T0213271

Lecture Notes in Computer Science 9988

Commenced Publication in 1973
Founding and Former Series Editors:
Gerhard Goos, Juris Hartmanis, and Jan van Leeuwen

More information about this series at http://www.springer.com/series/7407

Jukka Suomela (Ed.)

Structural Information and Communication Complexity

23rd International Colloquium, SIROCCO 2016
Helsinki, Finland, July 19–21, 2016
Revised Selected Papers

 Springer

Editor
Jukka Suomela
Aalto University
Espoo
Finland

ISSN 0302-9743   ISSN 1611-3349   (electronic)
Lecture Notes in Computer Science
ISBN 978-3-319-48313-9   ISBN 978-3-319-48314-6   (eBook)
DOI 10.1007/978-3-319-48314-6

Library of Congress Control Number: 2016955506

LNCS Sublibrary: SL1 – Theoretical Computer Science and General Issues

Printed on acid-free paper

This Springer imprint is published by Springer Nature
The registered company is Springer International Publishing AG
The registered company address is: Gewerbestrasse 11, 6330 Cham, Switzerland

Preface

This volume contains the papers presented at SIROCCO 2016, the 23rd International Colloquium on Structural Information and Communication Complexity, held during July 19–21, 2016, in Helsinki, Finland.

This year we received 50 submissions in response to the call for papers. Each submission was reviewed by at least three reviewers; we had a total of 18 Program Committee members and 57 external reviewers. The Program Committee decided to accept 25 papers: 24 normal papers and one survey-track paper. Fabian Kuhn, Yannic Maus, and Sebastian Daum received the SIROCCO 2016 Best Paper Award for their work "Rumor Spreading with Bounded In-Degree." Selected papers will also be invited to a special issue of the *Theoretical Computer Science* journal.

In addition to the 25 contributed talks, the conference program included a keynote lecture by Yoram Moses, invited talks by Keren Censor-Hillel, Adrian Kosowski, Danupon Nanongkai, and Thomas Sauerwald, and the award lecture by Masafumi (Mark) Yamashita, the recipient of the 2016 SIROCCO Prize for Innovation in Distributed Computing.

I would like to thank all authors for their high-quality submissions and all speakers for their excellent talks. I am grateful to the Program Committee and all external reviewers for their efforts in putting together a great conference program, to the Steering Committee chaired by Andrzej Pelc for their help and support, and to everyone who was involved in the local organization for making it possible to have SIROCCO 2016 in sunny Helsinki.

Finally, I would like to thank our sponsors for their support: the Federation of Finnish Learned Societies, Helsinki Institute for Information Technology HIIT, and Helsinki Doctoral Education Network in Information and Communications Technology (HICT) provided financial support, Springer not only helped with the publication of these proceedings but also sponsored the best paper award, Aalto University provided administrative support and helped with the conference venue, and EasyChair kindly provided a free platform for managing paper submissions and the production of this volume.

September 2016   Jukka Suomela

Organization

Program Committee

Leonid Barenboim	Open University of Israel
Jérémie Chalopin	LIF, CNRS and Aix Marseille Université, France
Yuval Emek	Technion, Israel
Paola Flocchini	University of Ottawa, Canada
Pierre Fraigniaud	CNRS and Université Paris Diderot, France
Janne H. Korhonen	Reykjavik University, Iceland
Evangelos Kranakis	Carleton University, Canada
Christoph Lenzen	MPI for Informatics, Germany
Friedhelm Meyer auf der Heide	Heinz Nixdorf Institute and University of Paderborn, Germany
Danupon Nanongkai	KTH Royal Institute of Technology, Sweden
Calvin Newport	Georgetown University, USA
Gopal Pandurangan	University of Houston, USA
Merav Parter	MIT, USA
Peter Robinson	Queen's University Belfast, UK
Thomas Sauerwald	University of Cambridge, UK
Stefan Schmid	Aalborg University, Denmark
Jukka Suomela	Aalto University, Finland
Przemysław Uznański	ETH Zürich, Switzerland

Additional Reviewers

Abu-Affash, A. Karim
Alistarh, Dan
Amram, Gal
Assadi, Sepehr
Augustine, John
Blin, Lelia
Burman, Janna
Censor-Hillel, Keren
Chatterjee, Soumyottam
Cseh, Ágnes
Das, Shantanu
Delporte-Gallet, Carole
Devismes, Stephane
Di Luna, Giuseppe
Friedrichs, Stephan

Förster, Klaus-Tycho
Geissmann, Barbara
Gelashvili, Rati
Gelles, Ran
Giakkoupis, George
Godard, Emmanuel
Graf, Daniel
Halldorsson, Magnus M.
Jung, Daniel
Karousatou, Christina
Kling, Peter
Konrad, Christian
Konwar, Kishori
Kuszner, Lukasz
Kuznetsov, Petr

Labourel, Arnaud
Lempiäinen, Tuomo
Malatyali, Manuel
Mamageishvili, Akaki
Medina, Moti
Molla, Anisur Rahaman
Musco, Cameron
Navarra, Alfredo
Pacheco, Eduardo
Pemmaraju, Sriram
Podlipyan, Pavel
Purcell, Christopher
Rabie, Mikaël
Rajsbaum, Sergio

Ravi, Srivatsan
Santoro, Nicola
Sardeshmukh, Vivek B.
Schneider, Johannes
Scquizzato, Michele
Setzer, Alexander
Sourav, Suman
Su, Hsin-Hao
Tonoyan, Tigran
Trehan, Chhaya
Tschager, Thomas
Yamauchi, Yukiko
Yu, Haifeng

Laudatio

It is a pleasure to award the 2016 SIROCCO Prize for Innovation in distributed computing to Masafumi (Mark) Yamashita. Mark has presented many original ideas and important results that have enriched the theoretical computer science community and the distributed computing community, such as his seminal work "Computing on Anonymous Networks" (with T. Kameda), which introduced the notion of "view" and has inspired all the subsequent investigations on computability in anonymous networks, as well as his work on coteries, on self-stabilization, and on polling games, among others.

The prize is awarded for his lifetime achievements, but especially for introducing the computational universe of autonomous mobile robots to the algorithmic community and to the distributed community in particular. This has opened a new and exciting research area that has now become an accepted mainstream topic in theoretical computer science (papers on "mobile robots" now appear in all major theory conferences and journals) and clearly in distributed computing. The fascinating new area of research it opened is now under investigation by many groups worldwide.

The introduction of this area to the theory community was actually made in his SIROCCO paper [1]. The full version was then published in the *SIAM Journal on Computing* [2]. (This paper currently has more than 500 citations.)

The paper deals with the problem of coordination among autonomous robots moving on a plane. This and subsequent papers on this topic provided the first indications about which tasks can be accomplished using multiple deterministic, autonomous, and identical robots in a collaborative manner. The formal model for mobile robots introduced in the paper (called the Suzuki–Yamashita or SYM model) provides a nice abstraction that makes it easy to analyze algorithms but still captures many of the difficulties of coordination between the robots. Many of the recent results on distributed robotics are based on either this model or extensions of it. The paper provided the characterization (in terms of geometric pattern formation) of all tasks that can be performed by such teams of deterministic robots and provided some fundamental impossibility results including the impossibility of gathering two oblivious robots. A more recent work [3] extends the characterization to the model where robots are memory-less, thus showing the exact difference between oblivious robots and robots having memory.

The 2015 Award Committee[1]:
Thomas Moscibroda (Microsoft)
Guy Even (Tel Aviv University)
Magnús Halldórsson (Reykjavik University)
Shay Kutten (Technion) – Chair
Andrzej Pelc (Université du Québec en Outaouais)

---

[1] We wish to thank the nominators for the nomination and for contributing greatly to this text.

Selected Publications Related to Masafumi (Mark) Yamashita's Contribution:

1. Suzuki, I., Yamashita, M.: Distributed anonymous mobile robots. In: Proceedings of the 3rd International Colloquium on Structural Information and Communication Complexity, Siena, Italy, 6–8 June, pp. 313–330 (1996)
2. Suzuki, I., Yamashita, M.: Distributed anonymous mobile robots. SIAM J. Comput. **28**(4), 1347–1363 (1999)
3. Yamashita, M., Suzuki, I.: Characterizing geometric patterns formable by oblivious anonymous mobile robots. Theor. Comput. Sci. **411**(26–28), 2433–2453 (2010)
4. Dumitrescu, A., Suzuki, I., Yamashita, M.: Motion planning for metamorphic systems: feasibility, decidability, and distributed reconfiguration. IEEE Trans. Robot. **20**(3), 409–418 (2004)
5. Souissi, S., Defago, X., Yamashita, M.: Using eventually consistent compasses to gather memory-less mobile robots with limited visibility. ACM Trans. Auton. Adapt. Syst. **4**(1), #9 (2009)
6. Das, S., Flocchini, P., Santoro, N., Yamashita, M.: Forming sequences of geometric patterns with oblivious mobile robots. Distrib. Comput. **28**(2), 131–145 (2015)
7. Fujinaga, N., Yamauchi, Y., Ono, H., Shuji, K., Yamashita, M.: Pattern formation by oblivious asynchronous mobile robots. SIAM J. Comput. **44**(3), 740–785 (2015)

Towards a Theory of Formal Distributed Systems
(SIROCCO Prize Lecture)

Masafumi Yamashita

Department of Informatics, Kyushu University, Fukuoka, Japan
mak@inf.kyushu-u.ac.jp

In the title, the word *towards* means incomplete, immature or not ready for presenting, and the word *formal* means unrealistic, imaginary or useless. Please keep them in mind.

One might find similarity between two phenomena, seabirds competing for good nesting places in a small island and cars looking (or fighting) for parking space. Regardless of whether conscious or unconscious, they are solving a conflict resolution problem, which is a well-known problem in distributed computing (in computer science). This suggests us there are many (artificial or natural) systems that are in the face of solving distributed problems.

Lamport and Lynch [1] claimed "although one usually speak of a distributed system, it is more accurate to speak of a distributed *view* of a system," after defining the word *distributed* to mean spread across space. This claim seems to imply that every system is a distributed system at least from the view of atoms or molecules, and may be in the face of solving a distributed problem, when we concentrate on the distributed view, like seabirds and cars in the example above.

An abstract distributed view, which we call a *formal distributed system* (FDS), describes how system elements interact logically. Our final goal is to understand a variety of FDSs and compare them in terms of the solvability of distributed problems.

We first propose a candidate for the model of FDS in such a way that it can describe a wide variety of FDSs, and explain that many of the models of distributed systems (including ones suitable to describe biological systems) can be described as FDSs. Compared with other distributed system models, FDSs have two features: First, the system elements are modeled by points in d-dimensional space, where d can be greater than 3. Second incomputable functions can be taken as transition functions (corresponding to distributed algorithms).

We next explain some of our ongoing works in three research areas, localization, symmetry breaking and self-organization. In localization, we discuss the simplest problem of locating a single element with limited visibility to the center of a line segment. In symmetry breaking, we observe how elements in 3D space can eliminate some symmetries. Finally in self-organization, we examine why natural systems appear to have richer autonomous properties than artificial systems, despite that the latter would have stronger interaction mechanisms, e.g., unique identifiers, memory, synchrony, and so on.

Reference

1. Lamport, L., Lynch, N.: Distributed computing: models and methods, In: van Leeuwen, J. (ed.) Handbook of Theoretical Computer Science. Formal Models and Semantics, Chap. 18, vol. B, pp. 1157–1199. MIT Press/Elsevier (1990)

A Principled Way of Designing Efficient Distributed Protocols

(Keynote Lecture)

Yoram Moses

Technion - Israel Institute of Technology, Technion City, Haifa 32000, Israel
moses@ee.technion.ac.il

The focus of this invited talk is a demonstration of how knowledge-based reasoning can be used to design an efficient protocol in a stepwise manner. The *Knowledge of Preconditions* principle, denoted by (**K**o**P**), can be formulated as a theorem that applies in all the various distributed systems models [2]. Intuitively, it states that if some condition φ is a necessary condition for process i to perform action α, then, under every protocol that satisfies this constraint, process i must *know* φ when it performs α. We denote i knowing something by 'K_i'. **K**o**P** thus states that if φ is a necessary condition for i performing α, then $K_i\varphi$ is also a necessary condition for i performing α. Thus, for example, a process the enters the critical section (CS) in a mutual exclusion protocol must know that the CS is empty when it enters. Similarly, if an ATM must only provide cash to a customer that has a sufficient positive balance, then the ATM must know that the customer has such a balance.

The talk illustrates the design of an unbeatable protocol for Consensus based on the **K**o**P**, along the lines of [1]. Based on the Validity property in the specification. In Consensus, a process can decide 0 only if some initial value is 0. The **K**o**P** immediately implies that following every correct protocol for Consensus, a process must *know* of an initial value of 0 when it decides 0. We consider binary Consensus, in which values are 0 or 1. We seek the optimal rule for deciding 1 in a protocol in which deciding on 0 is favored, by having every process that knows of a 0 decide 0. The Agreement property of Consensus implies that a process cannot decide 1 at a point when other processes decide 0. It follows by **K**o**P** that a process that decides 1 must know that nobody is deciding 0. In particular, it must know that no active process knows of a 0. A combinatorial analysis of when a process knows that nobody knows of a 0 is performed, yielding a natural condition that can be easily computed. The outcome is an elegant and efficient protocol that strictly dominates all known protocols for Consensus in the synchronous crash-failure model, which cannot be strictly dominated.

A video of a similar invited talk given in February 2016 appears in IHP talk.

References

1. Castañeda, A., Gonczarowski, Y.A., Moses, Y.: Unbeatable consensus. In: Kuhn, F. (ed.) DISC 2014. LNCS, vol. 8784, pp. 91–106. Springer, Heidelberg (2014). Full version available on arXiv
2. Moses,Y.: Relating knowledge and coordinated action: the knowledge of preconditions principle. In: Proceedings of the 15th Conference on Theoretical Aspects of Rationality and Knowledge, pp. 207–216 (2015)

Invited Talks

The Landscape of Lower Bounds
for the Congest Model

Keren Censor-Hillel

Technion, Department of Computer Science, Haifa, Israel
ckeren@cs.technion.ac.il

Introduction. We address the classic Congest model of distributed computing [8], in which n nodes of a network communicate in synchronous rounds, during each of which they send messages of $O(\log n)$ bits on the available links. We focus on solving *global* graph problems, which require $\Omega(D)$ rounds of communication even in the LOCAL model in which messages can be of unbounded size. While in the LOCAL model D rounds suffice for solving these problems by gathering all information at a single node and solving the problem on its local processor, the Congest model imposes additional bandwidth restrictions, making such problems harder. Below we discuss some known lower bounds for global problems in Congest, glimpse into some new results, and discuss open questions.

Computing the Diameter. One of the lead examples of a global graph problem is that of computing the diameter. In the Congest model, the diameter can be computed in $O(n)$ rounds [7, 9], and a beautiful lower bound of $\Omega(n/\log n)$, which we describe next, is known even for small values of D [5, 7].

In a nutshell, the lower bound is obtained through a reduction from the wellknown 2-party communication complexity problem of set-disjointness, in which Alice and Bob receive input vectors $\bar{x}, \bar{y}$ of length k, respectively, and need to output whether there is an index $1 \le i \le k$ for which $x_i = y_i = 1$. The reduction is obtained by constructing a graph of n nodes, with two sets of nodes that are connected by a complete matching and some additional edges within each set. Alice and Bob are each responsible for one of the two sets, in terms of simulating the distributed algorithm for the nodes within that set. Any message that needs to be sent within a set is simulated locally, and communication is only needed for messages that cross the cut between the two sets.

The crux is that Alice and Bob add edges within their sets according to their input vectors, where a 0 input for index i corresponds to adding the corresponding edge. This is done in a way that promises that the diameter of the resulting graph determines the answer to the set-disjointness problem. The parameters are taken such that $k = \Theta(n^2)$, and since set-disjointness is known to require $\Omega(k)$ bits of communication, and the size of the cut between the two sets of nodes is of size $\Theta(n)$ and the message size is of log n bits, the end result is a lower bound of $\Omega(n/\log n)$ rounds.

Keren Censor-Hillel—Supported in part by the Israel Science Foundation (grant 1696/14).

Recently, Abboud et al. [1] introduce a new construction that allows obtaining a similar near-linear lower bound for computing the diameter. The main technical contribution is a *bit-gadget*, which allows the cut between the sets of Alice and Bob to be of size only $\Theta(\log n)$ and allows taking $k = \Theta(n)$, giving a lower bound of $\Omega(n/\log^2 n)$. While this is worse than the previously mentioned bound by a logarithmic factor, the strength of the bit-gadget is in reducing the size of the cut and having a sparse construction, which then allows improving the state-of-the-art for additional problems: It gives the first near-linear lower bounds for a $(3/2 - \epsilon)$-approximation for the diameter, for computing or approximating the radius, for approximating all eccentricities, and for verifying certain types of spanners. These can also be made to work for constant degree graphs.

Constructing a Minimum Spanning Tree (MST). To exemplify another type of lower bounds for Congest that uses set-disjointness albeit in a different manner, consider the problem of finding an MST.

We next describe the key idea of the $\Omega(\sqrt{n/\log n} + D)$-round lower bound of [11]. This bound is given for the problem of subgraph connectivity, which can be easily be shown to reduce to finding an MST. A base graph is given and some of its edges are marked to be in the subgraph H, according to the inputs of Alice and Bob. It is shown that H is connected iff the inputs are not disjoint. To simulate the required distributed algorithm, Alice and Bob need to exchange information on certain edges of the graph in a dynamic way. That is, there is no static partition of the nodes between the 2 players which makes the complexity depend on the size of the cut, but rather the assignment of the nodes to be simulated changes from round to round and is not a partition. Thus, while the cut between Alice and Bob's nodes in each round is large, the *used cut* is $O(\log n)$, and choosing $k = O(n^{1/2})$ gives almost the claimed lower bound (for ease of description, this is a slightly weakened simplification of the lower bound). In our context, the interesting thing here is that although this is also a reduction from set-disjointness, the framework is entirely different from the distance computation lower bounds.

Constructing Additive Spanners. Recently, another type of Congest lower bounds has been introduced, for constructing additive spanners. Previous work obtains various spanners in the Congest model [2, 3, 10], and a lower bound of $\Omega(D)$ is given in [10].

A $+\beta$-pairwise spanner of G is a subgraph S for which, given $P \subseteq V$, for every $u, v \in P$, it holds that $d_S(u, v) \leq d_G(u, v) + \beta$. In addition to algorithms for purely additive spanners, [4] give lower bounds, of which we describe the $\Omega(p/n \log n)$ lower bound for constructing (+2)-pairwise spanners with $|P| = p$. Consider here $p = n^{3/2}$. Define the (p, m)-partial-complement problem as follows. Alice receives a set x of p elements in $\{1, \ldots, m\}$ and Bob needs to output a set y of $m/2$ elements in $\{1, \ldots, m\} \setminus x$. First, it is proven that (p, m)-partial-complement requires $\Omega(p)$ bits of communication. Then, a distributed algorithm for constructing a +2-spanner is simulated on the graph that consists of an Erdös graph with girth 6 and $\Theta(n^{3/2})$ edges that is simulated by Bob, whose nodes are connected by a complete matching to an equal size independent set of nodes that are simulated by Alice. The only unknown is the set P, given only to Alice. To decide on an edge of the graph to be omitted from the constructed spanner, Bob must know that the corresponding pair on Alice's side is not in P, otherwise its removal increases the distance between these nodes from 3 to 7, violating the +2 stretch requirement. Since Bob must remove $\Theta(n^{3/2})$ edges, this implies solving the (p, m)-partial-complement problem, hence requires $\Omega(p/n \log n)$ rounds. This gives a lower bound of a new flavor, where the graph is known to both players, and the uncertainty only comes from the unknown set of pairs.

Discussion. There are many additional lower bounds that are not described here.

Many specific questions are still open in the above various settings and problems. One example is that, while our lower bounds for distance computations apply to sparse graphs, they are far from being planar. It is known that an MST can be computed in $O(D \log D)$ rounds in planar graphs [6], which raises the question of whether distance computations can be performed faster than the general lower bound as well. Specifically, can the diameter of planar graphs be computed in $o(n/\text{polylog } n)$ rounds?

References

1. Abboud, A., Censor-Hillel, K., Khoury, S.: Near-linear lower bounds for distributed distance computations, even in sparse networks. In: Gavoille, C., Ilcinkas, D. (eds.) DISC 2016. LNCS, vol. 9888, pp. 29–42. Springer, Heidelberg (2016)
2. Baswana, S., Kavitha, T., Mehlhorn, K., Pettie, S.: Additive spanners and (alpha, beta)-spanners. ACM Trans. Algorithms **7**(1), 5 (2010)
3. Baswana, S., Sen, S.: A simple and linear time randomized algorithm for computing sparse spanners in weighted graphs. Random Struct. Algorithms **30**(4), 532–563 (2007)
4. Censor-Hillel, K., Kavitha, T., Paz, A., Yehudayoff, A.: Distributed construction of purelyadditive spanners. In: Gavoille, C., Ilcinkas, D. (eds.) DISC 2016. LNCS, vol. 9888, pp. 129–142. Springer, Heidelberg (2016)
5. Frischknecht, S., Holzer, S., Wattenhofer, R.: Networks cannot compute their diameter in sublinear time. In: SODA, pp. 1150–1162 (2012)
6. Ghaffari, M., Haeupler, B.: Distributed algorithms for planar networks II: low-congestion shortcuts, mst, and min-cut. In: SODA, pp. 202–219 (2016)
7. Holzer, S., Wattenhofer, R.: Optimal distributed all pairs shortest paths and applications. In: PODC, pp. 355–364 (2012)
8. Peleg, D.: Distributed computing: a locality-sensitive approach. In: Society for Industrial and Applied Mathematics (2000)
9. Peleg, D., Roditty, L., Tal, E.: Distributed algorithms for network diameter and girth. In: Czumaj, A., Mehlhorn, K., Pitts, A., Wattenhofer, R. (eds.) ICALP 2012. LNCS, vol. 7392, pp. 660–672. Springer, Heidelberg (2012)
10. Pettie, S.: Distributed algorithms for ultrasparse spanners and linear size skeletons. Distrib. Comput. **22**(3), 147–166, (2010)
11. Sarma, A.D., Holzer, S., Kor, L., Korman, A., Nanongkai, D., Pandurangan, G., Peleg, D., Wattenhofer, R.: Distributed verification and hardness of distributed approximation. SIAM J. Comput. **41**(5), 1235–1265 (2012)

What Makes a Distributed Problem Truly Local?

Adrian Kosowski

Inria and IRIF, CNRS — Université Paris Diderot, 75013 Paris, France
adrian.kosowski@inria.fr

Abstract. In this talk we attempt to identify the characteristics of a task of distributed network computing, which make it easy (or hard) to solve by means of fast local algorithms. We look at specific combinatorial tasks within the LOCAL model of distributed computation, and rephrase some recent algorithmic results in a framework of constraint satisfaction. Finally, we discuss the issue of efficient computability for relaxed variants of the LOCAL model, involving the so-called non-signaling property.

In distributed network computing, autonomous computational entities are represented by the nodes of an undirected *system graph*, and exchange information by sending messages along its edges. A major line of research in this area concerns the notion of *locality*, and asks how much information about its neighborhood a node needs to collect in order to solve a given computational task. In particular, in the seminal LOCAL model [19], the complexity of a distributed algorithm is measured in term of number of *rounds*, where in each round all nodes synchronously exchange data along network links, and subsequently perform individual computations. A t-round algorithm is thus one in which every node exchanges data with nodes at distance at most t (i.e., at most t hops away) from it.

Arguably, the most important class of local computational tasks concerns *symmetry breaking*, and several forms of such tasks have been considered, including the construction of proper *graph colorings* [3–9, 11, 15, 17, 18, 22], of *maximal independent sets* (MIS) [1, 4, 5, 14, 16, 18], as well as edge-based variants of these problems (cf. e.g. [21]). In this talk we address the following question: What makes some symmetry-breaking problems in the LOCAL model easier than others?

We note that the LOCAL model has two flavors, involving the design of deterministic and randomized algorithms, which are clearly distinct [8]. When considering randomized algorithms, for n-node graphs of maximum degree Δ, a hardness separation between the randomized complexities of the specific problems of MIS and $(\Delta + 1)$-coloring has recently been observed [11, 14]. No analogous separation is as yet known when considering deterministic solutions to these problems. We look at some partial evidence in this direction, making use of the recently introduced framework of *conflict coloring* representations [9] for local combinatorial problems. A conflict coloring representation captures a distributed task through a set of local constraints on edges

---

This talk includes results of joint work with: P. Fraigniaud, C. Gavoille, M. Heinrich, and M. Markiewicz.

of the system graph, thus constituting a special case of the much broader class of constraint satisfaction problems (CSP) with binary constraints. Whereas all local tasks are amenable to a conflict coloring formulation, one may introduce a natural *constraint density* parameter, which turns out to be inherently smaller for some problems than for others. For example, for the natural representation of the $(\Delta + 1)$-coloring task, the constraint density is $1/(\Delta + 1)$, while for any accurate representation of MIS, the constraint density is at least $1/2$. We discuss implications of how low constraint density (notably, much smaller than $1/\Delta$) may be helpful when finding solutions to a distributed task, especially when applying the so-called *shattering method* [20] in a randomized setting, and more directly, when designing faster deterministic algorithms through a direct attack on the conflict coloring representation of the task [9].

We close this talk with a discussion of relaxed variants of the LOCAL model, inspired by the physical concept of non-signaling. In a computational framework, the *non-signaling property* can be stated as the following necessary (but not sufficient) property of the LOCAL model: for any $t > 0$, given two subsets of nodes S_1 and S_2 of the system graph, such that the distance between the nearest nodes of S_1 and S_2 is greater than t, in any t-round LOCAL algorithm, the outputs of nodes from S_1 must be (probabilistically) independent of the inputs of nodes from S_2. We point out that for a number of symmetry breaking tasks in the LOCAL model, the currently best known asymptotic lower bounds can be deduced solely by exploiting the non-signaling property. This is the case for problems such as MIS [10, 14] or 2-coloring of the ring [10]. On the other hand, such an implication is not true for, e.g., the $\Omega(\log^* n)$ lower bound on the number of rounds required to 3-color the ring [15] — this lower bound follows from different (stronger) properties of the LOCAL model [12, 13]. This leads us to look at the converse question: How to identify conditions under which non-signaling solutions to a distributed task can be converted into an algorithm in the LOCAL model? We note some progress in this respect for quantum analogues of the LOCAL model [2].

References

1. Alon, N., Babai, L., Itai, A.: A fast and simple randomized parallel algorithm for the maximal independent set problem. J. Algorithms **7**(4), 567–583 (1986)
2. Arrighi, P., Nesme, V., Werner, R.F.: Unitarity plus causality implies localizability. J. Comput. Syst. Sci. **77**(2), 372–378 (2011)
3. Barenboim, L.: Deterministic $(\Delta + 1)$-coloring in sublinear (in Δ) time in static, dynamic and faulty networks. In: Proceedings of the 34th ACM Symposium on Principles of Distributed Computing (PODC), pp. 345–354 (2015)
4. Barenboim, L., Elkin, M.: Distributed $(\Delta + 1)$-coloring in linear (in Δ) time. In: Proceedings of the 41th ACM Symposium on Theory of Computing (STOC), pp. 111–120 (2009)
5. Barenboim, L., Elkin, M.: Distributed graph coloring: fundamentals and recent developments. In: Synthesis Lectures on Distributed Computing Theory. Morgan & Claypool Publishers (2013)
6. Barenboim, L., Elkin, M., Kuhn, F.: Distributed $(\Delta + 1)$-coloring in linear (in Δ) time. SIAM J. Comput. **43**(1), 72–95 (2014)

7. Barenboim, L., Elkin, M., Pettie, S., Schneider, J.: The locality of distributed symmetry breaking. In: Proceedings of the 53rd IEEE Symposium on Foundations of Computer Science (FOCS), pp. 321–330 (2012)
8. Chang, Y-J., Kopelowitz, T., Pettie, S.: An exponential separation between randomized and deterministic complexity in the LOCAL model, In: Proceedings 57th IEEE Symposium on Foundations of Computer Science (FOCS) (2016, to appear). http://arxiv.org/abs/1602.08166.
9. Fraigniaud, P., Heinrich, M., Kosowski, A.: Local conflict coloring, In: Proceedings of the 57th IEEE Symposium on Foundations of Computer Science (FOCS) (2016, to appear). http://arxiv.org/abs/1511.01287
10. Gavoille, C., Kosowski, A., Markiewicz, M.: What can be observed locally? In: Keidar, I. (ed.) DISC 2009. LNCS, vol. 5805, pp. 243–257. Springer, Heidelberg (2009)
11. Harris, D.G., Schneider, J., Su, H-H.: Distributed $(\Delta + 1)$-coloring in sublog-arithmic rounds, In: Proceedings of the 48th Annual Symposium on the Theory of Computing (STOC), pp. 465–478 (2016)
12. Holroyd, A.E., Liggett, T.M.: Finitely dependent coloring. Submitted preprint. http://arxiv.org/abs/1403.2448
13. Holroyd, A.E. Liggett, T.M.: Symmetric 1-dependent colorings of the integers. Electron. Commun. Probab. **20**(31) (2015)
14. Kuhn, F., Moscibroda, T., Wattenhofer, R.: What cannot be computed locally! In: Proceedings of the 23rd ACM Symposium on Principles of Distributed Computing (PODC), pp. 300–309 (2004)
15. Linial, N.: Locality in distributed graph algorithms. SIAM J. Comput. 21(1), 193–201 (1992)
16. Luby. M. A simple parallel algorithm for the maximal independent set problem. SIAM J. Comput. **15**, 1036–1053 (1986)
17. Naor. M.: A lower bound on probabilistic algorithms for distributive ring coloring. SIAM J. Discrete Math. **4**(3), 409–412 (1991)
18. Panconesi, A. Srinivasan, A.: Improved distributed algorithms for coloring and network decomposition problems. In: Proceedings of the 24th ACM Symposium on Theory of Computing (STOC), pp. 581–592 (1992)
19. Peleg, D.: Distributed computing: a locality-sensitive approach. SIAM (2000). Philadelphia, PA
20. Schneider, J., Wattenhofer, R.: A new technique for distributed symmetry breaking. In: Proceedings of the 29th ACM Symposium on Principles of Distributed Computing (PODC), pp. 257–266 (2010)
21. Suomela, J.: Survey of local algorithms. ACM Comput. Surv. **45**(2), 24 (2013)
22. Szegedy, M., Vishwanathan, S.: Locality based graph coloring. In: Proceedings of the 25th ACM Symposium on Theory of Computing (STOC), pp. 201–207 (1993)

Some Challenges on Distributed Shortest Paths Problems, A Survey

Danupon Nanongkai

KTH Royal Institute of Technology, Stockholm, Sweden
danupon@gmail.com
https://sites.google.com/site/dannanongkai/

Abstract. In this article, we focus on the time complexity of computing distances and shortest paths on distributed networks (the CONGEST model). We survey previous key results and techniques, and discuss where previous techniques fail and where major new ideas are needed. This article is based on the invited talk given at SIROCCO 2016. The slides used for the talk are available at the webpage of SIROCCO 2016 (http://sirocco2016.hiit.fi/programme/#invited).

Keywords: Shortest paths · Graph algorithms · Distributed algorithms

Our focus is on solving the *single-source shortest paths* problem on undirected weighted distributed networks. The network is modeled by the CONGEST model, and the goal is for every node to know its distance to a given source node. The algorithm should run with the least number of *rounds* possible (known as *time complexity*). (See, e.g., [8] for detailed descriptions.) Through a series of studies (e.g. [1, 3, 4, 8, 10, 11, 12, 14]), we now know that

1. any distributed algorithm with polynomial approximation ratio needs $\tilde{\Omega}(\sqrt{n} + D)$ rounds [3][1], and
2. there is a deterministic $(1 + \epsilon)$-approximation algorithm that takes $\tilde{O}(\epsilon^{O(1)}(\sqrt{n} + D))$ rounds [1, 8].

Here, n and D are the number of nodes and the network diameter, respectively, and $\tilde{\Omega}$ and $\tilde{O}$ hide $\log^{O(1)} n$ factors. The above results imply that we already know the best number of rounds an approximation algorithm can achieve, modulo some lower-order terms. The case of *exact* algorithm is, however, widely open. The best exact algorithm we know of takes $O(n)$ rounds, due to the distributed version of the Bellman-Ford algorithm. Beating this bound is the first open problem we highlight:

Open problem 1: Is there an algorithm that can solve the single-source shortest paths (or simply compute the distance between two given nodes) *exactly* in time that is *sublinear* in n, i.e. in $\tilde{O}(n^{1-\epsilon})$ rounds for some constant $\epsilon > 0$?

Note that whether we can solve graph problems exactly in sublinear time (in n) is interesting for many graph problems (e.g. the minimum cut problem [6, 13]).

---

[1] This lower bound holds for randomized algorithms and, in fact, even for quantum algorithms [5].

An equally interesting question is whether we can solve the *all-pairs* shortest paths problem exactly in linear-time (in n). We already know that we can get a $(1 + \epsilon)$-approximate solution in such running time.

One challenge in answering the above open problems is to avoid computing *k-source h-hop distances*. The h-hop distance between nodes u and v, denoted by $dist^h(u, v)$, is the (weighted) length of the shortest path among paths between u and v containing at most h edges. In the k-source h-hop distances problem, we are given k source nodes $s_1, s_2, \ldots, s_k$ and want to make every node u knows its distance to every source node s_i. An $\tilde{O}(k + n)$ distributed algorithm for solving this problem was presented in [12] and was an important subroutine in subsequent algorithms (e.g. [1, 8]). The drawback of this subroutine is that it only provides $(1 + \epsilon)$-approximate distances. Unfortunately, obtaining exact distances within the same running time is impossible, as Lenzen and Patt-Shamir [11] showed that such algorithm requires $\tilde{\Omega}(kh)$ rounds.

Another open problem (raised before in [12]) is the *directed case* (referred to as the *asymmetric case* in [12]). This is when we think of each edge (u, v) as two *directed* edges, one from u to v and the other from v to u, and the weight of the two edges might be different. (Note that the directions and edge weight do not affect the communication between u and v.) Obviously, the lower bound of $\tilde{\Omega}(\sqrt{n} + D)$ [3] for the undirected case also holds for this case. Using the techniques in [12], we can get a $(1 + \epsilon)$-approximation $\tilde{\Omega}(\sqrt{nD} + D)$-time algorithm. If we do not care about the approximation ratio, and simply want to know whether there is a directed path from the source to each node (this problem is called *single-source reachability*), then the running time can be slightly improved to $\tilde{\Omega}(\sqrt{n}D^{1/4} + D)$ [7]

Open problem 2: Is there an algorithm that can solve the *directed* single-source shortest paths (or just reachability) with any approximation ratio in $\tilde{O}(\sqrt{n} + D)$ rounds?

The main challenge in answering this open problem is to avoid the use of *sparse spanner* and related structures. A spanner is a subgraph that approximately preserves the distance between every pairs of nodes. Spanner and other relevant structures, such as emulator and hopset were used previously as the main tools to obtain tight upper bounds for the undirected case (see, e.g., [1, 8]). Unfortunately, similar structures do not exist on directed graphs. A sparse spanner, for example, do not exist for a complete bipartite graph with edges directed from left to right; removing any edge (u, v) from such graph will cause the distance from u to v to increase from one to infinity.

The last open problem we highlight is on *congested cliques*, i.e. when the network is fully-connected. For approximately solving the single-source shortest paths problem, we already have a satisfying algorithm with polylogarithmic time and $(1 + \epsilon)$-approximation ratio [1, 8]. The best $(1 + \epsilon)$-approximation algorithm for all-pairs shortest paths take $\tilde{O}(n^{0.15715})$ time [2]. For exact solutions, both single-source and all-pairs shortest paths have the best known running time of $\tilde{O}(n^{1/3})$ [2].

Open problem 3: Can we improve the running time of [2] for solving single-source shortest paths exactly and all-pairs shortest paths $(1 + \epsilon)$-approximately on congested cliques?

The above problem is interesting because of its connection to algebraic techniques. Its answer might lead us to understand these techniques better. See [2, 9] for algebraic tools developed so far on congested cliques.

References

1. Becker, R., Karrenbauer, A., Krinninger, S., Lenzen, C.: Approximate undirected trans-shipment and shortest paths via gradient descent. CoRR abs/1607.05127 (2016)
2. Censor-Hillel, K., Kaski, P., Korhonen, J.H., Lenzen, C., Paz, A., Suomela, J.: Algebraic methods in the congested clique. In: Symposium on Principles of Distributed Computing, PODC, pp. 143–152 (2015)
3. Das Sarma, A., Holzer, S., Kor, L., Korman, A., Nanongkai, D., Pandurangan, G., Peleg, D., Wattenhofer R.: Distributed verification and hardness of distributed approximation. SIAM J. Comput. **41**(5) (2012). Announced at STOC 2011
4. Elkin, M.: An unconditional lower bound on the time-approximation trade-off for the distributed minimum spanning tree problem. SIAM J. Comput. **36**(2) (2006). Announced at STOC 2004
5. Elkin, M., Klauck, H., Nanongkai, D., Pandurangan, G.: Can quantum communication speed up distributed computation? In: Symposium on Principles of Distributed Computing, PODC (2014)
6. Ghaffari, M., Kuhn, F.: Distributed minimum cut approximation. In: Afek, Y. (ed.) DISC 2013. LNCS, vol. 8205, pp. 1–15. Springer, Heidelberg (2013)
7. Ghaffari, M., Udwani, R.: Brief announcement: distributed single-source reachability. In: ACM Symposium on Principles of Distributed Computing, PODC, pp. 163–165 (2015)
8. Henzinger, M., Krinninger, S., Nanongkai, D: A deterministic almost-tight distributed algorithm for approximating single-source shortest paths. In: Symposium on Theory of Computing, STOC (2016)
9. Le Gall, F.: Further algebraic algorithms in the congested clique model and applications to graph-theoretic problems. In: Gavoille, C., Ilcinkas, D. (eds.) DISC 2016. LNCS, vol. 9888, pp. 57–70. Springer, Heidelberg (2016)
10. 10. Lenzen, C., Patt-Shamir, B.: Fast routing table construction using small messages. In: Symposium on Theory of Computing, STOC (2013)
11. Lenzen, C., Patt-Shamir, B.: Fast partial distance estimation and applications. In: ACM Symposium on Principles of Distributed Computing, PODC, pp. 153–162 (2015)
12. Nanongkai, D.: Distributed approximation algorithms for weighted shortest paths. In: ACM Symposium on Theory of Computing, STOC (2014)
13. Nanongkai, D., Su, H.: Almost-tight distributed minimum cut algorithms. In: Kuhn, F. (ed.) DISC 2014. LNCS, vol. 8784, pp. 439–453. Springer, Heidelberg (2014)
14. Peleg, D., Rubinovich, V.: A near-tight lower bound on the time complexity of distributed minimum-weight spanning tree construction. SIAM J. Comput. **30**(5) (2000). Announced at FOCS 1999

A Survey on Smoothing Networks

Thomas Sauerwald

Computer Laboratory, University of Cambridge, USA

Abstract. In this talk we will consider smoothing networks (a.k.a. balancing networks) that accept an arbitrary stream of tokens on input and routes them to output wires. Pairs of wires can be connected by balancers that direct arriving tokens alternately to its two outputs. We first discuss some classical results and relate smoothing networks to their siblings, including sorting and counting networks. Then we will present some results on randomised smoothing networks, where balancers are initialised randomly. Finally, we will explore stronger notions of smoothing networks including a model where an adversary can specify the input and the initialisation of all balancers.

References

1. Aspnes, J., Herlihy, M., Shavit, N.: Counting networks. J. ACM **41**(5), 1020–1048 (1994)
2. Batcher, K.E.: Sorting networks and their applications. In: American Federation of Information Processing Societies: AFIPS Conference Proceedings (1968) Spring Joint Computer Conference, pp. 307–314 (1968)
3. Busch, C., Mavronicolas, M.: A combinatorial treatment of balancing networks. J. ACM **43**(5), 794–839 (1996)
4. Dowd, M., Perl, Y., Rudolph, L., Saks, M.E.: The periodic balanced sorting network. J. ACM **36**(4), 738–757 (1989)
5. Friedrich, T., Sauerwald, T., Vilenchik. D.: Smoothed analysis of balancing networks. Random Struct. Algorithms **39**(1), 115–138 (2011)
6. Herlihy, M., Tirthapura, S.: Randomized smoothing networks. J. Parallel Distrib. Comput. **66**(5), 626–632 (2006)
7. Herlihy, M., Tirthapura, S.: Self-stabilizing smoothing and balancing networks. Distrib. Comput. **18**(5), 345–357 (2006)
8. Klugerman, M.: Small-depth counting networks and related topics, Ph.D. thesis, Massachusetts Institute of Technology, Department of Mathematics, September 1994
9. Klugerman, M., Plaxton, C.G.: Small-depth counting networks. In: Proceedings of the 24th Annual ACM Symposium on Theory of Computing (STOC 1992), pp. 417–428 (1992)
10. Mavronicolas, M. Sauerwald, T.: A randomized, o(log w)-depth 2 smoothing network. In: Proceedings of the 21st Annual ACM Symposium on Parallelism in Algorithms and Architectures (SPAA 2009), pp. 178–187 (2009)
11. Mavronicolas, M., Sauerwald, T.: The impact of randomization in smoothing networks. Distrib. Comput. **22**(5–6), 381–411 (2010)
12. Sauerwald, T., Sun, H.: Tight bounds for randomized load balancing on arbitrary network topologies. In: Proceedings of the 53rd Annual IEEE Symposium on Foundations of Computer Science (FOCS 2012), pp. 341–350 (2012)

Contents

Mobile Agents

Data Dissemination and Routing

Message Passing

How Many Cooks Spoil the Soup?

Othon Michail[1,2(✉)] and Paul G. Spirakis[1,2]

[1] Department of Computer Science, University of Liverpool, Liverpool, UK
{Othon.Michail,P.Spirakis}@liverpool.ac.uk
[2] Computer Technology Institute and Press "Diophantus" (CTI), Patras, Greece

Abstract. In this work, we study the following basic question: *"How much parallelism does a distributed task permit?"* Our definition of *parallelism* (or *symmetry*) here is not in terms of speed, but in terms of identical *roles* that processes have at the same time in the execution. We initiate this study in population protocols, a very simple model that not only allows for a straightforward definition of what a role is, but also encloses the challenge of isolating the properties that are due to the protocol from those that are due to the *adversary scheduler*, who controls the interactions between the processes. We (i) give a *partial characterization* of the set of predicates on input assignments that can be *stably computed with maximum symmetry*, i.e., $\Theta(N_{min})$, where N_{min} is the minimum multiplicity of a state in the initial configuration, and (ii) we turn our attention to the remaining predicates and prove a *strong impossibility result* for the *parity* predicate: the inherent symmetry of any protocol that stably computes it is *upper bounded by a constant that depends on the size of the protocol*.

1 Introduction

George Washington said *"My observation on every employment in life is, that, wherever and whenever one person is found adequate to the discharge of a duty by close application thereto, it is worse executed by two persons, and scarcely done at all if three or more are employed therein"*. The goal of the present paper is to investigate whether the analogue of this observation in simple distributed systems is true. In particular, we ask whether a task that can be solved when a single process has a crucial duty is still solvable when that (and any other) duty is assigned to more than one process. Moreover, we are interested in quantifying the degree of *parallelism* (also called *symmetry* in this paper) that a task is susceptible of.

Leader election is a task of outstanding importance for distributed algorithms. One of the oldest [Ang80] and probably still one of the most commonly used approaches [Lyn96, AW04, AAD+06, KLO10] for solving a distributed task in a given setting, is to execute a distributed algorithm that manages to elect

---

Supported in part by the School of EEE/CS of the University of Liverpool, NeST initiative, and the EU IP FET-Proactive project **MULTIPLEX** under contract no 317532. The full version can be found at: https://arxiv.org/abs/1604.07187.

© Springer International Publishing AG 2016
J. Suomela (Ed.): SIROCCO 2016, LNCS 9988, pp. 3–18, 2016.
DOI: 10.1007/978-3-319-48314-6_1

a unique leader (or *coordinator*) in that setting and then compose this (either sequentially or in parallel) with a second algorithm that can solve the task by assuming the existence of a unique leader. Actually, it is quite typical, that the tasks of electing a leader and successfully setting up the composition enclose the difficulty of solving many other higher-level tasks in the given setting.

Due to its usefulness in solving other distributed tasks, the leader election problem has been extensively studied, in a great variety of distributed settings [Lyn96, AW04, FSW14, AG15]. Still, there is an important point that is much less understood, concerning whether an election step is *necessary* for a given task and *to what extent* it can be avoided. Even if a task T can be solved in a given setting by first passing through a configuration with a unique leader, it is still valuable to know whether there is a correct algorithm for T that avoids this. In particular, such an algorithm succeeds without the need to ever have less than k processes in a given "role", and we are also interested in how large k can be without sacrificing solvability.

Depending on the application, there are several ways of defining what the "role" of a process at a given time in the execution is. In the typical approach of electing a unique leader, a process has the leader role if a *leader* variable in its local memory is set to true and it does not have it otherwise. In other cases, the role of a process could be defined as its complete local history. In such cases, we would consider that two processes have the same role after t steps iff both have the same local history after each one of them has completed t local steps. It could also be defined in terms of the external interface of a process, for example, by the messages that the process transmits, or it could even correspond to the branch of the program that the process executes. In this paper, as we shall see, we will define the role of a process at a given time in the execution, as the entire content of its local memory. So, in this paper, two processes u and v will be regarded to have the same role at a given time t iff, at that time, the local state of u is equal to the local state of v.

Understanding the parallelism that a distributed task allows, is of fundamental importance for the following reasons. First of all, usually, the more parallelism a task allows, the more efficiently it can be solved. Moreover, the less symmetry a solution for a given problem has to achieve in order to succeed, the more vulnerable it is to faults. For an extreme example, if a distributed algorithm elects in every execution a unique leader in order to solve a problem, then a single crash failure (of the leader) can be fatal.

1.1   Our Approach

We have chosen to initiate the study of the above problem in a very minimal distributed setting, namely in Population Protocols of Angluin *et al.* [AAD+06] (see Sect. 1.2 for more details and references). One reason that makes population protocols convenient for the problem under consideration, is that the role of a process at a given step in the execution can be defined in a straightforward way as the state of the process at the beginning of that step. So, for example, if we are interested in an execution of a protocol that stabilizes to the correct answer without ever electing a unique leader, what we actually require is an execution that, up to stability, never goes through a configuration in which a state q is the state of a single node, which implies that, in every configuration of the execution, every state q is either absent or the state of at least two nodes. Then, it is straightforward to generalize this to any symmetry requirement k, by requiring that, in every configuration, every state q is either absent or the state of at least k nodes.

What is not straightforward in this model (and in any model with adversarially determined events), is how to isolate the symmetry that is *only* due to the protocol. For if we require the above condition on executions to be satisfied *for every* execution of a protocol, then most protocols will fail trivially, because of the power of the adversary scheduler. In particular, there is almost always a way for the scheduler to force the protocol to break symmetry maximally, for example, to make it reach a configuration in which some state is the state of a single node, even when the protocol does not have an *inherent* mechanism of electing a unique state. Moreover, though for computability questions it is sufficient to assume that the scheduler selects in every step a single pair of nodes to interact with each other, this type of a scheduler is problematic for estimating the symmetry of protocols. The reason is that even fundamentally parallel operations, necessarily pass through a highly-symmetry-breaking step. For example, consider the rule $(a, a) \rightarrow (b, b)$ and assume that an even number of nodes are initially in state a. The goal is here for the protocol to convert all as to bs. If the scheduler could pick a perfect matching between the as, then in one step all as would be converted to bs, and additionally the protocol would never pass trough a configuration in which a state is the state of fewer than n nodes. Now, observe that the sequential scheduler can only pick a single pair of nodes in each step, so in the very first step it yields a configuration in which state b is the state of only 2 nodes. Of course, there are turnarounds to this, for example by taking into account only equal-interaction configurations, consisting of the states of the processes after all processes have participated in an equal number of interactions, still we shall follow an alternative approach that simplifies the arguments and the analysis.

In particular, we will consider schedulers that can be maximally parallel. Such a scheduler, selects in every step a matching (of any possible size) of the complete interaction graph, so, in one extreme, it is still allowed to select only one interaction but, in the other extreme, it may also select a perfect matching in a single step. Observe that this scheduler is different both from the sequential scheduler traditionally used in the area of population protocols and from the fully parallel scheduler which assumes that $\Theta(n)$ interactions occur in parallel in every step. Actually, several recent papers assume a fully parallel scheduler implicitly, by defining the model in terms of the sequential scheduler and then performing their analysis in terms of parallel time, defined as the sequential time divided by n.

Finally, in order to isolate the *inherent* symmetry, i.e., the symmetry that is only due to the protocol, we shall focus on those schedules[1] that achieve as high symmetry as possible for the given protocol. Such schedules may look into the protocol and exploit its structure so that the chosen interactions maximize parallelism. It is crucial to notice that this restriction does by no means affect correctness. Our protocols are still, as usual, required to stabilize to the correct answer in *any* fair execution (and, actually, in this paper against a more generic scheduler than the one traditionally assumed). The above restriction is only a convention for estimating the *inherent* symmetry of a protocol designed to operate in an adversarial setting. On the other hand, one does not expect this *measure of inherent symmetry* to be achieved by the majority of executions. If, instead, one is interested in some *measure of the observed symmetry*, then it would make more sense to study an *expected observed symmetry* under some probabilistic assumption for the scheduler. We leave this as an interesting direction for future research (see Sect. 5 for more details on this).

For a given initial configuration, we shall estimate the symmetry breaking performed by the protocol not in any possible execution but an execution in which the scheduler tries to maximize the symmetry. In particular, we shall define the symmetry of a protocol on a given initial configuration c_0 as the maximum symmetry achieved over all possible executions on c_0. So, in order to lower bound by k the symmetry of a protocol on a given c_0, it will be sufficient to present *a* schedule in which the protocol stabilizes without ever "electing" fewer than k nodes. On the other hand, to establish an upper bound of h on symmetry, we will have to show that *in every* schedule (on the given c_0) the protocol "elects" at most h nodes. Then we may define the symmetry of the protocol on a set of initial configurations as the minimum of its symmetries over those initial configurations. The symmetry of a protocol (as a whole) shall be defined as a function of some parameter of the initial configuration and is deferred to Sect. 2.

Observation 1. *The above definition leads to very strong impossibility results, as these upper bounds are also upper bounds on the observed symmetry. In particular, if we establish that the symmetry of a protocol $\mathcal{A}$ is at most h then, it is clear that under any scheduler the symmetry of $\mathcal{A}$ is at most h.*

Section 2 brings together all definitions and basic facts that are used throughout the paper. In Sect. 3, we give a set of positive results. The main result here is a partial characterization, showing that a wide subclass of semilinear predicates is computed with symmetry $\Theta(N_{min})$, which is asymptotically optimal. Then, in Sect. 4, we study some basic predicates that seem to require much symmetry breaking. In particular, we study the *majority* and the *parity* predicates. For majority we establish a constant symmetry, while for parity we prove a strong impossibility result, stating that the symmetry of any protocol that stably computes it, is upper bounded by an integer depending only on the size of the protocol (i.e., a constant, compared to the size of the system). The latter

---

[1] By "schedule" we mean an "execution" throughout.

implies that there exist predicates which can *only* be computed by protocols that perform some sort of leader-election (not necessarily a unique leader but at most a constant number of nodes in a distinguished leader role). In Sect. 5, we give further research directions that are opened by our work. All omitted details and proofs can be found in the full version.

1.2 Further Related Work

In contrast to static systems with unique identifiers (IDs) and dynamic systems, the role of symmetry in *static anonymous systems* has been deeply investigated [Ang80, YK96, Kra97, FMS98]. *Similarity* as a way to compare and contrast different models of concurrent programming has been defined and studied in [JS85]. One (restricted) type of symmetry that has been recently studied in systems with IDs is the existence of *homonyms*, i.e., processes that are initially assigned the same ID [DGFG+11]. Moreover, there are several standard models of distributed computing that do not suffer from a necessity to break symmetry globally (e.g., to elect a leader) like Shared Memory with Atomic Snapshots [AAD+93, AW04], Quorums [Ske82, MRWW01], and the LOCAL model [Pel00, Suo13].

Population Protocols were originally motivated by highly dynamic networks of simple sensor nodes that cannot control their mobility. The first papers focused on the computational capabilities of the model which have now been almost completely characterized. In particular, if the interaction network is complete (as is also the case in the present paper), i.e., one in which every pair of processes may interact, then the computational power of the model is equal to the class of the *semilinear predicates* (and the same holds for several variations) [AAER07]. Interestingly, the generic protocol of [AAD+06] that computes all semilinear predicates, elects a unique leader in every execution and the same is true for the construction in [CDS14]. Moreover, according to [AG15], all known generic constructions of semilinear predicates "fundamentally rely on the election of a single initial *leader* node, which coordinates phases of computation". Semilinearity of population protocols persists up to $o(\log \log n)$ local space but not more than this [CMN+11]. If additionally the connections between processes can hold a state from a finite domain, then the computational power dramatically increases to the commutative subclass of **NSPACE**(n^2) [MCS11a]. The formal equivalence of population protocols to *chemical reaction networks* (CRNs), which model chemistry in a *well-mixed solution*, has been recently demonstrated [Dot14]. Moreover, the recently proposed *Network Constructors* extension of population protocols [MS16] is capable of constructing arbitrarily complex stable networks. Czyzowicz *et al.* [CGK+15] have recently studied the relation of population protocols to antagonism of species, with dynamics modeled by discrete Lotka-Volterra equations. Finally, in [CCDS14], the authors highlighted the importance of executions that necessarily pass through a "bottleneck" transition (meaning a transition between two states that have only constant counts in the population, which requires $\Omega(n^2)$ expected number of steps to occur), by proving that protocols that avoid such transitions can only compute existence predicates. To the best of our knowledge, our type of approach, of computing predicates stably without *ever* electing a unique leader, has not been followed before in this area (according to [AG15], "[DH15] proposes a leader-less framework for population computation", but this should not be confused with what we do in this paper, as it only concerns the achievement of dropping the requirement for a *pre-elected* unique leader that was assumed in all previous results for that problem). For introductory texts to population protocols, the interested reader is encouraged to consult [AR09, MCS11b].

2   Preliminaries

A *population protocol* (PP) is a 6-tuple (X, Y, Q, I, O, δ), where X, Y, and Q are all finite sets and X is the *input alphabet*, Y is the *output alphabet*, Q is the set of *states*, $I: X \to Q$ is the *input function*, $O: Q \to Y$ is the *output function*, and $\delta: Q \times Q \to Q \times Q$ is the *transition function*.

If $\delta(a, b) = (a', b')$, we call $(a, b) \to (a', b')$ a *transition*. A transition $(a, b) \to (a', b')$ is called *effective* if $x \neq x'$ for at least one $x \in \{a, b\}$ and *ineffective* otherwise. When we present the transition function of a protocol we only present the effective transitions. The system consists of a population V of n distributed *processes* (also called *nodes*). In the generic case, there is an underlying *interaction graph* $G = (V, E)$ specifying the permissible interactions between the nodes. Interactions in this model are always pairwise. In this work, G is a *complete directed interaction graph*.

Let Q be the set of states of a population protocol $\mathcal{A}$. A configuration c of $\mathcal{A}$ on n nodes is an element of $\mathbb{N}_{\geq 0}^{|Q|}$, such that, for all $q \in Q$, $c[q]$ is equal to the number of nodes that are in state q in configuration c and it holds that $\sum_{q \in Q} c[q] = n$. For example, if $Q = \{q_0, q_1, q_2, q_3\}$ and $c = (7, 12, 52, 0)$, then, in c, 7 nodes of the $7 + 12 + 52 + 0 = 71$ in total, are in state q_0, 12 nodes in state q_1, and 52 nodes in state q_2.

Execution of the protocol proceeds in discrete steps and it is determined by an *adversary scheduler* who is allowed to be *parallel*, meaning that, in every step, it may select one or more pairwise interactions (up to a maximum matching) to occur at the same time. This is an important difference from classical population protocols where the scheduler could only select a single interaction per step. More formally, in every step, a non-empty matching $(u_1, v_1), (u_2, v_2), \ldots, (u_k, v_k)$ from E is selected by the scheduler and, for all $1 \leq i \leq k$, the nodes u_i, v_i interact with each other and update their states according to the transition function δ. A *fairness condition* is imposed on the adversary to ensure the protocol makes progress. An infinite execution is *fair* if for every pair of configurations c and c' such that $c \to c'$ (i.e., c can go in one step to c'), if c occurs infinitely often in the execution then so does c'.

In population protocols, we are typically interested in computing predicates on the inputs, e.g., $N_a \geq 5$, being true whenever there are at least 5 as in the input.[2] Moreover, computations are *stabilizing* and not terminating, meaning

---

[2] We shall use throughout the paper N_i to denote the number of nodes with input/state i.

that it suffices for the nodes to eventually converge to the correct output. We say that a protocol *stably computes* a predicate if, on any population size, any input assignment, and any fair execution on these, all nodes eventually stabilize their outputs to the value of the predicate on that input assignment.

We define the *symmetry* $s(c)$ *of a configuration* c as the *minimum multiplicity of a state that is present in* c (unless otherwise stated, in what follows by "symmetry" we shall always mean "inherent symmetry"). That is, $s(c) = \min_{q \in Q \,:\, c[q] \geq 1}\{c[q]\}$. For example, if $c = (0, 4, 12, 0, 52)$ then $s(c) = 4$, if $c = (1, \ldots)$ then $s(c) = 1$, which is the minimum possible value for symmetry, and if $c = (n, 0, 0, \ldots, 0)$ then $s(c) = n$ which is the maximum possible value for symmetry. So, the range of the symmetry of a configuration is $\{1, 2, \ldots, n\}$.

Let $\mathcal{C}_0(\mathcal{A})$ be the set of all *initial configurations* for a given protocol $\mathcal{A}$. Given an initial configuration $c_0 \in \mathcal{C}_0(\mathcal{A})$, denote by $\Gamma(c_0)$ the set of all fair executions of $\mathcal{A}$ that begin from c_0, each execution being truncated to its prefix *up to stability*.[3]

Given any initial configuration c_0 and any execution $\alpha \in \Gamma(c_0)$, define the *symmetry breaking of* $\mathcal{A}$ *on* α as the difference between the symmetry of the initial configuration of α and the minimum symmetry of a configuration of α, that is, the *maximum drop in symmetry* during the execution. Formally, $b(\mathcal{A}, \alpha) = s(c_0) - \min_{c \in \alpha}\{s(c)\}$. Also define the *symmetry of* $\mathcal{A}$ *on* α as $s(\mathcal{A}, \alpha) = \min_{c \in \alpha}\{s(c)\}$. Of course, it holds that $s(\mathcal{A}, \alpha) = s(c_0) - b(\mathcal{A}, \alpha)$. Moreover, observe that, for all $\alpha \in \Gamma(c_0)$, $0 \leq b(\mathcal{A}, \alpha) \leq s(c_0) - 1$ and $1 \leq s(\mathcal{A}, \alpha) \leq s(c_0)$. In several cases we shall denote $s(c_0)$ by N_{min}.

The *symmetry breaking of a protocol* $\mathcal{A}$ *on an initial configuration* c_0 can now be defined as $b(\mathcal{A}, c_0) = \min_{\alpha \in \Gamma(c_0)}\{b(\mathcal{A}, \alpha)\}$ and:

Definition 1. *We define the symmetry of* $\mathcal{A}$ *on* c_0 *as* $s(\mathcal{A}, c_0) = \max_{\alpha \in \Gamma(c_0)}\{s(\mathcal{A}, \alpha)\}$.

Remark 1. To estimate the *inherent* symmetry with which a protocol computes a predicate on a c_0, we execute the protocol against an *imaginary* scheduler who is a *symmetry maximizer*.

Now, given the set $\mathcal{C}(N_{min})$ of all initial configurations c_0 such that $s(c_0) = N_{min}$, we define the *symmetry breaking of a protocol* $\mathcal{A}$ *on* $\mathcal{C}(N_{min})$ as $b(\mathcal{A}, N_{min}) = \max_{c_0 \in \mathcal{C}(N_{min})}\{b(\mathcal{A}, c_0)\}$ and:

Definition 2. *We define the symmetry of* $\mathcal{A}$ *on* $\mathcal{C}(N_{min})$ *as* $s(\mathcal{A}, N_{min}) = \min_{c_0 \in \mathcal{C}(N_{min})}\{s(\mathcal{A}, c_0)\}$.

Observe again that $s(\mathcal{A}, N_{min}) = N_{min} - b(\mathcal{A}, N_{min})$ and that $0 \leq b(\mathcal{A}, N_{min}) \leq N_{min} - 1$ and $1 \leq s(\mathcal{A}, N_{min}) \leq N_{min}$.

---

[3] In this work, we only require protocols to preserve their symmetry *up to stability*. This means that a protocol is allowed to break symmetry arbitrarily after stability, e.g., even elect a unique leader, without having to pay for it. We leave as an interesting open problem the comparison of this convention to the apparently harder requirement of maintaining symmetry forever.

This means that, in order to establish that a protocol $\mathcal{A}$ is at least $g(N_{min})$ symmetric asymptotically (e.g., for $g(N_{min}) = \Theta(\log N_{min})$), we have to show that for every sufficiently large N_{min}, the symmetry breaking of $\mathcal{A}$ on $\mathcal{C}(N_{min})$ is at most $N_{min} - g(N_{min})$, that is, to show that for all initial configurations $c_0 \in \mathcal{C}(N_{min})$ there exists an execution on c_0 that drops the initial symmetry by at most $N_{min} - g(N_{min})$, e.g., by at most $N_{min} - \log N_{min}$ for $g(N_{min}) = \log N_{min}$, or that does not break symmetry at all in case $g(N_{min}) = N_{min}$. On the other hand, to establish that the symmetry is at most $g(N_{min})$, e.g., at most 1 which is the minimum possible value, one has to show a symmetry breaking of at least $N_{min} - g(N_{min})$ on infinitely many N_{min}s.

3   Predicates of High Symmetry

In this section, we try to identify predicates that can be stably computed with much symmetry. We first give an indicative example, then we generalize to arrive at a partial characterization of the predicates that can be computed with maximum symmetry, and, finally, we highlight the role of output-stable states in symmetric computations.

3.1   An Example: Count-to-x

Protocol. *Count-to-x*: $X = \{0, 1\}$, $Q = \{q_0, q_1, q_2, \ldots, q_x\}$, $I(\sigma) = q_\sigma$, for all $\sigma \in X$, $O(q_x) = 1$ and $O(q) = 0$, for all $q \in Q \backslash \{q_x\}$, and δ: $(q_i, q_j) \rightarrow (q_{i+j}, q_0)$, if $i + j < x$, $(q_i, q_j) \rightarrow (q_x, q_x)$, otherwise.

Proposition 1. *The symmetry of Protocol* Count-to-x, *for any $x = O(1)$, is at least $(2/3)\lfloor N_{min}/x \rfloor - (x - 1)/3$, when $x \geq 2$, and N_{min}, when $x = 1$; i.e., it is $\Theta(N_{min})$ for any $x = O(1)$.*

Proof. The scheduler[4] partitions the q_1s, let them be $N_1(0)$ initially and denoted just N_1 in the sequel, into $\lfloor N_1/x \rfloor$ groups of x q_1s each, possibly leaving an incomplete group of $r \leq x - 1$ q_1s residue. Then, in each complete group, it performs a sequential gathering of $x - 3$ other q_1s to one of the nodes, which will go through the states $q_1, q_2, \ldots, q_{x-1}$. The same gathering is performed in parallel to all groups, so every state that exists in one group will also exist in every other group, thus, its cardinality never drops below $\lfloor N_1/x \rfloor$. In the end, at step t, there are many q_0s, $N_{x-1}(t) = \lfloor N_1/x \rfloor$, and $N_1(t) = \lfloor N_1/x \rfloor + r$, where $0 \leq r \leq x - 1$ is the residue of q_1s. That is, in all configurations so far, the symmetry has not dropped below $\lfloor N_1/x \rfloor$.

Now, we cannot pick, as a symmetry maximizing choice of the scheduler, a perfect bipartite matching between the q_1s and the q_{x-1}s converting them all to the alarm state q_x, because this could possibly leave the symmetry-breaking residue of q_1s. What we can do instead, is to match in one step as many as

---

[4] Always meaning the *imaginary symmetry-maximizing scheduler* when lower-bounding the symmetry.

we can so that, after the corresponding transitions, $N_x(t') \geq N_1(t')$ is satisfied. In particular, if we match y of the (q_1, q_{x-1}) pairs we will obtain $N_x(t') = 2y$, $N_{x-1}(t') = \lfloor N_1/x \rfloor - y$, and $N_1(t') = \lfloor N_1/x \rfloor - y + r$ and what we want is

$$2y \geq \lfloor N_1/x \rfloor - y + r \Rightarrow 3y \geq \lfloor N_1/x \rfloor + r \Rightarrow y \geq \frac{\lfloor N_1/x \rfloor + r}{3},$$

which means that if we match approximately $1/3$ of the (q_1, q_{x-1}) pairs then we will have as many q_x as we need in order to eliminate all q_1s in one step and all remaining q_{x-1}s in another step.

The minimum symmetry in the whole course of this schedule is

$$N_{x-1}(t') = \lfloor N_1/x \rfloor - y = \lfloor N_1/x \rfloor - \frac{\lfloor N_1/x \rfloor + r}{3}$$
$$= \frac{2}{3} \lfloor N_1/x \rfloor - \frac{r}{3} \geq \frac{2}{3} \lfloor N_1/x \rfloor - \frac{x-1}{3}.$$

So, we have shown that if there are no q_0s in the initial configuration, then the symmetry breaking of the protocol on the schedule defined above is at most $N_{min} - ((2/3) \lfloor N_1/x \rfloor - (x-1)/3) = N_{min} - ((2/3) \lfloor N_{min}/x \rfloor - (x-1)/3)$. Next, we consider the case in which there are some q_0s in the initial configuration. Observe that in this protocol the q_0s can only increase, so their minimum cardinality is precisely their initial cardinality N_0. Consequently, in case $N_0 \geq 1$ and $N_1 \geq 1$, and if $N_{min} = \min\{N_0, N_1\}$, the symmetry breaking of the schedule defined above is $N_{min} - \min\{N_0, N_{x-1}(t')\}$. If, for some initial configuration, $N_0 \geq N_{x-1}(t')$ then the symmetry breaking is $N_{min} - N_{x-1}(t') \leq N_{min} - ((2/3) \lfloor N_1/x \rfloor - (x-1)/3)$. This gives again $N_{min} - ((2/3) \lfloor N_{min}/x \rfloor - (x-1)/3)$, when $N_1 \leq N_0$, and less than $N_{min} - ((2/3) \lfloor N_{min}/x \rfloor - (x-1)/3)$, when $N_1 > N_0 = N_{min}$. If instead, $N_0 < N_{x-1}(t') < N_1$, then, in this case, the symmetry breaking is $N_{min} - \min\{N_0, N_{x-1}(t')\} = N_0 - N_0 = 0$. Finally, if $N_0 = n$, then the symmetry breaking is 0. We conclude that for every initial configuration, the symmetry breaking of the above schedule is at most $N_{min} - N_{x-1}(t') \leq N_{min} - ((2/3) \lfloor N_{min}/x \rfloor - (x-1)/3)$, for all $x \geq 2$, and 0, for $x = 1$. Therefore, the symmetry of the *Count-to-x* protocol is at least $(2/3) \lfloor N_{min}/x \rfloor + (x-1)/3 = \Theta(N_{min})$, for $x \geq 2$, and N_{min}, for $x = 1$.   □

3.2 A General Positive Result

Theorem 1. *Any predicate of the form $\sum_{i \in [k]} a_i N_i \geq c$, for integer constants $k \geq 1$, $a_i \geq 1$, and $c \geq 0$, can be computed with symmetry more than $\lfloor N_{min}/(c/\sum_{j \in L} a_j + 2) \rfloor - 2 = \Theta(N_{min})$.*

Proof. We begin by giving a parameterized protocol (Protocol 1) that stably computes any such predicate, and then we shall prove that the symmetry of this protocol is the desired one.

Take now any initial configuration C_0 on n nodes and let $L \subseteq [k]$ be the set of indices of the initial states that are present in C_0. Let also q_{min} be the

---

Protocol 1. *Positive-Linear-Combination*

$Q = \{q_0, q_1, q_2, \ldots, q_c\}$
$I(\sigma_i) = q_{a_i}$, for all $\sigma_i \in X$
$O(q_c) = 1$ and $O(q) = 0$, for all $q \in Q \backslash \{q_c\}$
δ:

$$(q_i, q_j) \rightarrow (q_{i+j}, q_0), \text{ if } i + j < c$$
$$\rightarrow (q_c, q_c), \text{ otherwise}$$

---

state with minimum cardinality, N_{min}, in C_0. Construct $\lfloor N_{min}/x \rfloor$ groups, by adding to each group $x = \lceil c / \sum_{j \in L} a_j \rceil$ copies of each initial state. Observe that each group has total sum $\sum_{j \in L} a_j x = x \sum_{j \in L} a_j = \lceil c / \sum_{j \in L} a_j \rceil (\sum_{j \in L} a_j) \geq c$. Moreover, state q_{min} has a residue r_{min} of at most x and every other state q_i has a residue $r_i \geq r_{min}$. Finally, keep $y = \lceil (N_{min} + r_{min})/(x + 1) \rceil - 1$ from those groups and drop the other $\lfloor N_{min}/x \rfloor - y$ groups making their nodes part of the residue, which results in new residue values $r'_j = x(\lfloor N_{min}/x \rfloor - y) + r_j$, for all $j \in L$. It is not hard to show that $y \leq r'_j$, for all $j \in L$.

We now present a schedule that achieves the desired symmetry. The schedule consists of two phases, the *gathering* phase and the *dissemination* phase. In the dissemination phase, the schedule picks a node of the same state from every group and starts aggregating to that node the sum of its group sequentially, performing the same in parallel in all groups. It does this until the alarm state q_c first appears. When this occurs, the dissemination phase begins. In the dissemination phase, the schedule picks one after the other all states that have not yet been converted to q_c. For each such state q_i, it picks a q_c which infects one after the other (sequentially) the q_is, until $N_c(t) \geq N_i(t)$ is satisfied for the first time. Then, in a single step that matches each q_i to a q_c, it converts all remaining q_is to q_c.

We now analyze the symmetry breaking of the protocol in this schedule. Clearly, the initial symmetry is N_{min}. As long as a state appears in the groups, its cardinality is at least y, because it must appear in each one of them. When a state q_i first becomes eliminated from the groups, its cardinality is equal to its residue r'_i. Thus, so far, the minimum cardinality of a state is

$$\min\{y, \min_{j \in L} r'_j\} = y = \left\lceil \frac{N_{min} + r_{min}}{x + 1} \right\rceil - 1 > \left\lfloor \frac{N_{min}}{c / \sum_{j \in L} a_j + 2} \right\rfloor - 2.$$

It follows that the maximum symmetry breaking so far is less than $N_{min} - \left\lfloor \frac{N_{min}}{c / \sum_{j \in L} a_j + 2} \right\rfloor + 2$.

Finally, we must also take into account the dissemination phase. In this phase, the q_cs are $2y$ initially and can only increase, by infecting other states, until they become n and the cardinalities of all other states decrease until they all become 0. Take any state $q_i \neq q_c$ with cardinality $N_i(t)$ when the dissemination phase begins. What the schedule does is to decrement $N_i(t)$, until $N_c(t') \geq N_i(t')$ is first satisfied, and then to eliminate all occurrences of q_i in one step. Due to the fact that N_i is decremented by one in each step resulting in a corresponding increase by one of N_c, when $N_c(t') \geq N_i(t')$ is first satisfied, it holds that $N_i(t') \geq N_c(t') - 1 \geq N_c(t) - 1 \geq 2y - 1 \geq y$ for all $y \geq 1$, which implies that the lower bound of y on the minimum cardinality, established for the gathering phase, is not violated during the dissemination phase.

We conclude that the symmetry of the protocol in the above schedule is more than $\lfloor N_{min}/(c/\sum_{j \in L} a_j + 2) \rfloor - 2$. $\qquad\square$

3.3 Output-Stable States

Informally, a state $q \in Q$ is called *output-stable* if its appearance in an execution guarantees that the output value $O(q)$ must be the output value of the execution. More formally, if q is output-stable and C is a configuration containing q, then the set of outputs of C' must contain $O(q)$, for all C' such that $C \rightsquigarrow C'$, where '$\rightsquigarrow$' means *reaches in one or more steps*. Moreover, if all executions under consideration stabilize to an agreement, meaning that eventually all nodes stabilize to the same output, then the above implies that if an execution ever reaches a configuration containing q then the output of that execution is necessarily $O(q)$.

A state q is called *reachable* if there is an initial configuration C_0 and an execution on C_0 that can produce q. We can also define reachability just in terms of the protocol, under the assumption that if $Q_0 \subseteq Q$ is the set of initial states, then any possible combination of cardinalities of states from Q_0 can be part of an initial configuration. A *production tree* for a state $q \in Q$, is a directed binary in-tree with its nodes labeled from Q such that its root has label q, if a is the label of an internal node (the root inclusive) and b, c are the labels of its children, then the protocol has a rule of the form $\{b, c\} \rightarrow \{a, \cdot\}$ (that is, a rule producing a by an interaction between a b and a c in any direction)[5], and any leaf is labeled from Q_0. Observe now that if a path from a leaf to the root repeats a state a, then we can always replace the subtree of the highest appearance of a by the subtree of the lowest appearance of a on the path and still have a production tree for q. This implies that if q has a production tree, then q also has a production tree of depth at most $|Q|$, that is, a production tree having at most $2^{|Q|-1}$ leaves, which is a constant number, when compared to the population size n, that only depends on the protocol. Now, we can call a state q reachable (by a protocol $\mathcal{A}$) if there is a production tree for it. These are summarized in the following proposition.

Proposition 2. *Let $\mathcal{A}$ be a protocol, C_0 be any (sufficiently large) initial configuration of $\mathcal{A}$, and $q \in Q$ any state that is reachable from C_0. Then there is an initial configuration C_0' which is a sub-configuration of C_0 of size $n' \leq 2^{|Q|-1}$ such that q is reachable from C_0'.*

Proposition 2 is crucial for proving negative results, and will be invoked in Sect. 4.

---

[5] Whenever we use an unordered pair in a rule, like $\{b, c\}$, we mean that the property under consideration concerns both (b, c) and (c, b).

Proposition 3. *Let p be a predicate. There is no protocol that stably computes p (all nodes eventually agreeing on the output in every fair execution), having both a reachable output-stable state with output 0 and a reachable output-stable state with output 1.*

An output-stable state q is called *disseminating* if $\{x, q\} \to (q, q)$, for all $x \in Q$.

Proposition 4. *Let $\mathcal{A}$ be a protocol with at least one reachable output-stable state, that stably computes a predicate p and let $Q_s \subseteq Q$ be the set of reachable output-stable states of $\mathcal{A}$. Then there is a protocol $\mathcal{A}'$ with a reachable disseminating state that stably computes p.*

Theorem 2. *Let $\mathcal{A}$ be a protocol with a reachable disseminating state q and let $\mathcal{C}_0^d$ be the subset of its initial configurations that may produce q. Then the symmetry of $\mathcal{A}$ on $\mathcal{C}_0^d$ is $\Theta(N_{min})$.*

Theorem 2 emphasizes the fact that disseminating states can be exploited for maximum symmetry. We have omitted its proof, because it is similar to the proofs of Proposition 1 and Theorem 1. This lower bound on symmetry immediately applies to single-signed linear combinations (where passing a threshold can safely result in the appearance of a disseminating state, because there are no opposite-signed numbers to inverse the process), thus, it can be used as an alternative way of arriving at Theorem 1. On the other hand, the next proposition shows that this lower bound does not apply to linear combinations containing mixed signs, because protocols for them cannot have output-stable states.

Proposition 5. *Let p be a predicate of the form $\sum_{i \in [k]} a_i N_i \geq c$, for integer constants $k \geq 1$, a_i, and $c \geq 0$ such that at least two a_is have opposite signs. Then there is no protocol, having a reachable output-stable state, that stably computes p.*

4   Harder Predicates

In this section, we study the symmetry of predicates that, in contrast to single-signed linear combinations, do not allow for output-stable states. In particular, we focus on linear combinations containing mixed signs, like the *majority* predicate, and also on modulo predicates like the *parity* predicate. Recall that these predicates are not captured by the lower bound on symmetry of Theorem 2.

4.1   Bounds for Mixed Coefficients

We begin with a proposition stating that the majority predicate (also can be generalized to any predicate with mixed signs) can be computed with symmetry that depends on the difference of the state-cardinalities in the initial configuration.

Proposition 6. *The majority predicate* $N_a - N_b > 0$ *can be computed with symmetry* $\min\{N_{min}, |N_a - N_b|\}$, *where* $N_{min} = \min\{N_a, N_b\}$.

Remark 2. A result similar to Proposition 6 can be proved for any predicate $\sum_{i \in [k]} a_i N_i - \sum_{j \in [h]} b_j N'_j > c$, for integer constants $k, h, a_i, b_j \geq 1$ and $c \geq 0$.

Still, as we prove in the following theorem, it is possible to do better in the worst case, and achieve any desired constant symmetry.

Theorem 3. *For every constant* $k \geq 1$, *the majority predicate* $N_a - N_b > 0$ *can be computed with symmetry* k.

4.2   Predicates that Cannot be Computed with High Symmetry

We now prove a strong impossibility result, establishing that there are predicates that cannot be stably computed with much symmetry. The result concerns the *parity* predicate, defined as $n \bmod 2 = 1$. In particular, all nodes obtain the same input, e.g., 1, and, thus, all begin from the same state, e.g., q_1. So, in this case, $N_{min} = n$ in every initial configuration, and we can here estimate symmetry as a function of n. The parity predicate is true iff the number of nodes is odd. So, whenever n is odd, we want all nodes to eventually stabilize their outputs to 1 and, whenever it is even, to 0. If symmetry is not a constraint, then there is a simple protocol that solves the problem [AAD+06]. Unfortunately, not only that particular strategy, but any possible strategy for the problem, cannot achieve symmetry more than a constant that depends on the size of the protocol, as we shall now prove.

Theorem 4. *Let* $\mathcal{A}$ *be a protocol with set of states* Q, *that solves the parity predicate. Then the symmetry of* $\mathcal{A}$ *is less than* $2^{|Q|-1}$.

Proof. For the sake of contradiction, assume $\mathcal{A}$ solves parity with symmetry $f(n) \geq 2^{|Q|-1}$. Take any initial configuration C_n for any sufficiently large odd n (e.g., $n \geq f(n)$ or $n \geq |Q| \cdot f(n)$, or even larger if required by the protocol). By definition of symmetry, there is an execution α on C_n that reaches stability without ever dropping the minimum cardinality of an existing state below $f(n)$. Call C_{stable} the first output-stable configuration of α. As n is odd, C_{stable} must satisfy that all nodes are in states giving output 1 and that no execution on C_{stable} can produce a state with output 0. Moreover, due to the facts that $\mathcal{A}$ has symmetry $f(n)$ and that α is an execution that achieves this symmetry, it must hold that every $q \in Q$ that appears in C_{stable} has multiplicity $C_{stable}[q] \geq f(n)$.

Consider now the initial configuration C_{2n}, i.e., the unique initial configuration on $2n$ nodes. Observe that now the number of nodes is even, thus, the parity predicate evaluates to false and any fair execution of $\mathcal{A}$ must stabilize to output 0. Partition C_{2n} into two equal parts, each of size n. Observe that each of the two parts is equal to C_n. Consider now the following possible finite prefix β of a fair execution on C_{2n}. The scheduler simulates in each of the two parts the previous execution α up to the point that it reaches the configuration C_{stable}.

So, the prefix β takes C_{2n} to a configuration denoted by $2C_{stable}$ and consisting precisely of two copies of C_{stable}. Observe that $2C_{stable}$ and C_{stable} consist of the same states with the only difference being that their multiplicity in $2C_{stable}$ is twice their multiplicity in C_{stable}. A crucial difference between C_{stable} and $2C_{stable}$ is that the former is output-stable while the latter is not. In particular, any fair execution of $\mathcal{A}$ on $2C_{stable}$ must produce a state q_0 with output 0. But, by Proposition 2, q_0 must also be reachable from a sub-configuration C_{small} of $2C_{stable}$ of size at most $2^{|Q|-1}$. So, there is an execution γ restricted on C_{small} that produces q_0.

Observe now that C_{small} is also a sub-configuration of C_{stable}. The reason in that (i) every state in C_{small} is also a state that exists in $2C_{stable}$ and, thus, also a state that exists in C_{stable} and (ii) the multiplicity of every state in C_{small} is restricted by the size of C_{small}, which is at most $2^{|Q|-1}$, and every state in C_{stable} has multiplicity at least $f(n) \geq 2^{|Q|-1}$, that is, C_{stable} has sufficient capacity for every state in C_{small}. But this implies that if γ is executed on the sub-configuration of C_{stable} corresponding to C_{small}, then it must produce q_0, which contradicts the fact that C_{stable} is output-stable with output 1. Therefore, we conclude that $\mathcal{A}$ cannot have symmetry at least $f(n) \geq 2^{|Q|-1}$.   □

Remark 3. Theorem 4 constrains the symmetry of any correct protocol for parity to be upper bounded by a constant that depends on the size of the protocol. Still, it does not exclude the possibility that parity is solvable with symmetry k, for any constant $k \geq 1$. The reason is that, for any constant $k \geq 1$, there might be a protocol with $|Q| > k$ that solves parity and achieves symmetry k, because $k < 2^{|Q|-1}$, which is the upper bound on symmetry proved by the theorem. On the other hand, the $2^{|Q|-1}$ upper bound of Theorem 4 excludes any protocol that would solve parity with symmetry depending on N_{min}.

5   Further Research

In this work, we managed to obtain a first partial characterization of the predicates with symmetry $\Theta(N_{min})$ and to exhibit a predicate (parity) that resists any non-constant symmetry. The obvious next goal is to arrive at an exact characterization of the allowable symmetry of all semilinear predicates.

Some preliminary results of ours, indicate that constant symmetry for parity can be achieved if the initial configuration has a sufficient number of *auxiliary nodes* in a distinct state q_0. It seems interesting to study how is symmetry affected by auxiliary nodes and whether they can be totally avoided.

Another very challenging direction for further research, concerns networked systems (either static or dynamic) in which the nodes have memory and possibly also unique IDs. Even though the IDs provide an *a priori* maximum symmetry breaking, still, solving a task and avoiding the process of "electing" one of the nodes may be highly non-trivial. But in this case, defining the role of a process as its complete local state is inadequate. There are other plausible ways of defining the role of a process, but which one is best-tailored for such systems is still unclear and needs further investigation.

Finally, recall that in this work we focused on the *inherent* symmetry of a protocol as opposed to its *observed* symmetry. One way to study the observed symmetry would be to consider *random parallel schedulers*, like the one that selects in every step a maximum matching uniformly at random from all such matchings. Then we may ask *"What is the average symmetry achieved by a protocol under such a scheduler?"*. In some preliminary experimental results of ours, the expected observed symmetry of the *Count-to-5* protocol (i) if counted until the alert state q_5 becomes an absolute majority in the population, seems to grow faster than $\sqrt{n}$ and (ii) if counted up to stability, seems to grow as fast as $\log n$ (see the full paper for more details).

Acknowledgements. We would like to thank Dimitrios Amaxilatis for setting up and running experiments for the evaluation of the observed symmetry.

References

[AAD+93] Afek, Y., Attiya, H., Dolev, D., Gafni, E., Merritt, M., Shavit, N.: Atomic snapshots of shared memory. J. ACM (JACM) **40**(4), 873–890 (1993)

[AAD+06] Angluin, D., Aspnes, J., Diamadi, Z., Fischer, M.J., Peralta, R.: Computation in networks of passively mobile finite-state sensors. Distrib. Comput. **18**(4), 235–253 (2006)

[AAER07] Angluin, D., Aspnes, J., Eisenstat, D., Ruppert, E.: The computational power of population protocols. Distrib. Comput. **20**(4), 279–304 (2007)

[AG15] Alistarh, D., Gelashvili, R.: Polylogarithmic-time leader election in population protocols. In: Halldórsson, M.M., Iwama, K., Kobayashi, N., Speckmann, B. (eds.) ICALP 2015. LNCS, vol. 9135, pp. 479–491. Springer, Heidelberg (2015). doi:10.1007/978-3-662-47666-6_38

[Ang80] Angluin, D.: Local and global properties in networks of processors. In: Proceedings of the 12th Annual ACM Symposium on Theory of Computing (STOC), pp. 82–93. ACM (1980)

[AR09] Aspnes, J., Ruppert, E.: An introduction to population protocols. In: Garbinato, B., Miranda, H., Rodrigues, L. (eds.) Middleware for Network Eccentric and Mobile Applications, pp. 97–120. Springer, Heidelberg (2009)

[AW04] Attiya, H., Welch, J.: Distributed Computing: Fundamentals, Simulations, and Advanced Topics, vol. 19. Wiley-Interscience, Hoboken (2004)

[CCDS14] Chen, H.-L., Cummings, R., Doty, D., Soloveichik, D.: Speed faults in computation by chemical reaction networks. In: Kuhn, F. (ed.) DISC 2014. LNCS, vol. 8784, pp. 16–30. Springer, Heidelberg (2014). doi:10.1007/978-3-662-45174-8_2

[CDS14] Chen, H.-L., Doty, D., Soloveichik, D.: Deterministic function computation with chemical reaction networks. Nat. Comput. **13**(4), 517–534 (2014)

[CGK+15] Czyzowicz, J., Gąsieniec, L., Kosowski, A., Kranakis, E., Spirakis, P.G., Uznański, P.: On convergence and threshold properties of discrete Lotka-Volterra population protocols. In: Halldórsson, M.M., Iwama, K., Kobayashi, N., Speckmann, B. (eds.) ICALP 2015. LNCS, vol. 9134, pp. 393–405. Springer, Heidelberg (2015). doi:10.1007/978-3-662-47672-7_32

[CMN+11] Chatzigiannakis, I., Michail, O., Nikolaou, S., Pavlogiannis, A., Spirakis, P.G.: Passively mobile communicating machines that use restricted space. Theoret. Comput. Sci. **412**(46), 6469–6483 (2011)

[DGFG+11] Delporte-Gallet, C., Fauconnier, H., Guerraoui, R., Kermarrec, A.-M., Ruppert, E., Tran-The, H.: Byzantine agreement with homonyms. In: Proceedings of the 30th Annual ACM SIGACT-SIGOPS Symposium on Principles of Distributed Computing (PODC), pp. 21–30. ACM (2011)

[DH15] Doty, D., Hajiaghayi, M.: Leaderless deterministic chemical reaction networks. Nat. Comput. **14**(2), 213–223 (2015)

[Dot14] Doty, D.: Timing in chemical reaction networks. In: Proceedings of the 25th Annual ACM-SIAM Symposium on Discrete Algorithms (SODA), pp. 772–784 (2014)

[FMS98] Flocchini, P., Mans, B., Santoro, N.: Sense of direction: definitions, properties, and classes. Networks **32**(3), 165–180 (1998)

[FSW14] Förster, K.-T., Seidel, J., Wattenhofer, R.: Deterministic leader election in multi-hop beeping networks. In: Kuhn, F. (ed.) DISC 2014. LNCS, vol. 8784, pp. 212–226. Springer, Heidelberg (2014). doi:10.1007/978-3-662-45174-8_15

[JS85] Johnson, R.E., Schneider, F.B.: Symmetry and similarity in distributed systems. In: Proceedings of the 4th Annual ACM Symposium on Principles of Distributed Computing (PODC), pp. 13–22. ACM (1985)

[KLO10] Kuhn, F., Lynch, N., Oshman, R.: Distributed computation in dynamic networks. In: Proceedings of the 42nd ACM Symposium on Theory of Computing (STOC), pp. 513–522. ACM (2010)

[Kra97] Kranakis, E.: Symmetry, computability in anonymous networks: a brief survey. In: Proceedings of the 3rd International Conference on Structural Information and Communication Complexity, pp. 1–16 (1997)

[Lyn96] Lynch, N.A.: Distributed Algorithms, 1st edn. Morgan Kaufmann, Burlington (1996)

[MCS11a] Michail, O., Chatzigiannakis, I., Spirakis, P.G.: Mediated population protocols. Theoret. Comput. Sci. **412**(22), 2434–2450 (2011)

[MCS11b] Michail, O., Chatzigiannakis, I., Spirakis, P. G.: New models for population protocols. In: Lynch, N.A. (ed.) Synthesis Lectures on Distributed Computing Theory. Morgan & Claypool (2011)

[MRWW01] Malkhi, D., Reiter, M.K., Wool, A., Wright, R.N.: Probabilistic quorum systems. Inf. Comput. **170**(2), 184–206 (2001)

[MS16] Michail, O., Spirakis, P.G.: Simple and efficient local codes for distributed stable network construction. Distrib. Comput. **29**(3), 207–237 (2016)

[Pel00] Peleg, D.: Distributed Computing: A Locality-Sensitive Approach. Society for Industrial and Applied Mathematics, Philadelphia (2000). SIAM monographs on discrete mathematics and applications

[Ske82] Skeen, D.: A quorum-based commit protocol. Technical report, Cornell University (1982)

[Suo13] Suomela, J.: Survey of local algorithms. ACM Comput. Surv. (CSUR) **45**(2), 24 (2013)

[YK96] Yamashita, M., Kameda, T.: Computing on anonymous networks. I. Characterizing the solvable cases. IEEE Trans. Parallel, Distrib. Syst. **7**(1), 69–89 (1996)

Dynamic Networks of Finite State Machines

Yuval Emek[1] and Jara Uitto[2]($\boxtimes$)

[1] Technion, Haifa, Israel
[2] Comerge AG, Zürich, Switzerland
jara.uitto@comerge.net

Abstract. Like distributed systems, biological multicellular processes are subject to dynamic changes and a biological system will not pass the survival-of-the-fittest test unless it exhibits certain features that enable fast recovery from these changes. In most cases, the types of dynamic changes a biological process may experience and its desired recovery features differ from those traditionally studied in the distributed computing literature. In particular, a question seldomly asked in the context of distributed digital systems and that is crucial in the context of biological cellular networks, is whether the system can keep the changing components *confined* so that only nodes in their vicinity may be affected by the changes, but nodes sufficiently far away from any changing component remain unaffected.

Based on this notion of confinement, we propose a new metric for measuring the dynamic changes recovery performance in distributed network algorithms operating under the *Stone Age* model (Emek and Wattenhofer, PODC 2013), where the class of dynamic topology changes we consider includes inserting/deleting an edge, deleting a node together with its incident edges, and inserting a new isolated node. Our main technical contribution is a distributed algorithm for maximal independent set (MIS) in synchronous networks subject to these topology changes that performs well in terms of the aforementioned new metric. Specifically, our algorithm guarantees that nodes which do not experience a topology change in their immediate vicinity are not affected and that all surviving nodes (including the affected ones) perform $\mathcal{O}((C+1)\log^2 n)$ computationally-meaningful steps, where C is the number of topology changes; in other words, each surviving node performs $\mathcal{O}(\log^2 n)$ steps when amortized over the number of topology changes. This is accompanied by a simple example demonstrating that the linear dependency on C cannot be avoided.

1 Introduction

The biological form of close-range (juxtacrine) message passing relies on designated messenger molecules that bind to crossmembrane receptors in neighboring cells; this binding action triggers a signaling cascade that eventually affects gene expression, thus modifying the neighboring cells' states. This mechanism should feel familiar to members of the distributed computing community as it resembles the message passing schemes of distributed digital systems. In contrast to

© Springer International Publishing AG 2016
J. Suomela (Ed.): SIROCCO 2016, LNCS 9988, pp. 19–34, 2016.
DOI: 10.1007/978-3-319-48314-6_2

nodes in distributed digital systems, however, biological cells are not believed to be Turing complete, rather each biological cell is pretty limited in computation as well as communication. In attempt to cope with these differences, Emek and Wattenhofer [13] introduced the *Stone Age* model of distributed computing (a.k.a. networked finite state machines), where each node in the network is a very weak computational unit with limited communication capabilities and showed that several fundamental distributed computing problems can be solved efficiently under this model.

An important topic left outside the scope of [13] is that of *dynamic topology changes*. Just like distributed digital systems, biological systems may experience local changes and the ability of the system to recover from these changes is crucial to its survival. However, the desired recovery features in biological cellular networks typically differ from those traditionally studied in the distributed computing literature. In particular, a major issue in the context of biological cellular networks, that is rarely addressed in the study of distributed digital systems, is that of *confining* the topology changes: while nodes in the immediate vicinity of a topology change are doomed to be affected by it (hopefully, to a bounded extent), isolating the nodes sufficiently far away from any topology change so that their operation remains unaffected, is often critical. Indeed, a biological multicellular system with limited energy resources cannot afford every cell division (or death) to have far reaching effects on the cellular network.

In this paper, we make a step towards bringing the models from computer science closer to biology by extending the Stone Age model to accommodate four types of dynamic topology changes: (1) deleting an existing edge; (2) inserting a new edge; (3) deleting an existing node together with its incident edges; and (4) inserting a new isolated node. We also introduce a new method for measuring the performance of a network in recovering from these types of topology changes that takes into account the aforementioned confinement property. This new method measures the number of "computationally-meaningful" steps made by the individual nodes, which are essentially all steps in which the node participates (in the weakest possible sense) in the global computational process. An algorithm is said to be *effectively confining* if (i) the runtime of the nodes that are not adjacent to any topology change is $\log^{\mathcal{O}(1)} n$; and (ii) the global runtime (including all surviving nodes) is $(C+1)\log^{\mathcal{O}(1)} n$, where C is the number of topology changes throughout the execution. In other words, the global runtime is $\log^{\mathcal{O}(1)} n$ when amortized over the number of changes.

Following that, we turn our attention to the extensively studied *maximal independent set (MIS)* problem and design a randomized effectively confining algorithm for it under the Stone Age model extended to dynamic topology changes. This is achieved by carefully augmenting the MIS algorithm introduced in [13] with new components, tailored to ensure fast recovery from topology changes. Being a first step in the study of recovery from dynamic changes under the Stone Age model, our algorithm assumes a *synchronous* environment and it remains an open question whether this assumption can be lifted. Nevertheless, this assumption is justified by the findings of Fisher et al. [14] that model cellular networks as being subject to a *bounded asynchrony* scheduler, which is equivalent to a synchronous environment from an algorithmic perspective.

Paper's Organization. An extension of the Stone Age model of [13] to dynamic topology changes is presented in Sect. 2 together with our new method for evaluating the recovery performance of distributed algorithms. In Sect. 3, we describe the details of our MIS algorithm. Then, in Sect. 4.1, we show that each node not affected by a topology change will reach an output state in $\mathcal{O}(\log^2 n)$ rounds. In Sects. 4.2 and 4.3, we finish the analysis of our MIS algorithm by establishing an $\mathcal{O}((C+1)\log^2 n)$ upper bound on the global runtime. The runtime of the new MIS algorithm is shown to be near-optimal in Sect. 5 by proving that the global runtime of any algorithm is $\Omega(C)$. In the full version of the paper, we show that the runtime of any node u can be further bounded by $\mathcal{O}((C_u+1)\log^2 n)$, where C_u is the number of topology changes that occur within $\mathcal{O}(\log n)$ hops from u.

Related work. The standard model for a network of communicating devices is the message passing model [28,35]. There are several variants of this model, where the power of the network has been weakened. Perhaps the best-known variant of the message passing model is the congest model, where the message size is limited to size logarithmic in the size of the input graph [35]. A step to weaken the model further is to consider interference of messages, i.e., a node only hears a message if it receives a single message per round — cf. the *radio network model* [9]. In the *beeping model* [11,15], the communication capabilities are reduced further by only allowing to send beeps that do not carry information, where a listening node cannot distinguish between a single beep and multiple beeps transmitted by its neighbors.

The models mentioned above focus on limiting the communication but not the computation, i.e., the nodes are assumed to be strong enough to perform unlimited (local) Turing computations in each round. Networks of nodes weaker than Turing machines have been extensively studied in the context of *cellular automata* [16,33,41]. While the cellular automata model typically considers the deployment of finite state machines in highly regular graph structures such as the grid, the question of modeling cellular automata in arbitrary graphs was tackled by Marr and Hütt in [31], where a node changes its binary state according to the densities of its neighboring states. Another extensively studied model for distributed computing in a network of finite state machines is *population protocols* [4] (see also [5,32]), where the nodes communicate through a series of pairwise rendezvous. Refer to [13] for a comprehensive account of the similarities and differences between the Stone Age model and the models of cellular automata and population protocols.

Distributed computing in dynamic networks has been extensively studied [6, 19,23,26,40]. A classic result by Awerbuch and Sipser states that under the message passing model, any algorithm designed to run in a static network can be transformed into an algorithm that runs in a dynamic network with only a constant multiplicative runtime overhead [7]. However, the transformation of Awerbuch and Sipser requires storing the whole execution history and sending it around the network, which is not possible under the Stone Age model. Some dynamic network papers rely on the assumption that the topology changes are spaced in time so that the system has an opportunity to recover before another change occurs [20, 21, 30]. The current paper does not make this assumption.

The *maximal independent set (MIS)* problem has a long history in the context of distributed algorithms [3, 10, 28, 29, 34, 39]. Arguably the most significant breakthrough in the study of message passing MIS algorithms was the $\mathcal{O}(\log n)$ algorithm of Luby [29] (developed independently also by Alon et al. [3]). Later, Barenboim et al. [8] showed an upper bound of $2^{\mathcal{O}(\sqrt{\log\log n})}$ in the case of polylogarithmic maximum degree and an $\mathcal{O}(\log \Delta + \sqrt{\log n})$ bound for general graphs. For growth bounded graphs, it was shown by Schneider et al. [36] that MIS can be computed in $\mathcal{O}(\log^* n)$ time. In a recent work, Ghaffari studied the *local* complexity of computing an MIS, where the time complexity is measured from the perspective of a single node, instead of the whole network [17]. In a similar spirit, we provide, in addition to a global runtime bound, a runtime analysis from the perspective of a single node in the full version of the paper.

In the radio networks realm, with f channels, an MIS can be computed in $\Theta(\log^2 n/f) + \tilde{\mathcal{O}}(\log n)$ time [12], where $\tilde{\mathcal{O}}$ hides factors polynomial in $\log\log n$. The MIS problem was extensively studied also under the beeping model [1, 2, 37]. Afek et al. [1] proved that if the nodes are provided with an upper bound on n or if they are given a common sense of time, then the MIS problem can be solved in $\mathcal{O}(\log^{\mathcal{O}(1)} n)$ time. This was improved to $\mathcal{O}(\log n)$ by Scott et al. [37] assuming that the nodes can detect sender collision.

On the negative side, the seminal work of Linial [28] provides a runtime lower bound of $\Omega(\log^* n)$ for computing an MIS in an n-node ring under the message passing model [28]. Kuhn et al. [22] established a stronger lower bound for general graphs stating that it takes $\Omega\left(\sqrt{\frac{\log n}{\log\log n}}\right) + \Omega\left(\frac{\log \Delta}{\log\log \Delta}\right)$ rounds to compute an MIS, where Δ is the maximum degree of the input graph. For uniform algorithms in radio networks (and therefore, also for the beeping model) with asynchronous wake up schedule, there exists a lower bound of $\Omega(\sqrt{n/\log n})$ communication rounds [1].

Containing faults within a small radius of the faulty node has been studied in the context of *self-stabilization* [18]. An elegant MIS algorithm was developed under the assumption that the activation times of the nodes are controlled by a *central daemon* who activates the nodes one at a time [27, 38]. In contrast, we follow the common assumption that all nodes are activated simultaneously in each round. In the self-stabilization realm, the performance of an algorithm is typically measured as a function of some network parameter, such as the size of the network or the maximum degree, whereas in the current paper, the performance depends also on the number of failures.

With respect to the performance evaluation, perhaps the works closest to ours are by Kutten and Peleg [24, 25], where the concepts of *mending algorithms* and *tight fault locality* are introduced. The idea behind a fault local mending algorithm is to be able to recover a legal state of the network after a fault occurs, measuring the performance in terms of the number of faults. The term tight fault locality reflects the property that an algorithm running in time $\mathcal{O}(T(n))$ without faults is able to mend the network in time $\mathcal{O}(T(F))$, where F denotes the number of faults. The algorithm of Kutten and Peleg recovers an MIS in time $\mathcal{O}(\log F)$, but they use techniques that require nodes to count beyond constant numbers, which is not possible in the Stone Age model. Furthermore, they consider transient faults, whereas we consider permanent changes in the network topology.

2   Model

Consider some network represented by an undirected graph $G = (V, E)$, where the nodes in V correspond to the network's computational units and the edges represent bidirectional communication channels. Adopting the (static) Stone Age model of [13], each node $v \in V$ runs an algorithm captured by the 8-tuple $\Pi = \langle Q, q_0, q_{yes}, q_{no}, \Sigma, \sigma_0, b, \delta \rangle$, where Q is a fixed set of *states*; $q_0 \in Q$ is the *initial state* in which all nodes reside when they wake up for the first time; q_{yes} and q_{no} are the *output states*, where the former (resp., latter) represents membership (resp., non-membership) in the output MIS; Σ is a fixed *communication alphabet*; $\sigma_0 \in \Sigma$ is the *initial letter*; $b \in \mathbb{Z}_{>0}$ is a *bounding parameter*; and $\delta : Q \times \{0, 1, \ldots, b\}^{|\Sigma|} \rightarrow 2^{Q \times (\Sigma \cup \{\varepsilon\})}$ is the *transition function*. For convenience, we sometimes denote a state transition from q to q' by $(q \rightarrow q')$ and omit the rules associated with this transition from the notation.

Node v communicates with its neighbors by transmitting messages that consist of a single letter $\sigma \in \Sigma$ such that the same letter σ is sent to all neighbors. It is assumed that v holds a port $\phi_u(v)$ for each neighbor u of v in which the last message (a letter in Σ) received from u is stored. Transmitting the designated empty symbol ε corresponds to the case where u does not transmit any message. In other words, when node u transmits the ε letter, the letters in ports $\phi_u(v)$, for all neighbors v of u, remain unchanged. In the beginning of the execution, all ports contain the initial letter σ_0.

The execution of the algorithm proceeds in discrete synchronous rounds indexed by the positive integers. In each round $r \in \mathbb{Z}_{>0}$, node v is in some state $q \in Q$. Let $\sharp(\sigma)$ be the number of appearances of the letter $\sigma \in \Sigma$ in v's ports in round r and let $\langle \min\{\sharp(\sigma), b\} \rangle_{\sigma \in \Sigma}$ be a Σ-indexed vector whose σ-entry is set to the minimum between $\sharp(\sigma)$ and the bounding parameter b. Then the state q' in which v resides in round $r + 1$ and the message σ' that v sends in round r (appears in the corresponding ports of v's neighbors in round $r + 1$ unless $\sigma' = \varepsilon$) are chosen uniformly at random among the pairs in

$$\delta\left(q, \langle \min\{\sharp(\sigma), b\} \rangle_{\sigma \in \Sigma}\right) \subseteq Q \times (\Sigma \cup \{\varepsilon\}).$$

This means that v tosses an unbiased die (with $|\delta(q, \langle \min\{\sharp(\sigma), b\} \rangle_{\sigma \in \Sigma})|$ faces) when deciding on q' and σ'.

To ensure well-defined state transitions, we require that $|\delta(q, x)| \geq 1$ for all $q \in Q$ and $x \in \{0, 1, \ldots, b\}^{|\Sigma|}$ and say that a state transition is *deterministic* if $|\delta(q, x)| = 1$.

Topology changes. In contrast to the model presented in [13], our network model supports dynamic *topology changes* that belong to the following four classes:

1. *Edge deletion*: Remove a selected edge $e = \{u, v\}$ from the current graph. The corresponding ports $\phi_u(v)$ and $\phi_v(u)$ are removed and the messages stored in them are erased.
2. *Edge insertion*: Add an edge connecting nodes $u, v \in V$ to the current graph. New ports $\phi_u(v)$ and $\phi_v(u)$ are introduced storing the initial letter $\sigma_0 \in \Sigma$.
3. *Node deletion*: Remove a selected node $v \in V$ from the current graph with all its incident edges. The corresponding ports $\phi_v(u)$ of all v's neighbors u are eliminated and the messages stored in them are erased.
4. *Node insertion*: Add a new isolated (i.e., with no neighbors) node to the current graph. Initially, the node resides in the initial state q_0.

We assume that the schedule of these topology changes is controlled by an *oblivious adversary*. Formally, the strategy of the adversary associates a (possibly empty) set of topology changes with each round $r \in \mathbb{Z}_{>0}$ of the execution so that the total number of changes throughout the execution, denoted by C, is finite. This strategy may depend on the algorithm Π, but not on the random choices made during the execution.

To be precise, each round is divided into 4 successive steps as follows: (i) messages arrive at their destination ports; (ii) topology changes occur; (iii) the transition function is applied; and (iv) messages are transmitted. In particular, a message transmitted by node v in round r will not be read by node u in round $r+1$ if edge $\{u, v\}$ is inserted or deleted in round $r+1$. It is convenient to define the *adversarial graph sequence* $\mathcal{G} = G_1, G_2, \ldots$ so that G_1 is the initial graph and G_{r+1} is the graph obtained from G_r by applying to it the topology changes scheduled for round r. By definition, $\mathcal{G}$ is fully determined by the initial graph and the adversarial policy (and vice versa). The requirement that C is finite implies, in particular, that $\mathcal{G}$ admits an infinite suffix of identical graphs. Let n be the largest number of nodes that co-existed in the same round, i.e., $n = \max_r |V(G_r)|$, where $V(G_r)$ is the node set of G_r.

Correctness. We say that node u resides in state q in round r if the state of u is q at the beginning of round r. The algorithm is said to be in an *output configuration* in round r if every node $u \in G_r$ resides in an output state (q_{yes} or q_{no}). The output configuration is said to be *correct* if the states of the nodes (treated as their output) correspond to a valid MIS of the current graph G_r. An algorithm Π is said to be *correct* if the following conditions are satisfied for every adversarial graph sequence: (C1) If Π is in a non-output configuration in round r, then it will move to an output configuration in some (finite) round $r' > r$ w.p. 1.[1] (C2) If Π is in an output configuration in round r, then this output configuration is correct (with respect to G_r). (C3) If Π is in an output configuration in round r and $G_{r+1} = G_r$, then Π remains in the same output configuration in round $r + 1$.

---

[1] Throughout, we use w.p. to abbreviate "with probability" and w.h.p. to abbreviate "with high probability", i.e., with probability n^{-c} for any constant c.

Restrictions on the Output States. Under the model of [13], the nodes are not allowed to change their output once they have entered an output state. On the other hand, this model allows for multiple output states that correspond to the same problem output ("yes" and "no" in the MIS case). In other words, it is required that the output states that correspond to each problem output form a sink of the transition function. Since our model accommodates dynamic topology changes which might turn a correct output configuration into an incorrect one, we lift this restriction: our model allows transitions from an output state to a non-output state, thus providing the algorithm designer with the possibility to escape output configurations that become incorrect. Nevertheless, to prevent nodes in an output state from taking any *meaningful* part in the computation process, we introduce the following new (with respect to [13]) restrictions: (1) each possible output is represented by a unique output state; (2) a transition from an output state to itself is never accompanied by a letter transmission (i.e., it transmits ε); (3) all transitions originating from an output state must be deterministic; and (4) a transition from an output state must lead either to itself or to a non-output state.

Runtime. Fix some adversarial graph sequence $\mathcal{G} = G_1, G_2, \ldots$ and let η be an execution of a correct algorithm Π on $\mathcal{G}$ (determined by the random choices of Π). Round r in η is said to be *silent* if Π is in a correct output configuration and no topology change occurs in round r. The *global runtime* of η is the number of non-silent rounds. Node v is said to be *active* in round r of η if it resides in a non-output state, i.e., some state in $Q - \{q_{yes}, q_{no}\}$. The *(local) runtime* of v in η is the number of rounds in which v is active.

Let $N_r(v)$ be the (inclusive) neighborhood of v and let $E_r(v)$ be the edges incident to v in G_r. Node v is said to be *affected* under $\mathcal{G}$ if either v or one of its neighbors experienced an edge insertion/deletion, that is, there exists some round r and some node $u \in N_r(v)$ such that $E_r(u)$ is not identical to $E_{r+1}(u)$. Algorithm Π is said to be *effectively confining* if the following conditions are satisfied for every adversarial graph sequence $\mathcal{G}$: (1) the expected maximum local runtime of the non-affected nodes is $\log^{\mathcal{O}(1)} n$; and (2) the expected global runtime is $(C+1) \cdot \log^{\mathcal{O}(1)} n$, namely, $\log^{\mathcal{O}(1)} n$ when amortized over the number of topology changes. Notice that the bound on the global runtime directly implies the same bound on the local runtime of any affected node.

3 An MIS Algorithm

Our main goal is to design an algorithm under the Stone Age model for the maximal independent set (MIS) problem that is able to tolerate topology changes. For a graph $G = (V, E)$, a set of nodes $I \subseteq V$ is independent if for all $u, v \in I$, $\{u, v\} \notin E$. An independent set I is maximal if there is no other set $I' \subseteq V$ such that $I \subset I'$ and I' is independent.

Following the terminology introduced in the model section, we show that our algorithm is effectively confining. In Sect. 5, we provide a straightforward lower bound example that shows that under our model, our solution is within a polylogarithmic factor from optimal. In other words, the linear dependency on the number of changes is inevitable. Throughout, the bounding parameter of our algorithm is 1. This indicates that a node is able to distinguish between the cases of having a single appearance of a letter σ in some of its ports and not having any appearances of σ in any of its ports.

The basic idea behind our algorithm is that we first use techniques from [13] to come up with an MIS quickly and then we fix any errors that the topology changes might induce. In other words, our goal is to first come up with a *proportional* MIS, where the likelihood of a node to join the MIS is inversely proportional to the number of neighbors the node has that have not yet decided their output. We partition the state set of our algorithm into two components. One of the components contains the input state and is responsible for computing the proportional MIS. Once a node has reached an output state or detects that it has been affected by a topology change, it transitions to the second component, which is responsible for fixing the errors, and never enters an active state of the first component. One of the reasons behind dividing our algorithm into two seemingly similar components is that the first component has stronger runtime guarantees, but requires that the nodes start in a "nice" configuration, whereas the second component does not have this requirement while providing weaker runtime bounds.

3.1 The Proportional Component

The component that computes the proportional MIS, called `Proportional`, follows closely the design from [13]. The goal of the rest of the section is to introduce a slightly modified version of their algorithm that allows the non-affected nodes to ignore affected nodes while constructing the initial MIS.

`Proportional` consists of states $Q = \{S, D_1, D_2, U_0, U_1, U_2, W, L\}$, where $Q_a = Q - \{W, L\}$ are referred to as *active* states and $q_{yes} = W$ and $q_{no} = L$ as *passive* states. We set S as the initial state. The communication alphabet is identical to the set of states; the algorithm is designed so that node v transmits letter q in round r whenever it resides in state q in round $r + 1$, i.e., state q was returned by the application of the transition function in round r.

Let us denote a state transition between states q and q' by $q \to q'$ omitting the rules associated with this transition. To ease our notation and to make our illustrations more readable, we say that each state transition $q \to q'$ is *delayed* by a set $\mathcal{D} = \mathcal{D}(q \to q') \subseteq Q$ of delaying states. For $q \to q'$ the set of delaying states corresponds to the states in $Q - \{q\}$ from which there is a state transition to q. Transition $q \to q'$ being delayed by state q'' indicates that node v does not execute transition $q \to q'$ as long as there is at least one letter in its ports that corresponds to the state q''. We say that node v in state q is delayed if there is a neighbor w of v that resides in a state that delays at least one of the transitions from q, i.e., node v cannot execute some transition because of node w.

The main idea of `Proportional` is that each node v iteratively competes against its neighbors and the winner of a competition enters the MIS. Every competition, for an active node v and its active neighbors, begins by the nodes entering state U_0. During each round every node in state $U_j, j \in \{0,1,2\}$, assuming that it is not delayed, tosses a fair coin and proceeds to $U_{j+1 \mod 3}$ if the coin shows heads and to D_2 otherwise. Notice that a competition does not necessarily start in the same round for two neighboring nodes. The circular delaying logic of the U-states ensure that node v is always at most one coin toss ahead of its neighbors. If node v observes that it is the only node in its neighborhood in a U-state, it enters state W, which corresponds to joining the MIS. In the case where, due to unfortunate coin tosses, v moves to state D_2 along with all of its neighbors, node v restarts the process by entering state D_1.

We call a maximal contiguous sequence of rounds v spends in state $q \in Q_a$ a *q-turn* and a maximal contiguous sequence of turns starting from a D_1-turn and not including another D_1-turn a *tournament*. We index the tournaments and the turns within the tournament by positive integers. The correctness of `Proportional` in the case of no topology changes follows the same line of arguments as in [13].

We note that dynamically adding edges between nodes in states U_0, U_1, and U_2 could potentially cause a deadlock. To avoid this, we add a state transition, that is not delayed by any state, from each state $Q - \{S, L\}$ to state D' (that is part of the fixing component explained in Sect. 3.2) under the condition that a node reads the initial letter S in at least one of its ports. In other words, node v enters state D' if an incident edge is added in any round $r > 1$. Refer to Fig. 1 for a detailed, though somewhat cluttered, illustration. If node v transitions to state D' due to this condition being met or if v is deleted while being in an active state, we say that v is *excluded* from the execution of `Proportional`.

3.2 The Greedy Component

Now we extend `Proportional` to fix the MIS in the case that a topology change leaves the network in an illegal state. This extension of `Proportional` is referred to as the *greedy component* and denoted by `Greedy`.

Intuition spotlight: The basic idea is that nodes in states L and W verify locally that the configuration corresponds to a legal state of the network. If a node detects that the local configuration does not correspond to an MIS anymore, it revokes its output and tries again to join the MIS according to a slightly modified competition logic. The crucial difference is that we design Greedy without the circular delay logic. The reason behind the design is twofold: First, we cannot afford long chains of nodes being delayed. Second, since a dynamic change is only allowed to affect its 1-hop neighborhood, we cannot maintain the local synchrony similarly to Proportional.

More precisely, the state set of the algorithm is extended by $Q_2 = \{U', D'\}$, where states U' or D' are referred to as active. To detect an invalid configuration, we add the following state transitions from the output states W and L: a transition from state L to state D' in case that a node v resides in state L and does not have any letters W in its ports and a transition from state W to

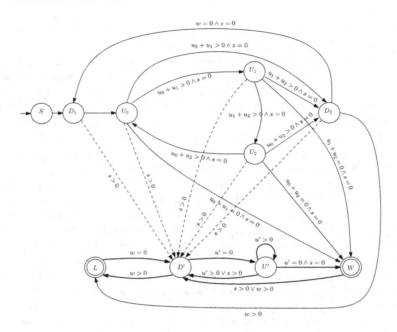

Fig. 1. The transition function of our MIS algorithm. The dashed lines correspond to the transitions that exclude nodes from the execution of `Proportional`. A state transition $q \to q'$ is delayed by state q'' if there is a transition indicated by a dashed edge or a thin edge from state q'' to q. A bold edge from q to q' implies that $q' \to q''$ is not delayed by q. As an example, the state transition from D' to U' is not delayed by state L and conversely, the state transition from D' to U' is delayed by state D_1. Similarly to the other state machine illustrations, the lower case letters in the transition rules correspond to the number of appearances of the corresponding letter and the rules associated with the delays are omitted for clarity. If a node v is in state q and none of the conditions associated with the transitions from q are satisfied (e.g., v is in state D', has a neighbor in state U', and no neighbors in state W), then v remains in state q. These self-loops are omitted from the picture for clarity.

state D' in case v resides in state W and reads a letter W in its ports. Finally, to prevent nodes in the active states of `Proportional` and `Greedy` from entering state W at the same time and thus, inducing an incorrect output configuration, we add a transition from W and U' to D' in case v reads the initial letter S.

The logic of the new states U' and D' is the following: node v in state D' goes into state U' if it does not read the letter U' in its ports. Then u and its neighbors that transitioned from D' to U' during the same round compete against each other: In every round, v tosses a fair coin and goes back to state D' if the coin shows tails. We emphasize that nodes in state U' are not delayed by nodes in state D'. Then, if v remains the only node in its neighborhood in state U', it declares itself as a winner and moves to state W. Conversely, if a neighbor of v wins, v goes into state L.

Observation 1 *(Proof deferred to the full version). A non-affected node is never in an active state of* Greedy.

4   Analysis

The authors of [13] analyze an algorithm very similar to the one induced by Proportional, showing that its runtime (on a static graph) is $O(\log^2 n)$ in expectation and w.h.p. In their analysis, they prove that with a positive constant probability, the number of edges decreases exponentially from one tournament to the next.

> **Intuition spotlight:** *The dynamic nature of the graph $\mathcal{G}$ prevents us from analyzing the execution of* Proportional *in the same manner. To overcome this obstacle, we introduce an auxiliary execution $\mathcal{E}$ of* Proportional *that turns out to be much easier to analyze, yet serves as a good approximation for the real execution.*

Consider some execution η of Proportional on $\mathcal{G}$ and recall that η is determined by the adversarial graph sequence $\mathcal{G}$ and by the coin tosses of the nodes. We associate with η the *auxiliary execution* $\mathcal{E} = \mathcal{E}(\eta)$ obtained from η by copying the same coin tosses (for each node) and modifying the adversarial policy by applying to it the following two rules for $i = 1, 2, \ldots$: (i) If node u is *excluded* during tournament i under η, then, instead, we *delete* node u immediately following its last U-turn under $\mathcal{E}(\eta)$. (ii) If node u becomes passive during tournament i under η and it would have remained active throughout tournament i under $\mathcal{E}(\eta)$, then we also *delete* node u immediately following its last U-turn under $\mathcal{E}(\eta)$.

Let $V_\eta(i)$ and $V(i)$ be the sets of nodes for which tournament i exists under η and $\mathcal{E}(\eta)$, respectively. Let $X^v(i)$ and $Y^v(i)$ denote the number of U-turns of node v in tournament i under η and $\mathcal{E}(\eta)$, respectively; to ensure that $X^v(i)$ and $Y^v(i)$ are well defined, we set $X^v(i) = 0$ if $v \notin V_\eta(i)$ and $Y^v(i) = 0$ if $v \notin V(i)$, i.e., if tournament i does not exist for v under η and $\mathcal{E}(\eta)$, respectively. Let $\Gamma_\eta(v, i)$ and $\Gamma(v, i)$ denote the (exclusive) neighborhood of v at the beginning of tournament i under η and $\mathcal{E}(\eta)$, respectively. Observe that by definition, $\Gamma_\eta(v, i) \subseteq V_\eta(i)$ and $\Gamma(v, i) \subseteq V(i)$. We say that node v *wins* in tournament i under η if $X^v(i) > X^w(i)$ for all $w \in \Gamma_\eta(v, i)$ and if v is not excluded in tournament i; likewise, we say that node v *wins* in tournament i under $\mathcal{E}(\eta)$ if $Y^v(i) > Y^w(i)$ for all $w \in \Gamma(v, i)$ and if v is not deleted in tournament i. The following lemma plays a key role in showing that the number of tournaments in $\mathcal{E}(\eta)$ is at least as large as that in η.

Lemma 2 *(Proof deferred to the full version). If $v \in V(i)$ wins in tournament i under $\mathcal{E}(\eta)$, then v wins in tournament i under η.*

4.1   Runtime of Proportional

Next, we analyze the number of tournaments in $\mathcal{E}(\eta)$. Once we have a bound on the number of tournaments executed in $\mathcal{E}(\eta)$, it is fairly easy to obtain the same bound for η. Similarly as before, the set of nodes for which tournament i exists according to $\mathcal{E}(\eta)$ is denoted by $V(i)$. Furthermore, let $\Gamma(v,i)$ be the set of neighbors of v for which tournament i exists in $\mathcal{E}(\eta)$ and let $E(i) = \bigcup_{v \in V(i)} \{\{v,w\} \mid w \in \Gamma(v,i)\}$ and $G(i) = (V(i), E(i))$. Notice that $G(i)$ is defined for the sake of the analysis and the topology of the underlying graph does not necessarily correspond to G_i in any round round i, where G_i is the ith element in the adversarial graph sequence.

According to the design of `Proportional`, no node that once enters a passive state can enter an active state of `Proportional`. Furthermore, if any edge is added adjacent to node v after v has transitioned out of state S, node v will not participate in any tournament after the addition. Therefore, we get that $V(i+1) \subseteq V(i)$ and $E(i+1) \subseteq E(i)$ for any i. The following lemma plays a crucial role in the runtime analysis of `Proportional`.

Lemma 3 *(Proof deferred to the full version). There are two constant $0 < p, \ell < 1$ such that $|E(i+1)| \le \ell |E(i)|$ with probability p.*

Theorem 4 *(Proof deferred to the full version). In expectation and w.h.p., any node participates in $\mathcal{O}(\log n)$ tournaments before becoming passive.*

The last step of the analysis of `Proportional` is to bound the number of rounds that any node v spends in an active state. Let $V_{\mathrm{MAX}}(v)$ denote the maximal connected component of nodes in active states of `Proportional` that contains node v. Consider the following modification of `Proportional`: before starting tournament $i+1$, node v waits for every other node in $V_{\mathrm{MAX}}(v)$ to finish tournament i or to become excluded. Notice that any node that gets inserted into the graph after node v has exited state S will not be part of $V_{\mathrm{MAX}}(v)$. We do not claim that we know how to implement such a modification, but clearly the modified process is not faster than the original one.

The length of tournament i is determined by $\max_v\{X^v(i)\}$. Given that the random variables $X^v(i)$ are independent and follow the $\mathrm{Geom}(1/2)$ distribution, we get that $\max_v\{X^v(i)\} \in \mathcal{O}(\log n)$ with high probability and in expectation. Combining with Observation 1, we get the following corollaries.

Corollary 5. *Let $t_v \ge 1$ be the round in which node v is inserted into the graph, where $t_v = 1$ if $v \in G_1$. If v is not deleted, it enters either state W or L by time $t_v + \mathcal{O}(\log^2 n)$ w.h.p. and in expectation.*

Corollary 6. *The maximum runtime of any non-affected node is $\mathcal{O}(\log^2 n)$ in expectation and w.h.p.*

4.2   The Quality of an MIS

Before we go to the analysis of `Greedy`, we want to point out that even though an invalid configuration can be detected locally, a single topology change can have an influence in an arbitrary subgraph.

We begin the analysis of `Greedy` by taking a closer look at the properties of non-affected nodes in the passive states and show that if a node has a high degree, it is either unlikely for this node to be in the MIS or that the neighbors of this node have more than one MIS node in their neighborhoods. Let i be the index of the tournament in which node v enters state W. We say that node v *covers* node $w \in \Gamma(v, i) \cup \{v\}$ if w entered state L in tournament i.

Definition 7. *The quality $q(v)$ of node v is given by $q(v) = |\{w \in \Gamma(v, i) \cup \{v\} \mid v \text{ covers } w\}|$. The quality $q(B)$ of any set of nodes B is defined as $\sum_{v \in B} q(v)$.*

Lemma 8 *(Proof deferred to the full version).* $\mathbb{E}[q(v)] \in \mathcal{O}(\log n)$ *for any node v.*

The quality of a node v gives an upper bound on the number of nodes in the neighborhood of v that are only covered by v. Let v be a node in state L. If t is the first round such that there are no nodes adjacent to v in state W, we say that v is *released* at time t. Similarly, we say that node v in state W is released if an adjacent edge is added. Every node can only be released once, i.e., node v counts as released even if it eventually again has a neighboring node in state W or if it enters state W.

Lemma 9 *(Proof deferred to the full version).* *Let H be the set of nodes that are eventually released. Then $\mathbb{E}[|H|] \in \mathcal{O}(C \log n)$.*

4.3   Fixing the MIS

We call a maximal contiguous sequence of rounds in which node v resides in state U' a *greedy tournament*. Unlike with `Proportional`, we index these greedy tournaments by the time this particular tournament starts. In other words, if node v resides in state D' in round $t - 1$ and in state U' in round t, we index this tournament by t. We denote the random variable that counts the number of transitions from U' to itself in greedy tournament t by node v by $X^v(t)$. Notice that $X^v(t)$ obeys $\text{Geom}(1/2)$ unless v is deleted or an adjacent edge is added during greedy tournament t. We say that a greedy tournament t is *active* if there is at least one node v in state U' of greedy tournament t.

Observation 10 *(Proof deferred to the full version).* *There are at most $\mathcal{O}(\log n)$ rounds in which greedy tournament t is active in expectation and w.h.p.*

Lemma 11 *(Proof deferred to the full version).* *The total expected number of greedy tournaments is $\mathcal{O}(C \log n)$.*

Observation 12 *(Proof deferred to the full version).* *Let k be the total number of greedy tournaments. There are at most $2k + 2C$ non-silent rounds without either (i) at least one node in an active state of `Proportional` or (ii) an active greedy tournament.*

Next we show that our MIS algorithm, that consists of `Proportional` and `Greedy`, fulfills the correctness properties given in the model section. Then, we establish that our MIS algorithm is indeed effectively confining.

Lemma 13 *(Proof deferred to the full version). Our MIS algorithm is correct.*

Theorem 14 *(Proof deferred to the full version). The global runtime of our MIS algorithm is $\mathcal{O}((C+1)\log^2 n)$ in expectation.*

5   Lower Bound

The runtime of our algorithm might seem rather slow at the first glance, since it is linear in the number of topology changes. In this section, we show that one cannot get rid of the linear dependency, i.e., there are graphs where the runtime of any algorithm grows at least linearly with the number of changes. In particular, we construct a graph where the runtime of any algorithm is $\Omega(C)$.

Let G^ℓ be a graph that consists of n nodes and $n = 3\ell$. The graph consists of ℓ components B_i, where B_i is a 3-clique for each $1 \leq i \leq \ell$. In addition, let $u_1, \ldots, u_\ell$ be a set of nodes such that $u_i \in B_i$ for every i.

Theorem 15. *There exists a graph G and an schedule of updates such that the runtime of any algorithm on G is $\Omega(C)$ in expectation and w.h.p.*

Proof. Consider graph G^ℓ and the following adversarial strategy. In round $2i - 1 \geq 1$, the adversary deletes node u_i, i.e., one of the nodes in component B_i. Our goal is to show that at least a constant fraction of the first $2C$ rounds are non-silent for any $C \leq n/3$. Consider round $2i$ and component B_i. Since the nodes in component B_i form a clique, their views are identical. Therefore, according to any algorithm that computes an MIS, their probability to join the MIS is equal, i.e., $1/3$. Thus, the probability that the nodes in $B_i - \{u_i\}$ are not in the MIS in any round $2j \leq 2i$ is at least $1/3$.

Let X_i then be the indicator random variable, where $X_i = 0$ if round i is silent and 1 otherwise and let $X = \sum_i^\infty X_i$ be the random variable that counts the number of non-silent rounds. Since any round $1 \leq 2i \leq 2C$ is non-silent with at least probability $1/3$, we get that $\mathbb{E}[X] \geq (1/3)C$. Now by applying a Chernoff bound, we get that $\mathbb{P}[X < 1/2\mathbb{E}[X]] = \mathbb{P}[X < (1/2) \cdot (1/3)C] \leq 2^{-C/12}$. Since $C = n/3$, we get that $\mathbb{P}[X < 1/2\mathbb{E}[X]] \in \mathcal{O}(n^{-k})$ for any constant k and thus, the claim follows.

References

1. Afek, Y., Alon, N., Bar-Joseph, Z., Cornejo, A., Haeupler, B., Kuhn, F.: Beeping a maximal independent set. In: Peleg, D. (ed.) DISC 2011. LNCS, vol. 6950, pp. 32–50. Springer, Heidelberg (2011). doi:10.1007/978-3-642-24100-0_3
2. Afek, Y., Alon, N., Barad, O., Hornstein, E., Barkai, N., Bar-Joseph, Z.: A biological solution to a fundamental distributed computing problem. Science **331**(6014), 183–185 (2011)
3. Alon, N., Babai, L., Itai, A.: A fast and simple randomized parallel algorithm for the maximal independent set problem. J. Algorithms **7**, 567–583 (1986)

4. Angluin, D., Aspnes, J., Diamadi, Z., Fischer, M.J., Peralta, R.: Computation in networks of passively mobile finite-state sensors. Distrib. Comput. **18**(4), 235–253 (2006)
5. Aspnes, J., Ruppert, E.: An introduction to population protocols. In: Garbinato, B., Miranda, H., Rodrigues, L. (eds.) Middleware for Network Eccentric and Mobile Applications, pp. 97–120. Springer, Heidelberg (2009)
6. Awerbuch, B., Patt-Shamir, B., Peleg, D., Saks, M.: Adapting to asynchronous dynamic networks (extended abstract). In: Proceedings of the 24th Annual ACM Symposium on Theory of Computing (STOC), pp. 557–570 (1992)
7. Awerbuch, B., Sipser, M.: Dynamic networks are as fast as static networks. In: 29th Annual Symposium on Foundations of Computer Science (FOCS), pp. 206–220 (1988)
8. Barenboim, L., Elkin, M., Pettie, S., Schneider, J.: The locality of distributed symmetry breaking. In: Proceedings of the 53rd Annual Symposium on Foundations of Computer Science (FOCS), pp. 321–330 (2012)
9. Chlamtac, I., Kutten, S.: On broadcasting in radio networks-problem analysis and protocol design. Commun., IEEE Trans. Commun. **33**(12), 1240–1246 (1985)
10. Cole, R., Vishkin, U.: Deterministic coin tossing with applications to optimal parallel list ranking. Inf. Control **70**(1), 32–53 (1986)
11. Cornejo, A., Kuhn, F.: Deploying wireless networks with beeps. In: Lynch, N.A., Shvartsman, A.A. (eds.) DISC 2010. LNCS, vol. 6343, pp. 148–162. Springer, Heidelberg (2010). doi:10.1007/978-3-642-15763-9_15
12. Daum, S., Ghaffari, M., Gilbert, S., Kuhn, F., Newport, C.: Maximal independent sets in multichannel radio networks. In: Proceedings of the ACM Symposium on Principles of Distributed Computing (PODC), pp. 335–344 (2013)
13. Emek, Y., Wattenhofer, R.: Stone age distributed computing. In: Proceedings of the ACM Symposium on Principles of Distributed Computing (PODC), pp. 137–146 (2013)
14. Fisher, J., Henzinger, T.A., Mateescu, M., Piterman, N.: Bounded asynchrony: concurrency for modeling cell-cell interactions. In: Fisher, J. (ed.) FMSB 2008. LNCS, vol. 5054, pp. 17–32. Springer, Heidelberg (2008). doi:10.1007/978-3-540-68413-8_2
15. Flury, R., Wattenhofer, R.: Slotted programming for sensor networks. In: IPSN (2010)
16. Gardner, M.: The fantastic combinations of John Conway's new solitaire game 'life'. Sci. Am. **223**(4), 120–123 (1970)
17. Ghaffari, M.: An improved distributed algorithm for maximal independent set. In: Proceedings of the 27th Annual ACM-SIAM Symposium on Discrete Algorithms (SODA), pp. 270–277 (2016)
18. Ghosh, S., Gupta, A., Herman, T., Pemmaraju, S.: Fault-containing self-stabilizing algorithms. In: Proceedings of the 15th Annual ACM Symposium on Principles of Distributed Computing (PODC), pp. 45–54 (1996)
19. Hayes, T., Saia, J., Trehan, A.: The forgiving graph: a distributed data structure for low stretch under adversarial attack. In: Proceedings of the 28th ACM Symposium on Principles of Distributed Computing (PODC), pp. 121–130 (2009)
20. König, M., Wattenhofer, R.: On local fixing. In: Baldoni, R., Nisse, N., Steen, M. (eds.) OPODIS 2013. LNCS, vol. 8304, pp. 191–205. Springer, Heidelberg (2013). doi:10.1007/978-3-319-03850-6_14
21. Korman, A.: Improved compact routing schemes for dynamic trees. In: Proceedings of the 27th Annual ACM Symposium on Principles of Distributed Computing (PODC), pp. 185–194 (2008)

22. Kuhn, F., Moscibroda, T., Wattenhofer, R.: What cannot be computed locally! In: Proceedings of the 23th Annual ACM Symposium on Principles of Distributed Computing (PODC), pp. 300–309 (2004)
23. Kuhn, F., Schmid, S., Wattenhofer, R.: A self-repairing peer-to-peer system resilient to dynamic adversarial churn. In: Castro, M., Renesse, R. (eds.) IPTPS 2005. LNCS, vol. 3640, pp. 13–23. Springer, Heidelberg (2005). doi:10.1007/11558989_2
24. Kutten, S., Peleg, D.: Fault-local distributed mending. J. Algorithms **30**(1), 144–165 (1999)
25. Kutten, S., Peleg, D.: Tight fault locality. SIAM J. Comput. **30**(1), 247–268 (2000)
26. Li, X., Misra, J., Plaxton, C.G.: Active and concurrent topology maintenance. In: Guerraoui, R. (ed.) DISC 2004. LNCS, vol. 3274, pp. 320–334. Springer, Heidelberg (2004). doi:10.1007/978-3-540-30186-8_23
27. Lin, J.C., Huang, T.: An efficient fault-containing self-stabilizing algorithm for finding a maximal independent set. IEEE Trans. Parallel Distrib. Syst. **14**(8), 742–754 (2003)
28. Linial, N.: Locality in distributed graph algorithms. SIAM J. Comput. **21**(1), 193–201 (1992)
29. Luby, M.: A simple parallel algorithm for the maximal independent set problem. SIAM J. Comput. **15**, 1036–1055 (1986)
30. Malpani, N., Welch, J.L., Waidya, N.: Leader election algorithms for mobile ad hoc networks. In: Proceedings of the 4th International Workshop on Discrete Algorithms and Methods for Mobile Computing and Communications (DIAL-M), pp. 96–103 (2003)
31. Marr, C., Hütt, M.T.: Outer-totalistic cellular automata on graphs. Phys. Lett. A **373**(5), 546–549 (2009)
32. Michail, O., Chatzigiannakis, I., Spirakis, P.G.: New Models for Population Protocols. Synthesis Lectures on Distributed Computing Theory. Morgan & Claypool Publishers, San Rafael (2011)
33. von Neumann, J.: Theory of Self-Reproducing Automata. University of Illinois Press, Champaign (1966)
34. Panconesi, A., Srinivasan, A.: On the complexity of distributed network decomposition. J. Algorithms **20**(2), 356–374 (1996)
35. Peleg, D.: Distributed Computing: A Locality-Sensitive Approach. Society for Industrial and Applied Mathematics, Philadelphia (2000)
36. Schneider, J., Wattenhofer, R.: A log-star distributed maximal independent set algorithm for growth-bounded graphs. In: Proceedings of the 27th ACM Symposium on Principles of Distributed Computing (PODC), pp. 35–44 (2008)
37. Scott, A., Jeavons, P., Xu, L.: Feedback from nature: an optimal distributed algorithm for maximal independent set selection. In: Proceedings of the 32nd Symposium on Principles of Distributed Computing (PODC), pp. 147–156 (2013)
38. Shukla, S., Rosenkrantz, D., Ravi, S.S.: Observations on self-stabilizing graph algorithms for anonymous networks (extended abstract). In: Proceedings of the Second Workshop on Self-Stabilizing Systems, pp. 1–15 (1995)
39. Valiant, L.G.: Parallel computation. In: 7th IBM Symposium on Mathematical Foundations of Computer Science (1982)
40. Walter, J., Welch, J.L., Vaidya, N.: A mutual exclusion algorithm for ad hoc mobile networks. Wireless Netw. **7**, 585–600 (1998)
41. Wolfram, S.: A New Kind of Science. Wolfram Media, Champaign (2002)

Setting Ports in an Anonymous Network: How to Reduce the Level of Symmetry?

Ralf Klasing[1], Adrian Kosowski[2], and Dominik Pajak[3(✉)]

[1] LaBRI, CNRS — Université de Bordeaux, 33400 Talence, France
ralf.klasing@labri.fr
[2] Inria and IRIF, CNRS — Université Paris Diderot, 75013 Paris, France
adrian.kosowski@inria.fr
[3] Faculty of Fundamental Problems of Technology, Institute of Informatics,
Wroclaw University of Technology, 50-370 Wroclaw, Poland
dominik.pajak@pwr.edu.pl

Abstract. A fundamental question in the setting of anonymous graphs concerns the ability of nodes to spontaneously break symmetries, based on their local perception of the network. In contrast to previous work, which focuses on symmetry breaking under arbitrary port labelings, in this paper, we study the following design question: Given an anonymous graph G without port labels, how to assign labels to the ports of G, in interval form at each vertex, so that symmetry breaking can be achieved using a message-passing protocol requiring as few rounds of synchronous communication as possible?

More formally, for an integer $l > 0$, the *truncated view* $\mathcal{V}_l(v)$ of a node v of a port-labeled graph is defined as a tree encoding labels encountered along all walks in the network which originate from node v and have length at most l, and we ask about an assignment of labels to the ports of G so that the views $\mathcal{V}_l(v)$ are distinct for all nodes $v \in V$, with the goal being to minimize l.

We present such efficient port labelings for any graph G, and we exhibit examples of graphs showing that the derived bounds are asymptotically optimal in general. More precisely, our results imply the following statements.

1. For any graph G with n nodes and diameter D, a uniformly random port labeling achieves $l = O(\min(D, \log n))$, w.h.p.
2. For any graph G with n nodes and diameter D, it is possible to construct in polynomial time a labeling that satisfies $l = O(\min(D, \log n))$.
3. For any integers $n \geq 2$ and $D \leq \log_2 n - \log_2 \log_2 n$, there exists a graph G with n nodes and diameter D which satisfies $l \geq \frac{1}{2}D - \frac{5}{2}$.

Research partially supported by the ANR project DISPLEXITY (ANR-11-BS02-014). This study has been carried out in the frame of the "Investments for the future" Programme IdEx Bordeaux - CPU (ANR-10-IDEX-03-02). Research partially supported by the National Science Centre, Poland - grant number 2015/17/B/ST6/01897.

© Springer International Publishing AG 2016
J. Suomela (Ed.): SIROCCO 2016, LNCS 9988, pp. 35–48, 2016.
DOI: 10.1007/978-3-319-48314-6_3

Keywords: Anonymous network · Port-labeled network · View · Level of symmetry

1   Introduction

Consider a network G in which nodes have no identifiers but ports at each node of degree d are uniquely labeled with the integers from 1 to d (in any order). A *view* from a node v in such a network is a rooted infinite tree defined recursively. This tree is composed of the views from all the neighbors of v connected to the root (corresponding to node v) using the edges with the same port labels as the edges connecting v to their children in graph G.

The concept of a view is useful for many applications as it allows for identification of the network topology and for breaking of symmetries between nodes. For example, the topological knowledge which can be gathered by a node of an anonymous graph running a deterministic algorithm, in a setup defined identically (uniformly) over nodes, corresponds to its view. More precisely, in an anonymous communication network, in k rounds, each node can learn up to k levels of its view. An important question for such applications is to determine the smallest possible integer l (also called the level of symmetry of G) such that for all the different views, their subtrees truncated at depth l are also different. As observed in [16] in the context of leader election in the $\mathcal{LOCAL}$ model, learning l levels of all the views is sufficient and necessary to determine whether leader election in the $\mathcal{LOCAL}$ model is feasible, hence l determines the number of communication rounds to elect a leader.

Another example of applications concerns walker-based models of computation. The information gathered by a walker traversing an anonymous graph, which does not have the ability to write to its environment, is simply a subtree of the view from its starting vertex. Hence, for example, rendezvous of deterministic walkers is only possible if they start from positions with different views.

A lot of the related work so far has considered the question of bounding the largest possible level of symmetry (taken over all labelings). By contrast, in this work, we would like to focus on the best case labelings, i.e., those having the smallest possible level of symmetry. Moreover, we also look at the case of uniformly random labelings to verify whether the average case is closer to the best or the worst case.

1.1   Overview of Results

For a graph G with labeling λ, we define the *level of symmetry* $l(G, \lambda)$ as the smallest integer l such that views from any two nodes, truncated at depth l, are different. If there exists a pair of nodes whose infinite views are equal, we say that the level of symmetry is infinite. For an unlabeled graph G, we refer to its level of symmetry as $l(G) = \min_\lambda l(G, \lambda)$, where the minimum is taken over all port labelings λ of G. We show the following results:

We first show that, for any graph G with n nodes and diameter D, the level of symmetry in the best case is $O(\min(D, \log n))$ by proposing an algorithm that labels any graph in such a way that the resulting labeled graph has level of symmetry $O(\min(D, \log n))$. Secondly, we show that a uniformly random labeling λ achieves level of symmetry $l(G, \lambda) = O(\min(D, \log n))$, w.h.p. Thirdly, we exhibit examples of graphs showing that these bounds are asymptotically tight in general.

1.2   Related Work

The notion of *view* was introduced and first studied by Yamashita and Kameda in [29] in the context of distributed message passing algorithms. Yamashita and Kameda proved that if views of two nodes truncated to depth n^2 are identical, then their infinite views are identical [29], where n is the number of nodes of the network. The bound has been improved to $n - 1$ by Norris [24]. Although this bound is asymptotically tight [1,24], it is far from being accurate for many networks. Hence, one may ask for bounds expressed as a function of different graph invariants. Fraigniaud and Pelc proved in [15] that if two nodes have the same views to depth $\hat{n} - 1$ then their views are the same, where $\hat{n}$ is the number of nodes having different views (or equivalently, $\hat{n}$ is the size of the quotient graph [29]). For some works on view computation see, e.g., [2,26]. Recently, Hendrickx [20] proved (for simple graphs with symmetric port labeling) an upper bound of $O(D \log(n/D))$ on the depth to which views need to be checked in order to be distinguished, where D is the diameter of the network, leaving the tightness of this bound as an open problem. A complementary bound of $\Omega(D \log(n/D))$ was shown in [9], and independently in [16] for the special case when $D = O(1)$.

View-based approaches have been successfully used when designing algorithms for various network problems, for example in leader election [4,7, 11,12,16,27,30]. In anonymous, port-labeled networks, the time to elect a leader or to declare the election infeasible is equal to $\Theta(D + l)$ [16] hence is $\Theta(D \log(n/D))$ [9,20] for the worst-case port labeling. Our results show that for the best- and random-case labeling the time of leader election reduces to $O(\min(D, \log n))$.

View-based approaches have also been used for map construction [3,10,12], rendezvous [6,8,19], and other tasks [14,28].

The result of Norris [24] is indeed more general as it works also for directed graphs and shows that in directed graphs the level of symmetry is at most $n - 1$. We may also consider the problem of distinguishing the views of two nodes from different graphs. We are interested in whether the isomorphism of two views truncated to a certain depth implies the isomorphism of the infinite views. Krebs and Verbitsky [23] showed that it is possible to construct two directed graphs, both with at most n nodes, such that the truncated views of two nodes from different graphs are isomorphic up to depth $(2 - o(1))n$.

In port-labeled graphs, the problem of setting the port labels to allow for quick exploration by a simple agent is well-studied. In most of the existing results, the considered agent is performing a so-called *basic walk* which consists in taking the port number $x + 1$ after entering a node via port number x (where $x + 1$ is taken modulo the degree of the node). It turns out that it is possible to find a port labeling such that an agent following the basic walk will explore the given graph in a number of rounds linear in n [5,13,17,18,21,22]. The question whether a port labeling allowing the basic walk to visit all the vertices can be assigned by a mobile agent making only local changes was posed in [13]. Steinová [25] answered affirmatively by exhibiting an agent with $O(\log n)$ bits of memory and one droppable pebble that assigns such a labeling.

1.3   Preliminaries and Notation

In this work, we consider anonymous port labeled (simple and undirected) networks $G = (V, E)$ (the terms graph and network are used interchangeably throughout) in which the nodes do not have identifiers and each edge $\{u, v\}$ has two integers assigned to its endpoints, called the *port numbers* at u and v, respectively. The port numbers are assigned in such a way that for each node v they are pairwise different and they form a consecutive set of integers $\{1, \ldots, d\}$, where d is the number of neighbors of v in G.

The number of neighbors of v in G is called the *degree* of v and is denoted by $\deg_G(v)$. The *maximum degree* of G (taken over all nodes v) is denoted by $\Delta(G)$. The *diameter* of G, denoted by D, is the maximum (taken over all pairs of nodes u and v) length of a shortest path between u and v in G. For a node $v \in V$, we write (G, v) to denote the rooted graph G with root v. For any $V' \subseteq V$, we denote by $G[V']$ the subgraph of G induced by V' and by $G[U]\backslash G[U']$ graph $G[U]$ without the edges that belong to $G[U']$. The set of nodes at distance at most l from a node $v \in V$ is denoted by $N_l(v)$, with $N(v) = N_1(v)$. A rooted subgraph $\mathcal{B}_l(v)$ of all nodes and edges reachable from v in a walk of at most l steps, $\mathcal{B}_l(v) = (G[N_l(v)]\backslash G[N_l(v)\backslash N_{l-1}(v)], v)$, will be called the *radius-l ball* around v. We recall the definition of a view [29]. Let G be a graph, v be a node of G and let λ be a port labeling for G. Given any $l \geq 0$, the *(truncated) view up to level l*, $\mathcal{V}_l(v)$, is defined as follows. $\mathcal{V}_0(v)$ is a tree consisting of a single node x_0. Then, $\mathcal{V}_{l+1}(v)$ is the port-labeled tree rooted at x_0 and constructed as follows. For every node v_i, $i \in \{1, \ldots, \deg_G(v)\}$, adjacent to v in G there is a child x_i of x_0 in $\mathcal{V}_{l+1}(v)$ such that the port number at x_0 corresponding to edge $\{x_0, x_i\}$ equals $\lambda(v, v_i)$, and the port number at x_i corresponding to edge $\{x_0, x_i\}$ equals $\lambda(v_i, v)$. For each $i \in \{1, \ldots, \deg_G(v)\}$ the node x_i is the root of the truncated view $\mathcal{V}_l(v_i)$. The *view* from v in G is the infinite port-labeled rooted tree $\mathcal{V}(v)$ such that $\mathcal{V}_l(v)$ is its truncation to level l, for each $l \geq 0$.

We will denote by $\mathcal{C}(G)$ the number of distinct labelings of a graph G. Notice that for each node v, its labeling can be defined independently of the other nodes in $\deg_G(v)!$ different ways, hence if the nodes of the graph were distinguishable then the number of distinct labelings would be equal to $\prod_{v \in V} \deg_G(v)!$. In our setting, nodes have no identifiers hence $\mathcal{C}(G) \leq \prod_{v \in V} \deg_G(v)!$.

Lemma 1. *The following bound holds for all $l > 1$ and all connected graphs G with more than 2 nodes:*

$$\mathcal{C}(\mathcal{B}_l(v)) \leq \mathcal{C}(G[N_l(v)]) \leq (\Delta(G[N_{l-1}(v)]))^{2|E(G[N_{l-1}(v)])|}$$
$$\leq 2^{2|E(\mathcal{B}_l(v))| \log_2 \Delta(\mathcal{B}_l(v))}. \tag{1}$$

Proof. From the definitions of $N_l(v)$ and $\mathcal{B}_l(v)$ we have, for any l, that $\mathcal{B}_{l-1}(v)$ is a subgraph of $G[N_{l-1}(v)]$, and $G[N_{l-1}(v)]$ is a subgraph of $\mathcal{B}_l(v)$. This shows the first and the last inequalities of (1).

Denote $H = G[N_{l-1}(v)]$. We have $\mathcal{C}(H) \leq \prod_{v \in H} \deg_H(v)!$. The middle inequality of (1) can be shown using the fact that $k! \leq k^k$, for all $k > 0$.

2   Lower Bounds

We start by providing a lower bound based on the following observation. Consider a graph G with labeling λ. If, for some integer $l > 0$ and for a pair of distinct nodes $u, v \in V$, their views truncated to depth l are distinct, but their radius-l balls are isomorphic (i.e., $\mathcal{V}_l(u) \neq \mathcal{V}_l(v)$ and $\mathcal{B}_l(u) \equiv \mathcal{B}_l(v)$), then the labeling λ around nodes u and v, restricted to their radius-l balls, must be distinct, i.e., $\lambda[\mathcal{B}_l(u)] \neq \lambda[\mathcal{B}_l(v)]$. It follows that if the number of isomorphic radius-l balls around nodes of G exceeds the number of possible labelings of such balls, we must have $l(G) > l$. (Here, isomorphism of rooted graphs (H_1, v_1) and (H_2, v_2) is understood as an isomorphism between H_1 and H_2 which maps v_1 into v_2.)

Lemma 2. *Given a graph G and $l > 0$, let $S \subseteq V$ be a subset of vertices such that all radius-l balls around nodes from S are isomorphic to some rooted graph B, $\mathcal{B}_l(v) \equiv B$. Then, $\mathcal{C}(B) < |S| \implies l(G) > l$.* □

The above lemma immediately implies the following two corollaries, the first of which applies inequality (1).

Corollary 1. *Given a connected graph G and $l > 0$, let $S \subseteq V$ be a subset of vertices such that all radius-l balls around nodes from S are isomorphic to some rooted graph B, $\mathcal{B}_l(v) \equiv B$. Then, $|E(B)| \log_2 \Delta(B) < \frac{1}{2} \log_2 |S| \implies l(G) \geq l$.* □

Corollary 2. *When G is a line or a ring, we have $l(G) \geq \frac{1}{2} \log_2 n - 2$.* □

Proposition 1. *For any integers $n \geq 2$ and $D \leq \log_2 n - \log_2 \log_2 n$, there exists a graph G with n nodes and diameter D which satisfies $l(G) \geq \frac{1}{2}D - \frac{5}{2}$.*

Proof. Assume that D is even. If $D < 6$ then the claim is trivial hence we assume that $D \geq 6$. It can be easily verified that for $D \leq \log_2 n - \log_2 \log_2 n$ and $n \geq 2$ we have $\lfloor \frac{n-1}{D-2} \rfloor \geq 4$. We construct a wheel graph G from a cycle with $n-1$ nodes and a node v connected to $x = \lfloor \frac{n-1}{D-2} \rfloor$ almost-equidistant nodes $s_1, s_2, \ldots, s_x$ on the cycle (see Fig. 1 for an example). We choose nodes $s_1, s_2, \ldots, s_x$ along the cycle in such a way that the distance between two consecutive nodes s_i and s_{i+1} (and between s_x and s_1) on the cycle is either $D-1$ or $D-2$. Let c_i be a middle

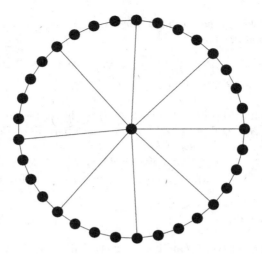

Fig. 1. Illustration of wheel graph for $n = 34$ and $D = 6$.

node of each such path and let c_x be a middle node on the path between s_x and s_1 (if the path has $D - 1$ edges then there are two middle nodes). Distance between any two (not consecutive) c_i and c_j is exactly D, because the shortest path goes via v and such not consecutive middle nodes exist because $x \geq 4$. Observe that the diameter of the graph is exactly D, because distance from v to any node is at most $D/2$. View from each c_i up to depth $D/2 - 2$ is a path with $D - 3$ nodes. There are 2^{D-4} possible labellings of such paths because at each vertex, different from c_i, we can either put port 1 towards c_i or away from c_i. Observe that

$$2^{D-4} \leq \frac{1}{16} 2^{\log_2 n - \log_2 \log_2 n} = \frac{n}{16 \log_2 n} < \left\lfloor \frac{n-1}{D-2} \right\rfloor.$$

Hence, by the pigeonhole principle, for at least two different nodes c_i, c_j their respective views up to depth $D/2 - 2$ are equal. Now observe that the view from the central node v is unique already at depth 1 hence by [20, Lemma 4], views from any pair of nodes is unique up do depth $D + 1$. This shows that $l(G) \geq D/2 - 2$.

If D is odd we construct the wheel graph on $n - 1$ nodes and diameter $D - 1$ and attach an extra node w to one of the middle nodes c. The construction of the wheel graph is possible because $\lfloor \frac{n-2}{D-3} \rfloor \geq 4$ holds true under our assumptions. We disregard the subpath to which we attach node w hence we obtain $\lfloor \frac{n-2}{D-3} \rfloor - 1$ paths of length $D - 4$. Similarly as in the previous case we have 2^{D-5} possible labelings of the paths and

$$2^{D-5} \leq \frac{1}{32} 2^{\log_2 n - \log_2 \log_2 n} = \frac{n}{32 \lg n} < \left\lfloor \frac{n-2}{D-3} \right\rfloor - 1.$$

Hence for odd D we obtain that $l(G) \geq (D-1)/2 - 2$.

3   Upper Bounds

In this section we want to propose algorithms for labeling of general graphs that will guarantee that all views up to certain depths will be distinct.

We will start by describing a simple procedure Greedy for labeling the ports of the graph. We start with a non-empty set X of nodes such that for any pair of nodes $u, v \in X$, if there is an edge between u and v then it has already been labeled (i.e., both ends of the edge have been labeled). Procedure Greedy labels all the remaining edges.

Procedure Greedy: *We pick any node $v \notin X$ and adjacent to some node from X. We label all the edges between v and the nodes from X (in any order) by choosing the smallest available label at both endpoints of such edges and we add v to X. Repeat until all edges are labeled.*

Let us also define two invariants. We will show that procedure Greedy maintains those invariants.

Invariants ($*$): *If at some step vertex $x \in X$ has α neighbors in X then it has port labels from 1 to α used to label the edges to these neighbors. The second invariant states that each $x \in X$ has at least one neighbor in X.*

Lemma 3. *If set X satisfies invariants ($*$) then procedure Greedy used with X as the initial set, labels all the remaining edges without using label pair $(1,1)$ on any newly labeled edge.*

Proof. Since graph G is connected then by repeating procedure Greedy we label all the edges of the graph.

In a single step we pick a vertex y not in X, but adjacent to some vertices from X, we label all the edges connecting y to vertices from X using the first available port numbers hence the invariant is maintained. From the invariant, port pair $(1,1)$ will not be used on any edge labeled by the procedure Greedy.

Theorem 1. *For any graph G of diameter D, we have $l(G) \leq D+1$. Moreover, a labeling λ such that $l(G, \lambda) \leq D+1$ can be constructed in polynomial time.*

Proof. We want to first show a procedure assigning labels in any graph in such a way that there always exists a node with a unique view up to depth 1. The rest follows from [20, Lemma 4] because then each view up do depth $D+1$ is unique.

Take any graph G with at least 2 edges. If G has a leaf (i.e., a node with degree 1) we assign label 1 to both endpoints of its unique incident edge. In this case we define set X to initially contain both endpoints of the edge.

On the other hand if G has no leaf then it has a cycle. We take this cycle (call it $\mathcal{C}$) and label both endpoints of its edges with port numbers 1 and 2 as in Fig. 2. We fix an arbitrary node v from cycle $\mathcal{C}$ and label its outgoing edges

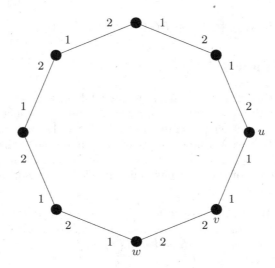

Fig. 2. Labeling of cycle $\mathcal{C}$.

with label pairs $(1, 1)$ and $(2, 2)$. All other edges of cycle $\mathcal{C}$ get labeled with label pairs $(1, 2)$. If there are other edges in G between the nodes of the cycle their labels are chosen by simply taking the first available label. In this case set X is defined to initially contain all nodes of the cycle. Observe that in both cases all edges with both endpoints in X have already been labeled.

The remaining labels are chosen by repeating procedure Greedy.

Since invariants $(*)$ are satisfied at the beginning of the procedure then procedure Greedy will not put port numbers $(1, 1)$ on any newly labeled edge.

Now observe that in the first case the view from the leaf is unique and in the second case the view from v is unique already at depth 1.

For a given orientation $\boldsymbol{G}$ of the edges of G, we denote by $\mathrm{indeg}_{\boldsymbol{G}}(v)$ (outdeg$_{\boldsymbol{G}}(v)$) the number of oriented edges entering (leaving) a vertex v. We also denote by $\boldsymbol{N}_l(v) \subseteq N_l(v)$ the subset of nodes reachable from v by following an outward-oriented path of at most l edges, starting from node v. Likewise, the radius-l out-ball of v is defined as $\boldsymbol{B}_l(v) = (\boldsymbol{G}[\boldsymbol{N}_l(v)]\backslash\boldsymbol{G}[\boldsymbol{N}_l(v)\backslash\boldsymbol{N}_{l-1}(v)], v)$.

Lemma 4. *Let λ be a port labeling of G chosen uniformly at random. Then, for any pair of distinct nodes $u, v \in V$ and any orientation $\boldsymbol{G}$ of G, we have under labeling λ:*

$$\Pr[\mathcal{V}_l(u) = \mathcal{V}_l(v)] \le \left(\prod_{w \in N_l(v)} \left((\mathrm{indeg}_{\boldsymbol{B}_l(v)}(w) - 1)! \deg_G(w) \right) \right)^{-1/2}.$$

Proof. We consider a process in which the labels of all ports at all nodes in λ are treated as initially covered cards with numbers written on their backs. Each node of degree d hands out a perfectly shuffled set of d cards with numbers $\{1, \ldots, d\}$ written on their backs to the d ports adjacent to it. We consider a process in which we sequentially consider edges of G (in some, possibly adaptive order), and in each time step uncover the cards at both ports of both endpoints of the edge. By $p_t(w)$ we denote the indegree of w with respect to covered incoming edges of $\boldsymbol{G}$, only, at the start of the t-th time step (initially, $p_1(w) = \text{indeg}_{\boldsymbol{B}_l(v)}(w)$). By $q_t(w)$ we denote a variable defined as $q_t(w) = \frac{\deg_G(w)}{\text{indeg}_{\boldsymbol{B}_l(v)}(w)}$ if all edges adjacent to w are covered at time t, and $q_t(w) = 1$ otherwise. For any pair of nodes u, v and integer i we define a subgraph $\text{VPM}_u(\boldsymbol{B}_i(v))$ as a "view preserving mapping" of $\boldsymbol{B}_i(v)$. We build $\text{VPM}_u(\boldsymbol{B}_i(v))$ in the following way. We take all directed paths of length i starting at v. For each such path we search for isomorphic with respect to λ path starting from u. Whenever such a path exists, we add its edges to $\text{VPM}_u(\boldsymbol{B}_i(v))$. If for some path P from v there is no isomorphic path from u then we will say that the path corresponding to P is empty (in this case we immediately distinguish views of u and v at depth i). We denote by S_t the event that after t steps of the process the views from u and v are equal.

The process proceeds in l stages, and we assume that at the beginning of the i-th stage, $i = (1, 2, \ldots, l)$, all edges from $\boldsymbol{B}_i(v) \cup \text{VPM}_u(\boldsymbol{B}_i(v))$ have been uncovered. In the i-th stage of the process, we sequentially consider edges connecting node pairs $\{w_{i-1}, w_i\}$, where $w_{i-1} \in \boldsymbol{N}_{i-1}(v)$, $w_i \in \boldsymbol{N}_i(v)$, such that the edge $\{w_{i-1}, w_i\} \in E(G)$ is oriented from w_{i-1} towards w_i in $E(G)$.

By the definition of our process, some shortest oriented path $\boldsymbol{P} = (v = w_0, w_1, \ldots, w_{i-1}, w_i)$, with $w_j \in \boldsymbol{N}_j(v) \backslash \boldsymbol{N}_{j-1}(v)$, has already been uncovered save for its last edge $\{w_{i-1}, w_i\}$, and so has the unoriented path P' originating at u and isomorphic to the unoriented version of $\boldsymbol{P}$. If edge $\{w_{i-1}, w_i\}$ is still uncovered, we uncover it together with its (possibly uncovered) counterpart $\{w'_{i-1}, w'_i\}$ in path P. (Remark that if $w_i = w'_i$, then we have an immediate distinction of the two considered views.) We observe that if two edges are uncovered in the current step t, then we have:

$$\Pr[S_t] \leq \frac{1}{\max\{p_t(w_i)q_t(w_i), p_t(w'_i)q_t(w'_i)\}} \Pr[S_{t-1}]$$

with $p_{t+1}(w_i) = p_t(w_i) - 1$, $p_{t+1}(w'_i) = p_t(w'_i) - 1$, and $p_{t+1}(x) = p_t(x)$ for all other nodes x; we also set $q_{t+1}(w_i) = q_{t+1}(w'_i) = 1$ and $q_{t+1}(x) = q_t(x)$ for all other nodes x. However, if edge $\{w'_{i-1}, w'_i\}$ was already previously uncovered, then we have:

$$\Pr[S_t] \leq \frac{1}{p_t(w_i)q_t(w_i)} \Pr[S_{t-1}]$$

with $p_{t+1}(w_i) = p_t(w_i) - 1$, and $p_{t+1}(x) = p_t(x)$ for all other nodes x; we also set $q_{t+1}(w_i) = 1$ and $q_{t+1}(x) = q_t(x)$ for all other nodes x. Denoting $\Pi_t = \prod_{w \in \boldsymbol{B}_l(v)}(p_t(w)! q_t(w))$, we notice that in both cases the following inequality holds:

$$\Pr[S_t] \leq \sqrt{\frac{\Pi_{t+1}}{\Pi_t}} \Pr[S_{t-1}]$$

By combining the above inequalities over all l phases of the process, in the final step m of the process we have $\Pi_m = 1$ and we eventually obtain:

$$\Pr[\mathcal{V}_l(u) = \mathcal{V}_l(v)] = \Pr[S_l] \le \Pi_1^{-1/2},$$

which, taking into account the definition of Π_1, p_1, and q_1, is exactly the claim of the lemma.

Choosing as G a BFS-out-orientation from vertex v (with edges within BFS levels arbitrarily oriented), we have $\boldsymbol{N}_l(v) = N_l(v)$, and we obtain the following corollary that follows directly from Lemma 4.

Corollary 3. *Let λ be a port labeling of G chosen uniformly at random. Then, for any pair of distinct nodes $u, v \in V$, we have under labeling λ:*

$$\Pr[\mathcal{V}_l(u) = \mathcal{V}_l(v)] \le \prod_{w \in N_l(v)} (\deg_G(w))^{-1/2}.$$

We use the above to show the following theorem.

Theorem 2. *There exists an absolute constant c such that, for any graph G with n nodes and diameter D, a uniformly random labeling λ satisfies $l(G, \lambda) \le c \min\{D, \log_2 n\}$, w.h.p.*[1]

Proof. Let us call any node with degree at least 2 a *non-leaf*. In our proof we will consider two cases. First assume that in G there are at least $6 \log_2 n$ non-leaves. Let $l = \min\{6 \log_2 n + 1, D\}$. For any v, in $\mathcal{B}_l(v)$ there are at least $6 \log_2 n$ non-leaves. Hence by Corollary 3, for any u, v we have $\Pr[\mathcal{V}_l(u) = \mathcal{V}_l(v)] \le n^{-3}$. By taking a union bound over all pairs of vertices we get that with probability at least n^{-1} all pairs of vertices are distinguished within radius l.

Now, consider the case where in G there are less than $6 \log_2 n$ non-leaves. Then the diameter of G is at most $6 \log_2 n + 1$. We will first compute the probability that all pairs of non-leaves can be distinguished within radius D. Let us fix any pair of different non-leaves u, v and compute the probability that, under random labeling, their views are equal up do depth D. We will, similarly as in the proof of Lemma 4 expose the port labels sequentially. If $\deg_G u \ne \deg_G v$ then u and v are distinguished within radius 1. Among non-leaves there exists a node u' (possibly equal to u, but different from v) with degree at least $n/(6 \log_2 n) - 1$. Take the shortest path P from u to u'. At every edge of P expose the port label of the port closer to u (the other port remains covered). Consider a path P' that has the same port numbers on the exposed ports but it starts at v. If the views of u and v are to be equal, P' must lead to another vertex with degree at least $n/(6 \log_2 n) - 1$, call it v'. If $u' = v'$, the views are already distinguished by the properties of the port labeling. Otherwise observe that the last ports on paths P and P' have to be equal, and since there are at least $n/(6 \log_2 n) - 1$ possibilities, then it is equal with probability at most $6 \log_2 n/(n - 6 \log_2 n)$. Hence

---

[1] With high probability means here with probability at least $1 - O(\text{polylog} n/n)$.

$\Pr[\mathcal{V}_l(u) = \mathcal{V}_l(v)] \leq 6\log_2 n/(n - 6\log_2 n)$. By taking the union bound over all non-leaves we get that with probability at least $1 - 108\log_2^3 n/(n - 6\log_2 n)$ all pairs of non-leaves are distinguished within radius D. This finishes the proof because the leaves can be distinguished using their unique neighbor (which is a non-leaf).

We remark that for a graph constructed from two equal stars connected with an edge, the probability that the centers are indistinguishable (to any depth) is $4/(n - 2)$. Hence the probability in our bound can be improved by at most a polylogarithmic factor.

The labelings given by Theorem 2 are non-constructive. By Theorem 1, we know how to constructively obtain a labeling λ satisfying $l(G, \lambda) = O(D)$. We now show how to constructively obtain $l(G, \lambda) = O(\log n)$. As a warmup exercise, we first perform the construction for the case when G is a path.

Proposition 2. *When G is a path, it is possible to construct in polynomial time a labeling λ such that $l(G, \lambda) = O(\log n)$.*

Proof. Encode the sequence of integers $1 * 2 * \ldots *$ on the ports of the path in binary, using a bit coding convention: $0 = aa$, $1 = ab$, $* = abbabbba$. Here, a means an edge with identical port labels at its endpoints, and b an edge with different ones.

We want to show that under such a labeling, within radius $l = 2\log_2 n + 12$, each node has a unique view. We can see the view from a node as a string of a and b. Observe that if the view has depth l, then it contains at least two complete sequences $*$. Moreover, this sequence $*$ is the only place in the string that contains at least 2 consecutive b-s hence it can be identified. Since, sequence $*$ is not a palindrome, then it allows to determine the orientation of the string. The substring between the two sequences $*$ can then be correctly decoded as a number. From the value of the number and the position of the node with respect to the first node encoding this number, we can uniquely determine the position of the node on the path. Hence the view up to depth l distinguishes all the nodes of the path.

Using the result for paths we can construct for any graph a labeling that yields logarithmic level of symmetry.

Proposition 3. *For any graph G, it is possible to construct in polynomial time a labeling λ such that $l(G, \lambda) = O(\log n)$.*

Proof. If $D = O(\log n)$, then the statement follows from Theorem 1. We assume here that $D \geq 4(\log_2 n + 1)$.

Let $x = 2\lceil \log_2 n \rceil + 15$ and let S be a set of vertices in G that form a maximal independent set in G^{2x}. Such a set can be constructed in polynomial time. For any vertex v we can then find a vertex $s \in S$ that is at distance at most $2x$ to v. Moreover any two vertices $s, s' \in S$ are at distance more than $2x$. Each $s \in S$ can be seen a cluster center and the ball of radius $2x$ around s is its cluster. Note, that our clusters are not necessarily disjoint but each cluster center belongs to exactly one cluster and balls of radius x from each cluster center are disjoint. Hence, for each $s \in S$ we can assign a simple path P_s (starting in s) of length x in such a way that for any $s, s' \in S$, their paths P_s and $P_{s'}$ are disjoint. For each s, we will label ports along path P_s using only port labels 1 and 2. We pick an arbitrary order $s_1, s_2, \ldots, s_k$ of the cluster centers. For each s_i, we write its identifier i on the path P_{s_i} using port labels 1 and 2 using the same coding as in the proof of Proposition 2. On P_{s_i} we write $*i*$, where $* = abbabbba$ and i is encoded in binary where $0 = aa$ and $1 = bb$. Here a means an edge with identical port endpoints, and b an edge with different ones. We label the path in such a way that the first a of each encoding is always a pair $(1, 1)$ (the encoding defines the remaining labels). If the encoding ends with pair $(2, 2)$, we label one more edge with port pair $(1, 1)$. The length of the whole encoding is then at most $2\lceil \log_2 n \rceil + 15$ hence the length of the path is sufficient. Let $\bar{P}_{s_i}$ be the subpath of P_{s_i} containing exactly the edges used to encode the identifier $*i*$ (possibly together with the additional edge).

The remaining edges are labeled as follows. We start with a set X of all the nodes whose at least one adjacent edge has been labeled. First we label all edges connecting two vertices from X, by picking the smallest available port number at both endpoints. We will not use port 1 because it has been used along the paths. Now all the remaining edges are labeled using procedure Greedy. Observe that now set X satisfies invariants $(*)$ hence by using procedure Greedy, we can label the remaining edges without putting port pair $(1, 1)$ on any newly labeled edge.

Now we want to show that under such labeling, all nodes have distinct views up to depth $3x$. To show that all the views are distinct, for any node v we will find a *tag* i.e. a labeled path in the view of v that will not appear in the view of any other node. In the view up to depth $3x$, vertex v has at least one cluster center $s \in S$ and its whole path $\bar{P}_s$. In labeled ball $\mathcal{B}_{3x}(v)$ we identify the set $\mathcal{P}$ of all paths induced by edges for which both endpoints were labeled using only ports 1 and 2. We keep in $\mathcal{P}$ only those paths which start and end with an edge labeled with $(1, 1)$ and remove the other paths. By the construction of our labeling, edge labeled with $(1, 1)$ can only appear on a path $\bar{P}_s$ of some cluster center s.

Now we remove from $\mathcal{P}$ all the paths that have more than x edges. Finally we remove from $\mathcal{P}$ the paths that are subpaths of other paths from $\mathcal{P}$. Each of the remaining paths $A \in \mathcal{P}$ is exactly $\bar{P}_s$ for some cluster center s. Path A cannot contain more than one $\bar{P}$ path because the length of A is at most x and the distance between the cluster centers is more than $2x$. Path A contains at least one $\bar{P}$ because A starts and ends with an edge labeled with $(1, 1)$ and A is not a subpath of any $\bar{P}$ path.

From each $A \in \mathcal{P}$ we can decode the identifier of the corresponding cluster center since $*$ is not a palindrome and hence we can deduce the direction of the path. If we obtained multiple identifiers of cluster centers for some starting vertex v, we simply pick the smallest one, call it s_i. The tag of v will then be the shortest path from v to s_i concatenated with $\bar{P}_{s_i}$ (if there are multiple shortest paths, we choose the one lexicographically smallest under the chosen labeling). We conclude the proof by observing that if for v, u their tags contain the same $\bar{P}_{s_i}$ then, by the properties of the port labeling, their paths to s_i have to be different.

4   Conclusion

In this paper, we have considered the problem of assigning labels to the ports of a graph G so that the views $\mathcal{V}_l(v)$ are distinct for all nodes $v \in V$, and such that l is minimized. We have shown that, for any graph G with n nodes and diameter D, a uniformly random labeling achieves $l = O(\min(D, \log n))$, w.h.p., and that it is possible to construct in polynomial time a labeling that satisfies $l = O(\min(D, \log n))$. In addition, we exhibited examples of graphs showing that these bounds are asymptotically tight in general.

An interesting direction of future work would be an analogue of the result of Steinová [25] in our setting. The goal is to propose a mechanism for the assignment (or reassignment) of the port labels by a mobile walker which starts in a graph with initially empty (or arbitrary) port labels, with the goal of obtaining a port labeling with small (best-case) level of symmetry. E.g., one could consider the implementation of the algorithm from Theorem 1 using a deterministic walker.

Acknowlegdements. The authors would like to thank Philippe Duchon and David Ilcinkas for proposing the problem, and for some initial discussions.

References

1. Boldi, P., Vigna, S.: Fibrations of graphs. Discrete Math. **243**(1–3), 21–66 (2002)
2. Boldi, P., Vigna, S.: Universal dynamic synchronous self-stabilization. Distrib. Comput. **15**(3), 137–153 (2002)
3. Chalopin, J., Das, S., Kosowski, A.: Constructing a map of an anonymous graph: applications of universal sequences. In: Lu, C., Masuzawa, T., Mosbah, M. (eds.) OPODIS 2010. LNCS, vol. 6490, pp. 119–134. Springer, Heidelberg (2010). doi:10.1007/978-3-642-17653-1_10
4. Chalopin, J., Métivier, Y.: An efficient message passing election algorithm based on Mazurkiewicz's algorithm. Fundamenta Informaticae **80**(1–3), 221–246 (2007)
5. Czyzowicz, J., Dobrev, S., Gasieniec, L., Ilcinkas, D., Jansson, J., Klasing, R., Lignos, I., Martin, R., Sadakane, K., Sung, W.: More efficient periodic traversal in anonymous undirected graphs. Theor. Comput. Sci. **444**, 60–76 (2012)
6. Czyzowicz, J., Kosowski, A., Pelc, A.: How to meet when you forget: log-space rendezvous in arbitrary graphs. Distrib. Comput. **25**(2), 165–178 (2012)
7. Das, S., Flocchini, P., Nayak, A., Santoro, N.: Effective elections for anonymous mobile agents. In: Asano, T. (ed.) ISAAC 2006. LNCS, vol. 4288, pp. 732–743. Springer, Heidelberg (2006). doi:10.1007/11940128_73
8. Das, S., Mihalák, M., Šrámek, R., Vicari, E., Widmayer, P.: Rendezvous of mobile agents when tokens fail anytime. In: Baker, T.P., Bui, A., Tixeuil, S. (eds.) OPODIS 2008. LNCS, vol. 5401, pp. 463–480. Springer, Heidelberg (2008). doi:10.1007/978-3-540-92221-6_29
9. Dereniowski, D., Kosowski, A., Pajak, D.: Distinguishing views in symmetric networks: a tight lower bound. Theor. Comput. Sci. **582**, 27–34 (2015)
10. Dereniowski, D., Pelc, A.: Drawing maps with advice. J. Parallel Distrib. Comput. **72**(2), 132–143 (2012)

11. Dereniowski, D., Pelc, A.: Leader election for anonymous asynchronous agents in arbitrary networks. Distrib. Comput. **27**(1), 21–38 (2014)
12. Dereniowski, D., Pelc, A.: Topology recognition and leader election in colored networks. Theor. Comput. Sci. **621**, 92–102 (2016)
13. Dobrev, S., Jansson, J., Sadakane, K., Sung, W.-K.: Finding short right-hand-on-the-wall walks in graphs. In: Pelc, A., Raynal, M. (eds.) SIROCCO 2005. LNCS, vol. 3499, pp. 127–139. Springer, Heidelberg (2005). doi:10.1007/11429647_12
14. Flocchini, P., Roncato, A., Santoro, N.: Computing on anonymous networks with sense of direction. Theor. Comput. Sci. **1–3**(301), 355–379 (2003)
15. Fraigniaud, P., Pelc, A.: Decidability classes for mobile agents computing. In: Fernández-Baca, D. (ed.) LATIN 2012. LNCS, vol. 7256, pp. 362–374. Springer, Heidelberg (2012). doi:10.1007/978-3-642-29344-3_31
16. Fusco, E.G., Pelc, A.: Knowledge, level of symmetry, and time of leader election. Distrib. Comput. **28**(4), 221–232 (2015)
17. Gasieniec, L., Klasing, R., Martin, R.A., Navarra, A., Zhang, X.: Fast periodic graph exploration with constant memory. J. Comput. Syst. Sci. **74**(5), 808–822 (2008)
18. Gąsieniec, L., Radzik, T.: Memory efficient anonymous graph exploration. In: Broersma, H., Erlebach, T., Friedetzky, T., Paulusma, D. (eds.) WG 2008. LNCS, vol. 5344, pp. 14–29. Springer, Heidelberg (2008). doi:10.1007/978-3-540-92248-3_2
19. Guilbault, S., Pelc, A.: Asynchronous rendezvous of anonymous agents in arbitrary graphs. In: Fernàndez Anta, A., Lipari, G., Roy, M. (eds.) OPODIS 2011. LNCS, vol. 7109, pp. 421–434. Springer, Heidelberg (2011). doi:10.1007/978-3-642-25873-2_29
20. Hendrickx, J.M.: Views in a graph: to which depth must equality be checked? IEEE Trans. Parallel Distrib. Syst. **25**(7), 1907–1912 (2014)
21. Ilcinkas, D.: Setting port numbers for fast graph exploration. Theor. Comput. Sci. **401**(1–3), 236–242 (2008)
22. Kosowski, A., Navarra, A.: Graph decomposition for memoryless periodic exploration. Algorithmica **63**(1–2), 26–38 (2012)
23. Krebs, A., Verbitsky, O.: Universal covers, color refinement, two-variable counting logic: lower bounds for the depth. In: Proceedings of 30th Annual ACM/IEEE Symposium on Logic in Computer Science (LICS 2015), pp. 689–700. IEEE (2015)
24. Norris, N.: Universal covers of graphs: isomorphism to depth $N − 1$ implies isomorphism to all depths. Discrete Appl. Math. **56**(1), 61–74 (1995)
25. Steinová, M.: On the power of local orientations. In: Shvartsman, A.A., Felber, P. (eds.) SIROCCO 2008. LNCS, vol. 5058, pp. 156–169. Springer, Heidelberg (2008). doi:10.1007/978-3-540-69355-0_14
26. Tani, S.: Compression of view on anonymous networks – folded view. IEEE Trans. Parallel Distrib. Syst. **23**(2), 255–262 (2012)
27. Tani, S., Kobayashi, H., Matsumoto, K.: Exact quantum algorithms for the leader election problem. ACM Trans. Comput. Theor. **4**(1), 1 (2012)
28. Yamashita, M., Kameda, T.: Computing functions on asynchronous anonymous networks. Math. Syst. Theor. **29**(4), 331–356 (1996)
29. Yamashita, M., Kameda, T.: Computing on anonymous networks: part I-characterizing the solvable cases. IEEE Trans. Parallel Distrib. Syst. **7**(1), 69–89 (1996)
30. Yamashita, M., Kameda, T.: Leader election problem on networks in which processor identity numbers are not distinct. IEEE Trans. Parallel Distrib. Syst. **10**(9), 878–887 (1999)

Fooling Pairs in Randomized Communication Complexity

Shay Moran[1,3,4]($\boxtimes$), Makrand Sinha[2], and Amir Yehudayoff[1]

[1] Technion, Israel Institute of Technology, Haifa 32000, Israel
shaymoran1@gmail.com
[2] Department of Computer Science and Engineering,
University of Washington, Seattle, USA
[3] Microsoft Research, Herzliya, Israel
[4] Max Planck Institute for Informatics, Saarbrücken, Germany

Abstract. The fooling pairs method is one of the standard methods for proving lower bounds for deterministic two-player communication complexity. We study fooling pairs in the context of randomized communication complexity. We show that every fooling pair induces far away distributions on transcripts of private-coin protocols. We use the above to conclude that the private-coin randomized ε-error communication complexity of a function f with a fooling set S is at least order $\log \frac{\log |S|}{\varepsilon}$. This relationship was earlier known to hold only for constant values of ε. The bound we prove is tight, for example, for the equality and greater-than functions.

As an application, we exhibit the following dichotomy: for every boolean function f and integer n, the $(1/3)$-error public-coin randomized communication complexity of the function $\bigvee_{i=1}^{n} f(x_i, y_i)$ is either at most c or at least n/c, where $c > 0$ is a universal constant.

1 Introduction

Communication complexity provides a mathematical framework for studying communication between two or more parties. It was introduced by Yao [12] and has found numerous applications since. We focus on the two-player case, and provide a brief introduction to it. For more details see the textbook by Kushilevitz and Nisan [8].

In this model, there are two players called Alice and Bob. The players wish to compute a function $f : \mathcal{X} \times \mathcal{Y} \to \mathcal{Z}$, where Alice knows $x \in \mathcal{X}$ and Bob knows $y \in \mathcal{Y}$. To achieve this goal, they need to communicate. The *communication complexity* of f measures the minimum number of bits the players must exchange in order to compute f. The communication is done according to a predetermined protocol. Protocols may be deterministic or use randomness that is either *public* (known to both players) or *private* (randomness held by one player

M. Sinha—Partially supported by BSF.
A. Yehudayoff—Horev fellow – supported by the Taub foundation. Supported by ISF and BSF.

© Springer International Publishing AG 2016
J. Suomela (Ed.): SIROCCO 2016, LNCS 9988, pp. 49–59, 2016.
DOI: 10.1007/978-3-319-48314-6_4

is not known to the other). In the case of deterministic protocols, we denote by $D(f)$ the minimum communication required to compute f correctly on all inputs. In the case of randomized protocols, we allow the protocol to err with a small probability. We denote by $R_\varepsilon(f)$ and $R_\varepsilon^{\mathsf{pri}}(f)$ the minimum communication required to compute f correctly with public and private-coin protocols with a probability of error at most ε on all inputs. We refer to Sect. 2.1 for formal definitions.

A fundamental problem in this context is proving lower bounds on the communication complexity of a given function f. Lower bounds methods for deterministic communication complexity are based on the fact that any protocol for f defines a partition of $\mathcal{X} \times \mathcal{Y}$ to f-monochromatic rectangles[1]. Thus, a lower bound on the size of a minimal partition of this kind readily translates to a lower bound on the communication complexity of f. Three basic bounds of this type are based on rectangle size, fooling sets, and matrix rank (see [8]). Both matrix rank and rectangle size lower bounds have natural and well-known analogues in the randomized setting: the approximate rank lower bound [7,9] and the discrepancy lower bound [8] respectively. In this paper we show that fooling sets also have natural counterparts in the randomized setting.

Although public-coin protocols are more general than private-coin ones, Newman [10] proved that for boolean functions every public-coin protocol can be efficiently simulated by a private-coin protocol: If $f : \mathcal{X} \times \mathcal{Y} \to \{0,1\}$ then for every $0 < \varepsilon < 1/2$,

$$R_{2\varepsilon}(f) \leq R_{2\varepsilon}^{\mathsf{pri}}(f) = O\left(R_\varepsilon(f) + \log \frac{\log(|\mathcal{X}||\mathcal{Y}|)}{\varepsilon}\right).$$

The additive logarithmic factor on the right-hand-side is often too small to matter, but it does make a difference in the bounds we prove below.

1.1 Fooling Pairs and Sets

Fooling sets are a well-known tool for proving lower bounds for $D(f)$. A pair $(x,y), (x',y') \in \mathcal{X} \times \mathcal{Y}$ is called a fooling pair for $f : \mathcal{X} \times \mathcal{Y} \to \mathcal{Z}$ if

- $f(x,y) = f(x',y')$, and
- either $f(x',y) \neq f(x,y)$ or $f(x,y') \neq f(x,y)$.

Observe that if (x,y) and (x',y') are a fooling pair then $x \neq x'$ and $y \neq y'$. When $\mathcal{Z} = \{0,1\}$ we distinguish between 0-fooling pairs (for which $f(x,y) = f(x',y') = 0$) and 1-fooling pairs (for which $f(x,y) = f(x',y') = 1$).

It is easy to see that if (x,y) and (x',y') form a fooling pair then there is no f-monochromatic rectangle that contains both of them. An immediate conclusion is the following:

---

[1] $R \subseteq \mathcal{X} \times \mathcal{Y}$ is an f-monochromatic rectangle if $R = A \times B$ for some $A \subseteq \mathcal{X}, B \subseteq \mathcal{Y}$ and f is constant over R.

Lemma 1 ([8]). *Let* $f : \mathcal{X} \times \mathcal{Y} \to \mathcal{Z}$ *be a function, let* (x, y) *and* (x', y') *be a fooling pair for* f *and let* π *be a deterministic protocol for* f. *Then*

$$\pi(x, y) \neq \pi(x', y').$$

A subset $\mathcal{S} \subseteq \mathcal{X} \times \mathcal{Y}$ is a *fooling set* if every $p \neq p'$ in $\mathcal{S}$ form a fooling pair. Lemma 1 implies the following basic lower bound for deterministic communication complexity.

Theorem 1 ([8]). *Let* $f : \mathcal{X} \times \mathcal{Y} \to \mathcal{Z}$ *be a function and let* S *be a fooling set for* f. *Then*

$$D(f) \geq \log_2(|\mathcal{S}|).$$

The same properties do not hold for randomized protocols, but one could expect their natural variants to hold. Let π be an ε-error private-coin protocol for f, and let $(x, y), (x', y')$ be a fooling pair for f. Then, one can expect that the probabilistic analogue of $\pi(x) \neq \pi(x')$ holds, *i.e.* $|\Pi(x, y) - \Pi(x', y')|$ is large, where $|\Pi(x, y) - \Pi(x', y')|$ denotes the statistical distance between the two distributions on transcripts.

Such a statement was previously only known for a specific type of fooling pair (that we call the AND fooling pair in Sect. 1.2) and was implicit in [2], where it is used as part of a lower bound proof for the randomized communication complexity of the disjointness function. Here, we prove that it holds for an arbitrary fooling pair.

Lemma 2 (Analogue of Lemma 1). *Let* $f : \mathcal{X} \times \mathcal{Y} \to \mathcal{Z}$ *be a function, let* (x, y) *and* (x', y') *be a fooling pair for* f, *and let* π *be an* ε-error private-coin *protocol for* f. *Then*

$$|\Pi(x, y) - \Pi(x', y')| \geq 1 - 2\sqrt{\varepsilon}.$$

Lemma 2 is not only an analogue of Lemma 1 but is actually a generalization of it. Indeed, plugging $\varepsilon = 0$ in Lemma 2 implies Lemma 1. Moreover, it implies that the bound from Theorem 1 holds also in the 0-error private-coin randomized case.

We use the above to prove an analogue of Theorem 1 as well.

Theorem 2 (Analogue of Theorem 1). *Let* $f : \mathcal{X} \times \mathcal{Y} \to \mathcal{Z}$ *be a function and let* S *be a fooling set for* f. *Let* $1/|\mathcal{S}| \leq \varepsilon < 1/3$. *Then,*

$$R_\varepsilon^{\mathsf{pri}}(f) = \Omega \left(\log \frac{\log |\mathcal{S}|}{\varepsilon} \right).$$

The lower bound provided by the theorem above seems exponentially weaker than the one in Theorem 1, but it is tight. The equality function EQ over n-bit strings has a large fooling set of size 2^n, but it is well-known (see [8]) that

$$R_\varepsilon^{\mathsf{pri}}(\mathsf{EQ}) = O \left(\log \frac{n}{\varepsilon} \right).$$

Theorem 2 therefore provides a tight lower bound on $R_\varepsilon^{\text{pri}}(\text{EQ})$ in terms of both n and ε. It also provides a tight lower bound for the greater-than function. Moreover, Theorem 2 is a generalization of Theorem 1 and basically implies it by choosing $\varepsilon = 1/|\mathcal{S}|$.

The proof of the lower bound uses a general lower bound on the rank of perturbed identity matrices by Alon [1]. Interestingly, although not every fooling set comes from an identity matrix (e.g. in the greater-than function), there is always some perturbed identity matrix in the background (the one used in the proof of Theorem 2).

We remark that for any constant $0 < \varepsilon < 1/3$, a version of Theorem 2 has been known for a long time. In particular, Håstad and Wigderson [5] give a proof of the following result[2] which appears in [12] without proof: for every function f with a fooling set $\mathcal{S}$ and for every $0 < \varepsilon < 1/3$,

$$R_\varepsilon^{\text{pri}}(f) = \Omega\left(\log \log |\mathcal{S}|\right). \tag{1}$$

The right-hand side above does not depend on ε. The same lower bound as in (1) also directly follows from Theorem 1 and from the following general result [8]: for every function f and for every $0 \leq \varepsilon < 1/2$,

$$R_\varepsilon^{\text{pri}}(f) = \Omega(\log D(f)).$$

1.2   Two Types of Fooling Pairs

Let $(x, y), (x', y')$ be a fooling pair for a boolean function f. There are two types of fooling pairs:

– The AND-type for which $f(x', y) \neq f(x, y')$.
– The XOR-type for which $f(x', y) = f(x, y')$.

A partial proof of Lemma 2 is implicit in [2]. The case considered in [2] corresponds to a 0-fooling pair of the AND-type. Let π be a private-coin ε-error protocol for f that is the AND of two bits. In this case, by definition it must hold that $\Pi(0, 0)$ is statistically far away from $\Pi(1, 1)$. The cut-and-paste property (see Corollary 1) implies that the same holds for $\Pi(0, 1)$ and $\Pi(1, 0)$, yielding Lemma 2 for the 0-fooling pair of the AND-type – $(0, 1), (1, 0)$.

The case of a pair of the XOR-type was not analyzed before. If π is a private-coin ε-error protocol for XOR of two bits, then it does not immediately follow that $\Pi(0, 0)$ is far away from $\Pi(1, 1)$, nor that $\Pi(0, 1)$ is far away from $\Pi(1, 0)$. Lemma 2 implies that in fact both are true, but the argument can not use the cut-and-paste property. Our argument actually gives a better quantitative result for the XOR function as compared to the AND function.

The importance of the special case of Lemma 2 from [2] is related to proving a lower bound on the randomized communication complexity of the disjointness

---

[2] In fact, the theorem in [5,12] is more general than the one stated here. We state the theorem in this form since it fits well the focus of this text.

function DISJ defined over $\{0,1\}^n \times \{0,1\}^n$: $\mathsf{DISJ}(x,y) = 1$ if for all $i \in [n]$ it holds that $x_i \wedge y_i = 0$. They reproved that $R_{1/3}(\mathsf{DISJ}) \geq \Omega(n)$. This lower bound is extremely important and useful in many contexts, and was first proved in [6].

On a high level, the proof of [2] can be summarized as follows. Let π be a private-coin protocol with $(1/3)$-error for DISJ. We want to show that $\mathsf{CC}(\pi) = \Omega(n)$. The argument has two different parts: The first part of the argument essentially relates the *internal information cost* (as was later defined in [3]) of computing one copy of the AND function with the communication of the protocol π for DISJ. This is a direct-sum-esque result. More concretely, if μ is a distribution on $\{0,1\}^2$ such that $\mu(1,1) = 0$ then

$$\mathsf{IC}_\mu(\mathsf{AND}) \leq \frac{\mathsf{CC}(\pi)}{n},$$

where $\mathsf{IC}_\mu(\mathsf{AND})$ is the infimum over all $(1/3)$-error private-coin protocols τ for AND of the internal information cost of τ. The second part of the argument shows that if μ is uniform on the set $\{(0,0),(0,1),(1,0)\}$ then $\mathsf{IC}_\mu(\mathsf{AND}) > 0$. The challenge in proving the second part stems from the fact that μ is supported on the zeros of AND, so it is trivial to compute AND on inputs from μ. However, the protocols τ in the definition of $\mathsf{IC}_\mu(\mathsf{AND})$ are guaranteed to succeed for every x, y and not only on the support of μ. The authors of [2] use the cut-and-paste property (see Corollary 1 below) to argue that indeed $\mathsf{IC}_\mu(\mathsf{AND}) > 0$.

Here we observe that these arguments can be cast into a more general fooling-pair based method. For example, consider the following function on a pair of n-tuples of elements:

$$f_k(x,y) = \bigvee_{i=1}^{n} \mathsf{EQ}_k(x_i, y_i),$$

where k is a positive integer and $\mathsf{EQ}_k : [k] \times [k] \to \{0,1\}$ denotes the equality function on elements of the set $[k]$.

The direct-sum reduction of [2] also works for the function f_3 and since EQ_3 contains a 0-fooling pair of the AND-type, we can straightaway conclude that the $(1/3)$-error randomized communication complexity and internal information cost of f_3 are $\Omega(n)$. However, for the seemingly similar function f_2, the direct sum reduction described above does not work (and all the fooling pairs are of the XOR-type). In fact, the $(1/3)$-error public-coin randomized communication complexity and internal information cost of f_2 are $O(1)$, since f_2 can be reduced to equality on n-bit strings.

The following theorem shows that this example is part of a general dichotomy. For example, there is no function f for which the randomized communication complexity of $\bigvee_{i=1}^{n} f(x_i, y_i)$ is $\Theta(\sqrt{n})$, when n tends to infinity.

Theorem 3. *There is a constant $c > 0$ so that for every boolean function f and integer n, the following holds:*

1. *If f contains a 0-fooling pair of the AND-type then the $(1/3)$-error public-coin randomized communication complexity of $\bigvee_{i=1}^{n} f(x_i, y_i)$ is at least n/c.*

2. *Else, the (1/3)-error public-coin randomized communication complexity of $\bigvee_{i=1}^{n} f(x_i, y_i)$ is at most c.*

A dual statement applies to the n-fold AND of f:

Theorem 4 (Dual of Theorem 3). *There is a constant $c > 0$ so that for every boolean function f and integer n, the following holds:*

1. *If f contains a 1-fooling pair of the AND-type then the (1/3)-error public-coin randomized communication complexity of $\bigwedge_{i=1}^{n} f(x_i, y_i)$ is at least n/c.*
2. *Else, the (1/3)-error public-coin randomized communication complexity of $\bigwedge_{i=1}^{n} f(x_i, y_i)$ is at most c.*

We provide a proof of Theorem 3. Theorem 4 can be derived by a similar argument, or alternatively by a reduction to Theorem 3 using the relation $\bigwedge_{i=1}^{n} f(x_i, y_i) = \neg \bigvee_{i=1}^{n} \neg f(x_i, y_i)$, which transforms 1-fooling pairs to 0-fooling pairs.

Proof (Proof of Theorem 3). To prove the first item, note that the sub-matrix corroponding to the 0-fooling pair of the AND-type can be mapped to the AND function and then taking the n-fold copy of it corresponds to computing the negation of the disjointness function on n bits. Applying the lower bound of [2] then proves that randomized communication complexity must be $\Omega(n)$.

For the second item, assume f does not contain any 0-fooling pair of the AND-type. Note that this implies that $\bigvee_{i=1}^{n} f(x_i, y_i)$ also does not contain any 0-fooling pair of the AND-type. Indeed, more generally, if f_1 and f_2 do not contain 0-fooling pairs of the AND-type then $f_1(x_1, y_1) \vee f_2(x_2, y_2)$ also does not contain such pairs.

So, it suffices to show that any function g that does not contain 0-fooling pairs of the AND type has public-coin randomized communication complexity $O(1)$. For any such g, the communication matrix of g does not contain a 2×2 sub-matrix with exactly three *zeros*. Without loss of generality, assume that the communication matrix contains no repeated rows or columns. We claim that this matrix contains at most one *zero* in each row and column. This will finish the proof since by permuting the rows and columns, we get the negation of the identity matrix with possibly one additional column of all ones or one additional row of all ones. Therefore, a simple variant of the $O(1)$ public-coin protocol for the equality function will compute g.

To see why there is at most one *zero* in each row and column, assume towards contradiction that it has two *zeros* in some row i, say in the first and second columns. Now, since the first and second *columns* differ, there must be some other row k on which they disagree. This means that the sub-matrix formed by rows i and k and columns 1 and 2 contains exactly three *zeros*, contradicting our assumption.

2   Preliminaries

2.1   Communication Complexity

A *private-coin communication protocol* for computing a function $f : \mathcal{X} \times \mathcal{Y} \to \mathcal{Z}$ is a binary tree with the following generic structure. Each node in the protocol is owned either by Alice or by Bob. For every $x \in \mathcal{X}$, each internal node v owned by Alice is associated with a distribution $P_{v,x}$ on the children of v. Similarly, for every $y \in \mathcal{Y}$, each internal node v owned by Bob is associated with a distribution $P_{v,y}$ on the children of v. The leaves of the protocol are labeled by $\mathcal{Z}$.

On input x, y, a protocol π is executed as follows.

1. Set v to be the root node of the protocol-tree defined above.
2. If v is a leaf, then the protocol outputs the label of the leaf. Otherwise, if Alice owns the node v, she samples a child according to the distribution $P_{v,x}$ and sends a bit to Bob indicating which child was sampled. The case when Bob owns the node is analogous.
3. Set v to be the sampled node and return to the previous step.

A protocol is *deterministic* if for every internal node v, the distribution $P_{v,x}$ or $P_{v,y}$ has support of size one. A *public-coin* protocol is a distribution over private-coin protocols defined as follows: Alice and Bob first sample a shared random r to choose a protocol π_r, and they execute a private protocol π_r as above.

For an input (x, y), we denote by $\pi(x, y)$ the sequence of messages exchanged between the parties. We call $\pi(x, y)$ the *transcript* of the protocol π on input (x, y). Another way to think of $\pi(x, y)$ is as a leaf in the protocol-tree. We denote by $L(\pi(x, y))$ the label of the leaf $\pi(x, y)$ in the tree. The *communication complexity* of a protocol π, denoted by $\mathsf{CC}(\pi)$ is the depth of the protocol-tree of π. For a private-coin protocol π, we denote by $\Pi(x, y)$ the distribution of the transcript of $\pi(x, y)$.

For a function f, the *deterministic* communication complexity of f, denoted by $D(f)$, is the minimum of $\mathsf{CC}(\pi)$ over all deterministic protocols π such that $L(\pi(x, y)) = f(x, y)$ for every x, y. For $\varepsilon > 0$, we denote by $R_\varepsilon(f)$ the minimum of $\mathsf{CC}(\pi)$ over all public-coin protocols π such that for every (x, y), it holds that $\mathbb{P}[L(\pi(x, y)) \neq f(x, y)] \leq \varepsilon$ where the probability is taken over all coin flips in the protocol π. We call $R_\varepsilon(f)$ the *ε-error public-coin randomized* communication complexity of f. Analogously we define $R_\varepsilon^{\mathrm{pri}}(f)$ as the *ε-error private-coin randomized communication complexity*.

2.2   Rectangle Property

In the case of deterministic protocols, the set of inputs reaching a particular leaf forms a rectangle (a product set inside $\mathcal{X} \times \mathcal{Y}$). In the case of private-coin randomized protocols, the following holds (see for example Lemma 6.7 in [2]).

Lemma 3 (Rectangle property for private-coin protocols). *Let π be a private-coin protocol over inputs from $\mathcal{X} \times \mathcal{Y}$, and let $\mathcal{L}$ denote the set of leaves of π. There exist functions $\alpha : \mathcal{L} \times \mathcal{X} \to [0,1]$, $\beta : \mathcal{L} \times \mathcal{Y} \to [0,1]$ such that for every $(x,y) \in \mathcal{X} \times \mathcal{Y}$ and every $\ell \in \mathcal{L}$,*

$$\mathbb{P}[\pi(x,y) \text{ reaches } \ell] = \alpha(\ell, x) \cdot \beta(\ell, y).$$

Here too the lemma is in fact a generalization of what happens in the deterministic case where α, β take values in $\{0,1\}$ rather than in $[0,1]$.

The next proposition immediately follows from the definitions.

Proposition 5. *Let $f : \mathcal{X} \times \mathcal{Y} \to \mathcal{Z}$ be a function and let (x,y) and (x',y') be such that $f(x,y) \neq f(x',y')$. Then, for any ε-error private-coin protocol π for f,*

$$|\Pi(x,y) - \Pi(x',y')| \geq 1 - 2\varepsilon.$$

2.3   Hellinger Distance and Cut-and-paste Property

The *Hellinger* distance between two distributions p, q over a finite set $\mathcal{U}$ is defined as

$$h(p,q) = \sqrt{1 - \sum_{u \in \mathcal{U}} \sqrt{p(u)q(u)}}.$$

Lemma 3 implies the following property of private-coin protocols that is more commonly known as the cut-and-paste property [4,11].

Corollary 1 (Cut-and-paste property). *Let (x,y) and (x',y') be inputs to a randomized private-coin protocol π. Then*

$$h(\Pi(x,y), \Pi(x',y')) = h(\Pi(x',y), \Pi(x,y')).$$

We also use the following relationship between Statistical and Hellinger Distances.

Proposition 6 (Statistical and Hellinger Distances). *Let p and q be distributions. Then,*

$$h^2(p,q) \leq |p - q| \leq \sqrt{h^2(p,q)(2 - h^2(p,q))}.$$

In particular, if $|p - q| \geq 1 - \varepsilon$ for $0 \leq \varepsilon \leq 1$. Then, $h^2(p,q) \geq 1 - \sqrt{2\varepsilon}$.

2.4   A Geometric Claim

We use the following technical claim that has a geometric flavor. For two vectors $\mathbf{a}, \mathbf{b} \in \mathbb{R}^m$, we denote by $\langle \mathbf{a}, \mathbf{b} \rangle$ the standard inner product between $\mathbf{a}, \mathbf{b}$. Denote by $\mathbb{R}_+$ the set of non-negative real numbers.

Claim 1. Let $\varepsilon_1, \varepsilon_2, \delta_1, \delta_2 > 0$ and let $\mathbf{a}, \mathbf{b}, \mathbf{c}, \mathbf{d} \in \mathbb{R}_+^m$ be vectors such that

$$\langle \mathbf{a}, \mathbf{b} \rangle \geq 1 - \varepsilon_1, \qquad\qquad \langle \mathbf{c}, \mathbf{d} \rangle \geq 1 - \varepsilon_2,$$
$$\langle \mathbf{a}, \mathbf{c} \rangle \leq \delta_1, \qquad\qquad \langle \mathbf{b}, \mathbf{d} \rangle \leq \delta_2.$$

Then,

$$\sum_{i \in [m]} |\mathbf{a}(i)\mathbf{b}(i) - \mathbf{c}(i)\mathbf{d}(i)| \geq 2 - (\varepsilon_1 + \varepsilon_2 + \delta_1 + \delta_2).$$

Proof.

$$\sum_{i \in [m]} |\mathbf{a}(i)\mathbf{b}(i) - \mathbf{c}(i)\mathbf{d}(i)|$$

$$\geq \sum_{i \in [m]} \left(\sqrt{\mathbf{a}(i)\mathbf{b}(i)} - \sqrt{\mathbf{c}(i)\mathbf{d}(i)} \right)^2 \qquad (\forall t, s \geq 0 \;\; |t - s| \geq (\sqrt{t} - \sqrt{s})^2)$$

$$= \langle \mathbf{a}, \mathbf{b} \rangle + \langle \mathbf{c}, \mathbf{d} \rangle - \sum_{i \in [m]} 2\sqrt{\mathbf{a}(i)\mathbf{b}(i)\mathbf{c}(i)\mathbf{d}(i)}$$

$$= \langle \mathbf{a}, \mathbf{b} \rangle + \langle \mathbf{c}, \mathbf{d} \rangle - \sum_{i \in [m]} 2\sqrt{\mathbf{a}(i)\mathbf{c}(i) \cdot \mathbf{b}(i)\mathbf{d}(i)}$$

$$\geq \langle \mathbf{a}, \mathbf{b} \rangle + \langle \mathbf{c}, \mathbf{d} \rangle - \sum_{i \in [m]} (\mathbf{a}(i)\mathbf{c}(i) + \mathbf{b}(i)\mathbf{d}(i)) \qquad \text{(AM-GM inequality)}$$

$$= \langle \mathbf{a}, \mathbf{b} \rangle + \langle \mathbf{c}, \mathbf{d} \rangle - (\langle \mathbf{a}, \mathbf{c} \rangle + \langle \mathbf{b}, \mathbf{d} \rangle)$$

$$\geq 2 - (\varepsilon_1 + \varepsilon_2 + \delta_1 + \delta_2). \qquad\qquad\qquad\qquad \square$$

3 Fooling Pairs and Sets

3.1 Fooling Pairs Induce Far Away Distributions

Proof (Proof of Lemma 2). Let the fooling pair be (x, y) and (x', y') and assume without loss of generality that $f(x, y) = f(x', y') = 1$. We distinguish between the following two cases.

(a) $f(x', y) \neq f(x, y')$.
(b) $f(x', y) = f(x, y') = z$ where $z \neq 1$.

In the first case, Proposition 5 implies that $|\Pi(x', y) - \Pi(x, y')| \geq 1 - 2\varepsilon$. Proposition 1 implies that $h(\Pi(x, y), \Pi(x', y')) = h(\Pi(x', y), \Pi(x, y'))$. Proposition 6 thus implies that $|\Pi(x, y) - \Pi(x', y')| \geq 1 - 2\sqrt{\varepsilon}$.

Let us now consider the second case. Let $\mathcal{L}$ be the set of all leaves of π and let $\mathcal{L}_1$ denote those leaves which are labeled by 1. For $x \in \mathcal{X}$, $y \in \mathcal{Y}$, define the vectors $\mathbf{a}_x \in \mathbb{R}_+^{\mathcal{L}_1}$ as $\mathbf{a}_x(\ell) = \alpha(\ell, x)$, and the vectors $\mathbf{b}_y \in \mathbb{R}_+^{\mathcal{L}_1}$ as $\mathbf{b}_y(\ell) = \beta(\ell, y)$ where α and β are the functions from Lemma 3. Since $f(x, y) = 1$ and π is an ε-error protocol for f,

$$\langle \mathbf{a}_x, \mathbf{b}_y \rangle = \sum_{\ell \in \mathcal{L}_1} \alpha(\ell, x) \cdot \beta(\ell, y) = \mathbb{P}[L(\pi(x, y)) = 1] \geq 1 - \epsilon.$$

Similarly, we have $\langle \mathbf{a}_{x'}, \mathbf{b}_{y'} \rangle \geq 1 - \varepsilon$, $\langle \mathbf{a}_x, \mathbf{b}_{y'} \rangle \leq \varepsilon$ and $\langle \mathbf{a}_{x'}, \mathbf{b}_y \rangle \leq \varepsilon$. Observe

$$2|\Pi(x, y) - \Pi(x', y')| \geq \sum_{\ell \in \mathcal{L}_1} |\mathbf{a}_x(\ell)\mathbf{b}_y(\ell) - \mathbf{a}_{x'}(\ell)\mathbf{b}_{y'}(\ell)|.$$

Applying Claim 1 with the vectors $\mathbf{a}_x, \mathbf{b}_y, \mathbf{a}_{x'}, \mathbf{b}_{y'}$ yields that $|\Pi(x, y) - \Pi(x', y')| \geq 1 - 2\varepsilon$.

3.2 A Lower Bound Based on Fooling Sets

The following result of Alon [1] on the rank of perturbed identity matrices is a key ingredient.

Lemma 4. *Let $\frac{1}{2\sqrt{m}} \leq \varepsilon < \frac{1}{4}$. Let M be an $m \times m$ matrix such that $|M(i, j)| \leq \varepsilon$ for all $i \neq j$ in $[m]$ and $|M(i, i)| \geq \frac{1}{2}$ for all $i \in [m]$. Then,*

$$\mathsf{rank}(M) = \Omega\left(\frac{\log m}{\varepsilon^2 \log(\frac{1}{\varepsilon})}\right).$$

Proof (Proof of Theorem 2). Let $\mathcal{L}$ denote the set of leaves of π. Let $A \in \mathbb{R}^{\mathcal{S} \times \mathcal{L}}$ be the matrix defined by

$$A_{(x,y),\ell} = \sqrt{\mathbb{P}[\pi(x, y) = \ell]}.$$

Let

$$M = AA^T$$

where A^T is A transposed. First,

$$M_{(x,y),(x,y)} = 1.$$

Second, if $(x, y) \neq (x', y')$ in $\mathcal{S}$ then by Lemma 2 we know $|\Pi(x, y) - \Pi(x', y')| \geq 1 - 2\sqrt{\varepsilon}$ so by Proposition 6

$$h^2(\Pi(x, y), \Pi(x', y')) \geq 1 - 2\varepsilon^{1/4}$$

which implies

$$M_{(x,y),(x',y')} = 1 - h^2(\Pi(x, y), \Pi(x', y')) \leq 2\varepsilon^{1/4}.$$

Lemma 4 implies that[3] the rank of M is at least $\Omega\left(\frac{\log|\mathcal{S}|}{\sqrt{\varepsilon}\log\left(\frac{1}{\varepsilon^{1/4}}\right)}\right) = \Omega\left(\left(\frac{\log|\mathcal{S}|}{\varepsilon}\right)^{1/4}\right)$. On the other hand,

$$2^{CC(\pi)} \geq |\mathcal{L}| \geq \mathsf{rank}(M).$$

---

[3] We may assume that say $\varepsilon < 2^{-12}$ by repeating the given randomized protocol a constant number of times.

References

1. Alon, N.: Perturbed identity matrices have high rank: proof and applications. Comb. Probab. Comput. **18**(1–2), 3–15 (2009). http://dx.doi.org/10.1017/S0963548307008917
2. Bar-Yossef, Z., Jayram, T.S., Kumar, R., Sivakumar, D.: An information statistics approach to data stream and communication complexity. In: FOCS, pp. 209–218 (2002)
3. Barak, B., Braverman, M., Chen, X., Rao, A.: How to compress interactive communication. SIAM J. Comput. **42**(3), 1327–1363 (2013)
4. Chor, B., Kushilevitz, E.: A zero-one law for Boolean privacy. SIAM J. Discrete Math. **4**(1), 36–47 (1991)
5. Hastad, J., Wigderson, A.: The randomized communication complexity of set disjointness. Theor. Comput. **3**(1), 211–219 (2007)
6. Kalyanasundaram, B., Schnitger, G.: The probabilistic communication complexity of set intersection. SIAM J. Discrete Math. **5**(4), 545–557 (1992)
7. Krause, M.: Geometric arguments yield better bounds for threshold circuits and distributed computing. Theor. Comput. Sci. **156**(1–2), 99–117 (1996)
8. Kushilevitz, E., Nisan, N.: Communication Complexity. Cambridge University Press, New York (1997)
9. Lee, T., Shraibman, A.: Lower bounds in communication complexity. Founda. Trends Theoret. Comput. Sci. **3**(4), 263–398 (2009)
10. Newman, I.: Private vs. common random bits in communication complexity. Inf. Process. Lett. **39**(2), 67–71 (1991)
11. Paturi, R., Simon, J.: Probabilistic communication complexity. J. Comput. Syst. Sci. **33**(1), 106–123 (1986)
12. Yao, A.C.C.: Some complexity questions related to distributive computing (preliminary report). In: STOC, pp. 209–213 (1979)

Public vs. Private Randomness in Simultaneous Multi-party Communication Complexity

Orr Fischer$^{(\boxtimes)}$, Rotem Oshman, and Uri Zwick

Computer Science Department, Tel-Aviv University, Tel Aviv, Israel
{orrfischer,roshman}@mail.tau.ac.il, zwick@tau.ac.il

Abstract. In simultaneous number-in-hand multi-party communication protocols, we have k players, who each receive a private input, and wish to compute a joint function of their inputs. The players simultaneously each send a single message to a referee, who then outputs the value of the function. The cost of the protocol is the total number of bits sent to the referee.

For two players, it is known that giving the players a public (shared) random string is much more useful than private randomness: public-coin protocols can be unboundedly better than deterministic protocols, while private-coin protocols can only give a quadratic improvement on deterministic protocols.

We extend the two-player gap to multiple players, and show that the private-coin communication complexity of a k-player function f is at least $\Omega(\sqrt{D(f)})$ for any $k \geq 2$. Perhaps surprisingly, this bound is tight: although one might expect the gap between private-coin and deterministic protocols to grow with the number of players, we show that the All-Equality function, where each player receives n bits of input and the players must determine if their inputs are all the same, can be solved by a private-coin protocol with $\tilde{O}(\sqrt{nk} + k)$ bits. Since All-Equality has deterministic complexity $\Theta(nk)$, this shows that sometimes the gap scales only as the square root of the number of players, and consequently the number of bits each player needs to send actually decreases as the number of players increases. We also consider the Exists-Equality function, where we ask whether there is a pair of players that received the same input, and prove a nearly-tight bound of $\tilde{\Theta}(k\sqrt{n})$ for it.

1 Introduction

In his seminal '79 paper introducing the notion of two-party communication complexity [18], Yao also briefly considered communication between more than two players, and pointed out "one situation that deserves special attention": two players receive private inputs, and send randomized messages to a third player, who then produces the output. Yao asked what is the communication complexity of the Equality function (called "the identification function" in [18]) in this model: in the Equality function EQ_n, the two players receive vectors $\{0,1\}^n$, and the goal is to determine whether $x = y$.

© Springer International Publishing AG 2016
J. Suomela (Ed.): SIROCCO 2016, LNCS 9988, pp. 60–74, 2016.
DOI: 10.1007/978-3-319-48314-6_5

Yao showed in [18] that EQ_n requires $\Omega(n)$ bits for determistic communi-cation protocols, even if the players can communicate back-and-forth. Using a *shared* random string, the complexity reduces to $O(1)$, and using *private* randomness, but more than a single round, the complexity is $\Theta(\log n)$. In modern nomenclature, the model described above is called *the 2-player simultaneous model*, and the third player (who announces the output) is called the *referee*. Yao's question is then: what is the communication complexity of EQ_n using private randomness in the simultaneous model of communication complexity?

Some seventeen years later, Yao's question was answered: Newman and Sezegy showed in [16] that EQ_n requires $\Theta(\sqrt{n})$ bits to compute in the model above, if the players are allowed only private randomness. (Using shared randomness the complexity reduces to $O(1)$, even for simultaneous protocols.) Moreover, Babai and Kimmel showed in [3] that for *any* function f, if the deterministic simultaneous complexity of f is $D(f)$, then the private-coin simultaneous communication complexity of f is $\Omega(\sqrt{D(f)})$, so in this sense private randomness is of only limited use for simultaneous protocols.

In this paper we study multi-player simultaneous communication complex-ity[1], and ask: how useful are private random coins for more than two players? Intuitively, one might expect that as the number of players grows, the utility of private randomness should decrease. We first extend the $\Omega(\sqrt{D(f)})$ lower bound of [3] to the multi-player setting, and show that for any k-player function f, the private-coin simultaneous communication complexity of f is $\Omega(\sqrt{D(f)})$. We then show, perhaps contrary to expectation, that the extended lower bound is still tight in some cases.

To see why this may be surprising, consider the function $\mathrm{ALLEQ}_{k,n}$, which generalizes EQ_n to k players: each player i receives a vector $x_i \in \{0,1\}^n$, and the goal is to determine whether all players received the same input. It is easy to see that the deterministic communication complexity of $\mathrm{ALLEQ}_{k,n}$ is $\Omega(nk)$ (not just for simultanoues protocols), and each player must send n bits to the referee in the worst case. From the lower bound above, we obtain a lower bound of $\Omega(\sqrt{nk})$ for the private-coin simultaneous complexity of $\mathrm{ALLEQ}_{k,n}$. It is easy to see that $\Omega(k)$ is also a lower bound, as each player must send at least one bit, so together we have a lower bound of $\Omega(\sqrt{nk} + k)$. If this lower bound is tight, then the average player only needs to send $O(\sqrt{n/k} + 1)$ bits to the referee in the worst-case, so in some sense we even *gain* from having more players, and indeed, if $k = \Omega(n)$, then the per-player cost of $\mathrm{ALLEQ}_{k,n}$ with private coins is constant, just as it would be with shared coins.

Nevertheless, our lower bound *is* nearly tight, and we are able to give a simultaneous private-coin protocol for $\mathrm{ALLEQ}_{k,n}$ where each players sends only $O(\sqrt{n/k} + \log(k))$ bits to the referee, for a total of $O(\sqrt{nk} + k \log \min\{k,n\})$ bits. This matches the lower bound of $\Omega(\sqrt{nk})$ when $k = O(n/\log^2 n)$. We also

---

[1] We consider the *number-in-hand* model, where each player receives a private input, rather than the perhaps more familiar *number-on-forehead* model, where each player can see the input of all the other players but not its own.

show that $\text{ALLEQ}_{k,n}$ requires $\Omega(k \log n)$ bits, so in fact our upper bound for ALLEQ is tight.

We then turn our attention to a harder class of k-player problems: those obtained by taking a 2-player function f and asking "do there exist two players on whose inputs f returns 1?". An example for this class is the function $\text{EXISTSEQ}_{k,n}$, which asks whether there exist two players that received the same input. We show that $\text{EXISTSEQ}_{k,n}$ requires $\tilde{\Theta}(k\sqrt{n})$ bits for private-coin simultaneous protocols, and moreover, any function in the class above has private-coin simultaneous complexity $\tilde{O}(kR(f))$, where $R(f)$ is the private-coin simultaneous complexity of f (with constant error).

1.1   Related Work

As we mention above, two-player simultaneous communication complexity was first considered by Yao in [18], and has received considerable attention since. The Equality problem was studied in [3,7,16], and another optimal simultaneous protocol is given in [2], using error-correcting codes. In [12], a connection is established between simultaneous and one-round communication complexity and the VC-dimension. [8,11] consider the question of simultaneously solving multiple copies of Equality and other functions, and in particular, [8] shows that solving m copies of Equality requires $\Omega(m\sqrt{n})$ bits for private-coin simultaneous 2-player protocols.

Multi-player communication complexity has also been extensively studied, but typically in the *number-on-forehead* model, where each player can see the inputs of all the other players but not its own. This model was introduced in [9]; sufficiently strong lower bounds on protocols in this model, even under restricted (but not simultaneous) communication patterns, would lead to new circuit lower bounds. Simultaneous communication complexity for number-on-forehead is considered in [4].

In contrast, in this paper we consider the *number-in-hand* model, where each player knows only its own input. This model is related to distributed computing and streaming (see, e.g., [17], which gives a lower bound for a promise version of Set Disjointness in our model).

An interesting "middle case" between the number-in-hand and number-on-forehead models is considered in [1,5,6]: there the input to the players is an undirected graph, and each player represents a node in the graph and receives the edges adjacent to this node as its input. This means that each edge is known to *two* players. This gives the players surprising power; for example, in [1] it is shown that graph connectivity can be decided in a total of $O(n \log^3 n)$ bits using public randomness. The power of private randomness in this model remains a fascinating open question and is part of the motivation for our work.

The functions ALLEQ and EXISTSEQ considered in this paper were also studed in, e.g., [10], but not in the context of simultaneous communication; the goal there is to quantify the communication cost of the network topology on communication complexity, in a setting where not all players can talk directly with each other.

2 Preliminaries

Notation. For a vector x of length n, we let x_{-i} denote the vector of length $n-1$ obtained by dropping the i-th coordinate of x (where $i \in [n]$).

Simultaneous Protocols. Fix input domains $\mathcal{X}_1, \ldots, \mathcal{X}_k$ of sizes $m_1, \ldots, m_k$ (respectively). A *private-coin k-player simultaneous communication protocol Π* on $\mathcal{X}_1 \times \ldots \times \mathcal{X}_k$ is a tuple of functions $(\pi_1, \ldots, \pi_k, O)$, where each π_i maps the inputs $\mathcal{X}_i$ of player i to a distribution on a finite set of messages $\mathcal{M}_i \subseteq \{0,1\}^*$, and O is the referee's output function, mapping each tuple of messages in $\mathcal{M}_1 \times \ldots \times \mathcal{M}_k$ to a distribution on outputs $\{0,1\}$.

We say that Π *computes a function* $f : \mathcal{X}_1 \times \ldots \times \mathcal{X}_k \to \{0,1\}$ *with error ϵ* if for each $(x_1, \ldots, x_k) \in \mathcal{X}_1 \times \ldots \times \mathcal{X}_k$ we have:

$$\Pr_{\{m_i \sim \pi_i(x_i)\}_{i \in [k]}} [O(m_1, \ldots, m_k) \neq f(x_1, \ldots, x_k)] \leq \epsilon.$$

A *deterministic* protocol is defined as above, except that instead of distributions on messages, the protocol maps each player's input to a deterministic message, and the referee's output is also a deterministic function of the messages it receives from the players.

Communication Complexity. The *communication complexity* of a protocol Π (randomized or deterministic), denoted by $CC(\Pi)$, is defined as the maximum total number of bits sent by the players to the referee in any execution of the protocol on any input.[2]

For a function f, the *deterministic communication complexity of f* is defined as

$$D(f) = \min_{\Pi} CC(\Pi),$$

where the minimum is taken over all deterministic protocols that compute f with no errors. The *private-coin ϵ-error communication complexity of f* is defined as

$$R_\epsilon(f) = \min_{\Pi : \Pi \text{ computes } f \text{ with error } \epsilon} CC(\Pi).$$

Individual Communication Complexity of a Player. We let $CC_i(\Pi)$ denote the maximum number of bits sent by player i to the referee in any execution. For general communication protocols, it could be that the players never simultaneously reach their worst-case message sizes — that is, we could have $CC(\Pi) < \sum_{i=1}^{k} CC_i(\Pi)$. However, with *simultaneous* protocols this cannot happen:

---

[2] Another reasonable definition for randomized protocols is to take the maximum over all inputs of the *expected* total number of bits sent. For two players this is asymptotically equivalent to the definition above [13]. For $k > 2$ players, the expectation may be smaller than the maximum by a factor of $\log(k)$.

Observation 1. *For any private-coin (or deterministic) simultaneous protocol* Π *we have* $\mathrm{CC}(\Pi) = \sum_{i=1}^{k} \mathrm{CC}_i(\Pi)$.

Proof. For each $i \in [k]$, let x_i be some input on which player i sends $\mathrm{CC}_i(\Pi)$ bits with non-zero probability. Then on joint input $(x_1, \ldots, x_k)$, there is a non-zero probability that each player i sends $\mathrm{CC}_i(\Pi)$ bits, for a total of $\sum_{i=1}^{k} \mathrm{CC}_i(\Pi)$ bits. Therefore $\mathrm{CC}(\Pi) \geq \sum_{i=1}^{k} \mathrm{CC}_i(\Pi)$. The inequality in the other direction is immediate, as there cannot be an execution of the protocol in which more than $\sum_{i=1}^{k} \mathrm{CC}_i(\Pi)$ bits are sent.

In the sequel we assume for simplicity that all players always send the same number of bits, that is, each player has a fixed message size. By the observation above, this does not change the communication complexity.

Maximal Message Complexity of a Protocol. The *maximal message complexity* of a protocol Π is the maximum individual communication complexity over all players. The deterministic maximum message complexity is $D^{\infty} = \min_{\Pi} \max_i \mathrm{CC}_i(\Pi)$, and the *private-coin ϵ-error maximal message complexity of* f is defined as

$$R_{\epsilon}^{\infty} = \min_{\Pi \text{ computes } f \text{ with error } \epsilon} \max_i \mathrm{CC}_i(\Pi)$$

Problem Statements. The two main problems we consider in this paper are:

- $\mathrm{ALLEQ}_{k,n}(x_1, \ldots, x_k) = 1$ iff $x_1 = \ldots = x_k$, where $x_1, \ldots, x_k \in \{0,1\}^n$;
- $\mathrm{EXISTSEQ}_{k,n}(x_1, \ldots, x_k) = 1$ iff for some $i \neq j$ we have $x_i = x_j$, where $x_1, \ldots, x_k \in \{0,1\}^n$.

We often omit the subscript when the number of players and the input size are clear from the context.

3   Lower Bound

In this section we extend the lower bound from [3] to multiple players, and show that for any k-player function f and constant error probability $\epsilon \in (0, 1/2)$ we have $R_{\epsilon}(f) = \Omega(\sqrt{D(f)})$.

When proving two-party communication complexity lower bounds, it is helpful to view the function being computed as a matrix, where the rows are indexed by Alice's input, the columns are indexed by Bob's input, and each cell contains the value of the function on the corresponding pair of inputs. The natural extension to k players is a "k-dimensional matrix" (or tensor) where the i-th dimension is indexed by the inputs to the i-th player, and the cells again contain the values of the function on that input combination. For conciseness we refer to this representation as a "matrix" even for $k > 2$ players.

In [3] it is observed that the deterministic simultaneous communication complexity of a function is exactly the sum of the logarithms of the number of unique rows and the number of unique columns in the matrix (rounded up to an integer). We generalize the notion of "rows and columns" to multiple players as follows.

Definition 1 (Slice). *Fix a k-dimensional matrix $M \in \{0,1\}^{m_1 \times \ldots \times m_k}$ For a player i and an input (i.e., index) $x_i \in [m_i]$, we define the (i, x_i)-th slice of M to be the projection of M onto a $(k-1)$-dimensional matrix $M|_{(i,x_i)} \in \{0,1\}^{m_1 \times \ldots \times m_{i-1} \times m_{i+1} \times \ldots \times m_k}$ obtained by fixing player i's input to x_i. That is, for each $x \in \mathcal{X}_1 \times \ldots \times \mathcal{X}_k$ we have $M|_{(i,x_i)}(x_{-i}) = M(x)$.*

Note that for $k = 2$ and a 2-dimensional matrix M, the $(1, x)$-th slice of M is simply the row indexed by x, and the $(2, y)$-th slice is the column indexed by y.

We assume that the matrices we deal with have *no redundant slices*: there does not exist a pair $(i, x_i), (i, x_i')$ (where $x_i \neq x_i'$) such that $M|_{(i,x_i)} = M|_{(i,x_i')}$. If there are redundant slices, we simply remove them; they correspond to inputs to player i on which the function value is the same, for any combination of inputs to the other players. Such inputs are "different in name only" and we can eliminate the redundancy without changing the communication complexity of the function being computed.

Let $\dim_i(M)$ denote the length of M in the i-th direction: this is the number of possible inputs to player i, after redundant slices are removed (i.e., the number of unique slices for player i in M). We rely upon the following observation, which generalizes the corresponding observation for two players from [3]:

Observation 2. *Let $f : \mathcal{X}_1 \times \ldots \times \mathcal{X}_k \to \{0,1\}$ be a k-player function, and let M_f be the matrix representing f. Then in any deterministic protocol for f, each player i sends at least $\log \dim_i(M_f)$ bits in the worst case, and $D(f) = \sum_{i=1}^{k} \lceil \log \dim_i(M_f) \rceil$.*

Proof. Suppose for the sake of contradiction that there is a deterministic protocol Π for f where some player i that always sends fewer than $\lceil \log \dim_i(M_f) \rceil$ bits in Π. For this player there exist two slices (i.e., inputs to player i) $M|_{(i,x_i)}$ and $M|_{(i,x_i')}$, with $x_i \neq x_i'$, on which the player sends the same message. Because we assumed that there are no redundant slices, there exists an input x_{-i} to the other players such that $M|_{(i,x_i)}(x_{-i}) \neq M|_{(i,x_i')}(x_{-i})$. But all players send the same messages to the referee on inputs (x_i, x_{-i}) and (x_i', x_{-i}), which means that on one of the two inputs the output of the referee is incorrect.

This shows that each player i must send at least $\lceil \log \dim_i(M_f) \rceil$ bits in the worst-case. This number of bits from each player is also sufficient to compute f, as the players can simply send the referee their input (after removing redundant slices, the number of remaining inputs is the number of unique slices). Therefore by Observation 1, $D(f) = \sum_{i=1}^{k} \lceil \log \dim_i(M_f) \rceil$.

In [3], Babai and Kimmel prove the following for two players:

Lemma 1 ([3]). *For any 2-player private-coin protocol Π with constant error $\epsilon < 1/2$,*

$$\mathrm{CC}_1(\Pi) \cdot \mathrm{CC}_2(\Pi) \geq \Omega(\log \dim_1(M_f) + \log \dim_2(M_f)).$$

Using this property of 2-player protocols, we can show:

Lemma 2. *Let Π be a k-player private-coin protocol for $f : \mathcal{X}_1 \times \ldots \times \mathcal{X}_k \to \{0,1\}$ with constant error $\epsilon \in (0, 1/2)$. Then for each $i \in [k]$:*

$$\mathrm{CC}_i(\Pi) \cdot \left(\sum_{j \neq i} \mathrm{CC}_j(\Pi) \right) = \Omega(\log \dim_i(M_f)).$$

Proof. Fix a player $i \in [k]$. The k-player protocol Π induces a 2-player protocol Π', where Alice plays the role of player i, and Bob plays the role of all the other players. We have $\mathrm{CC}_1(\Pi') = \mathrm{CC}_i(\Pi)$ and $\mathrm{CC}_2(\Pi') = \sum_{j \neq i} \mathrm{CC}_j(\Pi)$ (recall that we assume the message size of each player is fixed).

The 2-player function computed by Π' is still f, but now we view it as a 2-player function, represented by a 2-dimensional matrix M_f' with rows indexed by $\mathcal{X}_i$ and columns indexed by $\mathcal{X}_1 \times \ldots \times \mathcal{X}_{i-1} \times \mathcal{X}_{i+1} \times \ldots \times \mathcal{X}_k$. Note that $\dim_1(M_f') \geq \dim_i(M_f)$: if $M_f|_{(i,x_i)}$ and $M_f|_{(i,x_i')}$ are slices of M_f that are not equal, then the corresponding rows of M_f', indexed by x_i and x_i', differ as well. Thus, by Lemma 1,

$$\mathrm{CC}_i(\Pi) \cdot \left(\sum_{j \neq i} \mathrm{CC}_j(\Pi) \right) = \mathrm{CC}_1(\Pi') \cdot \mathrm{CC}_2(\Pi')$$

$$= \Omega(\log \dim_1(M_f')) = \Omega(\log \dim_i(M_f)).$$

We can now show:

Theorem 1. *For any k-player function f and constant error $\epsilon < 1/2$ we have $R_\epsilon(f) = \Omega(\sqrt{D(f)})$.*

Proof. Let Π be an ϵ-error private-coin simultaneous protocol for f. By the lemma, for each $i \in [k]$ we have

$$\mathrm{CC}_i(\Pi) \cdot \left(\sum_{j=1}^{n} \mathrm{CC}_j(\Pi) \right)$$

$$\geq \mathrm{CC}_i(\Pi) \cdot \left(\sum_{j \neq i} \mathrm{CC}_j(\Pi)) \right) = \Omega \left(\log \dim_i(M_f) \right).$$

Summing across all players, we obtain

$$\left(\sum_{i=1}^{n} \mathrm{CC}_i(\Pi) \right) \cdot \left(\sum_{j=1}^{n} \mathrm{CC}_j(\Pi) \right) = \Omega \left(\sum_{i=1}^{n} \log \dim_i(M_f) \right),$$

that is, by Observations 1 and 2,

$$\mathrm{CC}(\Pi)^2 = \Omega \left(D(f) \right).$$

The theorem follows.

From the theorem above we see that the *average* player must send $\Omega(\sqrt{D(f)/k})$ bits. But what is the relationship between the *maximum* number of bits sent by any player in a private-coin protocol and a deterministic protocol for f? This question is mainly of interest for non-symmetric functions, since for symmetric functions all players must send the same number of bits in the worst-case.

Theorem 2. *For any k-player function f and constant error ϵ, we have $R_\epsilon^\infty(f) = \Omega(\sqrt{D^\infty(f)/k})$.*

Proof. Recall that by Observation 2, $D^\infty(f) = \max_i \log \dim_i(M_f)$. Let i be a player maximizing $\log \dim_i(M_f)$. As we showed in the proof of Theorem 1, for this player we have in any private-coin simultaneous protocol Π:

$$\mathrm{CC}_i(\Pi) \cdot \left(\sum_{j=1}^n \mathrm{CC}_j(\Pi) \right) = \Omega\left(\log \dim_i(M_f) \right) = \Omega(D^\infty(f)).$$

Now let ℓ be the player with the maximum communication complexity in Π, that is, $\mathrm{CC}_j(\Pi) \le \mathrm{CC}_\ell(\Pi)$ for each $j \in [k]$. We then have

$$\mathrm{CC}_i(\Pi) \cdot \left(\sum_{j=1}^n \mathrm{CC}_j(\Pi) \right) \le \mathrm{CC}_\ell(\Pi) \cdot (k-1)\mathrm{CC}_\ell(\Pi) < k\mathrm{CC}_\ell^2(\Pi).$$

Combining the two, we obtain $\mathrm{CC}_\ell(\Pi) = \Omega\left(\sqrt{D^\infty(f)/k} \right)$, which proves the theorem.

Lower Bound of $\Omega(k \log n)$ for $\mathrm{AllEq}_{k,n}$

We next show that in the specific case of ALLEQ, each player needs to send at least $\Omega(\log n)$ bits, yielding a lower bound of $\Omega(k \log n)$. This improves on the lower bound of Theorem 1 when $k = \Omega(n/\mathrm{polylog}(n))$, and will show that the protocol in the next section is optimal.

Theorem 3. *For any constant $\epsilon < 1/2$, $R_\epsilon(\mathrm{ALLEQ}_{k,n}) = \Omega(k \log n)$.*

Proof. Fix a player $i \in [k]$. To show that player i must send $\Omega(\log n)$ bits, we reduce from EQ_n, but this time our reduction constructs a *one-way protocol*, where Alice, taking the role of player i, sends a message to Bob, representing all the other players and the referee; and Bob then outputs the answer. It is known that EQ_n requires $\Omega(\log n)$ bits of communication for private-coind protocols — this is true even with unrestricted back-and-forth communication between the two players [13]. The lower bound follows.

Let Π be a simultaneous private-coin protocol for $\mathrm{ALLEQ}_{k,n}$. We construct a one-way protocol for EQ_n as follows: on input (x, y), Alice sends Bob the message that player i would send on input x in Π. Bob computes the messages each player $j \ne i$ would send on input y, and then computes the output of the referee; this is the output of the one-way protocol. Clearly, $\mathrm{ALLEQ}_{k,n}(x, y, \ldots, y) = \mathrm{EQ}_n(x, y)$, so the one-way protocol succeeds whenever Π succeeds.

The lower bounds above use a series of 2-player reductions; they do not seem to exploit the full "hardness" of having k players with their own individual private randomness. This makes it more surprising that the lower bounds are tight, as we show in the next section.

4   Tight Upper Bound for AllEQ

In this section, we show that the lower bound proven in Sect. 3 is tight for $\text{ALLEQ}_{k,n}$. This is done by showing a protocol with maximal message of size $O(\sqrt{\frac{n}{k}} + \log{(\min(n, k))})$ bits per player, and $O(\sqrt{nk} + k\log{(\min(n, k))})$ bits of communication overall.

Theorem 4. *There exists a private-coin one sided error randomized simultaneous protocol for* $\text{ALLEQ}_{k,n}$ *with maximal message of size* $O(\sqrt{\frac{n}{k}} +$ $\log{(\min(n, k))}) = O(\sqrt{\frac{D^\infty(\text{ALLEQ}_{k,n})}{k}} + \log{(\min(n, k))})$ *bits per player.*

Corollary 1. *There exists a private-coin one sided error randomized simultaneous protocol for* $\text{ALLEQ}_{n,k}$ *of cost* $O(\sqrt{nk} + k\log{(\min(n, k))}) =$ $O(\sqrt{D(\text{ALLEQ}_{k,n})} + k\log{(\min(n, k))})$.

We note that the *deterministic* communication complexity of $\text{ALLEQ}_{n,k}$ is $\Theta(nk)$, and hence also $D^\infty(\text{ALLEQ}_{k,n}) = \Theta(n)$. This follows immediately from Observation 2.

Our randomized private-coin protocol is as follows.

Error-Correcting Codes. In the first step of the protocol, each player encodes its input using a predetermined error correcting code, and uses the encoded string as the new input. We review the definition of an error correcting code. In the definition below, n and k are the standard notation for error correcting codes, which we keep for the sake of consistency with the literature in coding; they are unrelated to the parameters n, k of the communication complexity problem and will be used in this context in the following definition only.

Definition 2. ([14]). $M \subseteq \{0,1\}^n$ *is called an* $[n, k, d]$*-code if it contains* 2^k *elements (that is,* $|M| = 2^k$*) and* $d_H(x, y) \geq d$ *for every two distinct* x, y*, where* d_H *is the Hamming distance. For a* $[n, k, d]$ *code, let* $\delta = \frac{d}{n}$ *denote the* relative distance *of the code.*

An $[n, k, d]$-code maps each of 2^k inputs to a code word of n bits, such that any two distinct inputs map to code words that have large relative distance. We use a simple error-correcting code (see [14]), which was also used in [2]:

Lemma 3 ([14], Theorem 17.30[3]). *For each* $m \geq 1$ *there is a* $[3m, m, \frac{m}{2}]$*-code.*

---

[3] The theorem in [14] gives a general construction for any distance up to $1/2$; here we use distance $1/6$.

The relative distance of the code in Lemma 3 is $\delta = \frac{(1/2)m}{3m} = \frac{1}{6}$.

When the players use the code from Lemma 3 to encode their inputs, each player's input grows by a constant factor (3), while the relative Hamming distance of any two differing inputs becomes at least δ. Let $N = 3n$ denote the length of the encoded inputs, and let $\bar{x}_i$ denote the encoding of player i's input x_i.

Partitioning into Blocks. After computing the encoding of their inputs, each player splits its encoded input into blocks of $L = \lceil \frac{N}{k} \rceil$ bits each, except possibly the last block, which may be shorter. For simplicity we assume here that all blocks have the same length, that is, L divides n. Let $b = N/L$ be the resulting number of blocks; we note that $b \leq \min(3n, k)$. Let $B_i(\ell) \in \{0,1\}^L$ denote the ℓ-th block of player i.

Because any two differing inputs have encodings that are far in Hamming distance, we can show that two players with different inputs will also disagree on many *blocks*:

Lemma 4. *For any two players i, j such that $x_i \neq x_j$, we have $|\{\ell \in [b] \mid B_i(\ell) \neq B_j(\ell)\}| \geq \delta b$.*

Proof. Assume by contradiction that $|\{\ell \in [b] | B_i(\ell) \neq B_j(\ell)\}| < \delta b$.

Let $\Delta = \{s \in [N] \mid \bar{x}_i(s) \neq \bar{x}_j(s)\}$ be the set of coordinates on which players i, j disagree. By the properties of the error correcting code, $|\Delta| \geq \delta N$.

Now partition Δ into disjoint sets $\Delta_1, \ldots, \Delta_b$, where each Δ_ℓ contains the locations inside block ℓ on which the encoded inputs disagree. Each Δ_ℓ contains between 0 and N/b coordinates, as the size of each block is $L = N/b$. By our assumption, there are fewer than δb blocks that contain any differences, so the number of non-empty sets Δ_ℓ is smaller than δb. It follows that $|\Delta| < \delta b \cdot (N/b) = \delta N$, which contradicts the relative distance of the code.

Comparing Blocks. Our goal now is to try to find two players that disagree on some block. We know that if there are two players with different inputs, then they will disagree on many different blocks, so choosing a random block will expose the difference with good probability. In order to compare the blocks, we use an optimal 2-player private-coin simultaneous protocol for EQ:

Theorem 5 ([3] Theorem 1.5). *There exists a private-coin one-sided error simultaneous protocol for the two player function EQ_m of cost $\Theta(\sqrt{m})$. If the inputs are equal, the protocol always outputs "Equal". If the inputs are not equal, then the protocol outputs "Equal" with probability $< 1/3$.*

Remark 1. We refer here to the symmetric variant of the equality protocol in Remark 3.3 of [3], in which both Alice and Bob use the same algorithm to compute their outputs.

We proceed as follows. Each player i chooses a block $\ell \in [b]$ at random. The player applies Alice's algorithm from [3]'s 2-player equality protocol on the chosen block $B_i(\ell)$, and sends the output to the referee, along with the index

ℓ of the block. In this process each player sends $O(\sqrt{\frac{n}{k}} + \log{(\min(n,k))})$ bits, because the length of a block is $L = O(n/k)$, and $b \leq \min(3n, k)$.

The referee receives the player's outputs $o_1, ..., o_k$, and for each pair that chose the same block index, it simulates [3]'s 2-player equality referee. If for all such pairs the output is 1 then the referee also outputs 1, otherwise it outputs 0. Let us denote by $Ref(o_1, \ldots, o_k)$ the referee's output function.

Analysis of the Error Probability. Note that if all inputs are equal, then our protocol always outputs 1: the EQ_L protocol from [3] has one-sided error, so in this case it will output 1 for any pair of blocks compared. On the other hand, if there exist two different inputs, we will only detect this if (a) the two corresponding players choose a block on which their encoded inputs differ, and (b) the EQ_L protocol from [3] succeeds and outputs 0. We show that this does happen with good probability:

Lemma 5. *If* $\mathrm{ALLEQ}(x_1, ..., x_k) = 0$, *then the protocol outputs* 0 *with probability at least* $\frac{2}{3}\delta(1 - e^{-\frac{1}{2}})$.

Proof. Since there are at least two distinct input strings, there exists an input string received by at most half the players. Let i be a player with such a string, and let $j'_1, ..., j'_{\frac{k}{2}}$ be $\frac{k}{2}$ players that disagree with player i's input.

Let A_t be the event that player j'_t chose the same block index as player i. Then

$$\Pr[Ref(o_1, ..., o_k) = 0] \geq \Pr\left[Ref(o_1, ..., o_k) = 0 \,\middle|\, \bigcup_{t=1}^{k/2} A_t\right] \cdot \Pr\left[\bigcup_{t=1}^{k/2} A_t\right].$$

We bound each of these two factors individually.

Since all A_t's are independent, and for a fixed t we have $Pr[A_t] = \frac{1}{b}$ then

$$\Pr\left[\bigcup_{t=1}^{k/2} A_t\right] = 1 - \left(1 - \frac{1}{b}\right)^{k/2} \geq 1 - \left(1 - \frac{1}{k}\right)^{k/2}$$

$$\geq 1 - \left(e^{-1/k}\right)^{k/2} = 1 - e^{-1/2}.$$

Next, let us condition on the fact that some specific A_r occurred. Given that at least one of the A_t's occurred, let A_r be such an event, that is, player r chose the same block as player i.

Clearly, conditioning on A_r does not change the probability of each block being selected, because the blocks are chosen uniformly and independently: that is, for each $i, r \in [k]$ and $\ell \in [b]$,

$$\Pr[\text{player } i \text{ chose block } \ell \mid A_r] = \frac{1}{b}.$$

Therefore, by Lemma 4, given the event A_r, players i and r disagree on the block they sent with probability at least $(\delta b)/b = \delta$. Whenever i and r send a block they disagree on, the protocol from [3] outputs 0 with probability 2/3. So overall,

$$\Pr\left[Ref(o_1, ..., o_k) = 0 \,\middle|\, \bigcup_{t=1}^{\frac{k}{2}} A_t\right] \geq \frac{2}{3}\delta.$$

Combining the two yields $Pr[Ref(o_1, ..., o_k) = 0] \geq \frac{2}{3}\delta(1 - e^{-\frac{1}{2}})$.

Proof (Proof of Theorem 4). By Lemma 5, if $\text{ALLEQ}(x_1, ..., x_k) = 0$ the algorithm errs with constant probability. If $\text{ALLEQ}(x_1, ..., x_k) = 1$ then since $\forall_{i,p,p'} B_p(i) = B'_p(i)$, and the fact that [3]'s protocol is a one-sided error protocol, the global protocol will always output 1, which is the correct value. Since this is a one-sided error protocol with constant error probability, this protocol can be amplified by repeating the protocol in parallel a constant number of times, so that the error probability becomes an arbitrarily small constant, and the communication is only increased by a constant factor.

5   On EXISTSEQ

The upper bound of Sect. 4 reduces the ALLEQ problem to a collection of 2-player EQ problems, which can then be solved efficiently using known protocols (e.g., from [3]). This works because asking whether *all* inputs are equal, and finding any pair of inputs that are not equal is sufficient to conclude that the answer is "no". What is the situation for the EXISTSEQ problem, where we ask whether there *exists* a pair of inputs that are equal? Intuitively the approach above should not help, and indeed, the complexity of EXISTSEQ is higher:

Theorem 6. *If $k \leq 2^{n-1}$, then $R_\epsilon(\text{EXISTSEQ}_{k,n}) = \Omega(k\sqrt{n})$ for any constant $\epsilon < 1/2$.*

Proof. We show that each player i must send $\Omega(\sqrt{n})$ bits in the worst case, and the bound then follows by Observation 1. The proof is by reduction from 2-player EQ_{n-1} (we assume that $n \geq 2$).

Let Π be a private-coin simultaneous protocol for $\text{EXISTSEQ}_{k,n}$ with error $\epsilon < 1/2$. Consider player 1 (the proof for the other players is similar). Assign to each player $i \in \{3, ..., k\}$ a unique label $b_i \in \{0, 1\}^{n-1}$ (this is possible because $k \leq 2^{n-1}$.

We construct a 2-player simultaneous protocol Π' for EQ_{n-1} with error $\epsilon < 1/2$ as follows: on inputs $(x, y) \in \{0, 1\}^{n-1}$, Alice plays the role of player 1 in Π, feeding it the input $1x$ (that is, the n-bit vector obtained by prepending '1' to x); Bob plays the role of player 2 with input $1y$; and the referee in Π' plays the role of all the other players and the referee in Π, feeding each player i the input $0b_i$, where b_i is the unique label assigned to player i.

After receiving messages from Alice and Bob, and sampling the messages players $3, \ldots, k$ would send in Π when given the inputs described above, the referee computes the output of Π and that is also the output of Π'.

Because we prefixed the true inputs x, y with 1, and players $3, \ldots, k$ received inputs beginning with 0, we have $\text{EXISTSEQ}(1x, 1y, 0b_3, \ldots, 0b_k) = \text{EQ}(x, y)$. Therefore Π' succeeds whenever Π does, and in particular it has error at most ϵ. By the lower bound of [3], then, player 1 must send $\Omega(\sqrt{n})$ bits in the worst case.

We note that if $k \geq 2^n$ then EXISTSEQ is trivial, as there must always exist two players with the same input. (The depencence of k on n in Theorem 6 can be improved to $k \leq (1 - o(1)) \cdot 2^n$ by assigning inputs to player $3, \ldots, k$ more cleverly.)

The lower bound above is tight up to $O(\log k)$, and indeed we can state something more general: for any 2-player function $f : \mathcal{X} \times \mathcal{Y} \to \{0, 1\}$, let $\exists_k f$ be the k-player function that outputs 1 on input $(x_1, \ldots, x_k)$ iff for some $i, j \in [k]$ we have $f(x_i, x_j) = 1$.

Lemma 6. *For any 2-player function f and $k \geq 2$ we have $R_{k^2 \epsilon}(\exists_k f) = O(k \cdot R_\epsilon(f))$.*

Proof. Let $\Pi = (\Pi_A, \Pi_B, O)$ be a 2-player private-coin simultaneous protocol for f with communication complexity $R_\epsilon(f)$.

We construct a protocol Π' for $\exists_k f$ as follows: on input $(x_1, \ldots, x_k)$, each player i samples two messages, $M_A^i \sim \Pi_A(x_i)$ and $M_B^i \sim \Pi_B(x_i)$, and sends them to the referee. The referee samples outputs $Z_{i,j} \sim O(M_A^i, M_B^j)$ for each $(i, j) \in [k]^2$ $(i \neq j)$ and outputs "1" iff for some pair $Z_{i,j} = 1$.

If $\exists_k f(x_1, \ldots, x_k) = 1$, then there exist i, j such that $f(x_i, x_j) = 1$, and for this pair of players we have $\Pr[Z_{i,j} = 0] \leq \epsilon$. Therefore the referee outputs "1" except with probability ϵ.

On the other hand, if $\exists_k f(x_1, \ldots, x_k) = 0$, then for every pair i, j of players we have $f(x_i, x_j) = 0$, so $\Pr[Z_{i,j} = 1] \leq \epsilon$. By union bound, in this case the probability that the referee outputs "1" is bounded by $\binom{k}{2} \epsilon < k^2 \epsilon$.

To handle the increased error of the protocol for $\exists_k f$, we can use a protocol for f that has error $O(1/k^2)$; this is achieved by taking a constant-error protocol for f, executing it $O(\log k)$ independent times, and taking the majority output [13]. We obtain the following:

Theorem 7. *For any 2-player function f, $k \geq 2$, and constant $\epsilon < 1/2$ we have $R_\epsilon(\exists_k f) = O(k \log k \cdot R_\epsilon(f))$.*

Corollary 2. *For EXISTSEQ we have $R_\epsilon(\text{EXISTSEQ}) = O(k \log k \sqrt{n})$, matching our lower bound up to a logarithmic factor.*

6   Conclusion

In this paper we extended the classical results of Babai and Kimmel [3] to the multi-player setting, and gave a tight bound for the gap between private-coin and deterministic communication complexity in the simultaneous setting. We showed that contrary to our initial expectations, the gap does not grow larger with the number of players, and indeed the per-player gap can shrink as the number of players grows. We also addressed a class of functions defined by taking a two-party function and asking whether there are two players whose inputs cause it to output 1.

Our work leaves open the interesting question of simultaneous lower bounds for the model considered in [1,5,6], where each player represents a node in a graph and is given the edges adjacent to that node. Our techniques do not apply to this scenario because of the sharing of edges between players. Indeed, the connectivity problem for this model (see [15]) cannot be addressed by reductions from two-player communication complexity, because it is easy for two players: we can simply have Alice and Bob compute spanning forests for their part of the input and send them to each other, at a total cost of $O(n \log n)$. Thus, further multi-party techniques need to be developed to address the hardness of connectivity.

References

1. Ahn, K.J., Guha, S., McGregor, A.: Graph sketches: sparsification, spanners, and subgraphs. In: Proceedings of the 31st Symposium on Principles of Database Systems, PODS 2012, pp. 5–14 (2012)
2. Ambainis, A.: Communication complexity in a 3-computer model. Algorithmica **16**(3), 298–301 (1996)
3. Babai, L., Kimmel, P.G.: Randomized simultaneous messages: solution of a problem of yao in communication complexity. In: Proceedings of the 12th Annual IEEE Conference on Computational Complexity, CCC 1997, p. 239. IEEE Computer Society (1997)
4. Babai, L., Gál, A., Kimmel, P.G., Lokam, S.V.: Communication complexity of simultaneous messages. SIAM J. Comput. **33**(1), 137–166 (2004)
5. Becker, F., Matamala, M., Nisse, N., Rapaport, I., Suchan, K., Todinca, I.: Adding a referee to an interconnection network: what can(not) be computed in one round. In: 25th IEEE International Symposium on Parallel and Distributed Processing, IPDPS 2011, pp. 508–514 (2011)
6. Becker, F., Montealegre, P., Rapaport, I., Todinca, I.: The simultaneous number-in-hand communication model for networks: private coins, public coins and determinism. In: Halldórsson, M.M. (ed.) SIROCCO 2014. LNCS, vol. 8576, pp. 83–95. Springer, Heidelberg (2014). doi:10.1007/978-3-319-09620-9_8
7. Bourgain, J., Wigderson, A.: Personal communication (see [3])
8. Chakrabarti, A., Shi, Y., Wirth, A., Yao, A.: Informational complexity and the direct sum problem for simultaneous message complexity. In: Proceedings of the Fourty-Second Annual Symposium on Foundations of Computer Science, FOCS 2001, pp. 270–278 (2001)

9. Chandra, A.K., Furst, M.L., Lipton, R.J.: Multi-party protocols. In: Proceedings of the Fifteenth Annual ACM Symposium on Theory of Computing, STOC 1983, pp. 94–99 (1983)
10. Chattopadhyay, A., Radhakrishnan, J., Rudra, A.: Topology matters in communication. In: 55th IEEE Annual Symposium on Foundations of Computer Science, FOCS 2014, Philadelphia, PA, USA, 18–21 October 2014, pp. 631–640 (2014)
11. Jain, R., Klauck, H.: New results in the simultaneous message passing model via information theoretic techniques. In: Proceedings of the Twenty-Fourth Annual IEEE Conference on Computational Complexity, CCC 2009, pp. 369–378 (2009)
12. Kremer, I., Nisan, N., Ron, D.: On randomized one-round communication complexity. In: Proceedings of the Twenty-seventh Annual ACM Symposium on Theory of Computing, STOC 1995, pp. 596–605 (1995)
13. Kushilevitz, E., Nisan, N.: Communication Complexity. Cambridge University Press, Cambridge (1997)
14. MacWilliams, F.J., Sloane, N.J.A.: The Theory of Error Correcting Codes. North-Holland, Amsterdam (1981)
15. McGregor, A.: Open problem 65. List of Open Problems in Sublinear Algorithms. http://sublinear.info/index.php?title=Open_Problems:65
16. Newman, I., Szegedy, M.: Public vs. private coin flips in one round communication games (extended abstract). In: Proceedings of the Twenty-Eighth Annual ACM Symposium on Theory of Computing, STOC 1996, pp. 561–570 (1996)
17. Weinstein, O., Woodruff, D.P.: The simultaneous communication of disjointness with applications to data streams. In: Halldórsson, M.M., Iwama, K., Kobayashi, N., Speckmann, B. (eds.) ICALP 2015. LNCS, vol. 9134, pp. 1082–1093. Springer, Heidelberg (2015). doi:10.1007/978-3-662-47672-7_88
18. Yao, A.C.-C.: Some complexity questions related to distributive computing (preliminary report). In: Proceedings of the Eleventh Annual ACM Symposium on Theory of Computing, STOC 1979, pp. 209–213 (1979)

Message Lower Bounds via Efficient Network Synchronization

Gopal Pandurangan[1]([✉]), David Peleg[2]([✉]), and Michele Scquizzato[1]([✉])

[1] University of Houston, Houston, USA
gopalpandurangan@gmail.com, michele@cs.uh.edu
[2] The Weizmann Institute of Science, Rehovot, Israel
david.peleg@weizmann.ac.il

Abstract. We present a uniform approach to derive message-time tradeoffs and message lower bounds for synchronous distributed computations using results from communication complexity theory.

Since the models used in the classical theory of communication complexity are inherently asynchronous, lower bounds do not directly apply in a synchronous setting. To address this issue, we show a general result called *Synchronous Simulation Theorem (SST)* which allows to obtain *message* lower bounds for synchronous distributed computations by leveraging lower bounds on communication complexity. The SST is a by-product of a new efficient synchronizer for complete networks, called σ, which has simulation overheads that are only logarithmic in the number of synchronous rounds with respect to both time and message complexity in the CONGEST model. The σ synchronizer is particularly efficient in simulating synchronous algorithms that employ silence. In particular, a curious property of this synchronizer, which sets it apart from its predecessors, is that it is *time-compressing*, and hence in some cases it may result in a simulation that is faster than the original execution.

While the SST gives near-optimal message lower bounds up to large values of the number of allowed synchronous rounds r (usually polynomial in the size of the network), it fails to provide meaningful bounds when a very large number of rounds is allowed. To complement the bounds provided by the SST, we then derive message lower bounds for the synchronous message-passing model that are *unconditional*, that is, independent of r, via direct reductions from multi-party communication complexity.

We apply our approach to show (almost) tight message-time tradeoffs and message lower bounds for several fundamental problems in the synchronous message-passing model of distributed computation. These include sorting, matrix multiplication, and many graph problems. All these lower bounds hold for any distributed algorithms, including randomized Monte Carlo algorithms.

G. Pandurangan—Supported, in part, by NSF grants CCF-1527867 and CCF-1540512.

© Springer International Publishing AG 2016
J. Suomela (Ed.): SIROCCO 2016, LNCS 9988, pp. 75–91, 2016.
DOI: 10.1007/978-3-319-48314-6_6

1   Introduction

Message complexity, which refers to the total number of messages exchanged during the execution of a distributed algorithm, is one of the two fundamental complexity measures used to evaluate the performance of algorithms in distributed computing [29]. Even when time complexity is the primary consideration, message complexity is significant. In fact, in practice the performance of the underlying communication subsystem is influenced by the load on the message queues at the various sites, especially when many distributed algorithms run simultaneously. Consequently, as discussed e.g. in [13], optimizing the message (as well as the time) complexity in some models for distributed computing has direct consequences on the time complexity in other models. Moreover, message complexity has also a considerable impact on the auxiliary resources used by an algorithm, such as energy. This is especially crucial in contexts, such as wireless sensor networks, where processors are powered by batteries with limited capacity. Besides, from a physical standpoint, it can be argued that energy leads to more stringent constraints than time does since, according to a popular quote by Peter M. Kogge, "You can hide the latency, but you can't hide the energy."

Investigating the message complexity of distributed computations is therefore a fundamental task. In particular, proving lower bounds on the message complexity for various problems has been a major focus in the theory of distributed computing for decades (see, e.g., [24,29,32,34]). Tight message lower bounds for several fundamental problems such as leader election [20,21], broadcast [4,20], spanning tree [17,20,24,34], minimum spanning tree [13,18,20,24,28,34], and graph connectivity [13], have been derived in various models for distributed computing.

One of the most important distinctions among message passing systems is whether the mode of communication is synchronous or asynchronous. In this paper we focus on proving lower bounds on the message complexity of distributed algorithms in the synchronous communication setting. Many of the message lower bounds mentioned above (e.g., [4,13,17,18,20]) use *ad hoc* (typically combinatorial) arguments, which usually apply only to the problem at hand. In this paper, on the other hand, the approach is to use *communication complexity* [19] as a uniform tool to derive message lower bounds for a variety of problems in the synchronous setting.

Communication complexity, originally introduced by Yao [38], is a subfield of complexity theory with numerous applications in several, and very different, branches of computer science (see, e.g., [19] for a comprehensive treatment). In the basic two-party model, there are two distinct parties, usually referred to as Alice and Bob, each of whom holds an n-bit input, say x and y. Neither knows the other's input, and they wish to collaboratively compute a function $f(x, y)$ by following an agreed-upon protocol. The cost of this protocol is the number of bits communicated by the two players for the worst-case choice of inputs x and y. It is important to notice that this simple model is inherently asynchronous, since it does not provide the two parties with a common clock. Synchronicity, however, makes the model subtly different, in a way highlighted by the following simple example (see, e.g., [33]). If endowed with a common clock, the two parties could agree upon a protocol in which time-coding is used to convey information: for instance, an n-bit message can be sent from one party to the other by encoding it with a single bit sent in one of 2^n possible synchronous rounds (keeping silent throughout all the other rounds). Hence, in a synchronous setting, *any* problem can be solved (deterministically) with communication complexity of one bit. This is a big difference compared to the classical (asynchronous) model! Likewise, as observed in [13], k bits of communication suffice to solve *any* problem in a complete network of k parties that initially agree upon a leader (e.g., the node with smallest ID) to whom they each send the bit that encodes their input. However, the low message complexity comes at the price of a high number of synchronous rounds, which has to be at least exponential in the size of the input that has to be encoded, as a single bit within time t can encode at most $\log t$ bits of information. The above observation raises many intriguing questions: (1) If one allows only a small number of rounds (e.g., polynomial in n) can such a low message complexity be achieved? (2) More generally, how can one show message lower bounds in the synchronous distributed computing model vis-a-vis the time (round) complexity? This paper tries to answer these questions in a general and comprehensive way.

Our approach is based on the design of a new and efficient *synchronizer* that can efficiently simulate synchronous algorithms that use (a lot of) silence, unlike previous synchronizers. Recall that a synchronizer ν transforms an algorithm S designed for a synchronous system into an algorithm $A = \nu(S)$ that can be executed on an asynchronous system. The goal is to keep T_A and C_A, the time and communication complexities of the resulting asynchronous algorithm A, respectively, close to T_S and C_S, the corresponding complexities of the original synchronous algorithm S. The synchronizers appearing in the literature follow a methodology (described, e.g., in [29]) which resulted in bounding the complexities T_A and C_A of the asynchronous algorithm A for every input instance $\mathcal{I}$ as

$$T_A(\mathcal{I}) \leq T_{\text{init}}(\nu) + \Psi_T(\nu) \cdot T_S(\mathcal{I}),$$
$$C_A(\mathcal{I}) \leq C_{\text{init}}(\nu) + C_S(\mathcal{I}) + \Psi_C(\nu) \cdot T_S(\mathcal{I}),$$

where $\Psi_T(\nu)$ (resp., $\Psi_C(\nu)$) is the time (resp., communication) overhead coefficient of the synchronizer ν, and $T_{\text{init}}(\nu)$ (resp., $C_{\text{init}}(\nu)$) is the time (resp., communication) initialization cost. In particular, the early synchronizers, historically named α [3], β [3], γ [3], and δ [31] (see also [29]), handled each synchronous round separately, and incurred a communication overhead of at least $O(k)$ bits per synchronous round, where k is the number of processors in the system. The synchronizer μ of [5] remedies this limitation by taking a more global approach, and its time and communication overheads $\Psi_T(\mu)$ and $\Psi_C(\mu)$ are both $O(\log^3 k)$, which is at most a polylogarithmic factor away from optimal under the described methodology.

Note, however, that the dependency of the asynchronous communication complexity $C_A(\mathcal{I})$ on the synchronous time complexity $T_S(\mathcal{I})$ might be problematic in situations where the synchronous algorithm takes advantage of synchronicity in order to exploit *silence*, and uses time-coding for conveying information while transmitting fewer messages (e.g., see [13,14]). Such an algorithm, typically having low communication complexity but high time complexity, translates into an asynchronous algorithm with high time and communication complexities. Hence, we may prefer a simulation methodology that results in a communication dependency of the form $C_A(\mathcal{I}) \leq C_{\text{init}}(\nu) + \Psi_C(\nu) \cdot C_S(\mathcal{I})$, and where $\Psi_C(\nu)$ is at most polylogarithmic in the number of rounds T_S of the synchronous algorithm.

1.1 Our Contributions

We present a uniform approach to derive message lower bounds for synchronous distributed computations by leveraging results from the theory of communication complexity. In this sense, this can be seen a companion paper of [6], which leverages the connection between communication complexity and distributed computing to prove lower bounds on the time complexity of synchronous distributed computations.

A New and Efficient Synchronizer. Our approach, developed in Sect. 3, is based on the design of a new and efficient synchronizer for complete networks, which we call synchronizer σ,[1] and which is of independent interest. The new attractive feature of synchronizer σ, compared to existing ones, is that it is *time-compressing*. To define this property, let us denote by T_S^c the number of *communicative* (or *active*) synchronous rounds, in which at least one node of the network sends a message. Analogously, let T_S^q denote the number of *quiet* (or *inactive*) synchronous rounds, in which all processors are silent. Clearly, $T_S = T_S^c + T_S^q$. Synchronizer σ compresses the execution time of the simulation by essentially discarding the inactive rounds, and remaining only with the active ones. This is in sharp contrast to all previous synchronizers, whereby every single round of the synchronous execution is simulated in the asynchronous network. A somewhat surprising consequence of this feature is that synchronizer σ may in certain situations result in a simulation algorithm whose execution time is *faster* than the original synchronous algorithm. Specifically, T_A can be strictly smaller than T_S when the number of synchronous rounds in which no node communicates is sufficiently high. (In fact, we observe that time compression may occur even when simulating the original synchronous algorithm on another *synchronous* network, in which case the resulting simulation algorithm may yield faster, albeit more communication-intensive, synchronous executions.)

Table 1 compares the complexities of various synchronizers when used for complete networks.

Synchronous Simulation Theorem. As a by-product of synchronizer σ, we show a general theorem, the *Synchronous Simulation Theorem (SST)*, which shows how

---

[1] We use σ as it is the first letter in the Greek word which means "silence".

Table 1. Comparison among different synchronizers for k-node complete networks. The message size is assumed to be $O(\log k)$ bits, and C_A is expressed in number of messages. (Note that on a complete network, synchronizers γ [3] and δ [31] are out-performed by β, hence their complexities are omitted from this table.)

Synchronizer	Time complexity T_A	Message complexity C_A
α [3]	$O(T_S)$	$O(k^2) + O(C_S) + O(T_S\, k^2)$
β [3]	$O(k) + O(T_S)$	$O(k \log k) + O(C_S) + O(T_S\, k)$
μ [5]	$O(k \log k) + O(T_S \log^3 k)$	$O(k \log k) + O(C_S) + O(T_S \log^3 k)$
σ [this paper]	$O(k) + O(T_S^c \log_k T_S)$	$O(k \log k) + O(C_S \log_k T_S)$

message lower bounds for synchronous distributed computations can be derived by leveraging communication complexity results obtained in the asynchronous model. More precisely, the SST provides a tradeoff between the message complexity of the synchronous computation and the maximum number of synchronous rounds T_S allowed to the computation. This tradeoff reveals that message lower bounds in the synchronous model are worse by at most logarithmic factors (in T_S and k, where T_S is the number of rounds taken in the synchronous model, and k is the network size) compared to the corresponding lower bounds in the asynchronous model.

Applications: Message-Time Tradeoffs. In Sect. 4 we apply the SST to obtain message-time tradeoffs for several fundamental problems. These lower bounds assume that the underlying communication network is complete, which is the case in many computational models [15,23]; however, the same lower bounds clearly apply also when the network topology is arbitrary. The corresponding bounds on communication complexity are tight up to polylogarithmic factors (in the input size n of the problem and network size k) when the number of rounds is at most *polynomial* in the input size. This is because a naive algorithm that sends all the bits to a leader via the time encoding approach of [14, Theorem 3.1] is optimal up to polylogarithmic factors. We next summarize our lower bound results for various problems for a precise statement of these results. All the lower bounds in this paper hold even for randomized protocols that can return the wrong answer with a small constant probability.

Our lower bounds assume that the underlying topology is complete and the input is partitioned (in an adversarial way) among the k nodes. We assume that at most T_S rounds of synchronous computation are allowed. (We will interchangeably denote the number of rounds with T_S and r.) For sorting, where each of the k nodes have $n \geq 1$ input numbers, we show a message lower bound[2] of $\tilde{\Omega}(nk/\log r)$. This result immediately implies that Lenzen's $\tilde{O}(1)$-round sorting algorithm for the Congested Clique model [22] has also optimal (to within log factors) message complexity. For the Boolean matrix multiplication of two

---

[2] Throughout this paper, the notation $\tilde{\Omega}$ hides polylogarithmic factors in k and n, i.e., $\tilde{\Omega}(f(n,k))$ denotes $\Omega(f(n,k)/(\text{polylog}\, n\, \text{polylog}\, k))$.

Boolean $n \times n$ matrices we show a lower bound of $\Omega(n^2/\log rk)$. For graph problems, there is an important distinction that influences the lower bounds: whether the input graph is initially partitioned across the nodes in an *edge-partitioning* fashion or in *vertex-partitioning* fashion. In the former, the edges of the graph are arbitrarily distributed across the k parties, while in the latter each vertex of the graph is initially held by one party, together with the set of its incident edges. In the edge-partitioning setting, using the results of [37] in conjunction with the SST yields non-trivial lower bounds of $\tilde{\Omega}(kn/\log rk)$, where n is the number of vertices of the input graph, for several graph problems such as graph connectivity, testing cycle-freeness, and testing bipartiteness. For testing triangle-freeness and diameter the respective bounds are $\tilde{\Omega}(km)$ and $\tilde{\Omega}(m)$. (In the vertex-partitioning setting, on the other hand, many graph problems such as graph connectivity can be solved with $O(n\,\text{polylog}\,n)$ message complexity [37].)

Unconditional Lower Bounds. While the SST gives essentially tight lower bounds up to very large values of T_S (e.g., polynomial in n), they become trivial for larger values of T_S (in particular, when T_S is exponential in n). To complement the bounds provided by the SST, in Sect. 5 we derive message lower bounds in the synchronous message-passing model which are *unconditional*, that is, independent of time. These lower bounds are established via direct reductions from *multi-party* communication complexity. They are of the form $\tilde{\Omega}(k)$, and this is almost tight since every problem can be solved with $O(k)$ bits of communication by letting each party encode its input in just one bit via time encoding. We point out that the unconditional lower bounds cannot be shown by reductions from 2-party case, as typically done for many reductions for these problems. A case in point are the reductions to establish the lower bounds for connectivity and diameter in the vertex-partitioning model. To show unconditional lower bounds for connectivity and diameter we define a new multi-party problem called *input-distributed disjointness (ID-DISJ)* (see Sect. 5) and establish a lower bound for it. We note that, unlike in the asynchronous setting, reduction from a 2-party setting will not yield the desired lower bound of $\Omega(k)$ in the synchronous setting (since 2-party problems can be solved trivially, exploiting clocks, using only one bit, as observed earlier).

1.2   Further Related Work

The first paper that showed how to leverage lower bounds on communication complexity in a synchronous distributed setting is [30], which proves a near-tight lower bound on the time complexity of distributed minimum spanning tree construction in the CONGEST model [29]. Elkin [9] extended this result to approximation algorithms. The same technique was then used to prove a tight lower bound for minimum spanning tree verification [16]. Later, Das Sarma et al. [6] explored the connection between the theory of communication complexity and distributed computing further by presenting time lower bounds for a long list of problems, including inapproximability results. For further work on time lower bounds via communication complexity see, e.g., [8,11,15,25,27], as well as [26] and references therein.

Researchers also investigated how to leverage results in communication complexity to establish lower bounds on the message complexity of distributed computations. Tiwari [35] shows communication complexity lower bounds for deterministic computations over networks with some specific topologies. Woodruff and Zhang [37] study the message complexity of several graph and statistical problems in complete networks. Their lower bounds are derived through a common approach that reduces those problem from a new meta-problem whose communication complexity is established. However, the models considered in these two papers are inherently asynchronous, hence their lower bounds do not hold if a common clock is additionally assumed.

Hegeman et al. [13] study the message complexity of connectivity and MST in the (synchronous) congested clique. However, their lower bounds are derived using ad hoc arguments. To the best of our knowledge, the first connection between the classical communication complexity theory and the message complexity in a synchronous setting has been established by Impagliazzo and Williams [14]. They show almost tight bounds for the case with two parties for deterministic algorithms, by efficiently simulating a synchronous protocol in an asynchronous model (like we do). Ellen et al. [10] claim a simulation result for $k \geq 2$ parties. Their results are similar to our synchronous simulation theorem. However, their simulation does not consider time, whereas ours is time-efficient as well.

2 Preliminaries: Models, Assumptions, and Notation

The *message-passing model* is one of the fundamental models in the theory of distributed computing, and many variations of it have been studied. We are given a complete network of k nodes, which can be viewed as a complete undirected simple graph where nodes correspond to the processors of the network and edges represent bidirectional communication channels. Each node initially holds some portion of the input instance I, and this portion is known only to itself and not to the other nodes. Each node can communicate directly with any other node by exchanging messages. Nodes wake up spontaneously at arbitrary times. The goal is to jointly solve some given problem Π on input instance I.

Nodes have a unique identifier of $O(\log k)$ bits. Before the computation starts, each node knows its own identifier but not the identifiers of any other node. Each link incident to a node has a unique representation in that node. All messages received at a node are stamped with the identification of the link through which they arrived. By the number of its incident edges, every node knows the value of k before the computation starts. All the local computation performed by the processors of the network happens instantaneously, and each processor has an unbounded amount of local memory. It is also assumed that both the computing entities and the communication links are fault-free.

A key distinction among message-passing systems is whether the mode of communication is synchronous or asynchronous. In the *synchronous* mode of communication, a global clock is connected to all the nodes of the network. The time interval between two consecutive pulses of the clock is called a *round*. The computation proceeds in rounds, as follows. At the beginning of each synchronous round, each node sends (possibly different) messages to its neighbors. Each node then receives all the messages sent to it in that round, and performs some local computation, which will determine what messages to send in the next round. In the *asynchronous* mode of communication, there is no global clock. Messages over a link incur finite but arbitrary delays (see, e.g., [12]). This can be modeled as each node of the network having a queue where to place outgoing messages, with an adversarial global scheduler responsible of dequeuing messages, which are then instantly delivered to their respective recipients. Communication complexity, the subfield of complexity theory introduced by Yao [38], studies the asynchronous message-passing model.

We now formally define the complexity measure studied in this paper. Most of these definitions can be found in [19]. The *communication complexity of a computation* is the total number of bits exchanged across all the links of the network during the computation (or, equivalently, the total number of bits sent by all parties). The *communication complexity of a distributed algorithm A* is the maximum number of bits exchanged during the execution of A over all possible inputs of a particular size. The *communication complexity of a problem Π* is the minimum communication complexity of any algorithm that solves Π. *Message complexity* refers to the total number of messages exchanged, where the message size is bounded by some value B of bits.

In this paper we are interested in lower bounds for *Monte Carlo* distributed algorithms. A Monte Carlo algorithm is a randomized algorithm whose output may be incorrect with some probability. Formally, *algorithm A solves a problem Π with ϵ-error* if, for every input I, A outputs $\Pi(I)$ with probability at least $1 - \epsilon$, where the probability is taken only over the random strings of the players. The *communication complexity of an ϵ-error randomized protocol/algorithm A on input I* is the maximum number of bits exchanged for any choice of the random strings of the parties. The *communication complexity of an ϵ-error randomized protocol/algorithm A* is the maximum, over all possible inputs I, of the communication complexity of A of input I. The *randomized ϵ-error communication complexity of a problem Π* is the minimum communication complexity of any ϵ-error randomized protocol that solves Π. In a model with $k \geq 2$ parties, this is denoted with $R_{k,\epsilon}(\Pi)$. The same quantity can be defined likewise for a synchronous model, in which case it is denoted with $SR_{k,\epsilon}(\Pi)$. Throughout the paper we assume ϵ to be a small constant and therefore, for notational convenience, we will drop the ϵ in the notation defined heretofore.

We say that a randomized distributed algorithm uses a *public coin* if all parties have access to a common random string. In this paper we are interested in lower bounds for public-coin randomized distributed algorithms. Clearly, lower bounds of this kind also hold for *private-coin* algorithms, in which parties do not share a common random string.

We now define a second complexity measure for a distributed computation, the *time complexity*. In the synchronous mode of communication, it is defined as the (worst-case) number of synchronous rounds that it comprises. It is additionally referred to as *round complexity*. Following [6], we define the *randomized ϵ-error r-round randomized communication complexity* of a problem Π in a synchronous model to be the minimum communication complexity of any protocol that solves Π with error probability ϵ when it runs in at most r rounds. We denote this quantity with $SR_{k,\epsilon,r}(\Pi)$. A lower bound on $SR_{k,\epsilon,r}(\Pi)$ holds also for Las Vegas randomized algorithms as well as for deterministic algorithms. In the asynchronous case, the time complexity of a computation is the (worst-case) number of time units that it comprises, assuming that each message incurs a delay of at most one time unit [29, Definition 2.2.2]. Thus, in arguing about time complexity, a message is allowed to traverse an edge in any fraction of the time unit. This assumption is used only for the purpose of time complexity analysis, and does not imply that there is a bound on the message transmission delay in asynchronous networks.

Throughout this paper, we shall use interchangeably node, party, or processor to refer to elements of the network, while we will use vertex to refer to a node of the input graph when the problem Π is specified on a graph.

3 Efficient Network Synchronization and the Synchronous Simulation Theorem

3.1 The Simulation

We present synchronizer σ, an efficient (deterministic) simulation of a synchronous algorithm S designed for a complete network of k nodes in the corresponding asynchronous counterpart. The main ideas underlying the simulation are the exploitation of inactive nodes and inactive rounds, via the use of the concept of *tentative time*, in conjunction with the use of acknowledgments as a method to avoid congestion and thus reduce the time overhead in networks whose links have limited bandwidth. It is required that all the possible communication sequences between any two nodes of the network are self-determining, i.e., no one is a prefix of another.

One of the k nodes is designated to be a *coordinator*, denoted with $\mathcal{C}$, which organizes and synchronizes the operations of all the processors. The coordinator is determined before the actual simulation begins, and this can be done by executing a leader election algorithm for asynchronous complete networks, such as the one in [1]. (The coordinator should not be confused with the notion of coordinator in the variant of the message-passing model introduced in [7]. In the latter, (1) the coordinator is an additional party, which has no input at the beginning of the computation, and which must hold the result of the computation at the end of the computation, and (2) nodes of the network are not allowed to communicate directly among themselves, and therefore they can communicate only with the coordinator.) After its election, the coordinator sends to each node a message START(1) instructing them to start the simulation of round 1.

At any given time each node v maintains a *tentative time* estimate $\mathbf{TT}(v)$, representing the next synchronous round on which v plans to send a message to one (or more) of its neighbors. This estimate may change at a later point, i.e., v may send out messages earlier than time $\mathbf{TT}(v)$, for example in case v receives a message from one of its neighbors, prompting it to act. However, assuming no such event happens, v will send its next message on round $\mathbf{TT}(v)$. (In case v currently has no plans to send any messages in the future, it sets its estimate to $\mathbf{TT}(v) = \infty$.) The coordinator $\mathcal{C}$ maintains k local variables, which store, at any given time, the tentative times of all the nodes of the network.

We now describe the execution of phase t of the simulation, which simulates the actions of the processors in round t of the execution ξ_S of algorithm S in the synchronous network. Its starting point is when the coordinator realizes that the current phase, simulating some round $t' < t$, is completed, in the sense that all messages that were supposed to be sent and received by processors on round t' of ξ_S were sent and received in the simulation on the asynchronous network. The phase proceeds as follows.

(1) The coordinator $\mathcal{C}$ determines the minimum value of $\mathbf{TT}(v)$ over all processors v, and sets t to that value. (In the first phase, the coordinator sets $t = 1$ directly.) If $t = \infty$ then the simulation is completed and it is possible to halt. If $t' + 1 < t$, then the synchronous rounds $t' + 1, \ldots, t - 1$ are inactive rounds, that is, in which all processors are silent. Thus, the system conceptually skips all rounds $t' + 1, \ldots, t - 1$, and goes straight to simulating round t. Since only the coordinator can detect the halting condition, it is also responsible for informing the remaining $k-1$ nodes by sending, in one time unit, $k-1$ additional HALT messages to each of them.

(2) The coordinator (locally) determines the set of *active* nodes, defined as the set of nodes whose tentative time is t, that is,

$$\mathcal{A}(t) = \{v \mid \mathbf{TT}(v) = t\},$$

and sends to each of them a message $\texttt{START}(t)$ instructing them to start round t. (In the first phase, all nodes are viewed as active, i.e., $\mathcal{A}(1) = V$.)

(3) Upon the receipt of this message, each active node v sends all the messages it is required by the synchronous algorithm to send on round t, to the appropriate subset $\mathcal{N}(v, t)$ of its neighbors. This subset of neighbors is hereafter referred to as v's *clan* on round t, and we refer to v itself as the *clan leader*. We stress that these messages are sent directly to their destination; they must not be routed from v to its clan via the coordinator, as this might cause congestion on the links from the coordinator to the members of $\mathcal{N}(v, t)$.

(4) Each neighbor $w \in \mathcal{N}(v, t)$ receiving such a message immediately sends back an acknowledgement directly to v.

Note that receiving a message from v may cause w to want to change its tentative time $\mathbf{TT}(w)$. However, w must wait for now with determining the new value of $\mathbf{TT}(w)$, for the following reason. Note that w may belong to more than one clan. Let $\mathcal{A}_w(t) \subseteq \mathcal{A}(t)$ denote the set of active nodes which are required to send a message to w on round t (namely, the clan leaders to whose clans w

belongs). At this time, w does not know the set $\mathcal{A}_w(t)$, and therefore it cannot be certain that no additional messages have been sent to it from other neighbors on round t. Such messages might cause additional changes in $\mathbf{TT}(w)$.

(5) Once an active node v has received acknowledgments from each of its clan members $w \in \mathcal{N}(v, t)$, v sends a message SAFE(v, t) to the coordinator $\mathcal{C}$.

(6) Once the coordinator $\mathcal{C}$ has received messages SAFE(v, t) from all the active nodes in $\mathcal{A}(t)$, it knows that all the original messages of round t have reached their destinations. What remains is to require all the nodes that were involved in the above activities (namely, all clan members and leaders) to recalculate their tentative time estimate. Subsequently, the coordinator $\mathcal{C}$ sends out a message ReCalcT to all the active nodes of $\mathcal{A}(t)$ (which are the only ones $\mathcal{C}$ knows about directly), and each $v \in \mathcal{A}(t)$ forwards this message to its entire clan, namely, its $\mathcal{N}(v, t)$ neighbors, as well.

(7) Every clan leader or member $x \in \mathcal{A}(t) \cup \bigcup_{v \in \mathcal{A}(t)} \mathcal{N}(v, t)$ now recalculates its new tentative time estimate $\mathbf{TT}(x)$, and sends it directly to the coordinator $\mathcal{C}$. (These messages must not be forwarded from the clan members to the coordinator via their clan leaders, as this might cause congestion on the links from the clan leaders to the coordinator.) The coordinator immediately replies each such message by sending an acknowledgement directly back to x.

(8) Once a (non-active) clan member w has received such an acknowledgement, it sends all its clan leaders in $\mathcal{A}_w(t)$ a message DoneReCalcT. (Note that at this stage, w already knows the set $\mathcal{A}_w(t)$ of its clan leaders—it is precisely the set of nodes from which it received messages in step (3).)

(9) Once an active node v has received an acknowledgement from $\mathcal{C}$ as well as messages DoneReCalcT from every member $w \in \mathcal{N}(v, t)$ of its clan, it sends the coordinator $\mathcal{C}$ a message DoneReCalcT, representing itself along with all its clan.

(10) Once the coordinator $\mathcal{C}$ has received an acknowledgement from every active node, it knows that the simulation of round t is completed.

3.2 Analysis of Complexity

Theorem 1. *Synchronizer σ is a synchronizer for complete networks such that*

$$T_A = O\left(k \log k + \left(1 + \frac{\log T_S}{B}\right) T_S^c\right), \tag{1}$$

$$C_A = O\left(k \log^2 k + C_S \log T_S\right), \tag{2}$$

where T_S^c is the number of synchronous rounds in which at least one node of the network sends a message, k is the number of nodes of the network, and B is the message size of the network, in which at most one message can cross each edge at each time unit.

Proof. For any bit sent in the synchronous execution ξ_S, the simulation uses $\lceil \log_2 T_S \rceil$ additional bits to encode the values of the tentative times, and a constant number of bits for the acknowledgments and for the special messages

START(t), SAFE(v, t), ReCalcT, and DoneReCalcT. Observe, finally, that no congestion is created by the simulation, meaning that in each synchronous round being simulated each node sends and receives at most $O(1 + \lceil \log_2 T_S \rceil)$ bits in addition to any bit sent and received in the synchronous execution ξ_S.

The $O(k \log k)$ and $O(k \log^2 k)$ additive factors in the first and second equation are, respectively, the time and message complexity of the asynchronous leader election algorithm in [1] run in the initialization phase. This algorithm exchanges a total of $O(k \log k)$ messages of size $O(\log k)$ bits each, and takes $O(k)$ time. □

3.3 Message Lower Bound for Synchronous Distributed Computations

Theorem 2 (Synchronous Simulation Theorem (SST)). *Let $SCC^{\mathcal{D}}_{k,r}(\Pi)$ be the r-round communication complexity of problem Π in the synchronous message-passing complete network model with k nodes, where $\mathcal{D}$ is the initial distribution of the input bits among the nodes. Let $CC^{\mathcal{D}'}_{k'}(\Pi)$ be the communication complexity of problem Π in the asynchronous message-passing complete network model with $k' \leq k$ nodes where, given some partition of the nodes of a complete network of size k into sets $S_1, S_2, \ldots, S_{k'}$, $\mathcal{D}'$ is the initial distribution of the input bits whereby, for each $i \in \{1, 2, \ldots, k'\}$, node i holds all the input bits held by nodes in S_i under the distribution $\mathcal{D}$. Then,*

$$SCC^{\mathcal{D}}_{k,r}(\Pi) = \Omega \left(\frac{CC^{\mathcal{D}'}_{k'}(\Pi) - k \log^2 k}{1 + \log r + \lceil (k - k')/k \rceil \log k} \right).$$

Proof. We leverage the communication complexity bound of the σ synchronizer result to prove a lower bound on $SCC^{\mathcal{D}}_{k,r}(\Pi)$, synchronous communication complexity, by relating it to $CC^{\mathcal{D}'}_{k'}(\Pi)$, the communication complexity in the asynchronous setting. More precisely, we can use the σ synchronizer to simulate any synchronous algorithm for the problem Π to obtain an asynchronous algorithm for Π whose message complexity satisfies Eq. (2) of Theorem 1. We first consider the case when $k' = k$. Rearranging Eq. (2), and by substituting T_S with r, C_A with $CC^{\mathcal{D}}_k(\Pi)$, and C_S with $SCC^{\mathcal{D}}_k(\Pi)$, and by setting $B = 1$ (since $(S)CC$ is expressed in number of bits), we obtain the claimed lower bound on $SCC^{\mathcal{D}}_{k,r}(\Pi)$.

Next we consider $k' < k$. In this case, we need to do a minor modification to the σ synchronizer. Since we assume that messages do not contain the ID of the receiver and of the sender, when the network carrying the simulation has fewer nodes than the network to be simulated the ID of both source and the destination of any message has to be appended to the latter. This is the sole alteration needed for the simulation to handle this case. This entails $\lceil (k - k')/k \rceil \cdot 2 \lceil \log k \rceil$ additional bits to be added to each message. In this case, the communication complexity of the σ synchronizer is increased by a factor of $O(\lceil (k - k')/k \rceil \log k)$. This gives the claimed result. □

Clearly, a corresponding lower bound on the total number of messages follows by dividing the communication complexity by the message size B. Observe that

CC and SCC can be either both deterministic or both randomized. In the latter case, such quantities can be plugged in Theorem 2 according to the definition of ϵ-error r-round protocols given in Sect. 2.

4 Message-Time Tradeoffs for Synchronous Distributed Computations

We now apply the Synchronous Simulation Theorem to get lower bounds on the communication complexity of some fundamental problems in the synchronous message-passing model.

4.1 Sorting

In this section we give a lower bound to the communication complexity of comparison-based sorting algorithms. At the beginning each of the k parties holds n elements of $O(\log n)$ bits each. At the end, the i-th party must hold the $(i-1)k+1, (i-1)k+2, \ldots, i \cdot k$-th order statistics. We have the following result.

Theorem 3. *The randomized r-round ϵ-error communication complexity of sorting in the synchronous message-passing model with k parties is $\Omega(nk/\log k \log r)$.*

4.2 Matrix Multiplication

We now show a synchronous message lower bound for Boolean matrix multiplication, that is, the problem of multiplying two $n \times n$ matrices over the semiring $(\{0,1\}, \wedge, \vee)$.

In [36, Theorem 4] it is shown the following: Suppose Alice holds a Boolean $m \times n$ matrix A, Bob holds a Boolean $n \times m$ matrix B, and the Boolean product of these matrices has at most z nonzeroes. Then the randomized communication complexity of matrix multiplication is $\tilde{\Omega}(\sqrt{z} \cdot n)$. To apply Theorem 2 we then just have to consider an initial partition of the $2mn$ input elements among k parties such that there exists a cut in the network that divides the elements of A from those of B. Given such a partition, we immediately obtain the following.

Theorem 4. *The randomized r-round ϵ-error communication complexity of Boolean matrix multiplication in the synchronous message-passing model with k parties is $\Omega(\sqrt{z} \cdot n/\log rk)$.*

4.3 Statistical and Graph Problems

The generality of the SSTs allows us to directly apply any previous result derived for the asynchronous message-passing model. As an example, of particular interest are the results of Woodruff and Zhang [37], who present lower bounds on the communication complexity of a number of fundamental statistical and graph problems in the (asynchronous) message-passing model with k parties, all connected to each other. We shall seamlessly apply the SST for complete networks to all of their results, obtaining the following.

Theorem 5. *The randomized r-round ϵ-error communication complexity of graph connectivity, testing cycle-freeness, testing bipartiteness, testing triangle-freeness, and diameter of graphs with n nodes in the synchronous message-passing model with k parties, where the input graph is encoded in edges (u, v) which are initially (adversarially) distributed among the parties, is $\tilde{\Omega}(nk/\log rk)$.*

5   Unconditional Message Lower Bounds for Synchronous Distributed Computations

The bounds resulting from the application of the synchronous simulation theorem of Sect. 3 become vanishing as r increases, independently of the problem Π at hand. Hence it is natural to ask whether there are problems that can be solved by exchanging, e.g., only a constant number of bits when a very large number of rounds is allowed. In this section we discuss problems for which this cannot happen.

Specifically, we show that $\tilde{\Omega}(k)$ bits is an unconditional lower bound for several important problems in a synchronous complete network of k nodes. The key idea to prove unconditional $\tilde{\Omega}(k)$ bounds via communication complexity is to resort to multiparty communication complexity (rather than just to classical 2-party communication complexity), where by a simple and direct information-theoretic argument (i.e., without reducing from the asynchronous setting, as in Sect. 3) we can show that many problems in such a setting satisfy an $\Omega(k)$-bit lower bound, no matter how many synchronous rounds are allowed.

5.1   Graph Problems in the Vertex-Partitioning Model

To show unconditional lower bounds for graph problems in the vertex-partitioning model, we use a reduction from a new multiparty problem, called *input-distributed disjointness* (ID-DISJ) defined as follows. For the rest of this section, we assume $k = n$ and thus each party is assigned one vertex (and all its incident edges).

Definition 1. *Given n parties, each holding one input bit, partitioned in two distinct subsets $S_1 = \{1, 2, \ldots, n/2\}$ and $S_2 = \{n/2 + 1, n/2 + 2, \ldots, n\}$ of $n/2$ parties each, the* input-distributed disjointness *function* ID-DISJ(n) *is 0 if there is some index $i \in [n/2]$ such that both the input bits held by parties i and $i + n/2$ are 1, and 1 otherwise.*

Notice that this problem is, roughly speaking, "in between" the classical 2-party set disjointness and the n-party set disjointness: as in the latter, there are n distinct parties, and as in the former, the input can be seen as two vectors of $(n/2)$ bits. We have the following result.

Theorem 6. *The randomized ϵ-error communication complexity of* ID-DISJ(n) *in the synchronous message-passing model is $\Omega(n)$.*

We now leverage the preceding result to prove lower bounds on the communication complexity of graph connectivity and graph diameter.

Theorem 7. *The randomized ϵ-error communication complexity of graph connectivity in the synchronous message-passing model, with vertex-partitioning, is $\Omega(n)$.*

Theorem 8. *The randomized ϵ-error communication complexity of computing the diameter in the synchronous message-passing model, with vertex-partitioning, is $\Omega(n/\log n)$.*

References

1. Afek, Y., Gafni, E.: Time and message bounds for election in synchronous and asynchronous complete networks. SIAM J. Comput. **20**(2), 376–394 (1991)
2. Avin, C., Borokhovich, M., Lotker, Z., Peleg, D.: Distributed computing on core-periphery networks: axiom-based design. In: Esparza, J., Fraigniaud, P., Husfeldt, T., Koutsoupias, E. (eds.) ICALP 2014. LNCS, vol. 8573, pp. 399–410. Springer, Heidelberg (2014). doi:10.1007/978-3-662-43951-7_34
3. Awerbuch, B.: Complexity of network synchronization. J. ACM **32**(4), 804–823 (1985)
4. Awerbuch, B., Goldreich, O., Peleg, D., Vainish, R.: A trade-off between information and communication in broadcast protocols. J. ACM **37**(2), 238–256 (1990)
5. Awerbuch, B., Peleg, D.: Network synchronization with polylogarithmic overhead. In: Proceedings of the 31st Annual Symposium on Foundations of Computer Science (FOCS), pp. 514–522 (1990)
6. Sarma, A.D., Holzer, S., Kor, L., Korman, A., Nanongkai, D., Pandurangan, G., Peleg, D., Wattenhofer, R.: Distributed verification and hardness of distributed approximation. SIAM J. Comput. **41**(5), 1235–1265 (2012)
7. Dolev, D., Feder, T.: Determinism vs. nondeterminism in multiparty communication complexity. SIAM J. Comput. **21**(5), 889–895 (1992)
8. Drucker, A., Kuhn, F., Oshman, R.: On the power of the congested clique model. In: Proceedings of the 33rd ACM Symposium on Principles of Distributed Computing (PODC), pp. 367–376 (2014)
9. Elkin, M.: An unconditional lower bound on the time-approximation trade-off for the distributed minimum spanning tree problem. SIAM J. Comput. **36**(2), 433–456 (2006)
10. Ellen, F., Oshman, R., Pitassi, T., Vaikuntanathan, V.: Brief announcement: private channel models in multi-party communication complexity. In: Proceedings of the 27th International Symposium on Distributed Computing (DISC), pp. 575–576 (2013)
11. Frischknecht, S., Holzer, S., Wattenhofer, R.: Networks cannot compute their diameter in sublinear time. In: Proceedings of the 23rd Annual ACM-SIAM Symposium on Discrete Algorithms (SODA), pp. 1150–1162 (2012)
12. Gallager, R.G., Humblet, P.A., Spira, P.M.: A distributed algorithm for minimum-weight spanning trees. ACM Trans. Program. Lang. Syst. **5**(1), 66–77 (1983)
13. Hegeman, J.W., Pandurangan, G., Pemmaraju, S.V., Sardeshmukh, V.B., Scquizzato, M.: Toward optimal bounds in the congested clique: graph connectivity and MST. In: Proceedings of the 2015 ACM Symposium on Principles of Distributed Computing (PODC), pp. 91–100 (2015)

14. Impagliazzo, R., Williams, R.: Communication complexity with synchronized clocks. In: Proceedings of the 25th Annual IEEE Conference on Computational Complexity (CCC), pp. 259–269 (2010)
15. Klauck, H., Nanongkai, D., Pandurangan, G., Robinson, P.: Distributed computation of large-scale graph problems. In: Proceedings of the 26th Annual ACM-SIAM Symposium on Discrete Algorithms (SODA), pp. 391–410 (2015)
16. Kor, L., Korman, A., Peleg, D.: Tight bounds for distributed minimum-weight spanning tree verification. Theor. Comput. Syst. **53**(2), 318–340 (2013)
17. Korach, E., Moran, S., Zaks, S.: The optimality of distributive constructions of minimum weight and degree restricted spanning trees in a complete network of processors. SIAM J. Comput. **16**(2), 231–236 (1987)
18. Korach, E., Moran, S., Zaks, S.: Optimal lower bounds for some distributed algorithms for a complete network of processors. Theor. Comput. Sci. **64**(1), 125–132 (1989)
19. Kushilevitz, E., Nisan, N.: Communication Complexity. Cambridge University Press, Cambridge (1997)
20. Kutten, S., Pandurangan, G., Peleg, D., Robinson, P., Trehan, A.: On the complexity of universal leader election. J. ACM **62**(1), 7:1–7:27 (2015)
21. Kutten, S., Pandurangan, G., Peleg, D., Robinson, P., Trehan, A.: Sublinear bounds for randomized leader election. Theor. Comput. Sci. **561**, 134–143 (2015)
22. Lenzen, C.: Optimal deterministic routing and sorting on the congested clique. In: Proceedings of the 2013 ACM Symposium on Principles of Distributed Computing (PODC), pp. 42–50 (2013)
23. Lotker, Z., Patt-Shamir, B., Pavlov, E., Peleg, D.: Minimum-weight spanning tree construction in $O(\log \log n)$ communication rounds. SIAM J. Comput. **35**(1), 120–131 (2005)
24. Lynch, N.A.: Distributed Algorithms. Morgan Kaufmann Publishers Inc., San Francisco (1996)
25. Nanongkai, D., Sarma, A.D., Pandurangan, G.: A tight unconditional lower bound on distributed randomwalk computation. In: Proceedings of the 30th ACM Symposium on Principles of Distributed Computing (PODC), pp. 257–266 (2011)
26. Oshman, R.: Communication complexity lower bounds in distributed message-passing. In: Halldórsson, M.M. (ed.) SIROCCO 2014. LNCS, vol. 8576, pp. 14–17. Springer, Heidelberg (2014). doi:10.1007/978-3-319-09620-9_2
27. Pandurangan, G., Robinson, P., Scquizzato, M.: Fast distributed algorithms for connectivity and MST in large graphs. In: Proceedings of the 28th ACM Symposium on Parallelism in Algorithms and Architectures (SPAA), pp. 429–438 (2016)
28. Pandurangan, G., Robinson, P., Scquizzato, M.: A time- and message-optimal distributed algorithm for minimum spanning trees. CoRR, abs/1607.06883 (2016)
29. Peleg, D.: Distributed Computing: A Locality-Sensitive Approach. Society for Industrial and Applied Mathematics, Philadelphia (2000)
30. Peleg, D., Rubinovich, V.: A near-tight lower bound on the time complexity of distributed minimum-weight spanning tree construction. SIAM J. Comput. **30**(5), 1427–1442 (2000)
31. Peleg, D., Ullman, J.D.: An optimal synchronizer for the hypercube. SIAM J. Comput. **18**(4), 740–747 (1989)
32. Santoro, N.: Design and Analysis of Distributed Algorithms. Wiley, Hoboken (2006)
33. Schneider, J., Wattenhofer, R.: Trading bit, message, and time complexity of distributed algorithms. In: Peleg, D. (ed.) DISC 2011. LNCS, vol. 6950, pp. 51–65. Springer, Heidelberg (2011). doi:10.1007/978-3-642-24100-0_4

34. Tel, G.: Introduction to Distributed Algorithms, 2nd edn. Cambridge University Press, Cambridge (2001)
35. Tiwari, P.: Lower bounds on communication complexity in distributed computer networks. J. ACM **34**(4), 921–938 (1987)
36. Van Gucht, D., Williams, R., Woodruff, D.P., Zhang, Q.: The communication complexity of distributed set-joins with applications to matrix multiplication. In: Proceedings of the 34th ACM Symposium on Principles of Database Systems (PODS), pp. 199–212 (2015)
37. Woodruff, D.P., Zhang, Q.: When distributed computation is communication expensive. Distrib. Comput. (to appear)
38. Yao, AC.-C.: Some complexity questions related to distributive computing. In: Proceedings of the 11th Annual ACM Symposium on Theory of Computing (STOC), pp. 209–213 (1979)

Recent Results on Fault-Tolerant Consensus in Message-Passing Networks

Lewis Tseng[⊠]

Coordinated Science Laboratory, Department of Computer Science,
University of Illinois at Urbana-Champaign, Urbana, USA
ltseng3@illinois.edu

Abstract. Fault-tolerant consensus has been studied extensively in the literature, because it is one of the important distributed primitives and has wide applications in practice. This paper surveys important works on fault-tolerant consensus in message-passing networks, and the focus is on results from the past decade. Particularly, we categorize the results into two groups: new problem formulations and practical applications. In the first part, we discuss new ways to define the consensus problem, which include larger input domains, enriched correctness properties, different network models, etc. In the second part, we focus on real-world systems that use Paxos or Raft to reach consensus, and Byzantine Fault-Tolerant (BFT) systems. We also discuss Bitcoin, which can be related to solving Byzantine consensus in anonymous systems, and compare Bitcoin with BFT systems and Byzantine consensus algorithms.

Keywords: Consensus · Paxos · Bitcoin · BFT · Byzantine · Crash

1 Introduction

Fault-tolerant consensus has received significant attentions over the past three decades [14,50] since the seminal work by Lamport, Shostak, and Pease [43,67] – some important results include solving consensus in an optimal way and identifying bounds on time and communication complexity under different models – please refer to [14,50,70] for these fundamental results. In this paper, we survey recent efforts on fault-tolerant consensus in message-passing networks, with the focus on results from the past decade. References [18,26,69] presented early surveys on the topic. To complement theses prior surveys, our paper focus on the following two directions:

- *Exploration of new problem formulations*: Lots of different consensus problems have been proposed in the past ten years in order to solve more complicated tasks and accommodate different system and network requirements. New problem formulations include enriched correctness properties, different fault models, different communication networks, and different input/output domains. For this part, we focus on the comparison of recently proposed problem formulations and relevant techniques.

© Springer International Publishing AG 2016
J. Suomela (Ed.): SIROCCO 2016, LNCS 9988, pp. 92–108, 2016.
DOI: 10.1007/978-3-319-48314-6_7

– *Exploration of practical applications*: Consensus has been applied in many practical systems. Here, we focus on three types of applications: (i) crash-tolerant consensus algorithms (mainly Paxos [40] and Raft [63]) and their applications in real-world systems, (ii) Practical Byzantine Fault-Tolerance (PBFT) [20] and subsequent works on improving PBFT, and (iii) Bitcoin [2] and its comparison with Byzantine consensus algorithms and Byzantine Fault-Tolerance (BFT) systems.

For lack of space, discussions on some results are omitted here. Further details can be found in [77].

Classic Problem Formulations of Fault-tolerant Consensus. We consider the consensus problem in a point-to-point message-passing network, which is modeled as an undirected graph. Without specifically mentioning, the communication network is assumed to be *complete* in this survey, i.e., each pair of nodes can communicate with each other directly. In the fault-tolerant consensus problem [14,50], each node is given an *input*, and after a finite amount of time, each fault-free node should produce an *output* – consensus algorithms should satisfy the *termination* property. Additionally, the algorithms should also satisfy appropriate *validity* and *agreement* conditions. There are three main categories of consensus problems regarding different agreement properties:

– *Exact* [40,67]: fault-free nodes have to agree on exactly the *same* output.
– *Approximate* [30,32]: fault-free nodes have to agree on "roughly" the *same* output – the difference between outputs at any pair of fault-free nodes is bounded by a given constant ϵ ($\epsilon > 0$) of each other.
– *k-set* [22,68]: the number of distinct outputs at fault-free nodes is $\leq k$.

Validity property is also required for consensus algorithms to produce meaningful output(s), since the property defines the acceptable relationship between inputs and output(s). Some popular validity properties include: (i) *strong validity*: output must be an input at some fault-free node, (ii) *weak validity*: if all fault-free nodes have the same input v, then v is the output, and (iii) *validity* (for approximate consensus): output must be bounded by the inputs at fault-free nodes. A consensus algorithm is said to be correct if it satisfies termination, agreement and validity properties given that enough number of nodes are fault-free throughout the execution of the algorithm. In this paper, we focus on three types of node failures – Byzantine, crash, and omission faults.

The other key component of the consensus problem formulation is *system synchrony*, i.e., a model specifying the relative speed of nodes and the network delay. There are also three main categories [14,16,31,50]:

– *Synchronous*: each node proceeds in a lock-step fashion, and there is a known upper bound on the network delay.

- *Partially synchronous*: there exists a *partially synchronous* period from time to time. In such a period, fault-free nodes and the network stabilize and behave (more) synchronously.[1]
- *Asynchronous*: no known bound exists on nodes' processing speed or the network delay.

2   Exploration of New Problem Formulations

In the past decade, researchers proposed many new consensus problems to handle more complicated tasks and/or environments. We categorize these efforts into four groups: (i) input/output domain, (ii) communication network and synchrony assumptions, (iii) link fault models, and (iv) enriched correctness properties, such as early-stopping and one-step properties. In this survey, we focus on the discussion of works on different input/output domain and communication networks. Please refer to [77] for works in other groups. In this section, we assume that the system consists of n nodes, and up to f of them may crash or become Byzantine faulty. Byzantine faulty nodes may have an arbitrary behavior.

2.1   Input/Output Domain

Multi-valued Consensus. In the original *exact* Byzantine consensus problem [43,67], both input and output are binary values. Later, references [51,80] proposed the multi-valued version in which input may take more than two *real* values. Recently, multi-valued consensus received renewed attentions and researchers proposed algorithms that achieve asymptotically optimal communication complexity (number of bits transmitted) in both synchronous and asynchronous systems. Perhaps a bit surprisingly, for L-bit inputs, these algorithms achieve asymptotic communication complexity of $O(nL)$ bits when L is large enough.

In synchronous systems, Fitzi and Hirt proposed a Byzantine multi-valued algorithm with small error probability [35]. Their algorithm is based on the reduction technique and has the following steps: (i) hash the inputs to much smaller values using universal hash function, (ii) apply (classic) Byzantine consensus algorithm using these hash values as inputs, and (iii) achieve consensus by obtaining the input value from nodes that have the same hash values (if there is enough number of such nodes) [35]. Later, Liang and Vaidya combined a different reduction technique (that divides an input into a large number of small values) with novel coding technique to construct an error-free algorithm in synchronous systems [47]. One key contribution is to introduce a lightweight fault detection (or fault diagnosis) mechanism using coding [47]. Their coding-based fault diagnosis is efficient for large inputs because the inputs are divided into batches of small values, and in each batch, either consensus (on the small value

---

[1] Note that there are also other definitions of partial synchrony. We choose to present this particular definition, since many BFT systems only satisfy liveness under this particular definition. Please refer to [12,31] for more models on partial synchrony.

of this batch) can be achieved with small communication complexity or some faulty nodes will be identified. Once all faulty nodes are identified, then consensus on the remaining batches becomes trivial. Since number of faulty nodes is bounded, consensus on most batches can be achieved with small communication complexity [47].

Subsequently, variants of reduction technique were applied to solve consensus problems with large inputs in asynchronous systems. References [65, 66] provided multi-valued algorithms with small error probability. Afterwards, Patra improved the results and proposed an error-free algorithm [64]. These algorithms terminate with overwhelming probability; however, the expected time complexity is large because these algorithms first divide inputs to small batches and achieve consensus on each batch using variants of fault diagnosis mechanisms.

Typically, to achieve optimal communication complexity, the number of batches is in the same order of L. Consequently, the number of messages is large, since by assumption, L is a large value (compared with n). Instead of achieving optimal bit complexity, Mostéfaoui and Raynal focused on a different goal – minimizing number of messages in asynchronous systems [56, 58]. Their algorithm relies on two new all-to-all communication abstractions, which have an $O(n^2)$ message complexity (i.e., $O(n^2 L)$ bits) and a constant time complexity. The first communication abstraction allows the fault-free to reduce the number of input values to a small constant c, which ranges from 3 to 6 depending on the bound on the number of faulty nodes. The second abstraction allows each fault-free node to obtain a set of inputs such that, if the set at a fault-free node contains a single value, then this value belongs to the set at any other fault-free nodes. The algorithm in [56, 58] consists of four phases: (i) nodes exchange input values in the first three phases with the first phase based on the first communication abstraction, and the two subsequent phases based on the second, and (ii) nodes use binary consensus in the final phase to determine whether it is safe to agree on the value learned from phase 3.

Multi-valued consensus has also been studied under the crash fault model. Mostéfaoui et al. proposed multi-valued consensus algorithms in both synchronous and asynchronous systems [9]. Later, Zhang and Chen proposed a more efficient multi-valued consensus algorithm in asynchronous systems [90].

High-Dimensional Input/Output. In the Byzantine vector consensus (or multidimensional consensus) [52, 82], each node is given a d-dimensional vector of reals as its input ($d \geq 1$), and the output is also a d-dimensional vector. In complete networks, the recent papers by Mendes and Herlihy [52] and Vaidya and Garg [82] addressed approximate vector consensus in the presence of Byzantine faults. These papers yielded lower bounds on the number of nodes, and algorithms with optimal resilience in asynchronous [52, 82] as well as synchronous systems [82]. The algorithms in [52, 82] are generalizations of the optimal iterative approximate Byzantine consensus for scalar inputs in asynchronous systems [11]. The algorithms in [52, 82] require sub-routines for geometric computation in the d-dimensional space to obtain each node's local state in each iteration, whereas, a simple average operation suffices when $d = 1$ (i.e., classic approximate consensus) [11]. These two papers [52] and [82] independently addressed the same problem, and developed different algorithms – mainly on different geometric computation techniques – which also resulted in different proofs.

Subsequent work by Vaidya [81] explored the approximate vector consensus problem in incomplete *directed* graphs. Later, Tseng and Vaidya [78] proposed the convex hull consensus problem, in which fault-free nodes have to agree on "largest possible" polytope in the d-dimensional space that may not necessarily equal to a d-dimensional vector (a single point). The asynchronous algorithm in [78] bears some similarity to the ones in [11,52,82]; however, Tseng and Vaidya used a different communication abstraction to achieve the "largest possible" polytope. Moreover, Tseng and Vaidya introduced a new proof technique to show the correctness of iterative consensus algorithms when the output is a polytope [78].

2.2 Communication Network

The fault-tolerant consensus problem has been studied extensively in complete networks (e.g., [14,30,40,50,67]) and in undirected networks (e.g., [29,33]). In these works, any pair of nodes can communicate with each other reliably either directly or via at least $2f + 1$ node-disjoint paths (for Byzantine faults) or $f + 1$ node-disjoint paths (for crash faults). Recently, researchers revisited the assumptions on the communication network and enriched the problem space in four main directions: directed graphs, dynamic graphs, unknown/anonymous networks and partial synchrony. Here, we focus on the works on directed graphs. Please find the discussion on the later three directions in [77].

Directed Graphs. Researchers started to explore various consensus problems in arbitrary directed graphs, i.e., two pairs of nodes may not share a bi-directional communication channel, and not every pair of nodes may be able to communicate with each other directly or indirectly. Significant efforts have also been devoted on *iterative* algorithms in incomplete graphs. In iterative algorithms, (i) nodes proceed in iterations; (ii) the computation of new state at each node is based only on local information, i.e., nodes own state and states from neighboring nodes; and (iii) after each iteration of the algorithm, the state of each fault-free node must remain in the convex hull of the states of the fault-free nodes at the end of the previous iteration. Vaidya et al. [83] proved *tight* conditions for achieving approximate Byzantine consensus in synchronous and asynchronous systems using *iterative* algorithms. The tight condition for achieving approximate crash-tolerant consensus using iterative algorithms in asynchronous systems was also proved in [76].

A more restricted fault model – called "malicious" fault model – in which the faulty nodes are restricted to sending identical messages to their neighbors has also been explored extensively, e.g., [44–46,89]. LeBlanc and Koutsoukos [45] addressed a continuous time version of the consensus problem with malicious faults in complete graphs. LeBlanc et al. [44] have obtained *tight* necessary and sufficient conditions for tolerating up to f faults in the network.

The aforementioned approximate algorithms (e.g., [44,74,83]) are generalizations of the iterative approximate consensus algorithm in complete network [30,32]. However, to accommodate directed links, the proofs are more involved. Particularly, for the sufficiency part, one has to prove that all fault-free nodes must be able to receive a non-trivial amount of a state at some fault-free node in finite number of iterations. The necessity proofs in the work on directed graphs (e.g., [44,83]) are generalizations of the indistinguishability proof [13,33]. The main contributions are to identify how faulty nodes can block the information flow so that (i) fault-free nodes can be divided into several groups, and (ii) there exists certain faulty behaviors for up to f nodes such that different groups of fault-free nodes have to agree on different outputs.

There were also works on using general algorithms to achieve consensus – an algorithm is *general* if nodes are allowed to have topology knowledge and the ability to route messages (send and receive messages using multiple node-disjoint paths). Furthermore, unlike iterative algorithms (e.g., [11,30]), the state maintained at each node in general algorithms is not constrained to a single value. Tseng and Vaidya [79] proved *tight* necessary and sufficient conditions on the underlying communication graphs for achieving (i) exact crash-tolerant consensus in synchronous systems, (ii) approximate crash-tolerant consensus in asynchronous systems, and (iii) exact Byzantine consensus in synchronous systems using *general* algorithms. The tight condition for achieving approximate Byzantine consensus in asynchronous systems remains open. Lili and Vaidya [74] proved tight conditions for achieving approximate Byzantine consensus using general algorithms.

The exact consensus algorithms in [79] require that some "common information" has to be propagated to all fault-free nodes even if some nodes may fail. Generally speaking, the algorithms in [79] proceed in phases such that in each phase, a group of nodes try to send information to the remaining nodes. The algorithms are designed to maintain validity at all time. Additionally, if no failure occurs in a phase, then agreement can be achieved, because some "common information" are guaranteed to be received by all nodes that have not failed yet. The algorithm in [74] can be viewed as an extension of the iterative algorithm that tolerates Byzantine faults in directed networks [83], and it utilized the routing information and network knowledge to tolerate more failures than the algorithm in [83] does.

3   Exploration of Practical Applications

Fault-tolerant consensus has been adopted in many practical systems. We start with real-world systems that are designed to tolerate crash node faults, particularly, those based on two families of algorithms – Paxos [40] and Raft [63]. Then, we discuss efforts on designing BFT (Byzantine Fault-Tolerance) systems. Finally, we compare Bitcoin-related work [60] with BFT systems and Byzantine consensus. In [77], we also discuss systems tolerating "arbitrary state corruption faults".

3.1   Paxos and Raft

Here, we discus exact consensus algorithms developed for asynchronous systems. Consensus algorithm needs to satisfy validity, agreement and termination as discussed in Sect. 1. However, it is impossible to achieve exact consensus in asynchronous systems [34]. Hence, the termination property is relaxed – progress (or liveness) is only ensured when there exist some time periods that enough messages are received within time.

Paxos [40–42,54] is the well-known family of consensus protocols tolerating crash node faults in asynchronous systems. Since Paxos was first proposed by Lamport [40,41], variants of Paxos were developed and implemented in real-world systems, such as Chubby lock service used in many Google systems [17,25], and membership management in Windows Azure [19].[2] Yahoo! also developed ZaB [71], a protocol achieving atomic broadcast in network equipped with FIFO channels, and used ZaB to build the widely-adopted coordination service, ZooKeeper [38]. ZooKeeper is later used in many practical storage systems, like HBase [4] and Salus [86]. Recently, many novel mechanisms have been proposed to improve the performance of Paxos, including quorum lease [55], diskless Paxos [75], even load balancing [54], and time bubbling (for handling nondeterministic network input timing) [28]. While the original Paxos [40,41] is theoretically elegant, practitioners have found it hard to implement Paxos in practice [21]. One difficulty mentioned in [21] is that membership/configuration management is non-trivial in practice, especially, when Multi-Paxos, and disk corruptions are considered. (Multi-Paxos is a generalization of Paxos which is designed to optimize the performance when there are multiple inputs to be agree upon [21].)

In 2014, Ongaro and Ousterhout from Stanford proposed a new consensus algorithm – Raft [63]. Their main motivation was to simplify the design of consensus algorithm so that it is easier to understand and verify the design and implementation. One interesting (social) experiment by Ongaro and Ousterhout was mentioned in [63]: *"In an informal survey of attendees at NSDI 2012, we found few people who were comfortable with Paxos, even among seasoned researchers"*. To simplify the conceptual design, Raft integrates the consensus-solving element deeply with leader election protocol and membership/configuration management protocol [63]. After their publication, Raft has quickly gained popularity, and been used in practical key-value store systems such as etcd [3] and RethinkDB [7]. Please refer to their website [6] for a list of papers and implementations.

3.2   Byzantine Fault Tolerance (BFT)

Generally speaking, Byzantine Fault-Tolerance (BFT) systems implement deterministic state machines over different machines (or *replicas*) to tolerate Byzantine node failures. In other words, BFT systems realize the State Machine Replication systems [72] that tolerate Byzantine faults. The main challenge is to design

---

[2] We would like to thank the anonymous reviewer who pointed out that Windows Azure also uses ZooKeepr to manage virtual machines [1].

a system such that it behaves like a centralized server to the clients in the presence of Byzantine faults. More precisely, the system is given requests from the clients, and the goals of a BFT system are: (i) the fault-free replicas agree on the total order of the requests, and then the replicas execute the requests following the agreed order (safety); and (ii) clients learn the responses to their requests eventually (liveness). Usually, safety is guaranteed at all time, and liveness is guaranteed only in the *grace periods*, i.e., when messages are delivered in time.

Since Castro and Liskov published their seminal work PBFT (Practical Byzantine Fault-Tolerance) [20], significant efforts have been devoted to improving BFT systems. There were mainly two directions of the improvements: (i) reducing the overhead like communication costs, or replication costs, and (ii) providing higher throughput or lower latency (in the form of round complexity). Below, we focus on different techniques for improving the performance. Please refer to [77] for the discussions on other works in this area, including hardening existing crash-fault-tolerant systems, hardware-based BFTs, BFTs with relaxed properties, BFT storage systems, and BFTs over intercloud.

Improving Performance. Castro and Liskov's work on Practical Byzantine Fault-Tolerance (PBFT) showed for the first time that BFT system is useful in practice [20]. PBFT requires $3f + 1$ replicas, where f is the upper bound on the number of Byzantine nodes in the system. Subsequently, Quorum-based solutions Q/U [10] and HQ [27] have been proposed, which only require one round of communication in contention-free case by allowing clients directly interact with all the replicas to agree on an execution order. Contention-free case means the time when all the following conditions hold: (i) no replica fails, (ii) the network has stable performance, and (iii) there is no contention on the proposed input value. The quorum-based solutions reduce latency (number of rounds) in some cases, but was shown to be more expensive in other cases [39]. Hence, Zyzzyva [39] focuses on increasing performance in failure-free case (when no replica fails) by allowing speculative operations that increase throughput significantly and adopting a novel roll-back mechanism to recover operations when failures are detected. Zyzzyva requires $3f + 1$ replicas; however, a single crash failure would significantly reduce the performance by forcing Zyzzyva to run in the slow mode – where no speculative operation can be executed [39]. Thus, Kotla et al. also introduced Zyzzyva5, which can be executed in the fast mode even if there are crash failures, but Zyzzyva5 requires $5f + 1$ replicas [39]. Subsequently, Scrooge [73] reduces the replication cost to $4f$ by requiring the participation from clients which help detect replicas' misbehaviors. Moreover, Scrooge runs in the fast mode even if there are crashed nodes.

Clement et al. observed that a single Byzantine replica or client can significantly impact the performance of HQ, PBFT, Q/U and Zyzzyva [24]. Thus, they proposed a new system Aardvark, which provides good performance when Byzantine failures happen by sacrificing the performance in the failure-free case [24]. Later, Clement et al. also demonstrated how to combine Zyzzyva and Aardvark so that the new system, Zyzzyvark, not only tolerates faulty clients, but also enjoys fast performance in the failure-free case due to the integration of speculative operations [23].

The aforementioned BFT systems are designed to optimize performance for certain circumstances, e.g., HQ for contention-free case and Zyzzyva for failure-free case. Guerraoui et al. proposed a new type of BFT systems that can be constructed to have optimized performance under difference circumstances [37]. Their tunable design is useful, since it allows the system administrators to explore the performance tradeoff space. Their systems are based on three core concepts: (i) abortable requests, (ii) composition of (abortable) BFT instances, and (iii) dynamic switching among BFT instances. The tunable parameter specifies the progress condition under which a BFT instance should not abort. Some example conditions include contention, system synchrony or node failures. In [37], Guerraoui et al. showed how to construct new BFT systems with different parameters; particularly, they proposed (i) *AZyzzyva* which composes Zyzzyva and PBFT together to have more stable performance than Zyzzyva does and faster failure-free performance than PBFT's performance, and (ii) *Aliph* which has three components: PBFT, Quorum-based protocol optimized for contention-free case, and Chain-based protocol optimized for high-contention case without failures and asynchrony [37].

For computation-heavy workload, Yin et al. proposed a novel idea that separates agreement protocol from executions of clients' requests [88]. This separation mechanism reduces the replication cost to $2f + 1$. Note that the system still requires $3f + 1$ replicas to achieve agreement on the order of the clients' requests, but the executions of requests, and data storage only occur at $2f + 1$ replicas. Later, Wood et al. built a system, ZZ, which reduces the replication cost to $f + 1$ using virtualization technique [87]. The idea behind ZZ is that $f + 1$ active replicas are sufficient for fault detection, and when fault is detected, their virtualization technique allows ZZ to replace the faulty replica by waking up fresh replica and retrieving current system state with small overhead [87].

3.3 Bitcoin

Bitcoin is a digital currency system proposed by Satoshi Nakamoto [60] and later gained popularity due to its characteristics of anonymity and decentralized design [2]. Since Bitcoin is based on cryptography tools (Proof-of-Work mechanism), it can be viewed as a cryptocurrency. Even though Bitcoin has large latencies (on the order of an hour), and the theoretical peak throughput is up to 7 transactions per second [85], Bitcoin is still one of the most popular cryptocurrencies. Here, we briefly discuss the core mechanism of Bitcoin and compare it with Byzantine consensus and BFT systems.

Bitcoin Mechanism. The core of Bitcoin is called *Blockchain*, which is a peer-to-peer ledger system, and acts as a virtually centralized ledger that keeps track of all bitcoin transactions. A set of bitcoin transactions are recorded in blocks. Owners of bitcoins can generate new transactions by broadcasting signed blocks to the Bitcoin network.[3] Then, a procedure called *mining* confirms the transactions and includes the transactions to the Blockchain (the centralized ledger system). Essentially, *mining* is a randomized distributed consensus component that confirms pending transactions by including them in the Blockchain. To include a transaction block, a miner needs to solve a "proof-of-work" (POW) or "cryptographic puzzle". The main incentive mechanism for Bitcoin participants to maintain the Blockchain and to confirm new transactions is to reward the participants (or the miners) some bitcoins – the first miner that solves the puzzle receives a certain amount of bitcoins. The main reason that the mining procedure can be related to consensus is because each miner maintains the chain of blocks (Blockchain) at local storage, and the global state is consistent at all miners eventually – all fault-free miners will have the same Blockchain eventually [60]. That is, fault-free Bitcoin participants need to agree on the total order of the transactions.

One important feature of the cryptocurrency system is to prevent the *double-spending attacks*, i.e., spending the same unit of money twice. In Bitcoin, the consistent global state – the order of transactions – can be used to prevent double-spending attacks. Since the attackers have no ability to reorganize the order of blocks (i.e., modify the Blockchain, the ledger system), the money recipient can simply check whether the money has already been spent in the Blockchain and reject the money if it has already been used.[4] In [60], Satoshi Nakamoto presented a simple analysis that showed with high probability, Bitcoin's participants maintain a total order of the transactions if adversary's computation power is less than 1/3 of the total computation power in the Bitcoin network. As a result, no double-spending attack is possible with high probability if adversary's computation power is bounded. However, the models under consideration were not well-defined and the analysis was not rigorous in [60]. Thus, significant efforts have been devoted to formally proving the correctness of Bitcoin mechanism or improving the design and performance. Please refer to a nice textbook [61] for a thorough discussion. Below, we focus on the comparison of Bitcoin and Byzantine consensus/BFT systems.

Comparison with Byzantine Consensus. There are several differences between the problem formulation of Byzantine consensus (as described in Sect. 1) and the assumptions of Bitcoin [36,53,60]. For example, in Bitcoin: (i) the number of participants is dynamic; (ii) participants are anonymous, and the participants cannot authenticate each other; (iii) as a result of (ii), participants have no way to identify the source of a received message; and (iv) the Bitcoin network

---

[3] Here, we follow the convention: (i) Bitcoin network includes all the anonymous participants in the Bitcoin system and the network that supports the anonymous communication; and (ii) throughout the discussion, "Bitcoin" refers to the system/network, whereas, "bitcoin" refers to the basic unit of the cryptocurrency.

[4] One technical issue here is that the Blockchain has the "eventually consistent" feature. The exact mechanism to handle the issue is beyond the scope of this survey. Please refer to a nice textbook [61] for some mechanisms.

is synchronized enough, and there is a notion of a "round", i.e., the network communication delay is negligible compared to computation time.

It was first suggested by Nakamoto that Bitcoin's POW-based mechanism can be used to solve Byzantine consensus [8,59]. However, the discussion was quite informal [59]. To the best of our knowledge, Miller and LaViola were the first to formalize the suggestion and proposed a POW-based model to achieve Byzantine consensus when majority of participants are fault-free. However, the validity is only ensured with non-negligible probability (but not with over-whelming probability). Subsequently, Garay et al. [36] extracted and analyzed the core mechanism of Bitcoin [36], namely Bitcoin Backbone. They first identified and formalized two properties of Bitcoin Backbone: (i) *common prefix property*: fault-free participants will possess a large common prefix of the Blockchain, and (ii) *chain-quality property*: enough blocks in the Blockchain are contributed by fault-free participants. Then, they presented a simple POW-based Byzantine consensus algorithm which is a variation of Nakamoto's suggestion [59], but satisfy agreement and validity assuming that the adversary's computation power (puzzle-solving power) is bounded by 1/3. Their algorithm can also be used to solve Byzantine consensus with strong validity [62]. Finally, they proposed a more complicated consensus protocol, which was proved to be secure assuming high network synchrony and that the adversary's computation power is strictly less than 1/2. In [36], Garay et al. focused on how to use Bitcoin-inspired mechanism to solve Byzantine consensus.

Comparison with BFT Systems. Conceptually, BFT and Bitcoin have similar goals: (i) *BFT*: clients' requests are executed in a total order distributively; and (ii) *Bitcoin*: a total order of blocks are maintained by participants distributively. Therefore, it is interesting to compare BFT with Bitcoin as well. Below, we address fundamental differences between the two.

– *Environment*: As discussed above, assumptions for BFT are similar to the ones for Byzantine consensus, which are very different from the ones for Bitcoin. One major difference is the anonymous node identity. In BFT, the system environment is well-controlled, and replicas' IDs are maintained and managed by the system administrators. In contrast, Bitcoin is a decentralized system where all the participants are anonymous. As a result, BFT systems can use many well-studied tools from the literature, e.g., atomic broadcast, and quorum-based mechanism, whereas, Bitcoin-related systems usually rely on POW (proof-of-work) or variants of cryptographic tools.
– *Features*: In [85], Marko Vukolic mentioned that the features of BFT and Bitcoin are at two opposite ends of the scalability/performance spectrum due to different application goals. Generally speaking, BFT systems offer good performance (low latency and high throughput) for small number of replicas (≤ 20 replicas), whereas, Bitcoin scales well (≥ 1000 participants), but the latency is prohibitively high and throughput is limited.
– *Incentive*: In BFT system, every fault-free replica/client is programmed to follow the algorithm specification. However, in Bitcoin, participants may choose

not to spend their computation power on solving puzzles; thus, there is a mechanism in Bitcoin to reward the mining process [60].

- *Correctness property*: As addressed in Sect. 3.2, BFT systems satisfy safety in asynchronous network and satisfy liveness when network is synchronous enough (in grace period). As shown in [36,60], Bitcoin requires network synchronous enough for ensuring correctness (when network delay is negligible compared to computation time).

- *Applications*: Bitcoin or Blockchain-based systems inspire lots of exciting applications beyond cryptocurrency, e.g., smart contract, identity/ownership management, digital access/contents, etc. In contrasts, applications for BFT systems are more traditional in the sense that there already exist those applications (that tolerate only crash faults), and BFTs help improve the fault-tolerance level.

In [85], Marko Vukolic proposed an interesting research direction on finding the synergies between Bitcoin and BFT systems, since both systems have their limitations and advantages. On one hand, the poor performance of POW-based mechanism limits the applicability of Blockchain in other domains like smart contract application [15,85]. On the other hand, BFT systems are not widely adopted in practice due to their poorer scalability and lack of killer applications [48,84]. SCP is a recent system that utilizes hybrid POW/BFT architecture [49]. However, further exploration of the synergy between Bitcoin and BFT systems is an interesting research direction.

4   Summary and Future Directions

Conclusion. Fault-tolerant consensus is a rich topic. This paper is only managed to sample a subset of recent results. To augment previous surveys/textbooks on the same topic, e.g., [14,18,26,50,69], we survey prior works from two angles: (i) new consensus problem formulations, and (ii) practical applications. For the second part, we focus on the Paxos- and Raft-based systems, and BFT systems. We also discuss Bitcoin which has close relationship with Byzantine consensus and BFT systems.

Future Directions. The future research directions below focus on one theme: *bridging the gap between theory and practice.* As discussed in the first part of the paper, researchers have explored wide variety of different (theoretical) problem formulations; however, there is no consolidated or unified framework. As a result, it is often hard to compare different algorithms and models, and it is also difficult for practitioners to decide which algorithms are most appropriate to solve their problems. Thus, making these results more coherent and more practical (e.g., giving rule-of-thumbs for picking algorithms) would be an important and interesting task.

In the second part, we discuss the efforts of applying fault-tolerant consensus in real-world systems. Unfortunately, the difficulty in implementing or even understanding the consensus algorithms prevents wider applications of consensus algorithms. Therefore, simplifying the conceptual design and verifying the implementation is also a key task. Raft [63] is one good example of how simplified design and explanation could help gain popularity and practicability. Another major task is to understand and analyze more thoroughly the real-world distributed systems. As suggested in [36,85], BFT systems and Bitcoin are not yet well-understood. The models presented in [36,53] and other works mentioned in [85] were only the first step toward this goal. Only after enough research and understanding, could we improve the state-of-art mechanisms. For example, as mentioned in [61], Bitcoin's core mechanism depends on the incentive mechanism to reward miners; however, not much work has analyzed Bitcoin from the perspective of game theory.

Acknowledgment. We would like to thank the anonymous reviewers for encouragement and suggestions. We also acknowledge Nitin H. Vaidya for early feedback and Michel Raynal for pointers to several new works.

References

1. Apache zookeeper on windows azure. https://msopentech.com/opentech-projects/apache-zookeeper-on-windows-azure-2/#
2. Bitcoin.org. https://bitcoin.org/en/
3. etcd. https://github.com/coreos/etcd
4. HBase. http://hbase.apache.org/
5. Leslie Lamport - A.M. Turing award winner. http://amturing.acm.org/award_winners/lamport_1205376.cfm
6. Raft. https://raft.github.io/
7. Rethinkdb. https://www.rethinkdb.com/
8. Dugcampbell's blog, 07 2015. http://www.dugcampbell.com/byzantine-generals-problem/
9. Mostefaoui, A., Raynal, M., Tronel, F.: From binary consensus to multivalued consensus in asynchronous message-passing systems. IPL **73**(5), 207–212 (2000)
10. Abd-El-Malek, M., Ganger, G.R., Goodson, G.R., Reiter, M.K., Wylie, J.J.: Fault-scalable Byzantine fault-tolerant services. SOSP **39**(5), 59–74 (2005)
11. Abraham, I., Amit, Y., Dolev, D.: Optimal resilience asynchronous approximate agreement. In: Higashino, T. (ed.) OPODIS 2004. LNCS, vol. 3544, pp. 229–239. Springer, Heidelberg (2005). doi:10.1007/11516798_17
12. Aguilera, M.K., Delporte-Gallet, C., Fauconnier, H., Toueg, S.: Partial synchrony based on set timeliness. Distrib. Comput. **25**(3), 249–260 (2012)
13. Attiya, H., Ellen, F.: Impossibility results for distributed computing. Synth. Lect. Distrib. Comput. Theor. **5**(1), 1–162 (2014). Morgan & claypool
14. Attiya, H., Welch, J.: Distributed Computing: Fundamentals, Simulations, and Advanced Topics. Parallel and Distributed Computing. Wiley, Hoboken (2004)
15. Bonneau, J., Miller, A., Clark, J., Narayanan, A., Kroll, J.A., Felten, E.W.: Sok: research perspectives and challenges for bitcoin and cryptocurrencies. In: 2015 IEEE Symposium on Security and Privacy, pp. 104–121, May 2015
16. Bouzid, Z., Mostefaoui, A., Raynal, M.: Minimal synchrony for Byzantine consensus. In: Symposium on Principles of Distributed Computing, PODC (2015)

17. Burrows, M.: The chubby lock service for loosely-coupled distributed systems. In: Proceedings of the Operating Systems Design and Implementation, OSDI (2006)
18. Cachin, C.: State machine replication with Byzantine faults. In: Charron-Bost, B., Pedone, F., Schiper, A. (eds.) Replication. LNCS, vol. 5959, pp. 169–184. Springer, Heidelberg (2010). doi:10.1007/978-3-642-11294-2_9
19. Calder, B., et al.: Windows azure storage: a highly available cloud storage service with strong consistency. In: SOSP (2011)
20. Castro, M., Liskov, B.: Practical Byzantine fault tolerance. In: Proceedings of the Operating Systems Design and Implementation, OSDI (1099)
21. Chandra, T.D., Griesemer, R., Redstone, J.: Paxos made live: an engineering perspective. In: Symposium on Principles of Distributed Computing, PODC (2007)
22. Chaudhuri, S.: More choices allow more faults: set consensus problems in totally asynchronous systems. Inf. Comput. **105**(1), 132–158 (1993)
23. Clement, A., Kapritsos, M., Lee, S., Wang, Y., Alvisi, L., Dahlin, M., Riche, T.: Upright cluster services. In: SOSP (2009)
24. Clement, A., Wong, E., Alvisi, L., Dahlin, M., Marchetti, M.: Making Byzantine fault tolerant systems tolerate Byzantine faults. In: NSDI (2009)
25. Corbett, J.C., et al.: Spanner: Google's globally-distributed database. In: OSDI (2012)
26. Correia, M., Veronese, G.S., Neves, N.F., Veríssimo, P.: Byzantine consensus in asynchronous message-passing systems: a survey. IJCCBS **2**(2), 141–161 (2011)
27. Cowling, J., Myers, D., Liskov, B., Rodrigues, R., Shrira, L.: HQ replication: a hybrid quorum protocol for Byzantine fault tolerance. In: OSDI (2006)
28. Cui, H., Gu, R., Liu, C., Chen, T., Yang, J.: Paxos made transparent. In: SOSP (2015)
29. Dolev, D.: The Byzantine generals strike again. J. Algorithms **3**(1), 14–30 (1982)
30. Dolev, D., Lynch, N.A., Pinter, S.S., Stark, E.W., Weihl, W.E.: Reaching approximate agreement in the presence of faults. J. ACM **33**(3), 499–516 (1986)
31. Dwork, C., Lynch, N., Stockmeyer, L.: Consensus in the presence of partial synchrony. J. ACM **35**(2), 288–323 (1988)
32. Fekete, A.D.: Asymptotically optimal algorithms for approximate agreement. In: PODC (1986)
33. Fischer, M.J., Lynch, N.A., Merritt, M.: Easy impossibility proofs for distributed consensus problems. In: PODC (1985)
34. Fischer, M.J., Lynch, N.A., Paterson, M.S.: Impossibility of distributed consensus with one faulty process. J. ACM **32**, 374–382 (1985)
35. Fitzi, M., Hirt, M.: Optimally efficient multi-valued Byzantine agreement. In: PODC (2006)
36. Garay, J., Kiayias, A., Leonardos, N.: The bitcoin backbone protocol: analysis and applications. In: Oswald, E., Fischlin, M. (eds.) EUROCRYPT 2015. LNCS, vol. 9057, pp. 281–310. Springer, Heidelberg (2015). doi:10.1007/978-3-662-46803-6_10
37. Guerraoui, R., Knežević, N., Quéma, V., Vukolić, M.: The next 700 bft protocols. In: EuroSys (2010)
38. Hunt, P., Konar, M., Junqueira, F.P., Reed, B.: Zookeeper: wait-free coordination for internet-scale systems. In: USENIX ATC (2010)
39. Kotla, R., Alvisi, L., Dahlin, M., Clement, A., Wong, E.: Zyzzyva: speculative Byzantine fault tolerance. In: SOSP (2007)
40. Lamport, L.: The part-time parliament. ACM Trans. Comput. Syst. **16**(2), 133–169 (1998)
41. Lamport, L.: Paxos made simple. SIGACT News **32**(4), 51–58 (2001)

42. Lamport, L.: Fast Paxos. Distrib. Comput. **19**(2), 79–103 (2006)
43. Lamport, L., Shostak, R., Pease, M.: The Byzantine generals problem. ACM Trans. Program. Lang. Syst. **4**(3), 382–401 (1982)
44. LeBlanc, H., Zhang, H., Koutsoukos, X., Sundaram, S.: Resilient asymptotic consensus in robust networks. IEEE J. Sel. Areas Commun. **31**(4), 766–781 (2013). Special Issue on In-Network Computation
45. LeBlanc, H., Koutsoukos, X.: Consensus in networked multi-agent systems with adversaries. In: HSCC (2011)
46. LeBlanc, H., Koutsoukos, X.: Low complexity resilient consensus in networked multi-agent systems with adversaries. In: HSCC (2012)
47. Liang, G., Vaidya, N.: Error-free multi-valued consensus with Byzantine failures. In: PODC (2011)
48. Liu, S., Cachin, C., Quéma, V., Vukolic, M.: XFT: practical fault tolerance beyond crashes. CoRR abs/1502.05831 (2015)
49. Luu, L., et al.: Scp: a computationally-scalable Byzantine consensus protocol for blockchains. National University of Singapore, Technical report (2015)
50. Lynch, N.A.: Distributed Algorithms. Morgan Kaufmann, San Francisco (1996)
51. Lynch, N., Fischer, M., Fowler, R.: Simple and efficient Byzantine generals algorithm. In: Symposium on Reliability in Distributed Software and Database Systems (1982)
52. Mendes, H., Herlihy, M.: Multidimensional approximate agreement in Byzantine asynchronous systems. In: STOC (2013)
53. Miller, A., LaViola Jr., J.J.: Anonymous Byzantine consensus from anonymous Byzantine consensus from moderately-hard puzzles: a model for bitcoin. University of Central Florida, Technical report (2012)
54. Moraru, I., Andersen, D.G., Kaminsky, M.: There is more consensus in egalitarian parliaments. In: SOSP (2013)
55. Moraru, I., Andersen, D.G., Kaminsky, M.: Paxos quorum leases: fast reads without sacrificing writes. In: SOCC (2014)
56. Mostéfaoui, A., Raynal, M.: Signature-free asynchronous Byzantine systems: from multivalued to binary consensus with t < n/3, O(n2) messages, and constant time. In: Scheideler, C. (ed.) SIROCCO 2015. LNCS, vol. 9439, pp. 194–208. Springer, Switzerland (2015). doi:10.1007/978-3-319-25258-2_14
57. Mostéfaoui, A., Raynal, M.: Intrusion-tolerant broadcast and agreement abstractions in the presence of Byzantine processes. IEEE Trans. Parallel Distrib. Syst. **27**(4), 1085–1098 (2016). http://dx.doi.org/10.1109/TPDS.2015.2427797
58. Mostéfaoui, A., Raynal, M.: Signature-free asynchronous Byzantine systems: from multivalued to binary consensus with t < n/3, O(n2) messages, and constant time. Acta Informatica (2016). doi:10.1007/s00236-016-0269-y
59. Nakamoto, S.: the proof-of-work chain is a solution to the Byzantine generals' problem. In: The Cryptography Mailing List, November 2008. http://www.mail-archive.com/cryptography@metzdowd.com/msg09997.html
60. Nakamoto, S.: Bitcoin: A Peer-to-Peer Electronic Cash System (October 2008). bitcoin.org
61. Narayanan, A., Bonneau, J., Felten, E., Miller, A., Goldfeder, S.: Bitcoin and Cryptocurrency Technologies. Princeton University Presss, Princeton (2016)
62. Neiger, G.: Distributed consensus revisited. IPL **49**(4), 195–201 (1994)
63. Ongaro, D., Ousterhout, J.: In search of an understandable consensus algorithm. In: 2014 USENIX Annual Technical Conference (USENIX ATC 2014) (2014)

64. Patra, A.: Error-free multi-valued broadcast and Byzantine agreement with optimal communication complexity. In: Fernàndez Anta, A., Lipari, G., Roy, M. (eds.) OPODIS 2011. LNCS, vol. 7109, pp. 34–49. Springer, Heidelberg (2011). doi:10. 1007/978-3-642-25873-2_4

65. Patra, A., Rangan, C.P.: Communication optimal multi-valued asynchronous broadcast protocol. In: Abdalla, M., Barreto, P.S.L.M. (eds.) LATINCRYPT 2010. LNCS, vol. 6212, pp. 162–177. Springer, Heidelberg (2010). doi:10.1007/ 978-3-642-14712-8_10

66. Patra, A., Rangan, C.P.: Communication optimal multi-valued asynchronous Byzantine agreement with optimal resilience. In: Fehr, S. (ed.) ICITS 2011. LNCS, vol. 6673, pp. 206–226. Springer, Heidelberg (2011). doi:10.1007/ 978-3-642-20728-0_19

67. Pease, M., Shostak, R., Lamport, L.: Reaching agreement in the presence of faults. J. ACM **27**(2), 228–234 (1980)

68. de Prisco, R., Malkhi, D., Reiter, M.: On k-set consensus problems in asynchronous systems. IEEE Trans. Parallel Distrib. Syst. **12**(1), 7–21 (2001)

69. Raynal, M.: Consensus in synchronous systems: a concise guided tour. In: Pacific Rim International Symposium on Dependable Computing (2002)

70. Raynal, M.: Concurrent Programming: Algorithms, Principles, and Foundations. Springer, Heidelberg (2013)

71. Reed, B., Junqueira, F.P.: A simple totally ordered broadcast protocol. In: Proceedings of the 2nd Workshop on Large-Scale Distributed Systems and Middleware, LADIS (2008)

72. Schneider, F.B.: Implementing fault-tolerant services using the state machine approach: a tutorial. ACM Comput. Surv. **22**(4), 299–319 (1990)

73. Serafini, M., Bokor, P., Dobre, D., Majuntke, M., Suri, N.: Scrooge: reducing the costs of fast Byzantine replication in presence of unresponsive replicas. In: Dependable Systems and Networks (DSN) (2010)

74. Su, L., Vaidya, N.: Reaching approximate Byzantine consensus with multi-hop communication. In: Pelc, A., Schwarzmann, A.A. (eds.) SSS 2015. LNCS, vol. 9212, pp. 21–35. Springer, Heidelberg (2015). doi:10.1007/978-3-319-21741-3_2

75. Trencseni, M., Gazsó, A., Reinhardt, H.: Paxoslease: diskless Paxos for leases. CoRR abs/1209.4187 (2012)

76. Tseng, L.: Fault-tolerant consensus in directed graphs and convex hull consensus. Ph.D. thesis. University of Illinois at Urbana-Champaign (2016)

77. Tseng, L.: Recent results on fault-tolerant consensus in message-passing networks. CoRR abs/1608.07923 (2016)

78. Tseng, L., Vaidya, N.H.: Asynchronous convex hull consensus in the presence of crash faults. In: Proceedings of the 2014 ACM Symposium on Principles of Distributed Computing, PODC (2014)

79. Tseng, L., Vaidya, N.H.: Fault-tolerant consensus in directed graphs. In: PODC (2015)

80. Turpin, R., Coan, B.A.: Extending binary Byzantine agreement to multivalued Byzantine agreement. IPL **18**(2), 73–76 (1984)

81. Vaidya, N.H.: Iterative Byzantine vector consensus in incomplete graphs. In: Chatterjee, M., Cao, J., Kothapalli, K., Rajsbaum, S. (eds.) ICDCN 2014. LNCS, vol. 8314, pp. 14–28. Springer, Heidelberg (2014). doi:10.1007/978-3-642-45249-9_2

82. Vaidya, N.H., Garg, V.K.: Byzantine vector consensus in complete graphs. In: PODC (2013)

83. Vaidya, N.H., Tseng, L., Liang, G.: Iterative approximate Byzantine consensus in arbitrary directed graphs. In: PODC (2012)

84. Vukolić, M.: The Byzantine empire in the intercloud. SIGACT News **41**(3), 105–111 (2010)
85. Vukolić, M.: The quest for scalable blockchain fabric: proof-of-work vs. BFT replication. In: Camenisch, J., Kesḑoğan, D. (eds.) iNetSec 2015. LNCS, vol. 9591, pp. 112–125. Springer, Heidelberg (2016). doi:10.1007/978-3-319-39028-4_9
86. Wang, Y., Kapritsos, M., Ren, Z., Mahajan, P., Kirubanandam, J., Alvisi, L., Dahlin, M.: Robustness in the salus scalable block store. In: NSDI (2013)
87. Wood, T., Singh, R., Venkataramani, A., Shenoy, P., Cecchet, E.: ZZ and the art of practical BFT execution. In: EuroSys (2011)
88. Yin, J., Martin, J.P., Venkataramani, A., Alvisi, L., Dahlin, M.: Separating agreement from execution for Byzantine fault tolerant services. SOSP **37**(5), 253–267 (2003)
89. Zhang, H., Sundaram, S.: Robustness of complex networks with implications for consensus and contagion. In: Proceedings of the 51st IEEE Conference on Decision and Control, CDC (2012)
90. Zhang, J., Chen, W.: Bounded cost algorithms for multivalued consensus using binary consensus instances. Inf. Process. Lett. **109**(17), 1005–1009 (2009)

Shared Memory

Asynchronous Coordination Under Preferences and Constraints

Armando Castañeda[1(✉)], Pierre Fraigniaud[2], Eli Gafni[3], Sergio Rajsbaum[1], and Matthieu Roy[4]

[1] Instituto de Matemáticas, UNAM, 04510 México D.F., Mexico
{armando.castaneda,rajsbaum}@im.unam.mx
[2] CNRS, University Paris Diderot, Paris, France
pierref@liafa.univ-paris-diderot.fr
[3] Computer Science Department, UCLA, Los Angeles, USA
eli@cs.ucla.edu
[4] LAAS-CNRS, Université de Toulouse, CNRS, Toulouse, France
roy@laas.fr

Abstract. *Adaptive renaming* can be viewed as a *coordination task* involving a set of asynchronous agents, each aiming at grabbing a single resource out of a set of resources Similarly, *musical chairs* is also defined as a coordination task involving a set of asynchronous agents, each aiming at picking one of a set of available resources, where every agent comes with an a priori *preference* for some resource. We foresee instances in which some combinations of resources are allowed, while others are disallowed.

We model these *constraints* as an undirected graph whose nodes represent the resources, and an edge between two resources indicates that these two resources cannot be used simultaneously. In other words, the sets of resources that are allowed are those which form *independent sets*.

We assume that each agent comes with an a priori *preference* for some resource. If an agent's preference is not in conflict with the preferences of the other agents, then this preference can be grabbed by the agent. Otherwise, the agents must coordinate to resolve their conflicts, and potentially choose non preferred resources. We investigate the following problem: given a graph, what is the maximum number of agents that can be accommodated subject to non-altruistic behaviors of early arriving agents?

Just for cyclic constraints, the problem is surprisingly difficult. Indeed, we show that, intriguingly, the natural algorithm inspired from optimal solutions to adaptive renaming or musical chairs is sub-optimal for cycles, but proven to be at most 1 to the optimal. The main message of this paper is that finding optimal solutions to the *coordination with constraints and*

A. Castañeda and S. Rajsbaum are supported by UNAM-PAPIIT IA101015 and IN107714. A. Castañeda is also supported by the project CONACYT C394/2016/271602. S. Rajsbaum also received support from ECOS-CONACYT and LAISLA. P. Fraigniaud received support from the ANR project DISPLEXITY, and from the INRIA project GANG. M. Roy is supported by CPSLab project H2020-ICT-644400 at http://www.cpse-labs.eu.

J. Suomela (Ed.): SIROCCO 2016, LNCS 9988, pp. 111–126, 2016.
DOI: 10.1007/978-3-319-48314-6_8

preferences task requires to design "dynamic" algorithms, that is, algorithms of a completely different nature than the "static" algorithms used for, e.g., renaming.

1   Introduction

1.1   Context and Objective

In distributed computing, several tasks have their *adaptive* versions in which the quality of the solution must depend only on the number of processes that participate in a given execution, and not on the total number of processes that could be involved in this task. A typical example of an adaptive task is *adaptive renaming* [4]. In renaming, each process is aiming at acquiring a *name* taken from a small range of integers $[1, r]$, under the constraint that all acquired names must be pairwise distinct. The *quality* of a renaming algorithm is judged based on the range r of names, the smaller the better. In adaptive renaming, r must depend only on the number k of participating processes. In the *asynchronous* setting with *crash-prone processes* and *read/write registers*, the optimal value for the range is known to be $r = 2k - 1$ [5, 13].

Interestingly, adaptive renaming can also be viewed as a *task* by interpreting the integers $1, \ldots, r$ as a total order on the names, where name i is preferred to name j whenever $i < j$. Hence, adaptive renaming can be viewed as an abstraction of the problem in which asynchronous *agents* are competing for *resources* totally ordered by their desirability. In other words, adaptive renaming is an abstraction of a problem of *coordination* between agents under *preferences*. Coordination between agents under preferences has been recently investigated in [2, 3] where the *musical chairs* game has been formally defined and solved. In this game, a set of players (modeling the agents) must coordinate so that each player eventually picks one of the available chairs (modeling the resources). Each player initially comes with an a priori *preference* for one chair. In absence of conflict with other players, the player can pick the desired chair, otherwise the conflicting players must coordinate so that they pick different chairs. It is proved that the smallest number r of chairs for which musical chairs with k players has a solution is $r = 2k - 1$.

We foresee that neither adaptive renaming nor musical chairs fully capture typical scenarios of agents competing for resources. Indeed, both tasks only capture scenarios in which the constraint is that no two agents can acquire the same resource. In practice, resources may not be independent, and the literature on scheduling, partitioning, resource allocation, etc. (see, e.g., [6, 7, 11, 15, 16]) provide several examples of problems in which resources are inter-dependent, causing some resource a not being allowed to be used simultaneously with resource b. That is, using one resource disables others. In this paper, we consider the case in which constraints are modeled as an indirected graph whose nodes are resources, and every edge $\{a, b\}$ indicates that resources a and b cannot be both simultaneously acquired, i.e., acquiring a node disables all its neighbors. In other words, the sets of resources that are allowed are those which form *independent sets* in the graphs. In this framework, renaming as well as musical chairs correspond to the case where the graph of constraints is a stable one (i.e., a graph with no edges). We thus address an extension of renaming and musical chairs, targeting an abstraction of a problem of coordination between agents under *constraints*.

Our objective is to understand the power and limitation of coordination between agents competing for interdependent resources. We are focussing on a scenario inspired from musical chairs in which a resource is a priori assigned to each agent, and the agents have to coordinate between them so that to eventually acquire pairwise non conflicting resources. In particular, if the initial assignment forms an independent set, then the agents do not have to do anything. Alternatively, if they are initially assigned conflicting resources, then they have to spread out and coordinate themselves so that they eventually acquire a set of resources that form an independent set. In other words, each agent comes with an a priori *preference* for some resource—these preferences for the resources do not need to be different. If an agent's preference is not in conflict with the preference of another agent, then it can grab its preference. Otherwise, this agent must choose another resource.

The coordination task between agents under preferences and constraints is thus defined as follows. Given an n-node graph $G = (V, E)$ modeling the constraints between the resources, an input is a multiset M of k elements in V representing the preferences of k processes $p_1, \ldots, p_k$ modeling the k agents. Outputs are independent sets $I = \{u_1, \ldots, u_k\}$ in G, of size k representing the fact that process p_i acquires u_i, for $i = 1, \ldots, k$. The literature on renaming [5] and musical chair [3] taught us that, in an asynchronous system in which the processes are subject to crash failures, the task is not solvable for k larger than some bound, even for the stable graph G (the value of the bound on k for the stable graph is roughly half the number of nodes of the graph). We are interested in the impact of the constraints on this bound. That is, given a graph G, we are interested in the largest k for which the coordination with constraints and preferences task in G is solvable for every preference multiset M of size at most k. We focus on asynchronous systems in which an arbitrarily large number of processes are subject to crash failures. Each process has its own private registers, and the processes communicate via read/write accesses to a shared memory.

1.2   Our Results

We first focus on the problem for the n-node path P_n because it enables to prove a lower bound on the size of Hamiltonian graphs for which the coordination with constraints and preferences task is solvable. Interestingly, this lower bound is almost twice as large as the $2k - 1$ bound without constraints resulting from renaming or musical chairs. Specifically, we establish the following:

Theorem 1. *Let k be a positive integer. The smallest integer n for which the coordination with constraints and preferences task in P_n is solvable for k processes satisfies $n = 4k - 3$. As a consequence, if the coordination with constraints and preferences task in an n-node Hamiltonian graph G is solvable for k processes then $n \geq 4k - 3$.*

The lower bound on n is based on a reduction to impossibility results for musical chairs, i.e., renaming with initial preferences. The upper bound on n comes from a wait-free algorithm, inspired from an optimal adaptive renaming algorithm, whose main lines are: (1) fix a maximum independent set I in P_n, (2) index the vertices of I from 1 to $2k-1$, and (3) run an optimal (adaptive) renaming algorithm on these indexes.

From this preliminary result on P_n, one may think that solving the coordination with constraints and preferences task in a graph G boils down to classical renaming once a maximum independent set in G is fixed. We show that this is *not* the case. In fact, even for an instance as simple as the n-node ring C_n, the problem becomes highly non trivial.

Theorem 2. *Let k be a positive integer. The smallest n for which the coordination with constraints and preferences task in C_n is solvable for k processes satisfies $4k - 3 \leq n \leq 4k - 2$.*

The lower bound is a consequence of Theorem 1 since C_n is Hamiltonian. A quite intriguing fact is that the wait-free algorithm derived from an adaptation of an optimal algorithm for classical renaming run on a maximum independent set of C_n does not match the lower bound, and is off by an additive factor $+1$. In fact, we prove that the true answer is probably the lower bound $4k-3$, which is shown to be tight for $k=2$ and 3 agents, using ad hoc algorithms that are radically different from renaming algorithm.

We believe that the difference of 1 between the lower and upper bounds for C_n is certainly not anecdotal, but is the witness of a profound phenomenon that is not yet understood, with potential impact on classical renaming and musical chairs. The main outcome of this paper is probably the observation that "static" algorithms, i.e., algorithms based on *fixed* precomputed positions in the graph of constraints, might be sub-optimal by allocating less resources than the optimal. Our optimal ad hoc algorithms for coordinating two or three processes in the ring are not static, and the set of allocated resources output by these algorithms can form *any* independent set. The design of optimal "dynamic" (i.e., non static) algorithms for solving the coordination with constraints and preferences task appears to be a challenge, even in the specific case of the cycle C_n.

The enormous difficulty for asynchronous crash-prone processes to coordinate under constraints and preferences, even in graphs with arbitrarily large independent sets, is also illustrated by the case of the complete bipartite graph $K_{x,y}$ with $n = x + y$ nodes. We show that, although $K_{x,y}$ has very large independent sets (of size at least $\min\{x, y\}$), processes cannot coordinate *at all* in this graph.

Theorem 3. *Let x, y be positive integers. Coordination with constraints and preferences in the complete bipartite graph $K_{x,y}$ is unsolvable for more than one process.*

Finally, on the positive side, given any graph G, we can design an static algorithm ALG solving the coordination with constraints and preferences task in G. ALG is based on the novel notion of k-*admissible* independent sets, which may have its interest on its own: given $G = (V, E)$, an independent set I of G is k-admissible if for every $W \subseteq V$ of size at most $k-1$, we have $|I \setminus N[W]| \geq |I \cap W| + 1$ where $N[W]$ denotes the set of nodes at distance at most 1 from a node in W. We prove that among static algorithms, ALG is optimal, which completely closes the problem for static algorithms.

Theorem 4. *Let G be a graph, and k be a positive integer. Let I be a k-admissible independent set in G. Then, ALG instantiated with I solves the coordination with constraints and preferences task in G with k processes. Moreover, if G has no $(k + 1)$-admissible independent set, then no static algorithms can solve the coordination with constraints and preferences task in G with more than k processes.*

1.3   Related Work

Since its introduction, the renaming problem has been extensively studied (see for example [5,10,19]). It was initially introduced as a *non-adaptive* problem in which processes just need to pick distinct output names in the space $[1, \ldots, M]$, where M is on function only on the total number of processes that might participate [4]. Several algorithms were proposed (e.g. [4,9,14]), and, as far as we know, all those initial algorithms are adaptive. Then the adaptive version of renaming was coined. The study of lower bounds for renaming have inspired new developments in topology techniques (see [17] for a detailed description). As explained above, the variant of renaming that is closest to this paper is the adaptive renaming version. This is the version we use to solve the coordination with constraints and preferences task on a graph with no edges.

Musical chairs [2,3] is a coordination problem on a stable graph in which each process starts with a initial vertex (chair) and processes are required to decide distinct vertices (chairs). The problem is studied in a model where the only communication between processes is an indication when two processes propose the same vertex. It has been shown that k processes can solve the problem only if the stable graph has at least $2k - 1$ vertices. It has been also shown that musical chairs and adaptive renaming are equivalent problems.

Interestingly, the coordination with constraints and preferences is also related to *mobile computing* [12], where mobile entities (modeling physical robots, software agents, insects, etc.) must cooperate in order to solve tasks such as rendezvous, gathering, exploration, patrolling, etc. In particular, in the asynchronous look-compute-move model of mobile computation, the "look" operation is very similar to a "snapshot" operation in shared memory, and the "move" operation is very similar to the "write" operation. The major differences between the wait-free model of distributed computation and the look-compute-move model of mobile computation are (1) the presence of failures in the former, and (2) the fact that agents are moving in an anonymous graph in the latter.

2   Model and Examples

2.1   Computational Model

We consider the standard asynchronous wait-free read/write model with k processes, $p_1, \ldots, p_k$ [5,19]. Processes are asynchronous, communicate by writing and reading from a reliable shared memory, and any set of processes may crash. We assume, without loss of generality, that processes can read the whole shared memory in a single atomic snapshot [1].

Problems in the wait-free model are usually defined as *tasks*, where processes get *input values*, and decide after a bounded number of operations on *output values*, such that the decided values represent a valid configuration associated to the initial values of the execution for this task.

Without loss of generality, we can assume that algorithms solving tasks are in *normal form*, that is, are of the form of a loop consisting of (1) writing to the shared memory the local state of the process, (2) taking a snapshot, and (3) performing some local computation. The loop is executed until the process returns an output (i.e., decides).

2.2   Coordination with Constraints and Preferences (CCP)

The task *coordination with constraints and preferences* (or CCP for short) is instantiated by a fixed n-node graph $G = (V, E)$. The graph is modeling the *constraints*. It is supposed to be simple, i.e., without loops and multiple edges, but does not need to be connected. Each process p_i gets as input one vertex $u \in V$, called its initial private *preference*, and must eventually decide on a vertex $v \in V$. It is required that the decided vertices form an *independent set* of G, that is, no two processes decide the same vertex, and no two decided vertices belong to the same edge. It is also required that if the initial preferences form an independent set, then each process must decide its initial preference (enforcing the fact that processes cannot discard their preferences).

We are interested in computing, for every n-node graph G, the largest k such that CCP in G is wait-free solvable for k processes. Note that an algorithm solving CCP in G is given the full description of G a priori. Hence, there are no issues such as, e.g., breaking symmetry between nodes of G, even if G is vertex-transitive. (In particular, the nodes of G might be given labels from 1 to n, a priori, in a specific order which may facilitate the task for the processes).

2.3   Examples and Basic Observations

CCP is trivially solvable for one process in every graph, by selecting its initial preference as output vertex. Also, CCP is trivially not solvable in G for k processes if k exceeds the size of a maximum independent set. In fact, CCP is not solvable in G for k processes if k exceeds the size of the smallest maximal independent set. Indeed, let $I = \{u_1, \ldots, u_\ell\}$ be a smallest maximal independent set in G, and assume that ℓ processes $p_1, \ldots, p_\ell$ are given preferences in

I (u_i to p_i for $i = 1, \ldots, \ell$). In a wait-free execution in which only those ℓ processes participate, they must decide I. If another process $p_{\ell+1}$ "wakes up" after the ℓ processes have decided, there is no more room for $p_{\ell+1}$ to acquire a vertex, because I is maximal. This holds even if there exists another independent set I' larger than I, since the first ℓ processes have already terminated.

The following result is a direct consequence of [3] as Musical Chairs is exactly our problem on the stable graph.

Proposition 1. Let k be a positive integer. The smallest integer n for which the coordination with constraints and preferences task in the n-node stable graph is solvable for k processes satisfies $n = 2k - 1$.

Also, we have the following observation.

Proposition 2. Let $G = (V, E)$ be a graph, and $G' = (V, E')$ with $E' \subseteq E$ be a subgraph of G. If the coordination with constraints and preferences task is solvable for k processes in G, then it is solvable for k processes in G'.

As a consequence of the above two propositions, we get a general lower bound on the size of graphs in which the coordination with constraints and preferences task is solvable for k processes.

Corollary 1. *Let G be an n-node graph. If the coordination with constraints and preferences task in G is solvable for k processes then $2k - 1 \leq n$.*

3   The Case of Cyclic Constraints

Our first results concern simple non-trivial sets of constraints, namely the cases of the n-node paths and cycles, respectively denoted P_n and C_n. The case of the path is entirely solved by the following results, which establish Theorem 1:

Proposition 3. Let k be a positive integer. The smallest integer n for which the coordination with constraints and preferences task in the n-node path is solvable for k processes satisfies $n = 4k - 3$.

Proof. Let us assume, for the purpose of contradiction, that there is an algorithm $\mathcal{A}$ solving CCP in the n-node path for k processes with $n = 4k - 4$. Such a path has a maximum matching M of size $2k - 2$. $\mathcal{A}$ guarantees that, for every edge of M, at most one process acquires an extremity of that edge. $\mathcal{A}$ can be used as a subroutine to solve CCP on a stable graph of size $2k - 2$, as we show below.

We assume that the n-path is oriented from left to right and hence for each edge in M there is a *left* vertex and a *right* vertex (thus, in the path, not two left (right) vertices are adjacent). Also, each vertex v of the stable graph of size $2k-2$ is mapped to a unique edge $f(v)$ in M. To solve CCP on the stable graph, each process p_i with initial preference v, invokes $\mathcal{A}$ with the left vertex of $f(v)$ and decides $f^{-1}(e)$, where e is the edge in M containing the vertex that $\mathcal{A}$ outputs to p_i The resulting algorithm $\mathcal{A}$ solves CCP on the stable graph of size $2k - 2$

because if processes start with distinct vertices, then all of them invoke $\mathcal{A}$ with left vertices and hence, each process decide its initial preference. If processes start with conflicting initial preferences, then $\mathcal{A}$ outputs vertices that belong to distinct edges in M. This is a contradiction with Proposition 1 because M is of size $2k - 2$.

We now describe an algorithm solving CCP in the n-node path for k processes with $n = 4k - 3$. The algorithm is based on a maximum independent set I of size $2k - 1$ in P_{4k-3}. That is, the nodes of P_{4k-3} are labeled off line as $(v_1, v_2, v_3, \ldots, v_{4k-3})$, and we define $I = \{v_1, v_3, v_5, \ldots, v_{4k-3}\}$. Essentially, the algorithm runs the textbook renaming algorithm of [5] on I, adapted to handle initial preferences. Indeed, selecting a node $w \notin I$ may block two positions in I (the two neighboring nodes of w). Nevertheless, there is still enough room to perform renaming, and hence to solve CCP. Indeed, let $N[w]$ be the closed neighborhood of w in P_n, i.e., the at most three nodes at distance at most 1 from w, and, for a set of nodes W, let $N[W] = \cup_{w \in W} N[w]$. In classical renaming, if W is the multiset of currently chosen names, there remain at least $2k - 1 - |W| \geq k$ available names to choose from. In the path, if W is the multiset of currently chosen nodes, there are only $|I \backslash N[W]|$ available nodes in I to choose from, and this number of available nodes can be less than k. However, the crucial observation is that $|I \backslash N[W]|| > |I \cap W|$ in the path P_{4k-3}, for any set of nodes W of size at most $k - 1$. Hence, there are more free nodes in I than occupied nodes in I, and thus the idea is to perform the ranking of the renaming algorithm only on processes sitting on the occupied nodes of I. Since $|I \backslash N[W]|| > |I \cap W|$, this ranking is valid, that is, systematically provides a position in $I \backslash N[W]$. Termination follows from classical arguments by assuming, by way of contradiction, that some processes do not terminate, and then by considering the process p with lowest ID that does not terminate. Eventually, the rank r of p will remain forever the same, and no other processes that do not terminate will conflict with the rth node in the subset of nodes in I that are not conflicting with terminated processes. At this point, process p terminates.   □

As a consequence of this result combined with Proposition 2, we get a general lower bound on the size of Hamiltonian graphs in which the coordination with constraints and preferences task is solvable for k processes. Interestingly, this bound is roughly twice as big as the bound for arbitrary graphs (cf. Corollary 1).

Corollary 2. *Let G be an n-node Hamiltonian graph. If the coordination with constraints and preferences task in G is solvable for k processes then $4k - 3 \leq n$.*

The case of P_n has attracted our interest for it enables deriving bounds for Hamiltonian graphs. The case of the cycle C_n may seem to behave quite similarly as P_n. Surprisingly, this is not the case, as the wraparound constraint yields an interesting phenomenon, namely "static" solutions inspired from renaming algorithms such as the ones for the stable graph and the path are not anymore optimal in term of number of processes, and are off by an additive factor $+1$ from the optimal. More precisely, we show the following, which establishes Theorem 2:

Proposition 4. Let k be a positive integer. The smallest integer n for which the coordination with constraints and preferences task in C_n is solvable for k processes satisfies $4k - 3 \leq n \leq 4k - 2$.

Proof. The lower bound follows directly from Corollary 2. The upper bound is directly derived from the algorithm in the proof of Proposition 3 by fixing a maximum independent set of size $2k - 1$ in the cycle C_{4k-2}. The correctness of the algorithm follows from the same arguments as for the path P_{4k-3}. □

Interestingly, the lower bound $4k - 3$ is most probably the right answer, and not the upper bound $4k - 2$. At least, this is the case for small numbers of processes:

Proposition 5. The smallest integer n for which the coordination with constraints and preferences task in C_n is solvable for k processes satisfies $n = 4k - 3$ for $k = 2$ and $k = 3$.

Proof. The nodes of C_{4k-3} are sequentially labeled offline as $v_1, v_2, \ldots, v_{4k-3}$. This labeling induces a clockwise direction (increasing labels) and a counter-clockwise direction.

The algorithm for two processes in C_5 is depicted on Fig. 1, which represents the snapshot of a process, and the action to take (represented as arrows) based on this snapshot when 2 processes participate. There are three cases, depending on whether the two processes are currently occupying nodes at distance 0, 1, or 2. (Of course, if the snapshot reveals that the process is alone, then it decides the node that it currently occupies, i.e., its preferred node). If the snapshot reveals that the two processes occupy the same node, then the action depends on the ID: going clockwise for the process with smallest ID, and counterclockwise otherwise. If the snapshot reveals that the two processes occupy two neighboring nodes, then the action is: going away from the other node. Finally, if the snapshot reveals that the two processes occupy two nodes at distance 2, then the action is to decide the currently occupied node. One can check that this asynchronous algorithm terminates, and wait-free solves CCP.

A similar algorithm for three processes in C_9 can be derived. Due to lack of space, this algorithm is deferred to the full version of the paper. □

From these two cases, we conjecture that the smallest cycle C_n enabling to solve CCP is $n = 4k - 3$, for all $k \geq 2$. If this is correct, it means that optimality requires processes to coordinate in a more complex way than they do for renaming, in order to spread out optimally in the graph of constraints, and eventually occupy a large number of nodes. The independent set they will eventually agree on cannot be decided a priori, but the processes must agree *on line* in order to eventually decide an independent set that fits their initial preferences, the constraints and the uncertainty resulting for asynchrony and failures.

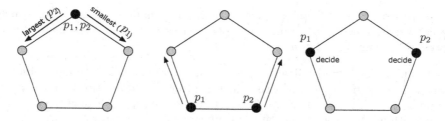

Fig. 1. CCP algorithm for two processes in C_5: rules when two processes are executing

4   A Generic Algorithm

The algorithms inspired by the original algorithm of [5] used in the proofs of Theorems 1 and 2 to establish the upper bounds on the smallest integer n for which the coordination with constraints and preferences task in the n-node path and in the n-node cycle are *static* in the sense that they are aiming at deciding within a fixed independent set. More precisely:

Definition 1. Let G be a graph, and I be an independent set in G. An algorithm $\mathcal{A}$ solving the coordination with constraints and preferences task in G is *static with respect to I* if, for every execution of $\mathcal{A}$, and for every process p, if p does not decide its initial input, then it decides a vertex in I.

To establish Theorem 4, we present a *generic* static algorithm to solve CCP on every graph $G = (V, E)$, and prove that this algorithm is the best possible static algorithm in the sense that it maximizes the number k of processes for which CCP is solvable in G. The generic algorithm is instantiated with an a priori ordered independent set I. That is, $I = \{w_1, \ldots, w_{|I|}\}$, and this ordering of the nodes in I is know a priori to every process. The processes proceed in a sequence of (asynchronous) rounds. At each round, every process p_i proposes a current vertex denoted cur_i (the first proposal is the input u_i of processes p_i). Then, p_i checks whether there is a *conflict* with other proposals. In absence of conflict, p_i decides its current proposal cur_i. If there is a conflict, p_i computes a new proposal in I, and repeats. Hence, in particular, if a process sees no conflict in its initial proposal, then it stays there. Otherwise it will try a new proposal in I. The new proposal of p_i is computed within the "free space" that is defined as the maximal subset of I such that there is no conflict with other processes' proposals.

Algorithm 1 is the pseudocode of the generic algorithm. The algorithm uses a shared array *view*, accessed with write and snapshot operations, where each entry is initially $\bot$. For convenience, it is easier to consider the array *view* as a multiset of nodes. The local variable cur_i stores the current proposal of process p_i. For a set $W \subseteq V$, we denote by $N[W]$ the *closed neighborhood* of W, that is, for $w \in V$, $N[w] = \{w\} \cup \{v \in V : \{v, w\} \in E\}$, and $N[W] = \cup_{w \in W} N[w]$.

Algorithm 1. $G = (V, E)$ is a graph, and I is an ordered independent set in G. Code for p_i,

```
function IndependentSet(u_i ∈ V: initial preference of p_i)
 1:  cur_i ← u_i
 2:  loop
 3:      write(cur_i)
 4:      snapshot memory to get view = {cur_{j_1}, ..., cur_{j_r}}    ▷ multiset of r elements, for some r
 5:      view' ← view\{cur_i}                                ▷ remove one occurrence of cur_i from view
 6:      if view' ∩ N[cur_i] = ∅ then                                         ▷ check for conflicts
 7:          return cur_i                                          ▷ no conflict ⇒ decide cur_i
 8:      else                                       ▷ conflict detected ⇒ compute a new position
 9:          free ← I\N[view']                               ▷ rule out conflicting vertices from I
10:          ℓ ← |{s : cur_{j_s} ∈ I and j_s < i}| + 1  ▷ ranking on the currently occupied vertices of I
11:          cur_i ← ℓ^{th} element in free                  ▷ try the ℓ^{th} free node for the next round
12:      end if
13:  end loop
```

Note that if the initial preferences of participating processes are distinct and form an independent set, then Algorithm 1 guarantees that each process decides on its initial preference. Note also that if two processes decide on vertices v_1, v_2, then $v_1 \neq v_2$ and $\{v_1, v_2\} \notin E$. However, for Algorithm 1 to function appropriately, we use it for a specific kind of independent sets, namely *admissible* independent sets, defined hereafter (see Fig. 2 for an illustration):

Definition 2. Let k be a positive integer, and $G = (V, E)$ be a graph. An independent set I of G is k-*admissible* if for every $W \subseteq V$ of size at most $k - 1$, we have $|I \setminus N[W]| \geq |I \cap W| + 1$.

Notice that any k-admissible independent set I satisfies $|I| \geq 2k - 1$ (instantiate Definition 2 with $W \subseteq I$ of size $k - 1$). To establish Theorem 4, we first prove the following result.

Proposition 6. Let G be a graph, and k be a positive integer. Let I be a k-admissible independent set in G. Then, Algorithm 1 instantiated with I solves the coordination with constraints and preferences task in G with k processes.

Proof. We have seen that the safety conditions (i.e., respect of the preferences, and take decisions on an independent set) are satisfied. It just remains to show that the algorithm is valid (i.e., whenever a process detects a conflict on Line 6, it is able to compute a new consistent preference), and terminate. We first show validity.

Claim. For any process that is about to execute Line 11, $|free| \geq \ell$.

Consider a process p_i that is about to execute Line 11. Such a process p_i must have detected a conflict Line 6. Let $view$ be the snapshot of p_i associated with this conflict, and $W = view'$. When p_i is about to execute Line 11, we have $free = I \setminus N[W]$. As there are at most k participating processes, it follows that $|W| < k$. Since I is k-admissible, we have $|I \setminus N[W]| \geq |W \cap I| + 1$, which implies $|free| \geq |W \cap I| + 1$. Moreover, p_i's ranking computed in line 10 is at most $|W \cap I| + 1$ because W does not contain p_i's preference, i.e., one ignores the nodes not in I when ranking. Hence $|free| \geq \ell$, and thus the algorithm is valid.

Claim. Algorithm 1 terminates.

Assume, for the sake of contradiction, that there is a run α in which some processes take an infinite number of steps without deciding a vertex. In this execution, let P be the set of all processes taking infinitely many steps, and let $p \in P$ be the process with minimum ID in P. Consider a suffix α' of α in which (1) all processes that do not run forever have stopped, and (2) all processes in P have already tried once to get a vertex in I, namely, they have executed Line 11 at least once. Note that such a suffix exists because every process that takes an infinite number of steps in α will see an infinite number of conflicts, and thus will eventually always execute Line 3 with its current vertex in I.

In α', the rank of p is fixed because, from there on, the set of processes that occupy vertices in I is fixed. Let r be the rank of p among the processes of α' with proposals in I, and let *good* be the set of vertices in I that do not conflict with preferences of stopped processes in α'. Eventually, there are no processes in P that are proposing one of the first $r - 1$ elements of *good*. Indeed, the rank of every process in P is at least r, and when a process $q \in P$ takes a snapshot, the set *free* that it computes satisfies *free* $\subseteq$ *good*. Thus, q proposes the x-th element in *free*, where $x \geq r$. Hence, this element cannot be one of the first $r - 1$ elements in *good*.

It follows that, in α', as soon as all running processes have seen each other, and have written at least twice, when these processes compute *free*, the first r elements are all elements of *good*, and only p will try to get the r-th element in *free*. When it does, it detects no conflict. Thus p can terminate the algorithm on Line 3, which yields to a contradiction.   □

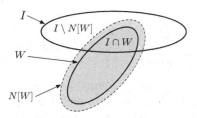

Fig. 2. To be k-admissible, I must satisfy $|I \backslash N[W]| \geq |I \cap W| + 1$ for every W of size $\leq k - 1$.

We now show the second part of Theorem 4, that is, Algorithm 1 is optimal on the number of processes, among static CCP algorithms. This is established thanks to the following result.

Proposition 7. Let G be a graph, k be a positive integer, and assume that G has no $(k + 1)$-admissible independent set. Then, no static algorithm can solve the coordination with constraints and preferences task in G with more than k processes.

The proof is based on the following claim, which is an interesting consequence of the Wait-free Computability Theorem of [18].

Claim. If there is a k-process CCP algorithm for G, then there exists a k-process CCP algorithm for G in which whenever a process sees only itself in its first snapshot, it immediately decides.

Proof (Sketch). The Asynchronous Computability Theorem ACT [18] states that a task is wait-free read/write solvable if and only if there is a chromatic subdivision of the input complex of the task (the simplicial complex represents all possible input configurations) with a coloring that agrees with the specification of the task.

Assume there is a k-process CCP algorithm for G, then let S be such a chromatic subdivision. Herlihy and Shavit proved that any chromatic subdivision can be approximated by an M-th standard chromatic subdivision [18], for some big enough M. This M-th standard chromatic subdivision is crucial because it directly implies a wait-free read/write protocol that solves the task that the coloring of S respects.

Now, the particular algorithms in the claim change the class of subdivisions we are dealing with. These subdivisions are very similar to the M-th standard chromatic subdivision with the difference that that after the 1-st standard chromatic subdivision, the corners of the subdivision are not changed, while the rest of the simplexes are subdivided in the same way. Let us call these subdivisions *M-solo chromatic subdivisions*. Considering this class of subdivisions, in which the diameter of these subdivisions shrinks as M grows, we can observe that, for a big enough M, say M_S, the M_S-solo chromatic subdivision approximates S. As explained above, this M_S-solo chromatic subdivision implies an algorithm that solves CCP for k processes for G in which whenever a process sees only itself in its first snapshot, it immediately decides. $\square$

Proof (Proposition 7). For contradiction, assume there is a static CCP algorithm A on G for $k+1$ processes, with respect to a non $(k+1)$-admissible independent set I. By the claim above, we can assume that if a process sees only itself in its first snapshot, then it decides its initial value (recall that A can be assumed to be in normal form). Since I is not $(k+1)$-admissible, there exists a set $W \subseteq V$ with at most k vertices such that $|I \backslash N[W]| < |W \cap I| + 1$.

Claim. $W \cap I \neq \emptyset$.

Proof. To prove the claim, suppose by contradiction that $W \cap I$ is empty. It follows that $|W \cap I| + 1 = 1$, and thus $|I \backslash N[W]| = 0$, i.e., $I \backslash N[W]$ is empty as well. Thus, $I \cap N[W] = I$. Now, let us consider an execution α of A in which processes $p_1, \ldots, p_{|W|}$ start with preferences for distinct vertices in W, and just write their input in shared memory, i.e., they only perform one write operation in shared memory. Then, consider any extension β of α in which a process q starts with a preference to a vertex in I, runs alone (the p_i's from α do not take steps), and decides. Note that such a process exists since $|W| \leq k$ and A is for

$k+1$ processes. Since A is static with respect to I, q decides a vertex $v \in I$. Let $v' \in W$ that is adjacent to v, whose existence is guaranteed by $I \cap N[W] = I$, and let q' be the process in α with input v'. Let α' be a reordering of α in which q' executes solo, decides, then all other processes in α do their write operation. q' sees only itself in its first snapshot, and will thus decide its input. As q cannot distinguish α from α', β is also a valid extension of α'. In $\alpha'\beta$, process q decides v. This is a contradiction, since q' decides v' in this execution, which is in conflict with v. Thus, $W \cap I \neq \emptyset$, which completes the proof of the claim.

Let $\rho = |W \cap I|$. As proved above, $\rho > 0$. We have $|I \setminus N[W]| < |W \cap I| + 1 = \rho + 1$, and thus $|(I \setminus N[W]) \cup (W \cap I)| < 2\rho - 1$. Let us consider an execution α of A in which processes $p_1, \ldots, p_{|W \setminus I|}$ start with preferences for distinct vertices in $W \setminus I$, and they just write their input in the shared memory. Using the same indistinguishability argument as in the proof of the claim above, we can show that that there is no extension of α in which a process $q \notin \{p_1, \ldots, p_{|W \setminus I|}\}$ decides a vertex in $I \cap N[W]$. Thus, in every extension of α, such a process q decides its initial preference vertex, or a vertex in $(I \setminus N[W]) \cup (W \cap I)$. Let us then consider processes $q_1, \ldots, q_{\rho+1}$, all distinct from $p_1, \ldots, p_{|W \setminus I|}$ (note that these processes exist because $|W| \leq k$). In every extension of α in which $q_1, \ldots, q_{\rho+1}$ start with preferences in $(I \setminus N[W]) \cup (W \cap I)$, and these processes decide, they necessarily decide vertices in $(I \setminus N[W]) \cup (W \cap I)$. This implies that $|(I \setminus N[W]) \cup (W \cap I)| \geq \rho + 1$.

Now, from A and execution α, we construct a p-process algorithm A' that solves CCP in the stable graph with too few vertices as follows. In A', the shared memory has the state in α, and each process q_i, $1 \leq i \leq \rho + 1$, follows the same code as in A. To see that A' is correct, note that (1) if processes start with distinct vertices of $(I \setminus N[W]) \cup (W \cap I)$, then each process decides its input, since A is correct, and (2) if processes start with inputs in conflict, then they decide distinct vertices in $(I \setminus N[W]) \cup (W \cap I)$, as shown before. Now, A' solves CCP for $\rho+1$ processes over the stable graph with vertex set $(I \setminus N[W]) \cup (W \cap I)$ (and no edges), which is impossible since this set has at most $2\rho - 2$ vertices, and, by Proposition 1, CCP is unsolvable for so few vertices. Therefore, assuming the existence of A yields a contradiction, which completes the proof.   □

The following is an interesting consequence of Proposition 7 to the case of CCP in cycles.

Corollary 3. *If $\mathcal{A}$ is a wait-free algorithm solving the coordination with constraints and preferences task in C_{4k-3} for k processes then $\mathcal{A}$ cannot be static.*

A natural guess is that CCP can be solved in G for a number of processes that grows with the size of the smallest maximal independent set in G. Having this in mind, Theorem 4 may appear to be rather weak. Indeed, for instance, in complete bipartite graphs $K_{x,y} = (X \cup Y, X \times Y)$, with $x = |X|$, and $y = |Y|$, there are no 2-admissible independent sets. Thus our static algorithm, although optimal among static algorithms, cannot do better than solving CCP in $K_{x,y}$ for just one process! The truth is that our algorithm is not to be blamed. Indeed, the intuition that the number of processes that can be accommodated should grow with the size of the smallest maximal independent set is wrong, and *any* algorithm, not just static ones, cannot do better that solving CCP in $K_{x,y}$ for a single process only, as shown below (which establishes Theorem 3).

Proposition 8. Let x, y be positive integers. Coordination with constraints and preferences in $K_{x,y}$ is unsolvable for more than one process.

Proof. Recall that, in the *s-set agreement* task, each process proposes a value, and each correct process decides a proposed value so that the number of distinct decisions is at most s. We show that if algorithm $\mathcal{A}$ solves CCP in $K_{x,y}$ for k processes (hence either x or y is at least k), then there is an algorithm $\mathcal{B}$ that solves $\lceil \frac{k}{2} \rceil$-set agreement for k processes – yet, it is known that s-set agreement on k processes is unsolvable for any $1 \leq s \leq k - 1$ and $k \geq 2$ (see [8,18, 20]), a contradiction. In $K_{x,y} = (X \cup Y, X \times Y)$, X and Y are the unique two maximal independent sets. We say that processes $p_1, \ldots, p_{\lceil \frac{k}{2} \rceil}$ *belong* to X, and the remaining processes, $p_{\lceil \frac{k}{2} \rceil + 1}, \ldots, p_k$, *belong* to Y, in the following sense: to the processes $p_1, \ldots, p_{\lceil \frac{k}{2} \rceil}$, we assign pairwise distinct vertices $v_i \in X$ as preferences, $i = 1, \ldots, \lceil \frac{k}{2} \rceil$, and to the processes $p_{\lceil \frac{k}{2} \rceil + 1}, \ldots, p_k$, we assign pairwise distinct vertices $v_i \in Y$ as preferences, $i = \lceil \frac{k}{2} \rceil + 1, \ldots, k$. We use algorithm $\mathcal{A}$ solving CCP on this instance for construction an algorithm $\mathcal{B}$ solving $\lceil \frac{k}{2} \rceil$-set agreement.

A process p_i that belongs to X first announces its preference (considered as its proposal), and then invokes $\mathcal{A}$ using $v_i \in X$ as its input. If p_i gets assigned to a vertex of X in $\mathcal{A}$, then it decides v_i in $\mathcal{B}$, otherwise it decides any vertex $v_j \in Y$ with $j \in \{\lceil \frac{k}{2} \rceil + 1, \ldots, k\}$. A process p_i that belongs to Y proceeds similarly. That is, if p_i gets assigned to a vertex of Y in $\mathcal{A}$, then it decides its preference v_i in $\mathcal{B}$, otherwise it decides any vertex $v_j \in X$ with $j \in \{1, \ldots, \lceil \frac{k}{2} \rceil\}$. Since $\mathcal{A}$ is correct, and since there are no independent sets in $K_{x,y}$ that include vertices of both X and Y simultaneously, it follows that if a process gets assigned to a vertex of X (resp., Y) in $\mathcal{A}$, then every other process gets assigned to a vertex of X (resp., Y) as well. Therefore, in every execution of $\mathcal{B}$, processes either decide the preferences of the processes that belongs to X, or the preferences of the processes that belongs to Y, hence at most $\lceil \frac{k}{2} \rceil$ proposals are decided in $\mathcal{B}$. □

5   Conclusion

We have considered a generalization of renaming in graphs, in which deciding a node forbids others to use neighboring nodes. We proved a lower bound for Hamiltonian graphs, and provided optimal algorithms for 2 processes on a pentagon, and 3 processes on a nonagon. For the case where processes agree beforehand on a given maximal independent set, we designed optimal *static* algorithms for solving this problem. Static algorithms are however sub-optimal, as illustrated in the case of the rings. The design of optimal *dynamic* algorithms for solving the coordination with preferences and constraints tasks in graphs remains an open problem, even for rings.

References

1. Afek, Y., Attiya, H., Dolev, D., Gafni, E., Merritt, M., Shavit, N.: Atomic snapshots of shared memory. J. ACM **40**(4), 873–890 (1993)
2. Afek, Y., Babichenko, Y., Feige, U., Gafni, E., Linial, N., Sudakov, B.: Oblivious collaboration. In: Peleg, D. (ed.) DISC 2011. LNCS, vol. 6950, pp. 489–504. Springer, Heidelberg (2011). doi:10.1007/978-3-642-24100-0_45
3. Afek, Y., Babichenko, Y., Feige, U., Gafni, E., Linial, N., Sudakov, B.: Musical chairs. SIAM J. Discrete Math. **28**(3), 1578–1600 (2014)
4. Attiya, H., Bar-Noy, A., Dolev, D., Peleg, D., Reischuk, R.: Renaming in an asynchronous environment. J. ACM **37**(3), 524–548 (1990)
5. Attiya, H., Welch, J.: Distributed Computing Fundamentals, Simulations, and Advanced Topics, 2nd edn. Wiley, New York (2004)
6. Baker, B.S., Coffman Jr., E.G.: Mutual exclusion scheduling. Theor. Comput. Sci. **162**(2), 225–243 (1996)
7. Bodlaender, H.L., Jansen, K.: Restrictions of graph partition problems. Part I. Theor. Comput. Sci. **148**(1), 93–109 (1995)
8. Borowsky, E., Gafni, E.: Generalized FLP impossibility result for t-resilient asynchronous computations. In: Proceedings of the Twenty-Fifth Annual ACM Symposium on Theory of Computing, STOC 1993, pp. 91–100. ACM, New York (1993)
9. Borowsky, E., Gafni, E.: Immediate atomic snapshots and fast renaming. In: Proceedings of the Twelfth Annual ACM Symposium on Principles of Distributed Computing, PODC 1993, pp. 41–51. ACM, New York (1993)
10. Castañeda, A., Rajsbaum, S., Raynal, M.: The renaming problem in shared memory systems: an introduction. Comput. Sci. Rev. **5**(3), 229–251 (2011)
11. Even, G., Halldórsson, M.M., Kaplan, L., Ron, D.: Scheduling with conflicts: online and offline algorithms. J. Sched. **12**(2), 199–224 (2009)
12. Flocchini, P., Prencipe, G., Santoro, N.: Distributed Computing by Oblivious Mobile Robots. Synthesis Lectures on Distributed Computing Theory. Morgan & Claypool Publishers, San Rafeal (2012)
13. Gafni, E., Mostéfaoui, A., Raynal, M., Travers, C.: From adaptive renaming to set agreement. Theor. Comput. Sci. **410**(14), 1328–1335 (2009). Structural Information and Communication Complexity (SIROCCO 2007)
14. Gafni, E., Rajsbaum, S.: Recursion in distributed computing. In: Dolev, S., Cobb, J., Fischer, M., Yung, M. (eds.) SSS 2010. LNCS, vol. 6366, pp. 362–376. Springer, Heidelberg (2010). doi:10.1007/978-3-642-16023-3_30
15. Garey, M.R., Graham, R.L.: Bounds for multiprocessor scheduling with resource constraints. SIAM J. Comput. **4**(2), 187–200 (1975)
16. Halldórsson, M.M., Kortsarz, G., Proskurowski, A., Salman, R., Shachnai, H., Telle, J.A.: Multicoloring trees. Inf. Comput. **180**(2), 113–129 (2003)
17. Herlihy, M., Kozlov, D., Rajsbaum, S.: Distributed Computing Through Combinatorial Topology. Morgan Kaufmann, San Francisco (2013)
18. Herlihy, M., Shavit, N.: The topological structure of asynchronous computability. J. ACM **46**(6), 858–923 (1999)
19. Raynal, M.: Concurrent Programming: Algorithms, Principles, and Foundations. Springer, Heidelberg (2013)
20. Saks, M., Zaharoglou, F.: Wait-free k-set agreement is impossible: the topology of public knowledge. SIAM J. Comput. **29**(5), 1449–1483 (2000)

Concurrent Use of Write-Once Memory

James Aspnes[1], Keren Censor-Hillel[2(✉)], and Eitan Yaakobi[2]

[1] Department of Computer Science, Yale University, New Haven, USA
aspnes@cs.yale.edu
[2] Technion, Department of Computer Science, Haifa, Israel
{ckeren,yaakobi}@cs.technion.ac.il

Abstract. We consider the problem of implementing general shared-memory objects on top of write-once bits, which can be changed from 0 to 1 but not back again. In a sequential setting, write-once memory (WOM) codes have been developed that allow simulating memory that support multiple writes, even of large values, setting an average of $1 + o(1)$ write-once bits per write. We show that similar space efficiencies can be obtained in a concurrent setting, though at the cost of high time complexity and fixed bound on the number of write operations. As an alternative, we give an implementation that permits unboundedly many writes and has much better amortized time complexity, but at the cost of unbounded space complexity. Whether one can obtain both low time complexity and low space complexity in the same implementation remains open.

1 Introduction

Write-once memory (WOM) is a storage medium with memory elements, called cells, that can only *increase* their value. These media can be represented as a collection of binary cells, each of which initially represents a bit value 0 that can be *irreversibly* overwritten with a bit value 1. WOM codes, first introduced by Rivest and Shamir [33], enable to record data multiple times without violating the asymmetry writing constraint in a WOM. The goal in the design of a WOM code is to maximize the total number of bits that can be written to the memory in t writes, while preserving the property that cells can only increase their level.

These codes were first motivated by storage media such as punch cards and optical storage. However, in the last decade, a wide study of these codes re-emerged due to their connection to *Flash memories*. Flash memories contain floating gate cells which are electrically charged with electrons to represent the cell level. While it is fast and simple to increase a cell level, reducing its level requires a long and cumbersome operation of first erasing its entire containing block and only then programming relevant cells. Applying a WOM code enables additional writes before having to physically erase the entire block.

This paper provides the first study of concurrency in write-once shared memory. We investigate concurrent write-once memory from a theoretical viewpoint, which, in particular, means that we consider the memory impossible to erase

© Springer International Publishing AG 2016
J. Suomela (Ed.): SIROCCO 2016, LNCS 9988, pp. 127–142, 2016.
DOI: 10.1007/978-3-319-48314-6_9

(as opposed to considering it to be expensive). We show that any problem that can be solved in a standard shared-memory model can be solved in a write-once memory model, at the cost of some overhead. Our goal is to provide an analysis of this cost, both in terms of step complexity and space complexity.

Motivation: In addition to our interest in WOM as a computing model, our study is motivated by two observations. First, WOM is not subject to the **ABA problem**, in which memory can change back and forth going unnoticed, which is proven to be hard to overcome [1].

The second reason is that several known concurrent algorithms are already implemented using write-once bits. In other words, for some specific problems, the overhead of using WOM can be reduced compared to the general case. Examples of such implementations are the **sifters** constructed by Alistarh and Aspnes [2] and by Giakkoupis and Woelfel [17], and some variants of the **conflict detectors** of Aspnes and Ellen [5].[1] A **max register** [3] is another example of an object that can be implemented using write-once bits (see overview in Sect. 3). Interestingly, the covering arguments used to prove lower bounds on max registers [3] imply that no historyless primitive can give a better implementation than write-once bits.

Yet these specific solutions do not immediately give a general implementation of arbitrary shared-memory objects, and the question arises whether the space efficiencies obtained by WOM codes in a sequential setting can transfer to a concurrent setting as well.

The challenge: To give a flavor of the challenge in adopting known WOM codes to concurrent use, we explain a simple example in Table 1, introduced by Rivest and Shamir [33], which enables the recording of two bits of information in three cells twice. It is possible to verify that after the first 2-bit data vector is encoded into a 3-bit codeword, if the second 2-bit data vector is different from the first, the 3-bit codeword into which it is encoded does not change any code bit 1 into a code bit 0, ensuring that it can be recorded in the write-once medium.

Table 1. A WOM code example

Data bits	First write	Second write
00	000	111
10	100	011
01	010	101
11	001	110

---

[1] This does not include the $\Theta(\log m / \log\log m)$-step m-valued conflict detector that appears in [5], but does include a simpler $\Theta(\log m)$-step conflict detector in which a write of a value whose bits are $x_{k-1}, \ldots, x_0$ is done by setting to 1 the corresponding bits $A[i][x_i]$ in a $k \times 2$ array A.

Suppose now that the above code is used in a concurrent WOM system, and that two processes p_1 and p_2 invoke write operations with input data bits 10 and 01, respectively. This means that p_1 needs to write 100 into the memory, and p_2 needs to write 010 to it. In other words, p_1 needs to set the first of the three bits to 1, while p_2 needs to set the second. Consider a schedule in which p_1, p_2 set their respective bit in some order, and afterwards another process p_3 reads the shared memory. The bits that it sees are 110, but these correspond to the input 11, which was never written into the memory, violating the specification of the memory.

The difficulty above is amplified by the fact that since more than a single process is writing and reading the content of the memory, it is not known what the value of t is, that is, how many writes have occurred so far. This is needed in the above example for both writing and reading.

We emphasize that there is a significant amount of fundamental simulations of different types of registers in the literature of distributed computing (see, e.g., [7, Chapter 10] and [21, Chapter 4]). The above WOM example satisfies the definition of a **single-writer-multi-reader (SWMR) safe register** [28,29], in which a read that is not concurrent with a write returns a correct value. Known simulations can use this object to construct **multi-writer-multi-reader (MWMR) atomic registers**. However, these simulations do not comply with the restrictions that arise from WOM, and hence different solutions must be sought.

1.1 Our Contribution

We first show that with one additional bit that indicates to read operations that a write operation has been completed, we can easily implement a write-once m-bit register. Then, we show how to support t writes, still for a single writer, within a space complexity of $2m + t$ bits. This appears in the full version. After these toy examples, our goal is to get closer to the $t(1 + o(1))$-space WOM code constructions for the non-concurrent setting. Carefully adapting the tabular code of [33] to our concurrent setting, allows us to obtain a SWMR m-bit register that supports t writes, with the following properties.

Theorem 1. *There is an algorithm that implements an n-process SWMR m-bit register supporting up to t writes, using space complexity of $(1 + o(1))t$ when $t = \omega(m2^m)$, and with amortized step complexity $O(n2^m)$ for a write and $O(2^m)$ for a read.*

We then extend our tabular construction to support multiple writers, with the aid of a reduction from MWMR registers to SWMR registers due to [24], and with incorporating safe-agreement objects [10] in order to efficiently share space. Our result is summarized as follows.

Theorem 2. *There is an algorithm that implements an n-process MWMR m-bit register that supports up to t writes, using space complexity of $(2 + o(1))t$ when $t = \omega((m + \log n)n^6 2^m)$, and with amortized step complexity $O(n^2 2^m)$ for both write and read operations.*

The drawback of the above implementation is its large step complexity. At the cost of increased space complexity, we show (most details appear in the full version) how to build a WOM code on top of a max register, which allows drastically reduced step complexities, as stated next. Here, a unique timestamp t is guaranteed to be associated by the algorithm with each write operation.

Theorem 3. *There is an algorithm that implements an n-process MWMR register of m bits with unbounded space, where the* amortized *step complexity of a write operation that gets associated with a timestamp t is $O(\log t + m + \log n)$ and the step complexity of a read operation that reads a value associated with a timestamp t is $O(\log t + m + \log n)$.*

Whether it is possible to obtain both low time complexity and low space complexity in the same implementation remains an intriguing open question.

1.2   Additional Related Work

In their pioneering work, Rivest and Shamir also reported on more WOM code constructions, including tabular WOM codes and linear WOM codes. Since then, several more constructions were studied in the 1980's and 1990's [13,15,18], and more interest to these codes was given in the past seven years; see e.g. [9,11,12, 14,34,35,37–40].

The capacity of a WOM was also rigorously investigated. The maximum sum-rate as well as the capacity regions were studied in [19,33,36] with extensions to the non-binary case in [16]. The implementation of WOM codes in several applications such flash memories and phase-change memories was recently explored in [26,30,31,41,42]. These works were motivated by the system implementation on WOM codes in these memories, while taking into account the hardware and architecture limitations when implementing these codes into the system.

Write-once memory should not be confused with **sticky registers** as defined by Plotkin [32], which in some recent systems literature (e.g. [8]) have been described as registers with **write-once semantics**. Sticky registers initially hold a default "empty" value, and any write after the first has no effect. Such registers are equivalent to consensus objects, and thus significantly more powerful than standard shared memory. In contrast, write-once memory as considered here and in the WOM code literature is weaker than standard shared memory.

1.3   Model

We use a standard asynchronous shared memory system, restricted by the assumption that registers hold only a single bit and `write` operations can only write 1. We assume that all registers are initially 0, as any register initialized to 1 conveys no information and can safely be omitted. Asynchrony is modeled by interleaving according to a schedule chosen by an adversary. As we consider only deterministic algorithms, it is reasonable to assume that the adversary has unrestricted knowledge of the state of the system at all times, and can choose the schedule to make things as difficult for the algorithm as possible. The computational power of the adversary is unlimited; indeed, the adversary is essentially just a personification of the universal quantifier applied to schedules.

When implementing an object, our goal is **linearizability** [22]; given an execution of S of the implemented object, there should exist a sequential execution S' of the same object with the same operations, such that whenever some operation π finishes in S before another operation π' starts, π precedes π' in S'.

Time complexity: We use the standard notion of **step complexity**. The worst-case **individual step complexity** of an operation is the maximum number of **steps** (read or write operations applied to a write-once bit) carried out by the process executing the operation between when it starts and finishes. The **total step complexity** of a collection of operations is the maximum number of steps taken by all processes in any execution involving these operations.

We say that an operation π has **amortized step complexity** $C(\pi)$ if, in any execution, the total step complexity is bounded by the sum of the amortized step complexities of all operations in the execution. Note that the amortized step complexity of an operation is not uniquely determined, and there may be more than one way to trade off the amortized step complexities of operations.

Space complexity and storage: The straightforward measure of **space complexity** for write-once memory is the number of objects (in our case, write-once bits) that can be accessed during the execution of an algorithm or implementation. In traditional shared-memory models, this quantity is fixed throughout the execution of the algorithm.

However, for one of our implementations, we will assume that the space is unbounded, in order to exemplify its property of obtaining good step complexity despite supporting an unbounded number of writes. A simple argument (see also Sect. 4) shows that infinite space is inherent for supporting an unbounded number of writes in a write-once medium.

2   Registers Based on the Tabular WOM Code

Here we give a family of register implementations based on the tabular WOM code of Rivest *et al.* [33]. These allow up to t writes of m-bit values. For the single-writer case (see Sect. 2.2), the construction requires only $(1 + o(1))t$ write-once bits provided $t = \omega(m2^m)$, for an average of $1 + o(1)$ bits per write. For the multi-writer case (Sect. 2.3), it requires $(2 + o(2))t$ bits under the same conditions on t. In both cases the amortized time complexity of each operation is polynomial in n and 2^m, even for very large tables. An alternative implementation that sacrifices space for speed will be given later in Sect. 3.

2.1   The Tabular WOM Code

The tabular WOM code represents 2^m distinct values as an array of k rows of $m + \ell$ bits each, where k and ℓ are parameters selected to maximize the efficiency of the code. Each row $A[i]$ consists of an m-bit increment field $A[i]$.increment, interpreted as an element of $\mathbb{Z}_{2^m}$, together with an ℓ-bit unary counter $A[i]$.count. A row is **unused** if all bits in the counter are 0, and **full** if all are 1. A row that is neither unused nor full is **active**. The value stored in the array is given by

$$\left(\sum_{i=1}^{k} A[i].\text{increment}.A[i].\text{count} \right) \bmod 2^m. \tag{1}$$

To change the current value in A from x to y, the writer first checks for a used, non-full row that already has an increment value equal to $(y - x) \bmod 2^m$, and if so increments the counter in that row by one by writing an additional 1 bit. If there is no such row, the writer selects an unused row, writes $(y - x) \bmod 2^m$ to its increment field, and sets the count to 1 by writing a single one bit to the counter field. This process continues until the writer can no longer find an unused row when trying to write an increment that cannot be stored otherwise.

We would like to get the space needed for t write operations as close to t as possible. There are two sources of space overhead that prevent this in the tabular WOM code. The first is that each increment field adds m bits that must be amortized over the ℓ write operations handled by that row; this gives $1 + o(1)$ overhead provided $\ell = \omega(m)$. The second is that up to $2^m - 1$ rows may be only partially used (if more than this are unused, we have rows available for all possible increments and can perform any new write operation). This overhead also becomes $1 + o(1)$ provided $k = \omega(2^m)$. Setting both $\ell = \omega(m)$ and $k = \omega(2^m)$ gives $t = \omega(m2^m)$ and a space complexity of $(1 + o(1))t$.

2.2   Single-Writer Implementation

The tabular WOM code has the useful property that as long as the writer writes $A[i]$.increment in a new row before setting any of the bits in $A[i]$.count, the value stored in A changes atomically at the moment that the writer sets a bit in $A[i]$.count. This means that with a single writer, no special effort is needed to ensure linearizability, and we can treat the linearization point of a write operation as the moment it sets a bit in some count row.

On the other side, a read operation needs to obtain an atomic snapshot of the entire array to be able to compute the sum of the entries as given in (1). This can be done in a straightforward way using a double-collect snapshot, with some further optimizations possible by taking advantage of predicting which bits could be written next. Note that even with a snapshot, it is possible that a reader may observe an incomplete write of $A[i]$.increment for some i. However, this can only occur if the corresponding $A[i]$.count is still 0. So a read operation always returns the sum of the increments of all writes that linearize before it, giving correctness.

For $t = \omega(m2^m)$, the average time complexity of a write is $1 + o(1)$, though the cost of a specific write may range from 1 to $1 + m$, depending on whether it needs to set an increment field.

For read operations the cost may be much higher. Unlike the writer, a reader may need to read the same bit more than once to see if it has changed. Indeed, a naive implementation of the double-collect snapshot would force a reader to read all $k\,(m+\ell)$ bits at least twice during any read operation, and again for each write that occurs during the read. We can reduce the amortized cost by observing that the reader never needs to re-read a bit that is already 1, and by enforcing that the writer use new rows and write count bits in a specified order. This means that each new write might write to at most 2^m distinct locations in the count fields: one for each active row, plus at most one bit at the start of an unused row if the active rows do not span all 2^m possible increments. This reduces the cost imposed on each reader by a new write to at most $2^m + m$ operations (2^m count bits plus at most one increment field). If we multiply this by n potential readers, this raises the amortized cost of a write to $O(n2^m)$ bit operations, which is large but still independent of the table size. Shifting costs to the writers in this way still leaves the reader with an amortized cost of $O(2^m)$ to re-read zero bits to confirm that no new writes have occurred.

Pseudocode for an implementation that applies these optimizations is given in Algorithm 1. The above discussion essentially proves the following.

Theorem 1. There is an algorithm that implements an n-process SWMR m-bit register supporting up to t writes, using space complexity of $(1 + o(1))t$ when $t = \omega(m2^m)$, and with amortized step complexity $O(n2^m)$ for a write and $O(2^m)$ for a read.

2.3   Multi-writer Extension

We can extend the single-writer construction to multiple writers using a construction of Israeli and Shaham [25, Sect. 4]. This construction implements a multi-writer multi-reader (MWMR) register from n single-writer multi-reader (SWMR) registers, one for each writer. Each MWMR write operation requires $O(n)$ SWMR read operations and 2 SWMR write operations. MWMR read operations require only $O(n)$ SWMR read operations. Each SWMR register must be large enough to store the contents of the MWMR register, plus an addition $6 \lg n + O(1)$ bits for pointers used to determine the linearization order.

By implementing each SWMR register as in the preceding section, for sufficiently large t, each writer process can carry out up to t writes at an amortized space complexity of $2 + o(1)$ bits per write. However, both the bound on t to obtain this space complexity and the time complexity of both read and write operations becomes quite large: t must be $\omega((m+\log n)n^6 2^m)$ and the amortized cost of both read and write operations rises to $O(n^2 2^m)$. Whether one can retain low per-write space complexity while getting low time complexity in a MWMR setting remains open.

A further annoyance is that the low amortized space complexity applies only when each writer individually uses up its allotment of $t = \omega((m + \log n)n^6 2^m)$ writes. While this might be a reasonable assumption for some applications, in the worst case we can imagine a single writer using up its allotment while the other writers do nothing, giving a per-write space complexity of $\Theta(n)$.

shared data: Array A[0..$r-1$] of rows, where each row A[i] has fields
A[i].increment of m write-once bits and A[i].count[0..$\ell-1$] of ℓ
write-once bits;

local data: Array next[0..2^m-1] where each entry holds either $\perp$ or an
$\langle$index, position$\rangle$ where index is an index into A and position is an
index into A[index].count;

current, equal to the most recently computed value of the register;
Array MyA[0..$r-1$] of rows, where each row MyA[i] has fields MyA[i].increment
of m write-once bits and MyA[i].count[0..$\ell-1$] of ℓ write-once bits;

```
 1  procedure write(v)
 2  │   Let i = v − current (mod 2^m);
 3  │   if next[i] = ⊥ then
 4  │   │   next[i] ← ⟨r, 0⟩ where r is a newly-allocated row;
 5  │   └   A[next[i].index].increment ← i;
 6  │   A[next[i].index].count[next[i].position] ← 1;
 7  │   if next[i].position = ℓ − 1 then
 8  │   │   next[i] ← ⊥;
 9  │   else
10  │   └   next[i].position = next[i].position + 1;
11  └   current ← v;

12  procedure read()
13  │   repeat
14  │   │   foreach i such that MyA[i].count is not all 0 or all 1 do
15  │   │   └   copy(MyA[i].count, A[i].count);
16  │   │   Let i be the smallest index such that MyA[i].count is all 0;
17  │   │   if A[i].count[0] ≠ 0 then
18  │   │   │   MyA[i].increment ← A[i].increment;
19  │   │   └   copy(MyA[i].count, A[i].count);
20  │   until MyA is unchanged throughout an iteration;
21  └   return ∑_{i=0}^{r−1} ( MyA[i].increment · ∑_{j=0}^{ℓ−1} MyA[i].count[j] )  (mod 2^m) ;

    // Helper procedure for read
    // Copies bits to X from Y assuming Y contains no 0 to the left of
    //   a 1
22  procedure copy(X, Y)
23  │   Let j be the smallest index such that X[j] = 0;
24  │   while j < ℓ ∧ Y[j] = 1 do
25  │   │   X[j] ← 1;
26  └   └   j ← j + 1;
```

Algorithm 1. Single-writer register implemented using a tabular WOM code

2.4   Allocating Table Rows from a Common Pool

We solve this problem by allocating table rows from a common pool. In this section we describe a simple storage allocator, based on the safe-agreement objects of Borowsky et al. [10]. Our storage allocator guarantees that all but $n - 1$ rows in a k-row array are assigned to some writer.

A **safe-agreement object** provides a weak version of consensus that guarantees agreement and validity but not termination. Any process that accesses a safe-agreement object is guaranteed to obtain the id of a unique winner among the users of the object, provided no process halts during a special *unsafe* segment of its execution; if some process does halt, the object never returns. This means that, if we assign a safe-agreement object to control ownership of each of the k rows in our pool, at most $n - 1$ rows will never be allocated, assuming at least one process continues to run.

```
    // propose_i(v)
1   A[i] ← 001;
2   if snapshot(A) contains 101 for some j ≠ i then
        // Back off
3       A[i] ← 011;
4   else
        // Advance
5       A[i] ← 101;
    // safe_i
6   repeat
7   |   s ← snapshot(A);
8   until s[j] does not equal 001 for any j;
    // agree_i
9   return the smallest index j with s[j] = 101;
```
Algorithm 2. Safe agreement (adapted from [10])

Algorithm 2 shows how to implement a safe-agreement object using WOM. The mechanism is essentially the same as in the original Borowksy et al. algorithm, except that we encode the values 0 as 000 when it represents the initial value and 011 when it represents the result of a back-off, the value 1 as 001, and the value 2 as 101. The intuition is that a process first advances to level 1 (001), then backs off if it detects another process already at level 2 (101). If a snapshot includes no processes at level 1, it is safe for any process that sees that snapshot to agree on the smallest process at level 2, because any later process will back off before reaching level 2. Termination is also guaranteed as long as no process stays at level 1 forever.

To implement the storage allocator, we add a safe-agreement object to each row; this increases the size of each row by $3n$ bits. We also include a $\lceil \lg n \rceil$-bit field to allow a reader to quickly identify the owner of a row. Despite these additions, we still get $1 + o(1)$ amortized bits per write by making $\ell = \omega(m + n)$.

To allocate a new row, a writer interleaves attempts to win the safe-agreement objects for the next n rows for which it has not yet determined a winner. At least one of these safe-agreement objects will eventually return a value. If this is the id of the writer, it can claim the row by writing its id to the id field and proceed as in the single-writer construction. If not, it continues to attempt to acquire a row from the set obtained by throwing in the next row that it has not previously attempted to acquire. In either case the writer eventually acquires a row or reaches a state where all but $n - 1$ rows have been allocated.

The reader's task is largely unchanged from the basic MWMR construction: for each of the n SWMR registers, there are at most 2^m active rows it must check for updates, plus up to n additional rows it must check for new activity. This again gives an amortized cost from the readers of $O(n2^m)$ steps per write operation. In addition, each write operation may impose a cost of $O(n)$ bit operations from extra collects in the snapshot on each other writer, for a total of $O(n^2)$ bit operations, for each row it attempts to allocate. This gives a total cost over all writes of $O(kn^3)$ for an amortized cost of $O(n^3/\ell) = O(n^2)$ per write. So the total amortized cost per write is $O(n(n + 2^m))$. This gives:

Theorem 2. There is an algorithm that implements an n-process MWMR m-bit register that supports up to t writes, using space complexity of $(2 + o(1))t$ when $t = \omega((m + \log n)n^6 2^m)$, and with amortized step complexity $O(n^2 2^m)$ for both write and read operations.

3   An Unrestricted MWMR Implemention Based on Max Registers

The tabular WOM code constructions have two deficiencies: they have huge time complexity, and they are limited-use, permitting only a fixed maximum number t of write operations. In this section, we give a different construction (using unbounded space) that implements a wait-free m-bit MWMR register on top of a max register [3]. A max register provides WriteMax and ReadMax operations, where ReadMax returns the *largest* value written by any preceding WriteMax.

There are several known constructions of max registers [3,4,20], each of which has different goals. The basic structure we use here follows the tree implementation of [3], described in Sect. 3.1 for completeness. In Sect. 3.2 we construct our full MWMR m-bit register and prove its properties.

3.1   Tree-Based Max Register

The standard tree-based max register is any binary tree whose leaves correspond to the possible values of the max register. Each node represents a single-bit register that can hold a value in $\{0, 1\}$. The aim is to have the current value of the tree be the rightmost leaf that is set to 1. To implement this, a ReadMax operation travels down the tree starting from the root node, going to the left child of a node if it reads 0 and going to the right child if it reads 1. A WriteMax(v)

starts from the leaf that corresponds to the value v, and travels up the tree to the root, setting to 1 all bits to which it arrives from the right. An important technicality is that in order to make the above linearizable, before a `WriteMax` makes any change to a left subtree of a node, it checks that the bit at this node is 0. This allows, for example, implementing a b-bounded max register (supporting values in values in $\{0, ..., b-1\}$) using a balanced binary tree of depth $O(\log b)$.

However, we can also use an unbalanced binary tree with the property that each leaf v is at depth $O(\log v)$. Since the step complexity of any operation is proportional to the depth of the leaf it writes or returns, the latter gives an implementation with a step complexity of $O(\log v)$. This implementation also has the nice property that it can be extended to support an unbounded number of values. This is done by having a leaf at depth $O(n)$ point to a multi-writer snapshot object. This way the step complexity does not increase with the value v beyond limit, but is rather bounded by $O(\min\{\log v, n\})$, since there are linear-time implementations of snapshot objects [6,23]. The problem with having the step complexity increase beyond limit is not only a complexity problem—it is also a computational problem in the sense that the implementation is not wait-free if we keep the tree infinite, since a `ReadMax` operation can always be pushed farther down to the right side of the tree by a new `WriteMax` operation with a larger value.

Using WOM, the tree-based max register implementation has the nice property that only single-bit registers are used and their value can only be changed from 0 to 1. However, we cannot use the snapshot object that truncates the tree at depth $O(n)$, because its known implementations do not translate into the write-once model. Another approach that avoids the usage of the snapshot object is the randomized helping mechanism used in [4]. But this also does not translate to WOM, and hence we seek a different helping solution.

3.2   Adding the Helping Mechanism

For the sake of presentation, we start with describing an attempt for building a standard register out of a tree-based max register. This most basic approach only gives a non-blocking SWMR register. Then, we add a helping mechanism to obtain wait-freedom. This still only works for the case of a single-writer-single-reader (SWSR) implementation. We then explain the challenges in extending this to the multi-writer-multi-reader (MWMR) case. We keep the descriptions of the non-blocking and wait-free SWMR registers informal for clarity, and leave the pseudocode and formal proof for the presentation of our full MWMR construction with a more involved helping mechanism for all processes. Due to lack of space, the wait-free SWSR and MWMR registers are deferred to the full version.

A Non-blocking SWSR Write-Once Register. Suppose we have a single writing process p_W, and a single reading process p_R. We first describe an implementation of a SWSR register that is non-blocking but not wait-free, in order to give intuition for our framework.[2] We maintain an infinite unbalanced tree-based unbounded max register Max, and associate an m-bit register value(t) with each leaf t. The values of Max represent timestamps and value(t) represents the value written in the t-th operation, as follows. On its t-th write operation, p_W writes its input value into value(t) and then executes a WriteMax(t) operation on Max. Upon its read operation, p_R performs ReadMax on Max and then reads and returns value(t), where t is the timestamp returned from the ReadMax operation. The problem with this implementation is that operations of p_R are not wait-free because ReadMax may never return if p_W keeps invoking WriteMax operations and thus constantly pushes p_R down the rightmost infinite path of the tree.

Wait-Free SWSR and MWMR Registers. To make this implementation wait-free, we employ the following simple helping mechanism, which consists of an infinite array of bits HelpReq and an infinite array HelpData where each location has a 1-bit flag field and an unbounded register TS. When p_R starts its read operation, it first starts performing a ReadMax operation up to the first time at which it either returns the last value t it saw in previous invocations of read (or 0 if this is its first), or it discovers that a larger value was written. If t has not changed then p_R returns the same value value(t) that it returned for its previous read operation. Otherwise, p_R writes 1 into HelpReq[k], where k is an integer that increases by 1 every time that p_R accesses HelpReq. Then, p_R alternates between taking another step in its ReadMax operation and reading HelpData[k].flag. The operation completes either when p_R reads 1 from HelpData[k].flag, in which case it reads t' from HelpData[k].TS and returns value(t'), or when the ReadMax operation finishes and returns t', in which case p_R reads and returns value(t').

When p_W performs its t-th write operation, it firsts writes its input v to value(t) and then executes a WriteMax(t) operation on Max. Then, it checks whether p_R needs help by reading HelpReq[k], where k is greater by 1 compared with the last index at which p_W accessed HelpReq, and 0 if this is its first access. If HelpReq[k] is 1 then p_W writes t into HelpData[k].TS and 1 into HelpData[k].flag and returns.

The correctness of this SWSR implementation is deferred to the full version, which also contains our full MWMR register implementation and the proof of the following main result.

Theorem 3. There is an algorithm that implements an n-process MWMR register of m bits with unbounded space, where the *amortized* step complexity of a write operation that gets associated with a timestamp t is $O(\log t + m + \log n)$ and the step complexity of a read operation that reads a value associated with a timestamp t is $O(\log t + m + \log n)$.

---

[2] For the purpose of obtaining only a non-blocking SWSR write-once register, it is sufficient to construct an infinite array of m-bit locations to which the writer writes in increasing order and the reader searches for the last written location. However, we use here a max-register based implementation in order to build upon it when constructing our following wait-free SWSR and MWMR implementations.

4   Lower Bounds

For any implementation of a standard register from write-once memory, it is trivial to see that we must use an infinite amount of space in order to support an unbounded number of write operations. This holds even without concurrency, because a finite number of bits can encode a finite number of values, and because we cannot reset a 1 bit to 0. In this section, we provide two additional, non-trivial, lower bounds.

The first is an $\Omega(\log t)$ lower bound on the worst-case cost of a read operation in an execution with t write operations, even when implementing a one-bit register. This is an immediate consequence of Kraft's inequality [27]. Consider a family of executions $\Xi_0, \Xi_1, \ldots \Xi_t$, in which a single writer process alternates between writing a 0 value and a 1 value, with i writes in Ξ_i. In each execution, following these writes is a second reader process p that executes read.

Assume that the reader is deterministic. Let x_i be the sequence of bits read by p in Ξ_i. Observe that the x_i form a prefix-free code (where no codeword is a prefix of another), because the reader chooses to stop deterministically based on the bits it has read so far. Observe further that because write-once bits can never switch from 1 to 0, the x_i can only increase in lexicographic order: in particular this means they are all distinct. Kraft's inequality [27] then gives that $\sum_{i=0}^{t} 2^{-|x_i|} \leq 1$, implying that at least one (and indeed most) of the x_i have length $\Omega(\log t)$.

By treating a randomized reader as a mixture of deterministic readers, the same result applies to the expected worst-case cost of a read. Note that this holds even with an oblivious adversary, because the argument depends only on the information-theoretic properties of the possible sequences of bits observed by a reader, and not on any interaction between the reader and the schedule.

The previous lower bound assumes that the reader performs only one read operation. A reader that performs multiple reads may be able to save work by avoiding re-reading bits that it already knows to be 1. However, we can still show a second lower bound that is a trade-off between the number of bits written by a write operation that writes an m-bit value and the number of bits a read operation op has to look at to get new value, even if it observed the contents of memory immediately before the write.

Suppose that the read operation accesses at most r bits, and the write operation sets at most k bits. As in the previous bound we can consider each possible sequences of bits $x_0, \ldots x_{2^m-1}$ read by the reader, where x_i gives the sequence corresponding to the value i. Each such sequence is distinct, has length at most r and contains at most k ones, so we have $\sum_{i=1}^{k} \binom{r}{i} = 2^m$. For $k = 1$, this bound is reached (up to constants) by the construction of Sect. 2. It is an interesting question whether the trade-off can be realized in general for larger k.

5   Discussion

The present work initiates the study of write-once memory in a concurrent setting. Our results demonstrate that it is in principle possible to implement operations for standard, rewritable shared-memory using write-once memory with low space overhead and polynomial amortized time complexity. Several open questions remain:

1. Is it possible to combine low space overhead with low time overhead?
2. To what extent could a small amount of rewritable shared memory allow more efficiency in use of write-once shared memory?
3. What can one say about stronger write-once primitives, such as (non-resettable) test-and-set bits, either as a target or a base object for implementations?

Acknowledgments. Keren Censor-Hillel is supported in part by the Israel Science Foundation (grant 1696/14). The authors thank the anonymous reviewers for helpful comments and suggestions.

References

1. Aghazadeh, Z., Woelfel, P.: On the time and space complexity of ABA prevention and detection. In: Proceedings of the 2015 ACM Symposium on Principles of Distributed Computing, (PODC), pp. 193–202 (2015)
2. Alistarh, D., Aspnes, J.: Sub-logarithmic test-and-set against a weak adversary. In: Peleg, D. (ed.) DISC 2011. LNCS, vol. 6950, pp. 97–109. Springer, Heidelberg (2011). doi:10.1007/978-3-642-24100-0_7
3. Aspnes, J., Attiya, H., Censor-Hillel, K.: Polylogarithmic concurrent data structures from monotone circuits. J. ACM **59**(1), 2 (2012)
4. Aspnes, J., Censor-Hillel, K.: Atomic snapshots in $O(\log^3 n)$ steps using randomized helping. In: Proceedings of the 27th International Symposium on Distributed Computing, (DISC), pp. 254–268 (2013)
5. Aspnes, J., Ellen, F.: Tight bounds for adopt-commit objects. Theor. Comput. Syst. **55**(3), 451–474 (2014)
6. Attiya, H., Fouren, A.: Adaptive and efficient algorithms for lattice agreement and renaming. SIAM J. Comput. **31**(2), 642–664 (2001)
7. Attiya, H., Welch, J.: Distributed Computing: Fundamentals, Simulations and Advanced Topics, 2nd edn. Wiley, Chichester (2004)
8. Balakrishnan, M., Malkhi, D., Davis, J.D., Prabhakaran, V., Wei, M., Wobber, T.: CORFU: a distributed shared log. ACM Trans. Comput. Syst. **31**(4), 10:1–10:24 (2013)
9. Bhatia, A., Iyengar, A., Siegel, P.H.: Multilevel 2-cell t-write codes. In: IEEE Information Theory Workshop (ITW) (2012)
10. Borowsky, E., Gafni, E., Lynch, N.A., Rajsbaum, S.: The BG distributed simulation algorithm. Distrib. Comput. **14**(3), 127–146 (2001)
11. Burshtein, D., Strugatski, A.: Polar write once memory codes. IEEE Trans. Inf. Theory **59**(8), 5088–5101 (2013)
12. Cassuto, Y., Yaakobi, E.: Short (Q)-ary fixed-rate WOM codes for guaranteed rewrites and with hot/cold write differentiation. IEEE Trans. Inf. Theory **60**(7), 3942–3958 (2014)
13. Cohen, G.D., Godlewski, P., Merkx, F.: Linear binary code for write-once memories. IEEE Trans. Inf. Theory **32**(5), 697–700 (1986)

14. Gad, E.E., Wentao, H., Li, Y., Bruck, J.: Rewriting flash memories by message passing. In: IEEE International Symposium on Information Theory (ISIT) (2015)
15. Fiat, A., Shamir, A.: Generalized 'write-once' memories. IEEE Trans. Inf. Theory **30**(3), 470–479 (1984)
16. Fu, F.-W., Han Vinck, A.J.: On the capacity of generalized write-once memory with state transitions described by an arbitrary directed acyclic graph. IEEE Trans. Inf. Theory **45**(1), 308–313 (1999)
17. Giakkoupis, G., Woelfel, P.: On the time and space complexity of randomized test-and-set. In: Proceedings of the ACM Symposium on Principles of Distributed Computing, (PODC), pp. 19–28 (2012)
18. Godlewski, P.: WOM-codes construits à partir des codes de Hamming. Discrete Math. **65**(3), 237–243 (1987)
19. Heegard, C.: On the capacity of permanent memory. IEEE Trans. Inf. Theory **31**(1), 34–41 (1985)
20. Helmi, M., Higham, L., Woelfel, P.: Strongly linearizable implementations: possibilities and impossibilities. In: Proceedings of the ACM Symposium on Principles of Distributed Computing, (PODC), pp. 385–394 (2012)
21. Herlihy, M., Shavit, N.: The Art of Multiprocessor Programming. Morgan Kaufmann, San Francisco (2008)
22. Herlihy, M., Wing, J.M.: Linearizability: a correctness condition for concurrent objects. ACM Trans. Program. Lang. Syst. **12**(3), 463–492 (1990)
23. Inoue, M., Masuzawa, T., Chen, W., Tokura, N.: Linear-time snapshot using multi-writer multi-reader registers. In: Tel, G., Vitányi, P. (eds.) WDAG 1994. LNCS, vol. 857, pp. 130–140. Springer, Heidelberg (1994). doi:10.1007/BFb0020429
24. Israeli, A., Shaham, A.: Optimal multi-writer multi-reader atomic register. In: Proceedings of the Eleventh Annual ACM Symposium on Principles of Distributed Computing, PODC 1992, pp. 71–82, New York, NY, USA. ACM (1992)
25. Israeli, A., Shaham, A.: Time and space optimal implementations of atomic multi-writer register. Inf. Comput. **200**(1), 62–106 (2005)
26. Jacobvitz, A.N., Calderbank, R., Sorin, D.J.: Coset coding to extend the lifetime of memory. In: IEEE 19th International Symposium on High Performance Computer Architecture (HPCA) (2013)
27. Kraft, L.G.: A device for quantizing, grouping, and coding amplitude-modulated pulses. M.S. thesis, Department of Electrical Engineering, Massachusetts Institute of Technology (1949)
28. Lamport, L.: On interprocess communication part I: basic formalism. Distrib. Comput. **1**(2), 77–85 (1986)
29. Lamport, L.: On interprocess communication part II: algorithms. Distrib. Comput. **1**(2), 86–101 (1986)
30. Li, J., Mohanram, K.: Write-once-memory-code phase change memory. In: Design, Automation and Test in Europe Conference and Exhibition (DATE) (2014)
31. Margaglia, F., Brinkmann, A.: Improving MLC flash performance and endurance with extended P/E cycles. In: IEEE 31st Symposium on Mass Storage Systems and Technologies (MSST) (2015)
32. Plotkin, S.A.: Sticky bits and universality of consensus. In: Proceedings of the Eighth Annual ACM Symposium on Principles of Distributed Computing, PODC 1989, pp. 159–175, New York, NY, USA. ACM (1989)
33. Rivest, R.R., Shamir, A.: How to reuse a "write-once" memory. Inf. Control **55**, 1–19 (1982)
34. Shpilka, A.: New constructions of WOM codes using the Wozencraft ensemble. IEEE Trans. Inf. Theory **59**(7), 4520–4529 (2013)

35. Shpilka, A.: Capacity achieving multiwrite WOM codes. IEEE Trans. Inf. Theory **60**(3), 1481–1487 (2014)
36. Wolf, J.K., Wyner, A.D., Ziv, J., Korner, J.: Coding for a write-once memory. AT&T Bell Laboratories Tech. J. **63**(6), 1089–1112 (1984)
37. Yunnan, W.: Low complexity codes for writing a write-once memory twice. In: IEEE International Symposium on Information Theory (ISIT) (2010)
38. Wu, Y., Jiang, A.: Position modulation code for rewriting write-once memories. IEEE Trans. Inf. Theory **57**(6), 3692–3697 (2011)
39. Yaakobi, E., Kayser, S., Siegel, P.H., Vardy, A., Wolf, J.K.: Codes for write-once memories. IEEE Trans. Inf. Theory **58**(9), 5985–5999 (2012)
40. Yaakobi, E., Shpilka, A.: High sum-rate three-write and non-binary WOM codes. In: IEEE International Symposium on Information Theory (ISIT) (2012)
41. Yadgar, G., Yaakobi, E., Schuster, A.: Write once, get 50% free: saving SSD erase costs using WOM codes. In: 13th USENIX Conference on File and Storage Technologies (FAST) (2015)
42. Zhang, X., Jang, L., Zhang, Y., Zhang, C., Yang, J.: WoM-SET: low power proactive-SET-based PCM write using WoM code. In: IEEE International Symposium on Low Power Electronics and Design (ISLPED) (2013)

In the Search for Optimal Concurrency

Vincent Gramoli[1], Petr Kuznetsov[2], and Srivatsan Ravi[3(✉)]

[1] Data61-CSIRO and University of Sydney, Sydney, Australia
[2] Télécom ParisTech, Paris, France
petr.kuznetsov@telecom-paristech.fr
[3] Purdue University, West Lafayett, USA
srivatsanravi@purdue.edu

Abstract. It is common practice to use the epithet "highly concurrent" referring to data structures that are supposed to perform well in concurrent environments. But how do we measure the concurrency of a data structure in the first place? In this paper, we propose a way to do this, which allowed us to formalize the notion of a *concurrency-optimal* implementation.

The concurrency of a program is defined here as the program's ability to accept concurrent schedules, *i.e.*, interleavings of steps of its sequential implementation. To make the definition sound, we introduce a novel correctness criterion, *LS-linearizability*, that, in addition to classical linearizability, requires the interleavings of memory accesses to be *locally* indistinguishable from sequential executions. An implementation is then *concurrency-optimal* if it accepts *all* LS-linearizable schedules. We explore the concurrency properties of *search* data structures which can be represented in the form of directed acyclic graphs exporting insert, delete and search operations. We prove, for the first time, that *pessimistic* (*e.g.*, based on conservative locking) and *optimistic serializable* (*e.g.*, based on serializable transactional memory) implementations of search data-structures are incomparable in terms of concurrency. Thus, neither of these two implementation classes is concurrency-optimal, hence raising the question of the existence of concurrency-optimal programs.

Keywords: Concurrency · Search data structures · Lower bounds

1 Introduction

In the concurrency literature, it is not unusual to meet expressions like "highly concurrent data structures", used as a positive characteristics of their performance. Leaving aside the relation between performance and concurrency, the first question we should answer is what is the concurrency of a data structure in the first place. How do we measure it?

At a high level, concurrency is the ability to serve multiple requests in parallel. A data structure designed for the conventional sequential settings, when used *as is* in a concurrent environment, while being intuitively very concurrent, may face different kinds of inconsistencies caused by races on the shared data.

© Springer International Publishing AG 2016
J. Suomela (Ed.): SIROCCO 2016, LNCS 9988, pp. 143–158, 2016.
DOI: 10.1007/978-3-319-48314-6_10

To avoid these races, a variety of synchronization techniques have been developed [9]. Conventional pessimistic synchronization protects shared data with locks before reading or modifying them. Optimistic synchronization, achieved using transactional memory (TM) or conditional instructions such as CAS or LL/SC, optimistically executes memory accesses with a risk of aborting them in the future. A programmer typically uses these synchronization techniques to "wrap" fragments of a sequential implementation of the desired data structure, in order to preserve a correctness criterion. Therefore, intuitively, concurrency of a data structure is its ability to allow multiple processes to concurrently *make progress*, *i.e.*, to advance in their sequential operations on the shared data.

It is however difficult to tell in advance which of the techniques will provide more concurrency, *i.e.*, which one would allow the resulting programs to process more executions of concurrent operations without data conflicts. Implementations based on TMs [20,28], which execute concurrent accesses speculatively, may seem more concurrent than lock-based counterparts whose concurrent accesses are blocking. But TMs conventionally impose *serializability* [26] or even stronger properties [15] on operations encapsulated within transactions. This may prohibit certain concurrent scenarios allowed by a large class of dynamic data structures [10].

In this paper, we reason formally about the "amount of concurrency" one can obtain by turning a sequential program into a concurrent one. To enable fair comparison of different synchronization techniques, we (1) define what it means for a concurrent program to be correct regardless of the type of synchronization it uses and (2) define a metric of concurrency. These definitions allow us to compare concurrency properties offered by serializable optimistic and pessimistic synchronization techniques, whose popular examples are, respectively, transactions and conservative locking.

Correctness. Our novel consistency criterion, called *locally-serializable linearizability*, is an intersection of *linearizability* and a new *local serializability* criterion.

Suppose that we want to design a concurrent implementation of a data type T (*e.g.*, integer set), given its sequential implementation S (*e.g.*, based on a sorted linked list). A concurrent implementation of T is *locally serializable* with respect to S if it ensures that the local execution of *reads* and *writes* of each operation is, in precise sense, equivalent to *some* execution of S. This condition is weaker than serializability since it does not require the existence of a *single* sequential execution that is consistent with all local executions. It is however sufficient to guarantee that executions do not observe an inconsistent transient state that could lead to fatal arithmetic errors, *e.g.*, division-by-zero.

In addition, for the implementation of T to "make sense" globally, every concurrent execution should be *linearizable* [3,23]: the invocation and responses of high-level operations observed in the execution should constitute a correct sequential history of T. The combination of local serializability and linearizability gives a correctness criterion that we call *LS-linearizability*, where LS stands for "locally serializable". We show that LS-linearizability, just like linearizability, is *compositional* [21,23]: a composition of LS-linearizable implementations is also LS-linearizable.

Concurrency metric. We measure the amount of concurrency provided by an LS-linearizable implementation as the set of schedules it accepts. To this end, we define a concurrency metric inspired by the analysis of parallelism in database concurrency control [18,32] and transactional memory [11]. More specifically, we assume an external scheduler that defines which processes execute which steps of the corresponding sequential program in a dynamic and unpredictable fashion. This allows us to define concurrency provided by an implementation as the set of *schedules* (interleavings of reads and writes of concurrent sequential operations) it *accepts* (is able to effectively process).

Our concurrency metric is platform-independent and allows for measuring relative concurrency of LS-linearizable implementations using arbitrary synchronization techniques. The combination of our correctness and concurrency definitions provides a framework to compare the concurrency one can get by choosing a particular synchronization technique for a specific data type.

Measuring concurrency: pessimism vs. serializable optimism. We explore the concurrency properties of a large class of *search* concurrent data structures. Search data structures maintain data in the form of a rooted directed acyclic graph (DAG), where each node is a ⟨*key, value*⟩ pair, and export operations *insert*(*key, value*), *delete*(*key*), and *find*(*key*) with the natural sequential semantics. The class includes many popular data structures, such as linked lists, skiplists, and search trees, implementing various abstractions like sets, multisets and dictionaries.

In this paper, we compare the concurrency properties of two classes of search-structure implementations: *pessimistic* and *serializable optimistic*. Pessimistic implementations capture what can be achieved using classic conservative locks like mutexes, spinlocks, reader-writer locks. In contrast, optimistic implementations, however proceed speculatively and may roll back in the case of conflicts. Additionally, *serializable* optimistic techniques, *e.g.*, relying on conventional TMs, like TinySTM [8] or NOrec [5] allow for transforming any sequential implementation of a data type to a LS-linearizable concurrent one.

Main contributions. The main result of this paper is that synchronization techniques based on pessimism and serializable optimism, are not concurrency-optimal: we show that no one of their respective set of accepted concurrent schedules include the other.

On the one hand, we prove that there exist simple schedules that are not accepted by *any* pessimistic implementation, but accepted by a serializable optimistic implementation. Our proof technique, which is interesting in its own right, is based on the following intuitions: a pessimistic implementation has to proceed irrevocably and over-conservatively reject a potentially acceptable schedule, simply because it *may* result in a data conflict leading the data structure to an inconsistent state. However, an optimistic implementation of a search data structure may (partially or completely) restart an operation depending on the current schedule. This way even schedules that potentially lead to conflicts may be optimistically accepted.

On the other hand, we show that pessimistic implementations can be designed to exploit the semantics of the data type. In particular, they can allow operations updating disjoint sets of data items to proceed independently and preserving linearizability of the resulting history, even though the execution is not serializable. In such scenarios, pessimistic implementations carefully adjusted to the data types we implement can supersede the "semantic-oblivious" optimistic serializable implementations. Thus, neither pessimistic nor serializable optimistic implementations are concurrency-optimal.

Our comparative analysis of concurrency properties of pessimistic and serializable optimistic implementation suggests that combining the advantages of pessimism, namely its semantics awareness, and the advantages of optimism, namely its ability to restart operations in case of conflicts, enables implementations that are strictly better-suited for exploiting concurrency than any of these two techniques taken individually. To the best of our knowledge, this is the first formal analysis of the relative abilities of different synchronization techniques to exploit concurrency in dynamic data structures and lays the foundation for designing concurrent data structures that are concurrency-optimal.

Roadmap. We define the class of concurrent implementations we consider in Sect. 2. In Sect. 3, we define the correctness criterion and our concurrency metric. Section 4 defines the class of data structures for which our concurrency lower bounds apply. In Sect. 5, we analyse the concurrency provided by pessimistic and serializabile optimistic synchronization techniques to search data structures. We discuss the related work in Sect. 6 and conclude in Sect. 7. Full proofs are delegated to the accompanying tech report [12].

2   Preliminaries

Sequential types and implementations. An *type* τ is a tuple $(\Phi, \Gamma, Q, q_0, \delta)$ where Φ is a set of operations, Γ is a set of responses, Q is a set of states, $q_0 \in Q$ is an initial state and $\delta \subseteq Q \times \Phi \times Q \times \Gamma$ is a *sequential specification* that determines, for each state and each operation, the set of possible resulting states and produced responses [2].

Any type $\tau = (\Phi, \Gamma, Q, q_0, \delta)$ is associated with a *sequential implementation IS*. The implementation encodes states in Q using a collection of elements $X_1, X_2, \ldots$ and, for each operation of τ, specifies a sequential *read-write* algorithm. Therefore, in the implementation *IS*, an operation performs a sequence of *reads* and *writes* on $X_1, X_2, \ldots$ and returns a response $r \in \Gamma$. The implementation guarantees that, when executed sequentially, starting from the state of $X_1, X_2, \ldots$ encoding q_0, the operations eventually return responses satisfying δ.

Concurrent implementations. We consider an asynchronous shared-memory system in which a set of processes communicate by applying *primitives* on shared *base objects* [19].

We tackle the problem of turning the sequential algorithm *IS* of type τ into a *concurrent* one, shared by n processes $p_1, \ldots, p_n$ ($n \in \mathbb{N}$). The idea is that the concurrent algorithm essentially follows *IS*, but to ensure correct operation under concurrency, it replaces read and write operations on $X_1, X_2, \ldots$ in operations of *IS* with their base-object implementations.

Throughout this paper, we use the term *operation* to refer to high-level operations of the type. Reads and writes implemented by a concurrent algorithm are referred simply as *reads* and *writes*. Operations on base objects are referred to as *primitives*.

We also consider concurrent implementation that execute portions of sequential code *speculatively*, and restart their operations when conflicts are encountered. To account for such implementations, we assume that an implemented read or write may *abort* by returning a special response $\bot$. In this case, we say that the corresponding (high-level) operation is *aborted*.

Therefore, our model applies to all concurrent algorithms in which a high-level operation can be seen as a sequence of reads and writes on elements $X_1, X_2, \ldots$ (representing the state of the data structure), with the option of aborting the current operation and restarting it after. Many existing concurrent data structure implementations comply with this model as we illustrate below.

Executions and histories. An *execution* of a concurrent implementation is a sequence of invocations and responses of high-level operations of type τ, invocations and responses of read and write operations, and invocations and responses of base-object primitives. We assume that executions are *well-formed*: no process invokes a new read or write, or high-level operation before the previous read or write, or a high-level operation, resp., returns, or takes steps outside its operation's interval.

Let $\alpha|p_i$ denote the subsequence of an execution α restricted to the events of process p_i. Executions α and α' are *equivalent* if for every process p_i, $\alpha|p_i = \alpha'|p_i$. An operation π *precedes* another operation π' in an execution α, denoted $\pi \rightarrow_\alpha \pi'$, if the response of π occurs before the invocation of π'. Two operations are *concurrent* if neither precedes the other. An execution is *sequential* if it has no concurrent operations. A sequential execution α is *legal* if for every object X, every read of X in α returns the latest written value of X. An operation is *complete* in α if the invocation event is followed by a *matching* (non-$\bot$) response or aborted; otherwise, it is *incomplete* in α. Execution α is *complete* if every operation is complete in α.

The *history exported by an execution* α is the subsequence of α reduced to the invocations and responses of operations, reads and writes, except for the reads and writes that return $\bot$ (the abort response).

High-level histories and linearizability. A *high-level history* $\tilde{H}$ of an execution α is the subsequence of α consisting of all invocations and responses of *non-aborted* operations. A complete high-level history $\tilde{H}$ is *linearizable* with respect to an object type τ if there exists a sequential high-level history S equivalent to H such that (1) $\rightarrow_{\tilde{H}} \subseteq \rightarrow_S$ and (2) S is consistent with the sequential specification of type τ. Now a high-level history $\tilde{H}$ is linearizable if it can be *completed* (by adding matching responses to a subset of incomplete operations in $\tilde{H}$ and removing the rest) to a linearizable high-level history [3, 23].

Optimistic and pessimistic implementations. Note that in our model an implementations may, under certain conditions, abort an operation: some read or write return $\perp$, in which case the corresponding operation also returns $\perp$. Popular classes of such *optimistic* implementations are those based on "lazy synchronization" [17,21] (with the ability of returning $\perp$ and re-invoking an operation) or transactional memory (*TM*) [5,28].

In the subclass of *pessimistic* implementations, no execution includes operations that return $\perp$. Pessimistic implementations are typically *lock-based* or based on pessimistic TMs [1]. A lock provides exclusive (resp., shared) access to an element X through synchronization primitives lock(X) (resp., lock-shared(X)), and unlock(X) (resp., unlock-shared(X)). A process *releases* the lock it holds by invoking unlock(X) or unlock-shared(X). When lock(X) invoked by a process p_i returns, we say that p_i *holds a lock on* X (until p_i returns from the subsequent lock(X)). When lock-shared(X) invoked by p_i returns, we say that p_i *holds a shared lock on* X (until p_i returns from the subsequent lock-shared(X)). At any moment, at most one process may hold a lock on an element X. Note that two processes can hold a shared lock on X at a time. We assume that locks are *starvation-free*: if no process holds a lock on X forever, then every lock(X) eventually returns. Given a sequential implementation of a data type, a corresponding lock-based concurrent one is derived by inserting the synchronization primitives (lock and unlock) to protect read and write accesses to the shared data.

3   Correctness and Concurrency Metric

In this section, we define the correctness criterion of *locally serializable linearizability (LS-linearizability)* and introduce the framework for comparing the relative abilities of different synchronization technique in exploiting concurrency.

3.1   Locally Serializable Linearizability

Let H be a history and let π be a high-level operation in H. Then $H|\pi$ denotes the subsequence of H consisting of the events of π, except for the last aborted read or write, if any. Let *IS* be a sequential implementation of an object of type τ and Σ_{IS}, the set of histories of *IS*.

Definition 1 (LS-linearizability). *A history H is* locally serializable with respect to *IS if for every high-level operation π in H, there exists $S \in \Sigma_{IS}$ such that $H|\pi = S|\pi$.*

A history H is LS-linearizable with respect to (IS, τ) *(we also write H is (IS, τ)-LSL) if: (1) H is locally serializable with respect to IS and (2) the corresponding high-level history $\tilde{H}$ is linearizable with respect to τ.*

Observe that local serializability stipulates that the execution is seen as a sequential one by every operation. Two different operations (even when invoked by the same process) are not required to witness mutually consistent sequential executions.

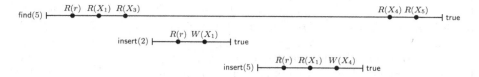

Fig. 1. A concurrency scenario for a set, initially $\{1, 3, 4\}$, where value i is stored at node X_i: insert(2) and insert(5) can proceed concurrently with find(5). The history is LS-linearizable but not serializable; yet accepted by HOH-find. (Not all read-write on nodes is presented here.)

A concurrent implementation I is *LS-linearizable with respect to* (IS, τ) (we also write I is (IS, τ)-LSL) if every history exported by I is (IS, τ)-LSL. Throughout this paper, when we refer to a concurrent implementation of (IS, τ), we assume that it is LS-linearizable with respect to (IS, τ).

We show in [12] that just as linearizability, LS-linearizability is *compositional* [21,23]: a composition of LSL implementations is also LSL. However, LS-linearizability is not non-blocking [21,23]: local serializability may prevent an operation in a finite LS-linearizable history from having a completion, e.g., because, it might read an inconsistent system state caused by a concurrent incomplete operation.

LS-linearizability and other consistency criteria. LS-linearizability is a two-level consistency criterion which makes it suitable to compare concurrent implementations of a sequential data structure, regardless of synchronization techniques they use. It is quite distinct from related criteria designed for database and software transactions, such as serializability [26,31] and multilevel serializability [30,31].

For example, serializability [26] prevents sequences of reads and writes from conflicting in a cyclic way, establishing a global order of transactions. Reasoning only at the level of reads and writes may be overly conservative: higher-level operations may commute even if their reads and writes conflict [29]. Consider an execution of a concurrent *list-based set* depicted in Fig. 1. We assume here that the set initial state is $\{1, 3, 4\}$. Operation find(5) is concurrent, first with operation insert(2) and then with operation insert(5). The history is not serializable: insert(5) sees the effect of insert(2) because $R(X_1)$ by insert(5) returns the value of X_1 that is updated by insert(2) and thus should be serialized after it. Operation find(5) misses element 2 in the linked list and must read the value of X_4 that is updated by insert(5) to perform the read of X_5, *i.e.*, the element created by insert(5). This history is, however, LSL since each of the three local histories is consistent with some sequential history of LL.

Multilevel serializability [30,31] was proposed to reason in terms of multiple semantic levels in the same execution. LS-linearizability, being defined for two levels only, does not require a global serialization of low-level operations as 2-level serializability does. LS-linearizability simply requires each process to observe a local serialization, which can be different from one process to another. Also, to make it more suitable for concurrency analysis of a concrete data structure, instead of semantic-based commutativity [29], we use the sequential specification of the high-level behavior of the object [23].

Linearizability [3, 23] only accounts for high-level behavior of a data structure, so it does not imply LS-linearizability. For example, Herlihy's universal construction [19] provides a linearizable implementation for any given object type, but does not guarantee that each execution locally appears sequential with respect to any sequential implementation of the type. Local serializability, by itself, does not require any synchronization between processes and can be trivially implemented without communication among the processes. Therefore, the two parts of LS-linearizability indeed complement each other. ·

3.2   Concurrency Metric

To characterize the ability of a concurrent implementation to process arbitrary interleavings of sequential code, we introduce the notion of a *schedule*. Intuitively, a schedule describes the order in which complete high-level operations, and sequential reads and writes are invoked by the user. More precisely, a schedule is an equivalence class of complete histories that agree on the *order* of invocation and response events of reads, writes and high-level operations, but not necessarily on the responses of read operations or of high-level operations. Thus, a schedule can be treated as a history, where responses of read and high-level operations are not specified.

We say that an implementation I *accepts* a schedule σ if it exports a history H such that $complete(H)$ exhibits the order of σ, where $complete(H)$ is the subsequence of H that consists of the events of the complete operations that returned a matching response. We then say that the execution (or history) *exports* σ. A schedule σ is (IS, τ)-LSL if there exists an (IS, τ)-LSL history exporting σ.

An (IS, τ)-LSL implementation is therefore *concurrency-optimal* if it accepts all (IS, τ)-LSL schedules.

4   Search Data Structures

In this section, we introduce a class $\mathcal{D}$ of *dictionary-search* data structures (or simply *search structures*), inspired by the study of *dynamic databases* undertaken by Chaudhri and Hadzilacos [4].

Data representation. At a high level, a search structure is a dictionary that maintains data in a directed acyclic graph (DAG) with a designated *root* node (or element). The vertices (or nodes) of the graph are key-value pairs and edges specify the *traversal function, i.e*, paths that should be taken by the dictionary operations in order to find the nodes that are going to determine the effect of the operation. Keys are natural numbers, values are taken from a set V and the outgoing edges of each node are locally labelled. By a light abuse of notation we say that G *find* both nodes and edges. Key values of nodes in a DAG G are related by a partial order $\prec_G$ that additionally defines a property $\mathbb{P}_G$ specifying if there is an outgoing edge from node with key k to a node with key k' (we say that G *respects* $\mathbb{P}_G$).

If G contains a node a with key k, the k-*relevant* set of G, denoted $V_k(G)$, is a plus all nodes b, such that G contains (b, a) or (a, b). If G contains no nodes with key k, $V_k(G)$ consists of all nodes a of G with the smallest $k \prec_G k'$ plus all nodes b, such that (b, a) is in G. The k-*relevant graph of* G, denoted $R_k(G)$, is the subgraph of G that consists of all paths from the root to the nodes in $V_k(G)$.

Sequential specification. Every data structure in $\mathcal{D}$ exports a sequential specification with the following operations: (i) *insert*(k, v) checks whether a node with key k is already present and, if so, returns false, otherwise it creates a node with key k and value v, *links* it to the graph (making it reachable from the root) and returns true; (ii) *delete*(k) checks whether a node with key k is already present and, if so, *unlinks* the node from the graph (making it unreachable from the root) and returns true, otherwise it returns false; (iii) *find*(k) returns the pointer to the node with key k or false if no such node is found.

Traversals. For each operation $op \in \{insert(k, v), delete(k), find(k)\}_{k \in \mathbb{N}, v \in V}$, each search structure is parameterized by a (possibly randomized) *traverse function* τ_{op}. Given the *last visited* node a and the DAG of already visited nodes G_{op}, the traverse function τ_{op} returns a new node b to be *visited*, i.e., accessed to get its *key* and the list of descendants, or $\emptyset$ to indicate that the search is complete.

Find, insert and delete operations. Intuitively, the traverse function is used by the operation op to explore the search structure and, when the function returns $\emptyset$, the sub-DAG G_{op} explored so far contains enough information for operation op to complete.

If $op = find(k)$, G_{op} either contains a node with key k or ensures that the whole graph does not contain k. As we discuss below, in *sorted* search structures, such as sorted linked-lists or skiplists, we can stop as soon as all outgoing edges in G_{op} belong to nodes with keys $k' \geq k$. Indeed, the remaining nodes can only contain keys greater than k, so G_{op} contains enough information for op to complete.

An operation $op = insert(k, v)$, is characterized by an *insert function* $\mu_{(k,v)}$ that, given a DAG G and a new node $\langle k, v \rangle \notin G$, returns the set of edges from nodes of G to $\langle k, v \rangle$ and from $\langle k, v \rangle$ to nodes of G so that the resulting graph is a DAG containing $\langle k, v \rangle$ and respects $\mathbb{P}_G$.

An operation $op = delete(k)$, is characterized by a *delete function* ν_k that, given a DAG G, gives the set of edges to be removed and a set of edges to be added in G so that the resulting graph is a DAG that respects $\mathbb{P}_G$.

Sequential implementations. We make the following natural assumptions on the sequential implementation of a search structure: (i) *Traverse-update*: Every operation op starts with the read-only *traverse* phase followed with a write-only *update* phase. The traverse phase of an operation op with parameter k completes at the latest when for the visited nodes G_{op} contains the k-relevant graph. The update phase of a *find(k)* operation is empty; (ii) *Proper traversals and updates*: For all DAGs G_{op} and nodes $a \in G_{op}$, the traverse function $\tau_{op}(a, G_{op})$ returns b such that $(a, b) \in G$. The update phase of an *insert(k)* or *delete(k)* operation modifies outgoing edges of k-relevant nodes; (iii) *Non-triviality*: There exist a key k and a state G such that (1) G contains no node with key k, (2) If G' is the state resulting after applying insert(k, v) to G, then there is exactly one edge (a, b) in G' such that b has key k, and (3) the shortest path in G' from the root to a is of length at least 2.

The non-triviality property says that in some cases the read-phase may detect the presence of a given key only at the last step of a traverse-phase. Moreover, it excludes the pathological DAGs in which all the nodes are always reachable in one hop from the root. Moreover, the traverse-update property and the fact that keys are natural numbers implies that every traverse phase eventually terminates. Indeed, there can be only finitely many vertices pointing to a node with a given key, thus, eventually a traverse operation explores enough nodes to be sure that no node with a given key can be found.

Examples of search data structures. A *sorted linked list* maintains a single path, starting at the *root* sentinel node and ending at a *tail* sentinel node, and any traversal with parameter k simply follows the path until a node with key $k' \geq k$ is located. The traverse function for all operations follows the only path possible in the graph until the two relevant nodes are located.

A *skiplist* [27] of n nodes is organized as a series of $O(\log n)$ sorted linked lists, each specifying shortcuts of certain length. The bottom-level list contains all the nodes, each of the higher-level lists contains a sublist of the lower-level list. A traversal starts with the top-level list having the longest "hops" and goes to lower lists with smaller hops as the node with smallest key $k' \geq k$ get closer.

A *binary search tree* represents data items in the form of a binary tree. Every node in the tree stores a key-value pair, and the left descendant of a non-leaf node with key k roots a subtree storing all nodes with keys less than k, while the right descendant roots a subtree storing all nodes with keys greater than k. Note that, for simplicity, we do not consider *rebalancing* operations used by balanced trees for maintaining the desired bounds on the traverse complexity. Though crucial in practice, the rebalancing operations are not important for our comparative analysis of concurrency properties of synchronization techniques.

Non-serializable concurrency. There is a straightforward LSL implementation of any data structure in $\mathcal{D}$ in which updates (*inserts* and *deletes*) acquire a lock on the root node and are thus sequential. Moreover, they take exclusive locks on the set of nodes they are about to modify (k-relevant sets for operations with parameter k).

A *find* operation uses *hand-over-hand shared* locking [29]: at each moment of time, the operation holds shared locks on all outgoing edges for the currently visited node a. To visit a new node b (recall that b must be a descendant of a), it acquires shared locks on the new node's descendants and then releases the shared lock on a. Note that just before a $find(k)$ operation returns the result, it holds shared locks on the k-relevant set.

This way updates always take place sequentially, in the order of their acquisitions of the root lock. A $find(k)$ operation is linearized at any point of its execution when it holds shared locks on the k-relevant set. Concurrent operations that do not contend on the same locks can be arbitrarily ordered in a linearization.

The fact that the operations acquire (starvation-free) locks in the order they traverse the directed acyclic graph implies:

Theorem 1. HOH-find *is a starvation-free LSL implementation of a search structure.*

As we show in Sect. 5, the implementation is however not (safe-strict) serializable.

5   Pessimism vs. Serializable Optimism

In this section, we show that, with respect to search structures, pessimistic locking and optimistic synchronization providing safe-strict serializability are *incomparable*, once we focus on LS-linearizable implementations.

5.1   Classes $\mathcal{P}$ and $\mathcal{SM}$

A *synchronization technique* is a set of concurrent implementations. We define below a specific optimistic synchronization technique and then a specific pessimistic one.

$\mathcal{SM}$: *serializable optimistic.* Let α denote the execution of a concurrent implementation and $ops(\alpha)$, the set of operations each of which performs at least one event in α. Let α^k denote the prefix of α up to the last event of operation π_k. Let $Cseq(\alpha)$ denote the set of subsequences of α that consist of all the events of operations that are complete in α. We say that α is *strictly serializable* if there exists a legal sequential execution α' equivalent to a sequence in $\sigma \in Cseq(\alpha)$ such that $\rightarrow_\sigma \subseteq \rightarrow_{\alpha'}$.

This paper focuses on optimistic implementations that are strictly serializable and whose operations (even aborted or incomplete) observes correct (serial) behavior. More precisely, an execution α is *safe-strict serializable* if (1) α is strictly serializable, and (2) for each operation π_k, there exists a legal sequential execution $\alpha' = \pi_0 \cdots \pi_i \cdot \pi_k$ and $\sigma \in Cseq(\alpha^k)$ such that $\{\pi_0, \cdots, \pi_i\} \subseteq ops(\sigma)$ and $\forall \pi_m \in ops(\alpha') : \alpha'|m = \alpha^k|m$.

Safe-strict serializability captures nicely both local serializability and linearizability. If we transform a sequential implementation IS of a type τ into a *safe-strict serializable* concurrent one, we obtain an LSL implementation of (IS, τ). Thus, the following lemma is immediate.

Lemma 1. *Let I be a safe-strict serializable implementation of (IS, τ). Then, I is LS-linearizable with respect to (IS, τ).*

Fig. 2. (a) a history of integer set (implemented as linked list or binary search tree) exporting schedule σ, with initial state $\{1,2,3\}$ (r denotes the root node); (b) a history exporting a problematic schedule σ', with initial state $\{3\}$, which should be accepted by any $I \in \mathcal{P}$ if it accepts σ

Indeed, we make sure that completed operations witness the same execution of IS, and every operation that returned $\bot$ is consistent with some execution of IS based on previously completed operations. Formally, $\mathcal{SM}$ denotes the set of optimistic, safe-strict serializable LSL implementations.

$\mathcal{P}$: *deadlock-free pessimistic.* Assuming that no process stops taking steps of its algorithm in the middle of a high-level operation, at least one of the concurrent operations return a matching response [22]. Note that $\mathcal{P}$ includes implementations that are not necessarily safe-strict serializable.

5.2   Suboptimality of Pessimistic Implementations

We show now that for any search structure, there exists a schedule that is rejected by *any* pessimistic implementation, but accepted by certain optimistic strictly serializable ones. To prove this claim, we derive a safe-strict serializable schedule that cannot be accepted by any implementation in $\mathcal{P}$ using the *non-triviality* property of search structures. It turns out that we can schedule the traverse phases of two *insert(k)* operations in parallel until they are about to check if a node with key k is in the set or not. If it is, both operations may safely return false (schedule σ). However, if the node is not in the set, in a pessimistic implementation, both operations would have to modify outgoing edges of the same node a and, if we want to provide local serializability, both return true, violating linearizability (schedule σ').

In contrast, an optimistic implementation may simply abort one of the two operations in case of such a conflict, by accepting the (correct) schedule σ and rejecting the (incorrect) schedule σ'.

Proof Intuition. We first provide an intuition of our results in the context of the *integer set* implemented as a *sorted linked list* or *binary search tree*. The set type is a special case of the dictionary which stores a set of integer values, initially empty, and exports operations *insert(v)*, *remove(v)*, *find(v)*; $v \in \mathbb{Z}$. The update operations, *insert(v)* and *remove(v)*, return a boolean response, true if and only if v is absent (for *insert(v)*) or present (for *remove(v)*) in the set. After *insert(v)* is complete, v is present in the set, and after *remove(v)* is complete, v is absent in the set. The *find(v)* operation returns a boolean, true if and only if v is present in the set.

An example of schedules σ and σ' of the set is given in Fig. 2. We show that the schedule σ depicted in Fig. 2(a) is not accepted by any implementation in $\mathcal{P}$. Suppose the contrary and let σ be exported by an execution α. Here α starts with three sequential insert operations with parameters 1, 2, and 3. The resulting "state" of the set is $\{1, 2, 3\}$, where value $i \in \{1, 2, 3\}$ is stored in node X_i. Suppose, by contradiction, that some $I \in \mathcal{P}$ accepts σ. We show that I then accepts the schedule σ' depicted in Fig. 2(b), which starts with a sequential execution of insert(3) storing value 3 in node X_1. We can further extend σ' with a complete find(1) (by deadlock-freedom of $\mathcal{P}$) that will return false (the node inserted to the list by insert(1) is lost)—a contradiction since I is linearizable with respect to set.

Theorem 2. *Any abstraction in $\mathcal{D}$ has a strictly serializable schedule that is not accepted by any implementation in $\mathcal{P}$, but accepted by an implementation in $\mathcal{SM}$.*

5.3   Suboptimality of Serializable Optimism

We show below that for any search structure, there exists a schedule that is rejected by *any* serializable implementation but accepted by a certain pessimistic one (*HOH-find*, to be concrete).

Proof Intuition. We first illustrate the proof in the context of the integer set. Consider a schedule σ_0 of a concurrent set implementation depicted in Fig. 1. We assume here that the set initial state is $\{1, 3, 4\}$. Operation find(5) is concurrent, first with operation insert(2) and then with operation insert(5). The history is not serializable: insert(5) sees the effect of insert(2) because $R(X_1)$ by insert(5) returns the value of X_1 that is updated by insert(2) and thus should be serialized after it. But find(5) misses node with value 2 in the set, but must read the value of X_4 that is updated by insert(5) to perform the read of X_5, *i.e.*, the node created by insert(5). Thus, σ_0 is not (safe-strict) serializable. This history though is LSL since each of the three local histories is consistent with some sequential history of the integer set. However, there exists an execution of our HOH-find implementation that exports σ_0 since there is no read-write conflict on any two consecutive nodes accessed.

To extend the above idea to any search structure, we use the *non-triviality* property of data structures in $\mathcal{D}$. There exist a state G' in which there is exactly one edge (a, b) in G' such that b has key k. We schedule a $op_f = find(k)$ operation concurrently with two consecutive delete operations: the first one, op_{d1}, deletes one of the nodes explored by op_f before it reaches a (such a node exists by the *non-triviality* property), and the second one, op_{d2} deletes the node with key k in G'. We make sure that op_f is not affected by op_{d1} (observes an update to some node c in the graph) but is affected by op_{d2} (does not observe b in the graph). The resulting schedule is not strictly serialializable (though linearizable). But our HOH-find implementation in $\mathcal{P}$ will accept it.

Theorem 3. *For any abstraction in $D \in \mathcal{D}$, there exists an implementation in $\mathcal{P}$ that accepts a non-strictly serializable schedule.*

Since any strictly serializable optimistic implementation only produces strictly serializable executions, from Theorem 3 we deduce that there is a schedule accepted by a pessimistic algorithm that no strictly serializable optimistic one can accept. Therefore, Theorems 2 and 3 imply that, when applied to search structures and in terms of concurrency, the strictly serializable optimistic approach is incomparable with pessimistic locking. As a corollary, none of these two techniques can be concurrency-optimal.

6   Related Work

Sets of accepted schedules are commonly used as a metric of concurrency provided by a shared memory implementation. For static database transactions, Kung and Papadimitriou [25] acknowledge that this metric may have "practical significance, if the schedulers in question have relatively small scheduling times as compared with waiting and execution times". Herlihy [18] implicitly considers a synchronization technique as highly concurrent, namely optimal, if no other technique accepts more schedules. By contrast, we focus here on a dynamic model where the scheduler cannot use the prior knowledge of all the shared addresses to be accessed.

Gramoli et al. [11] defined a concurrency metric, the *input acceptance*, as the ability of a TM to commit classes of input patterns of memory accesses without violating conflict-serializability. Guerraoui et al. [14] defined the notion of *permissiveness* as the ability for a TM to abort a transaction only if committing it would violate consistency. In contrast with these definitions, our framework for analyzing concurrency is independent of the synchronization technique. David et al. [6] consider that the closer the throughput of a concurrent algorithm is to that of its (inconsistent) sequential variant, the more concurrent the algorithm. In contrast, the formalism proposed in our paper allows for relating concurrency properties of various correct concurrent algorithms.

Our definition of search data structures is based on the paper by Chaudhri and Hadzilacos [4] who studied them in the context of dynamic databases. Safe-strict serializable implementations ($\mathcal{SM}$) require that every transaction (even aborted and incomplete) observes "correct" serial behavior. It is weaker than popular TM correctness conditions like opacity [15] and its relaxations like TMS1 [7] and VWC [24]. Unlike TMS1, we do not require the *local* serial executions to always respect the real-time order among transactions.

7   Concluding Remarks

In this paper, we presented a formalism for reasoning about the relative power of optimistic and pessimistic synchronization techniques in exploiting concurrency in search structures. We expect our formalism to have practical impact as the search structures are among the most commonly used concurrent data structures, including trees, linked lists, skip lists that implement various abstractions ranging from key-value stores to sets and multi-sets.

Our results on the relative concurrency of $\mathcal{P}$ and $\mathcal{SM}$ imply that none of these synchronization techniques might enable an optimally-concurrent algorithm. Of course, we do not claim that our concurrency metric necessarily captures efficiency, as it does not account for other factors, like cache sizes, cache coherence protocols, or computational costs of validating a schedule, which may also affect performance on multi-core architectures. In [13] we already described a *concurrency-optimal* implementation of the linked-list set abstraction that combines the advantages of $\mathcal{P}$, namely the semantics awareness, with the advantages of $\mathcal{SM}$, namely the ability to restart operations in case of conflicts. We recently observed empirically that this optimality can result in higher performance than state-of-the-art algorithms [16,17]. Therefore, our findings motivate the search for concurrency-optimal algorithms.

Acknowledgments. This research was supported under Australian Research Council's Discovery Projects funding scheme (project number 160104801) entitled "Data Structures for Multi-Core". Vincent Gramoli is the recipient of the Australian Research Council Discovery International Award. Petr Kuznetsov was supported by the Agence Nationale de la Recherche, under grant agreement N ANR-14-CE35-0010-01, project DISCMAT. Srivatsan Ravi acknowledges support from the National Science Foundation (NSF) grant CNS-1117065.

References

1. Afek, Y., Matveev, A., Shavit, N.: Pessimistic software lock-elision. In: Aguilera, M.K. (ed.) DISC 2012. LNCS, vol. 7611, pp. 297–311. Springer, Heidelberg (2012). doi:10.1007/978-3-642-33651-5_21
2. Aguilera, M.K., Frølund, S., Hadzilacos, V., Horn, S.L., Toueg, S.: Abortable and query-abortable objects and their efficient implementation. In: PODC, pp. 23–32 (2007)
3. Attiya, H., Welch, J.: Distributed Computing: Fundamentals, Simulations, and Advanced Topics. Wiley, New York (2004)
4. Chaudhri, V.K., Hadzilacos, V.: Safe locking policies for dynamic databases. J. Comput. Syst. Sci. **57**(3), 260–271 (1998)
5. Dalessandro, L., Spear, M.F., Scott, M.L.: NOrec: streamlining STM by abolishing ownership records. In: PPOPP, pp. 67–78 (2010)
6. David, T., Guerraoui, R., Trigonakis, V.: Asynchronized concurrency: the secret to scaling concurrent search data structures. In: ASPLOS, pp. 631–644 (2015)
7. Doherty, S., Groves, L., Luchangco, V., Moir, M.: Towards formally specifying and verifying transactional memory. Electron. Notes Theor. Comput. Sci. **259**, 245–261 (2009)
8. Felber, P., Fetzer, C., Riegel, T.: Dynamic performance tuning of word-based software transactional memory. In: PPoPP, pp. 237–246 (2008)
9. Gramoli, V.: More than you ever wanted to know about synchronization: synchrobench, measuring the impact of the synchronization on concurrent algorithms. In: PPoPP, pp. 1–10 (2015)
10. Gramoli, V., Guerraoui, R.: Democratizing transactional programming. Commun. ACM **57**(1), 86–93 (2014)

11. Gramoli, V., Harmanci, D., Felber, P.: On the input acceptance of transactional memory. Parallel Process. Lett. **20**(1), 31–50 (2010)
12. Gramoli, V., Kuznetsov, P., Ravi, S.: In the search of optimal concurrency. CoRR, abs/1603.01384 (2016)
13. Gramoli, V., Kuznetsov, P., Ravi, S., Shang, D.: Brief announcement: a concurrency-optimal list-based set. In: 29th International Symposium on Distributed Computing, DISC 2015, Tokyo, Japan, 7–9 October 2015. Technical report http://arxiv.org/abs/1502.01633
14. Guerraoui, R., Henzinger, T.A., Singh, V.: Permissiveness in transactional memories. In: Taubenfeld, G. (ed.) DISC 2008. LNCS, vol. 5218, pp. 305–319. Springer, Heidelberg (2008). doi:10.1007/978-3-540-87779-0_21
15. Guerraoui, R., Kapalka, M.: Principles of Transactional Memory: Synthesis Lectures on Distributed Computing Theory. Morgan and Claypool, San Rafael (2010)
16. Harris, T.L.: A pragmatic implementation of non-blocking linked-lists. In: Welch, J. (ed.) DISC 2001. LNCS, vol. 2180, pp. 300–314. Springer, Heidelberg (2001). doi:10.1007/3-540-45414-4_21
17. Heller, S., Herlihy, M., Luchangco, V., Moir, M., Scherer, W.N., Shavit, N.: A lazy concurrent list-based set algorithm. In: Anderson, J.H., Prencipe, G., Wattenhofer, R. (eds.) OPODIS 2005. LNCS, vol. 3974, pp. 3–16. Springer, Heidelberg (2006). doi:10.1007/11795490_3
18. Herlihy, M.: Apologizing versus asking permission: optimistic concurrency control for abstract data types. ACM Trans. Database Syst. **15**(1), 96–124 (1990)
19. Herlihy, M.: Wait-free synchronization. ACM Trans. Prog. Lang. Syst. **13**(1), 123–149 (1991)
20. Herlihy, M., Moss, J.E.B.: Transactional memory: architectural support for lock-free data structures. In: ISCA, pp. 289–300 (1993)
21. Herlihy, M., Shavit, N.: The Art of Multiprocessor Programming. Morgan Kaufmann, San Francisco (2008)
22. Herlihy, M., Shavit, N.: On the nature of progress. In: Fernàndez Anta, A., Lipari, G., Roy, M. (eds.) OPODIS 2011. LNCS, vol. 7109, pp. 313–328. Springer, Heidelberg (2011). doi:10.1007/978-3-642-25873-2_22
23. Herlihy, M., Wing, J.M.: Linearizability: a correctness condition for concurrent objects. ACM Trans. Program. Lang. Syst. **12**(3), 463–492 (1990)
24. Imbs, D., de Mendívil, J.R.G., Raynal, M.: Brief announcement: virtual world consistency: a new condition for STM systems. In: PODC, pp. 280–281 (2009)
25. Kung, H.T., Papadimitriou, C.H.: An optimality theory of concurrency control for databases. In: SIGMOD, pp. 116–126 (1979)
26. Papadimitriou, C.H.: The serializability of concurrent database updates. J. ACM **26**, 631–653 (1979)
27. Pugh, W.: Skip lists: a probabilistic alternative to balanced trees. Commun. ACM **33**(6), 668–676 (1990)
28. Shavit, N., Touitou, D.: Software transactional memory. In: PODC, pp. 204–213 (1995)
29. Weihl, W.E.: Commutativity-based concurrency control for abstract data types. IEEE Trans. Comput. **37**(12), 1488–1505 (1988)
30. Weikum, G.: A theoretical foundation of multi-level concurrency control. In: PODS, pp. 31–43 (1986)
31. Weikum, G., Vossen, G.: Transactional Information Systems: Theory, Algorithms, and the Practice of Concurrency Control and Recovery. Morgan Kaufmann, San Francisco (2002)
32. Yannakakis, M.: Serializability by locking. J. ACM **31**(2), 227–244 (1984)

The F-Snapshot Problem

Gal Amram[(⊠)]

Ben-Gurion University, Beer-Sheva, Israel
galamra@cs.bgu.ac.il

Abstract. Aguilera, Gafni and Lamport introduced the signaling problem in [3]. In this problem, two processes numbered 0 and 1 can call two procedures: update and Fscan. A parameter of the problem is a two-variable function $F(x_0, x_1)$. Each process p_i can assign values to variable x_i by calling update(v) with some data value v, and compute the value: $F(x_0', x_1)$ by executing an Fscan procedure. The problem is interesting when the domain of F is infinite and the range of F is finite. In this case, some "access restrictions" are imposed that limit the size of the registers that the Fscan procedure can access.

Aguilera et al. provided a non-blocking solution and asked whether a wait-free solution exists. A positive answer can be found in [5]. The natural generalization of the two-process signaling problem to an arbitrary number of processes turns out to yield an interesting generalization of the fundamental snapshot problem, which we call the F-snapshot problem. In this problem n processes can write values to an n-segment array (each process to its own segment), and can read and obtain the value of an n-variable function F on the array of segments. In case that the range of F is finite, it is required that only bounded registers are accessed when the processes apply the function F to the array, although the data values written to the segments may be taken from an infinite set. We provide here an affirmative answer to the question of Aguilera et al. for an arbitrary number of processes. Our solution employs only single-writer atomic registers, and its time complexity is $O(n \log n)$.

1   Introduction

In this paper we introduce a solution to the F-snapshot problem, which is a generalization of the well-studied snapshot problem (introduced independently by Afek et al. [1], by Anderson [6] and by Aspnes and Herlihy [7]). The snapshot object involves n asynchronous processes that share an array of n segments. Each process p_i can write values to the i-th segment by invoking an update procedure with a value taken from some range of values: *Vals*, and can scan the entire array by invoking an instantaneous scan procedure. For any function $F: Vals^n \to D$ (where D is any set and $Vals^n$ is the set of n-tuples of members of *Vals*) the F-snapshot variant differs from the snapshot problem in that the Fscan operation has to return the value $F(v_0, \ldots, v_{n-1})$ of the instantaneous segment

Research partially supported by the Israel Science Foundation.

J. Suomela (Ed.): SIROCCO 2016, LNCS 9988, pp. 159–176, 2016.
DOI: 10.1007/978-3-319-48314-6_11

values $v_0, \ldots, v_{n-1}$. That is in comparison to the standard scan operation, which returns the vector of values that the segments store at an instantaneous moment.

The F-snapshot problem is interesting only if we impose an additional requirement, without which it can be trivially implemented by applying the function F (assumed to be computable) to the values returned by the standard scan operation. This additional requirement, for the case $n = 2$, was suggested by Aguilera, Gafni and Lamport [3] (see also [2]) in what they called there the signaling problem. Thus, our F-snapshot problem is a generalization of both the standard snapshot problem and the signaling problem (generalizing this problem from the $n = 2$ case to the general case of arbitrary n).

In the signaling problem, the set *Vals* is possibly infinite and the range of F, D is finite (and small). It is required that an Fscan operation uses only bounded registers. That is, registers that can store only finitely many different values (the update operations may access unbounded registers). The signaling problem was formulated just for two processes in [3], and a wait-free solution for this problem was left there as an open problem. Thus, solving the general F-snapshot sets quite a challenge. A wait-free solution to the signaling problem is given in [5], and here we present a wait free solution to the general F-snapshot problem.

In [3], the signaling problem is justified for efficiency reasons. We consider a case in which the processes write values to their segments taken from an infinite range, but they are interested in some restricted data regarding these values (for example, which process invoked the largest value, how many different values there are etc.). An F-snapshot implementation may be more efficient in these cases than a snapshot implementation, since it is not necessary to scan the entire array for extracting the required information, and it suffices to read only bounded registers. Efficiency is mostly guaranteed when the Fscan operations are likely to be invoked much more frequently than the update procedures.

Moreover, the authors of [3] showed that a signaling algorithm can be used to solve the mailbox problem which is the main problem that [3] deals with. With a similar approach, a signaling algorithm can be also used to implement a solution to the N-buffer problem [22] (see also [5] for further discussion). At the mailbox problem, a processor and a device communicate through an interrupt controller and it is required that the processor will know if there are some unhandled requests, only by reading bounded registers. At the N-buffer problem, a producer sends messages to a consumer, through a message buffer of size N, and they need to check if the buffer is empty, full or neither-empty-nor-full by reading bounded registers.

In the same way, a solution to the F-snapshot problem can be used to implement a generalized mailbox algorithm, in which there are several devices, and the processor can check which devices are waiting for its response by reading only bounded registers. Similarly, an F-snapshot algorithm can be used to implement a generalized N-buffer in which there are possibly many producers and many consumers[1].

---

[1] Assuming that the queue of messages supports enqueue and dequeue operations by several processes.

Now we describe the F-snapshot problem formally. Let $P = \{p_0, \ldots, p_{n-1}\}$ be a set of n-asynchronous processes that communicate through shared registers and let $F\colon Vals^n \longrightarrow D$ be an n-variable computable function from a (possibly infinite) domain $Vals$, into $D = Rng(F)$. The problem is to implement two procedures:

1. update(v) - invoked with an element $v \in Vals$. This procedure writes v to the i-th segment of an n-array A, when invoked by p_i.
2. Fscan - returns a value $d \in D$. This procedure returns $F(A[0], \ldots, A[n-1])$, in contrast to a scan procedure which returns the values stored at the entire array: $(A[0], \ldots, A[n-1])$.

The implementation needs to satisfy the following requirements:

1. All procedures are wait free. That is, each procedure eventually returns, if the executing process keep taking steps.
2. If D is finite, then only bounded registers are accessed during Fscan operations.
3. Only single-writer multi-reader atomic registers are employed.

The reader may note that as the domain of F could be infinite, the update procedure must access also unbounded registers.

For correctness of F-snapshot implementations, we adapt the well known Linearizability condition, formulated by Herlihy and Wing [15]. Roughly speaking, an F-snapshot algorithm is correct if for any of its executions the following hold: Each procedure execution can be identified with a unique moment during its actual execution (named the linearization point), such that the resulting sequential execution belongs to a set of correct sequential executions, the sequential specification of the object. The sequential specification of the F-snapshot object includes all executions of the atomic implementation, presented in Fig. 1. The code uses an array $A[0 \ldots n-1]$.

```
1: procedure update (v)          1: procedure Fscan
2:     A[i] ← v                   2:     return F(A[0], ..., A[n])
3: end procedure                  3: end procedure
```

Fig. 1. F-snapshot atomic implementation

In this paper, we present a solution to the F-snapshot problem. Each operation in our algorithm consists of $O(n \log n)$ actions addressed to the shared registers. The rest of the paper is organized as follows: Preliminaries are given in Sect. 2. In Sect. 3 we present our algorithm and explain the ideas behind it. Correctness proof is given in Sect. 4, and Sect. 5 concludes.

Related Work

As been explained, the F-snapshot problem generalizes the signaling problem, from the case that there only two processes to an arbitrary number of processes. In [3], Aguilera et al. presented the signaling problem and gave a non-blocking solution. A wait free algorithm is given in [5].

Jayanti presented the f-array object which assumes n processes that write values to m multi-writer registers, $v_1, \ldots, v_m$ and compute an m-variable function f on the values that the registers store [20]. However, while in this paper we seek for a linearizable implementation for the F-snapshot object from single-writer registers in which the Fscan operation accesses bounded registers, the scope of [20] is different. Jayanti presents in [20] an implementation for the f-array object from registers and an LL/SC object. The motivation of [20] is to improve complexity measures, when assumptions on the function f implies that it is not necessary to scan the entire array for the computation of f.

2   Preliminaries

2.1   The Model of Computation

The model of computation in this paper is standard. We assume n asynchronous processes $p_1, \ldots, p_n$ that communicate through shared single-writer registers. Thus, each registers is "owned" by a unique process. At an individual step, a process may write to one of its single-writer registers and perform an internal computation, or read a register and perform an internal computation that depends on the value it read.

An F-snapshot implementation provides each process with a code for the update and Fscan procedures. At an execution, each process executes update and Fscan operations in some arbitrary order, and the update operations are invoked with arbitrary data values from the set *Vals*. Formally, an execution τ is a finite or infinite sequence of atomic actions (also named low-level events) that the processes perform while executing update and Fscan operations. As the processes are asynchronous, for each process p_i, there is no bound on the number of steps taken by other processes in-between two consecutive actions performed by p_i. In particular, a process may crash and take no additional steps at an execution.

At an execution τ, each atomic action is a part of a unique Fscan or update procedure execution, executed by some process. A set of actions that corresponds to a procedure execution is named an operation, or an high-level event. An high-level event may be complete if the executing process executed all the procedure instructions and returned, or pending otherwise.

The low-level events at an execution are linearly ordered by the precedence relation $<$. The precedence relation $<$ is naturally extended over high-level events: We write $A < B$ if A is complete and for every $a \in A$ and $b \in B$, $a < b$. Similarly, we relate high-level events with low-level events. For a low-level event e and high-level event A, we write $e < A$ if $e < a$ for every $a \in A$ and we write $A < e$ if A is complete and $a < e$ for every $a \in A$. Incomparable events are said to be concurrent.

2.2 Linearizability

An execution is linearizable if the high-level events can be identified with instantaneous moments during their executions so that this identification yields a correct sequential execution. More precisely, it is required to find such linearization points for all complete high-level events, where pending operations may be omitted or artificially completed.

In some cases the linearization points can be identified with an execution of some fixed instruction. When no such fixed linearization points exist, it is convenient to define the linearizability criterion in an equivalent manner. We say that an execution τ is linearizable, if there is a linear ordering $(\mathcal{H}, \prec)$ such that

1. $\mathcal{H}$ includes all complete high-level events and some pending high-level events.
2. $\prec$ extends $<$ over $\mathcal{H}$.
3. $(\mathcal{H}, \prec)$ belongs to the sequential specification. Namely, $\prec$ is a precedence relation over $\mathcal{H}$ obtained by some execution of the algorithm in Fig. 1.

An implementation is linearizable if all its executions are linearizable.

As an Fscan operation needs to return a value obtained by applying F on the values that the segments store, it is required to assume some initial value for each segment. For convenience, we assume that at the beginning of each execution, each process executes an initial update event invoked with an initial data value. These events write initial values to the registers and variables and they precede all other high-level events.

3 The F-Snapshot Algorithm

First we explain the main ideas behind the algorithm. The reader may want to consider our explanations, while examining the code of the algorithm and its local procedures in Fig. 3.

The crucial obstacle for solving the problem is that an Fscan procedure cannot access unbounded registers, but it is required to apply the function F on values that are stored in unbounded registers. To overcome this issue, we apply the function F on the values that the segments store, during an execution of an update operation. When process p_i performs an update operation invoked with a data value val, it writes val into a snapshot object V (line 4), scans this snapshot object (line 5), applies the function F on the view it obtained and stores the outcome in a local variable ans (line 6). Then, before it returns, the process writes the outcome it obtained into a snapshot object named $Flags$ (line 19). A process executing an Fscan operation, scans the snapshot object $Flags$ and it needs to choose the most up-to-date value among the values suggested by the processes. We need to provide the $Flags$ object with an additional information, so that the executing process could decide correctly which value to return. However, this additional information needs to be taken from a finite range due to the problem limitations.

As a first attempt, one may suggest to use bounded concurrent timestamps (see [10–12]). Bounded timestamps are used traditionally to label writes, but here we actually need to label scans. Namely, we need to know in what order the scans of the snapshot object V occurred. If we will try instead, for example, to use timestamps to label writes to $Flags$, then a process executing an Fscan operation can mistakenly rely on the order between the labeling operations and not on the order between the scan events addressed to the snapshot object V. Similarly, labeling the writes to V will not work either by the same reason. We see that the approach of using bounded timestamps, at least in its simplest form, will not succeed.

3.1   The Classify Mechanism

For determining the ordering between scan events addressed to V, we adapt the common technique of counting update events [8,9,18]. When a process performs an update operation, it increases a counter (line 2) and writes this counter to V together with the data value with which the update operation was invoked (line 3). When process p_i scans V (line 5), it sums these counters to obtain a natural number that reflects how recent its view is (line 12). This approach resembles the snapshot algorithm presented by Israeli, Shaham and Shirazi [18]. They used this technique to implement a snapshot algorithm in which the time complexity of the scan procedure is $O(n)$. In their construction, while executing a scan operation, the executing process returns the view of the process that presents the latest activity, reflected by the largest sum.

Since an Fscan operation cannot access unbounded registers, we cannot adopt the discussed approach as it was used in [18]. In our algorithm, the reading of these natural numbers is done within the update operation. The process writes the sum it obtained into a snapshot object named $ViewSum$ (lines 13, 15), and scans $ViewSum$ to compare its view with the views obtained by the other processes. Afterward, it classifies all other processes into two categories: $winners$ - the processes that possess a later view, reflected by a largest sum, and $losers$ - processes with outdated view. This is done by calling the local classify procedure, when processes id's are used for breaking symmetry. These sets of $winners$ and $losers$ are stored at the segment $Flags[i]$ (lines 18, 19).

3.2   The Coloring Mechanism

When process p_k executes an Fscan operation, it scans the $Flags$ array and it tries to extract the most up-to-date view, referring to the fields $Flags[i].winner$ and $Flags[i].losers$ for $i = 0, \ldots, n-1$. For any pair of processes p_i and p_j, p_k may want to know which process's view is more recent. The problem is that the processes may provide contradicting information. As an example, the process may find that $j \in Flags[i].winners$ (which means that p_i thinks that p_j's view is more up-to-date than its view), but it is possible that also $i \in Flags[j].winners$. Namely, it is possible that both p_i and p_j think that the other process knows better.

The coloring mechanism ensures that the problem described above can occur only in some "typical" executions (with which our next mechanism deals). The update events by each process alternate between 3 possible colors: 0, 1 or 2 (line 3). Each process holds a three-field variable, in correspondence to the three colors, named *myview*. After a process sums the counters it sees (lines 5, 12), it writes the sum it obtained into *myview*[*color*] (line 13) and deletes data obtained in its second-previous update operation (line 14) to erase confusing information. Then, it writes the value that *myview* stores into *ViewSum* (line 15). Now, when process p_i scans *ViewSum*, in each segment *ViewSum*[*j*], it finds two integers. These are the sums that p_j computed in its two previous update events. When p_i executes its local classify procedure, it also specifies the color it saw. For example, it writes (j, c) to *winners* for $c \in \{0, 1, 2\}$, if it reads from *ViewSum*[*j*][*c*] a number larger than the sum it obtained in line 9 (when id's are taken into account for breaking symmetry). Each process p_i writes the values of its local sets *winners* and *losers* to $Flags[i]$, together with the color of the update operation it is executing.

Coming back to our example, assume that process p_k executes an Fscan operation and it finds that $(j, c) \in Flags[i].winners$. Then, it understands that p_i saw in *ViewSum*[*j*][*c*] an integer larger than the number it obtained. However, if it sees that $Flags[j].color \neq c$, it just disregards p_i's information.

3.3   Adding Bounded Timestamps

The coloring mechanism does not prevent entirely the possibility that processes will provide contradicting information. Assume for example, that while executing an Fscan operation, p_k finds that $Flags[i].color = c_i$, $Flags[j].color = c_j$, $(j, c_j) \in Flags[i].winners$ and $(i, c_i) \in Flags[j].winners$. Thus, both p_i and p_j claim that the other process is more up-to-date. When such a situation occurs, one of the processes provides reliable information. This is the process that scanned *ViewSum* later before updating *Flags*.

When such a situation occurs, the processes use bounded timestamps to inform which process is trustworthy. We use a simple timestamping system in which the timestamps are vertices of a nine-vertices directed graph $G = (V_G, E_G)$. An illustration and a detailed explanation can be found in Chap. 2 of [14], or in [17]. The graph G consists of three cycles, each cycle includes three vertices. In addition, there is an edge from each vertex at the i-th cycle to each vertex at the $i - 1 \pmod 3$ cycle. Formally, $V_G = \{(i, j) : i, j \in \{0, 1, 2\}\}$, and there is an edge from $v = (i_1, j_1)$ to $u = (i_2, j_2)$ if $i_1 = i_2$ and $j_1 = j_2 + 1$ $\pmod 3$, or $i_1 = i_2 + 1 \pmod 3$. The vertices of G are named timestamps, and if $(v, u) \in E$ we say that v dominates u and we write $u <_{ts} v$. Intuitively, v dominates u means that the timestamp v represents a later moment than the timestamp u.

We note that there are no cycles of length two in G. In addition, for any two timestamps v, u, we can find a timestamp w that dominates both v and u. We take a function $next : V_G \times V_G \longrightarrow V_G$ that satisfies this property. That is, for any timestamps v, u: $v <_{ts} next(v, u)$ and $u <_{ts} next(v, u)$.

Any process p_i holds n pairs of timestamps. Each pair consists of a new timestamp and an old timestamp. These pairs are stored in a snapshot object named VTS. When p_i executes an update operation it scans VTS (line 7). Then, against each process p_j it chooses a timestamp that dominates the pair of timestamps it read from $VTS[j][i]$, using the function $next$. p_i stores the timestamp it obtained as its new timestamp, keeps its former new timestamp available as its old timestamp and updates VTS (consider lines 8–10 and the local procedure newts). Finally, p_i stores its n-vector of pairs of timestamps in $Flags[i]$ while updating the $Flags$ object (lines 18, 19).

Now, we consider again the situation in which process p_k executes an Fscan operation, and finds that two processes p_i and p_j, provide contradicting information as described earlier. In this case, the scanner checks the timestamps that the processes present. The process that its new timestamp dominates the other process's new timestamp is the reliable one. More precisely, the scanner considers the timestamps $Flags[i].vts[j].new$ and $Flags[j].vts[i].new$. The information provided by the process with the later timestamp is the correct information. These timestamps are used only when processes provide contradicting information. In other cases the timestamps do not necessarily reflect the right ordering between the processes' views.

3.4   The Code

In this subsection we present our F-snapshot algorithm. The algorithm uses four snapshot objects in addition to local variables. The type of the variables and the type of the snapshot objects segment's are specified at Fig. 2.

Note that there are two fixed values used for initialization, $x_0 \in Vals$ and $v_0 \in V_G$. Segments of V store pairs from $\mathbb{N} \times Vals$. For such a pair, $v = (m, val)$, we write $m = v.counter$ and $val = v.val$. Each entry $VTS[i]$ stores an n-tuple of pairs of timestamps. If $VTS[i][j] = (c, d)$ we write $c = VTS[i][j].old$ and $d = VTS[i][j].new$. Each segment of the $Flags$ snapshot object is of type $flag$ that consists of five fields: $flag.color \in \{0, 1, 2\}$, $flag.vts \in (V_G \times V_G)^n$, $flag.winners, flag.losers$ are sets that store elements from $\{0, \ldots, n-1\} \times \{0, 1, 2\}$, $flag.ans \in D$. The initial value of $Flags[i]$ is $flag_{0,i}$ where $flag_{0,i}.color = 0$, $flag_{0,i}.vts = ((v_0, v_0), \ldots, (v_0, v_0))$, $flag_{0,i}.winners = \{(j, 0): i < j\}$, $flag_{0,i}.losers = \{(j, 0): i > j\}$ and $flag_{0,i}.ans = F(x_0, x_0, \ldots, x_0)$.

Our snapshot implementation uses four local procedures, classify, newflag, newts and find_max. The implementation for process p_i, together with the code for the local procedures is presented in Fig. 3. In the next subsection we present the find_max local procedure.

snapshot-objects

name	segment-type	initial-value
V	$\mathbb{N} \times Vals$	$(0, x_0)$
VTS	$(V_G \times V_G)^n$	$((v_0, v_0), \ldots, (v_0, v_0))$
$ViewSum$	$(\mathbb{N} \cup \{null\})^3$	$(0, null, null)$
$Flags$	$flag$	$flag_{0,i}$

Local variables

name	type	initial-value
$Flags$	$flag$	$flag_{0,i}$
val	$Vals$	arbitrary
$counter, viewsum$	$\mathbb{N}$	0
$color$	$\{0, 1, 2\}$	0
$v_0, \ldots, v_{n-1}$	$\mathbb{N} \times Vals$	arbitrary
ans	D	arbitrary
$vts, \ldots, vts_{n-1}$	$(V_G \times V_G)^n$	arbitrary
$myview, vu_0, \ldots, vu_{n-1}$	$(\{0, \ldots, n-1\} \cup \{null\})^3$	$(0, null, null)$
$flg, flg_0, \ldots, flg_n$	$\{0, \ldots, n-1\}^3$	arbitrary
$winner$	$\{0, 1, \ldots, n-1\}$	arbitrary
$ts.old, ts.new$	V_G	arbitrary
$winners, losers$	sets over $\{0, 1, \ldots, n-1\} \times \{0, 1, 2\}$	arbitrary

Fig. 2. Snapshot objects and local variables

3.5   The Procedure find_max

This procedure is invoked during an execution of an Fscan event S, and it returns the id of the most up-to-date process. Thus, the process that executes S returns the value $flg_i.ans$ in case that find_max returns i.

The find_max procedure of an Fscan operation S gets n flags as arguments: $flags(S) := (flg_0, \ldots, flg_{n-1})$. The procedure returns a maximal element in relation $<_S \subseteq \{0, \ldots, n-1\} \times \{0, \ldots, n-1\}$ that we define here. Relation $<_S$ is defined by reference to $flags(S)$ in Definition 2.

Definition 1. *Let p_i and p_j be two processes and write: $flg_i.color = c_i$ and $flg_j.color = c_j$. We say that p_i and p_j are in conflict in S, if one of the following occurs:*

1. $(j, c_j) \in flg_i.winners$ and $(i, c_i) \in flg_j.winners$.
2. $(j, c_j) \in flg_i.losers$ and $(i, c_i) \in flg_j.losers$.

Definition 1 is important since, as we shall prove, for each two processes p_i, p_j and an Fscan event S, the *flag* of one of these processes determines correctly the ordering between p_i and p_j. That is, if p_i is the reliable process and if (for example) $(j, c_j) \in flg_i.winners$ and the color in p_j's *flag* is c_j, than p_j is indeed more up-to-date than p_i (more precisely, the *ans* field of p_j's *flag* is more up-to-date) as indicated by p_i's *flag*. The problem is that we do not know which

```
1: procedure update (val)                    1: procedure Fscan
2:     counter ← counter + 1                 2:     (flg_0, ..., flg_{n-1}) ←
3:     color := color (mod 3)                       Flags.scan
4:     V.update(counter, val)                3:     winner ←
5:     (v_0, ..., v_{n-1}) ← V.scan                 find_max(flg_0, ..., flg_{n-1})
6:     ans ← F(v_0.val, ..., v_{n-1}.val)    4:     return flg_{winner}.ans
7:     (vts_0, ..., vts_{n-1}) ← VTS.scan    5: end procedure
8:     for j = 0 to n − 1 do                 1: procedure newflag
9:         vts_i[j] ← newts(vts_j[i], vts_i[j])  2:     flg.color ← color
10:    end for                               3:     flg.vts ← vts_i
11:    VTS.update(vts_i)                      4:     flg.winners ← winners
12:    viewsum ← Σ_{j=0}^{n-1} v_j.counter   5:     flg.losers ← losers
13:    myview[color] ← viewsum               6:     flg.ans ← ans
14:    myview[color + 1 (mod 3)] ← null      7:     return flg
15:    ViewSum.update(myview)                8: end procedure
16:    (vu_0, ..., vu_{n-1}) ← ViewSum.scan  1: procedure newts((c, d), (c', d'))
17:    classify(vu_0, ..., vu_{n-1})         2:     ts.old ← d'
18:    flag ← newflag                        3:     ts.new ← (next(c, d))
19:    Flags.update(flag)                    4:     return (ts.old, ts.new)
20: end procedure                            5: end procedure
1: procedure classify((vu_0, ..., vu_{n-1}))
2:     winners ← {(j, c) : view_j[c] > viewsum}∪
           {(j, c) : view_j[c] = viewsum ∧ i < j}
3:     losers ← {(j, c) : view_j[c] < viewsum}∪
           {(j, c) : view_j[c] = viewsum ∧ i > j}
4: end procedure
```

Fig. 3. F-snapshot implementation

process provides correct information among any pair of processes. However, this problem does not arise when the processes are not in conflict. When processes provide contradicting information we use the processes' timestamps to find the trustworthy process.

Definition 2. *Let p_i and p_j be two processes and write: $flg_i.color = c_i$, $flg_j.color = c_j$. $i <_S j$ if one of the following occurs:*

1. *p_i and p_j are not in conflict in S and $(i, c_i) \in flg_j.losers$.*
2. *p_i and p_j are not in conflict in S and $(j, c_j) \in flg_i.winners$.*
3. *p_i and p_j are in conflict in S, $flg_i.vts[j].new <_{ts} flg_j.vts[i].new$ and $(i, c_i) \in flg_j.losers$.*
4. *p_i and p_j are in conflict in S, $flg_j.vts[i].new <_{ts} flg_i.vts[j].new$ and $(j, c_j) \in flg_i.winners$.*

An element $i \in \{0, ..., n-1\}$ is maximal in $<_S$ if there is no $j \neq i$ such that $i <_S j$. The procedure find_max($flg_0, ..., flg_{n-1}$) (line 3) returns a maximal element in $<_S$ (we shall prove that such a maximal element exists in any Fscan event). This procedure accesses only local variables and we omit the technical but easy implementation of this procedure.

4   Correctness

Fixing an execution τ of our algorithm, we need to show that the precedence relation defined over the high-level events in τ, $<$ can be extended into a linear ordering, $\prec$ that belongs to the sequential specification of the F-snapshot object. In our proof, we assume that τ is finite, that all the operations in τ are complete, and that every operation addressed to one of the snapshot objects is atomic. It is suffices to assume that τ is finite since linearizability of deterministic objects is a safety property, as Guerraoui and Ruppert proved [13]. It is suffices to assume that all operations are complete since the implementation is wait-free. Indeed, if there are any pending operation we can just let the processes take several additional steps and to complete these operations. Of course, it suffices to show that the resulting execution is linearizable. Finally, by taking a linearizable implementation for the snapshot objects, we may assume that all operations addressed to these object are atomic, since we can identify the executions of these operations with their linearization points. Further discussion about using linearizable implementation can be found in [4,15].

As explained in Subsect. 2.2, we assume n initial update events by the processes, where the initial update event by process p_j is denoted I_j. These initial high-level events write initial values to the variables and the snapshot objects, and they precede all other events.

Our algorithm employs several snapshot objects. Thus, for preventing confusion, we use the notation $A.\mathsf{update}$ and $A.\mathsf{scan}$ to denote invocations of update and scan procedures addressed to object A. We write: update and Fscan to denote high-level events in τ. Note that an $A.\mathsf{update}(x)$ invocation by p_i writes x to the i-th segment of A.

If e is an $A.\mathsf{update}$ event, $val(e)$ is the value with which e is invoked, and if e is an $A.\mathsf{scan}$ operation, $val(e)$ is the vector of values that e returns. If e is an $A.\mathsf{scan}$ operation and j a process id, $\mu_j(e)$ is the last $p_j\text{-}A.\mathsf{update}$ operation that precedes e. Hence, $val(e)[j] = val(\mu_j(e))$ in this case. For an $A.\mathsf{update}$ (respectively, $A.\mathsf{scan}$) event e, $[e]$ is the update (respectively, Fscan) operation that includes e.

For an operation X and a snapshot-object A, X includes at most one $A.\mathsf{update}$ (respectively, $A.\mathsf{scan}$) event. If such an event exists, it is denoted $A.\mathsf{update}(X)$ (respectively, $A.\mathsf{scan}(X)$). Similarly, if r is a local variable, X includes at most one write to r. If X includes a write to r, $r(X)$ is the value written to r in X. In particular, if I is an initial update operation that includes a write to r, $r(I)$ is the initial value of r.

For an Fscan operation S, $\beta_j(S)$ is the p_j-update operation that wrote to $Flags$ the value read in S. That is, $\beta_j(S) = [\mu_j(Flags.\mathsf{scan}(S))]$.

Recall that $<$ is the precedence relation defined over high and low level events in τ (see Sect. 2). For two pairs of integers (a, b), (x, y) we write $(a, b) <_{lex} (x, y)$ if (a, b) precedes (x, y) lexicographically. We write $x \leq y$ (respectively, $x \leq_{lex}$, $x \preceq y, x \leq_S y$) to denote that $x < y$ (respectively, $x <_{lex} y, x \prec y, x <_S y$) or $x = y$.

Before defining a total order $\prec$ over the high-level events in τ, we prove few technical lemmas. We fix an Fscan operation S. For process p_i, we write $m_i = viewsum(\beta_i(S))$, $c_i = color(\beta_i(S))$, $flg_i = flg(\beta_i(S))$ and $e_i = ViewSum.\text{update}(\beta_i(S))$ (line 15).

Lemma 1. *Assume that* $\beta_j(S) \neq I_j$ *and write* $e = ViewSum.\text{scan}(\beta_j(S))$ *(line 16). If* $e_i < e_j$, *then one of the following holds:*

1. $\mu_i(e) = e_i$ *or,*
2. $\mu_i(e) = e' > e_i$ *and there is no* p_i-update *event between* $\beta_i(S) = [e_i]$ *and* $[e']$.

Proof. Since $e_i < e_j$ and since (by the code) $e_j < e$, we see that $e_i < e$ and hence, $e_i \leq \mu_i(e)$. Thus, we need to show that there is at most one $ViewSum.\text{update}$ event by p_i between e_i and e.

Assume for a contradiction that e' and e'' are two $ViewSum.\text{update}$ events by p_i such that $e_i < e' < e'' < e$. Each $ViewSum.\text{update}$ event belongs to a unique update event so there are two different p_i-update operations $U' = [e']$ and $U'' = [e'']$. Recall that $[e_i] = \beta_i(S)$ and observe that $\beta_i(S) < U' < e'' < e$.

Now, the event $Flags.\text{scan}(S)$ occurs after e, so the value read in this event from $Flags[i]$ is the value written to $Flags[i]$ in U' or in a later event. We have: $\beta_i(S) < U' \leq [\mu_i(Flags.\text{update}(S))]$ in contradiction to the definition of $\beta_i(S)$.   □

The following two lemmas follows, and their proof is left as an exercise for the reader.

Lemma 2. *Assume that* $\beta_j(S) \neq I_j$. *If* $e_i < e_j$, *then* p_j *reads* m_i *from* $ViewSum[i][c_i]$ *at the event* $ViewSum.\text{scan}(\beta_j(S))$ *(line 16) and in addition:*

1. *If* $(m_i, i) <_{lex} (m_j, j)$, *then* $(i, c_i) \in losers(\beta_j(S))$.
2. *If* $(m_j, j) <_{lex} (m_i, i)$, *then* $(i, c_i) \in winners(\beta_j(S))$.

Lemma 3. *Assume that* p_i *and* p_j *are not in conflict in* S *(the definition is given in Subsect. 3.5). Then,* $(m_i, i) <_{lex} (m_j, j) \iff i <_S j$.

Our next goal is to prove the same for the case that the processes are in conflict. If the processes are in conflict, we know by Lemma 2 that the process that wrote later to $ViewSum$ provides reliable information. Recall that in this case, $<_S$ is determined according to the *flag* of the process that presents a later timestamp. We need to show that the process with the later timestamp is also the process that wrote later to $ViewSum$.

Lemma 4. *Assume that* $e_i < e_j$ *and let* ts_i *and* ts_j *the timestamps written to* $Flags[i].vts[j].new$ *and to* $Flags[j].vts[i].new$ *in* $\beta_i(S)$ *and in* $\beta_j(S)$ *respectively. If* p_i *and* p_j *are in conflict in* S, *then* $ts_i <_{ts} ts_j$.

Proof. First, note that $\beta_j(S) \neq I_j$. Indeed, if $\beta_j(S) = I_j$, then also $\beta_i(S) = I_i$ (since $e_i < e_j$) which implies that the processes are not in conflict.

For the rest of the proof we assume that also $\beta_i(S) \neq I_i$. If $\beta_i(S) = I_i$, then similar (and simpler) arguments can be applied. Write $s_i = ViewSum.\text{scan}(\beta_i(S))$. By Lemma 2, p_j reads m_i from $ViewSum[i][c_i]$ in $\beta_j(S)$, but since p_i and p_j are in conflict in S, we conclude that p_i reads some $m \neq m_j$ from $ViewSum[j][c_j]$ in s_i. Hence,

$$\mu_j(s_i) \neq e_j. \tag{1}$$

Since $e_i < e_j$ and (by the code) $e_i < s_i$, either $e_i < s_i < e_j$ or $e_i < e_j < s_i$. We claim that the former occurs and $s_i < e_j$. Assume otherwise, and use Eq. 1 to conclude that $e_j < \mu_j(s_i)$. Note that there can be at most one $ViewSum.\text{update}$ event by p_j that follows e_j and precedes s_i, since otherwise we would have $\beta_i(S) < \mu_i(Flags.\text{scan}(S))$ which is impossible. Hence, s_i reads from $ViewSum[j]$ the value of this event. However, by the code of the update procedure, the update operation by p_j that follows $\beta_j(S)$ also writes m_j to $ViewSum[j][c_j]$. Thus, if $e_j < s_i$, then p_i reads m_j from $ViewSum[j][c_j]$ in s_i, in contradiction to the assumption that the processes are in conflict. We conclude that $e_i < s_i < e_j$.

Now we claim that there is a p_j-$ViewSum.\text{update}$ event between s_i and e_j. Indeed, assume not and let e' be the last $ViewSum.\text{update}$ event by p_j that precedes e_j. By our assumption we have $\mu_j(s_i) = e'$. $e' \in U'$, the p_j-update event that precedes $\beta_j(S)$ and hence, $color(U') = c_j - 1 \pmod 3$. Therefore, $val(e')[c_j] = null$. We conclude that p_i reads $null$ from $ViewSum[j][c_j]$ in s_i and this contradicts the fact that p_i and p_j are in conflict.

We see that there is a $ViewSum.\text{update}$ event by p_j between s_i and e_j. Let e'_j denotes this event. Hence, $s_i < e'_j < e_j$. Write $[e'_j] = U'$, and note that $U' < \beta_j(S)$. Therefore, $e'_j < \beta_j(S)$.

Now, write $t_i = VTS.\text{update}(\beta_i(S))$ (line 11 in the code) and write $val(t_i)[j] = (x,y)$. Observe that since $t_i \in \beta_i(S)$, (x,y) is also the value of $flg_i.vts[j]$. Write $s_j = VTS.\text{scan}(\beta_j(S))$ (line 7) and note that since $e'_j < \beta_j(S)$, $e'_j < s_j$. By the code and by our conclusions we have, $t_i < s_i < e'_j < s_j$. Therefore, $\mu_i(s_j) \geq t_i$. However, there could be most one $VTS.\text{update}$ event by p_i between t_i and s_j (by definition of $\beta_i(S)$). Furthermore, if there is such an event, it writes to $VTS[i][j]$: (y,z) for some vertex $z \in V_G$ (consider the newts code). Let (a,b) denotes the value of $VTS[j][i]$ before the execution of $\beta_j(S)$.

Case 1. $\mu_i(s_j) = t_i$ and hence p_j reads in s_j from $VTS[i][j]$: (x,y). Thus, p_j writes in $\beta_j(S)$ to $Flags[j].vts[i]$: $\text{newts}((x,y),(a,b)) = (b, next(x,y))$. Since $(next(x,y),y) \in E_G$, $flg_i.vts[j].new <_{ts} flg_j.vts[i].new$ as required.

Case 2. $\mu_i(s_j) > t_i$ and hence p_j reads in s_j from $VTS[i][j]$: (y,z). In this case, p_j writes in $\beta_j(S)$ to $Flags[j].vts[i]$: $\text{newts}((y,z),(a,b)) = (b, next(y,z))$. Since also $(next(y,z),y) \in E_G$, $flg_i.vts[j].new <_{ts} flg_j.vts[i].new$. We see that the lemma holds in this case as well. $\square$

The previous lemma shows that if two processes are in conflict and their *flags* provide contradicting information, the *flag.vts* fields determine correctly which among the two processes is the reliable one. The conclusion is that relation $<_S$ determines correctly which process presents the most up-to-date view in its *flag.ans* field.

Lemma 5. *Let p_i and p_j be two processes. Then, $(m_i, i) <_{lex} (m_j, j) \Longleftrightarrow i <_S j$. In particular, $<_S$ admits a maximal element.*

Proof. First assume that $(m_i, i) <_{lex} (m_j, j)$. If p_i and p_j are not in conflict, then this is the case of Lemma 3. If p_i and p_j are in conflict, assume w.l.o.g. that $e_i < e_j$ and note that $\beta_j(S) \neq I_j$. By Lemma 2, $(i, c_i) \in flg_j.losers$. By the previous lemma $flg_i.vts[j].new <_{ts} flg_j.vts[i].new$ thus $i <_S j$.

Now, for the other direction, we leave for the reader to verify that $<_S$ is an a-symmetric total order over $\{0, 1, \ldots, n-1\}$, and to conclude that if $i <_S j$, then $(m_i, i) <_{lex} (m_j, j)$. □

Before showing that τ is linearizable, we add few notations. For a non-initial update event U, $\alpha_j(U)$ is the p_j-update operation that wrote to V (line 4) the value read form V in U (line 5). Namely, $\alpha_j(U) = [\mu_j(V.\text{scan}(U))]$. If U is the initial update operation, $\alpha_j(U) = I_j$. Assume that S is an Fscan event in which $winner(S) = k$ (the find_max procedure execution in S returns k) and write $U_k = \beta_k(S)$. Then, we define $\alpha_j(S) = \alpha_j(U_k)$. The next lemma follows and its strait-forward proof is left for the reader.

Lemma 6. – *if U is a p_i-update event, then $\alpha_i(U) = U$.*
- *If $U < U'$ are two update operations and i a process id, then $\alpha_i(U) \leq \alpha_i(U')$.*
- *Let U, U' be two update operations by p_j and p_k respectively. If $(viewsum(U), j) <_{lex} (viewsum(U'), k)$, then for each process id i, $\alpha_i(U) \leq \alpha_i(U')$.*
- *Let S be an Fscan event, and for each process id i, assume that $\alpha_i(S)$ is invoked with value val_i. Then, S returns $F(val_0, \ldots, val_{n-1})$.*

Now we are ready to show that τ is a linearizable by defining a linear ordering $\prec$ over the high-level events in τ. First, we define $\prec$ over the update events. For two update events U, U', we set $U \prec U'$ if $V.\text{scan}(U) < V.\text{scan}(U')$. Now we linearize also Fscan operations by choosing for each Fscan operation an update operation to linearized after it. For an Fscan S, we linearize S after the $\prec$-maximal update event among $\{\alpha_0(S), \ldots, \alpha_{n-1}(S)\}$. Finally, if we linearized several Fscan events (say, $S_1, \ldots, S_l$) immediately after the same update event, we extend $\prec$ over these events in some arbitrary way that extends $<$ over $S_1, \ldots, S_l$.

It is easy to verify that $\prec$ is a linear ordering. To complete our proof, the next two lemmas show that $\prec$ extends $<$, and that $\prec$ belongs to the sequential specification of the F-snapshot object.

Lemma 7. $\prec$ *extends* $<$.

Proof. Assume that $A < B$ are two operations in τ. We need to show that $A \prec B$. The claim is obvious when both are update operations, and it is left to deal with all other cases.

First, assume that $A = U$ is an update operation, say by p_i and $B = S$ is an Fscan operation. For each process p_j, write $U_j = \beta_j(S)$. Note that since $U < S$, $U \leq U_i$. Assume that the $winner(S) = k$ thus $i \leq_S k$. By Lemma 5,

$(viewsum(U_i), i) \leq_{lex} (viewsum(U_k), k)$. Therefore, by Lemma 6 we conclude: $U \leq U_i = \alpha_i(U_i) \leq \alpha_i(U_k) = \alpha_i(S)$. Recall that S was linearized after $\alpha_i(S)$ thus, since $\prec$ extends $<$ over update events, $U \preceq U_i \preceq (\alpha_i(S)) \prec S$.

Now we deal with the case that $A = S$ is an Fscan operation and $B = U$ is an update operation. For showing that $S \prec U$, we prove that for each process id j, $\alpha_j(S) \prec U$. Assume that $winner(S) = k$, and write $U_k' = \beta_k(S)$. For process p_j, write $U_j = \alpha_j(S) = \alpha_j(U_k')$. If $U_k' = I_k$, then $U_j = I_j$ and then it is clear that $U_j \prec U$ as required. Otherwise, write $e_j = V.\text{update}(U_j)$, $s = V.\text{scan}(U_k')$ and $e = V.\text{update}(U)$. Note that $e_j = \mu_j(s)$ thus $e_j < s$. Since $s \in U_k' = \beta_k(S)$, $\neg(S < s)$. But since $S < U$, we conclude that $s < e$. As a result, $e_j < s < e$ which implies that $U_j \prec U$ as required.

Finally, assume that $A = S$ and $B = S'$ are two Fscan events. It suffices to show that for each process id i, $\alpha_i(S) \leq \alpha_i(S)$. Assume that $winner(S) = j$, and write $U_j = \beta_j(S)$. Then, $\alpha_i(S) = \alpha_i(U_j)$. Write $winner(S') = k$, and $U_k' = \beta_k(S')$. Hence, $\alpha_i(S') = \alpha_i(U_k')$. Since $S < S'$, $\beta_j(S) \leq \beta_j(S')$. Thus, by Lemmas 5 and 6,

$$(viewsum(U_j), j) \leq_{lex} (viewsum(\beta_j(S')), j) \leq_{lex} (viewsum(U_k', k)).$$

Therefore, by Lemma 6, $\alpha_i(S) = \alpha_i(U_j) \leq \alpha_i(U_k') = \alpha_i(S')$ as required.   □

Lemma 8. *Let S be an* Fscan *operation. For each process id i, let U_i be the maximal p_i-update operation that precedes S in $\prec$, and assumes that U_i is invoked with value val_i. Then, S returns $F(val(val_0), \ldots, val(val_{n-1}))$.*

Proof. By Lemma 6, it suffices to show that for each process id i, $\alpha_i(S) = U_i$. Since S was linearized after the events $\alpha_0(S), \ldots, \alpha_{n-1}(S)$, clearly $\alpha_i(S) \preceq U_i$. We need to show that $U_i \preceq \alpha_i(S)$.

Toward a contradiction, assume that $\alpha_i(S) \prec U_i \prec S$. We conclude that there is some process p_j such that $\alpha_i(S) \prec U_i \prec \alpha_j(S)$ since otherwise, S would have been linearized before U_i. Note that $\alpha_j(S) \neq I_j$. Assume that $winner(S) = k$ and write $U_k' = \beta_k(S)$. Thus, $\alpha_j(S) = \alpha_j(U_k')$. Since $\alpha_j(S)$ is not the initial p_j-event, also $U_k' \neq I_k$. As a result, we get:

$$V.\text{update}(\alpha_i(S)) < V.\text{update}(U_i) < V.\text{update}(\alpha_j(S)) < V.\text{scan}(U_k')$$

in contradiction to $\alpha_i(S) = [\mu_i(V.\text{scan}(U_k'))]$.   □

5   Conclusions

We present the F-snapshot problem which generalizes the signaling problem introduced in [3], from the two-process case to an arbitrary number of processes. We described a wait-free F-snapshot algorithm that employs only single-writer registers, and proved its correctness. Our algorithm uses four snapshot objects. For efficiency, we can use the snapshot implementation by Attiya and Rachman [9] for these objects. Thus, the time complexity of our algorithm is $O(n \log n)$. Since the *Flags* object is accessed during Fscan operations, it is required to use the bounded version of the algorithm in [9] (described in Sect. 4.4). As the F-snapshot problem generalizes the snapshot problem, where F is chosen to be the identity function, the F-snapshot problem inherits the linear time lower bound of the snapshot problem. The $O(n)$ lower bound holds for the Fscan procedure [21] and also for the update procedure [19].

It is known that the snapshot object can be implemented with time complexity $O(n)$ when multi-writer registers are allowed as Inoue, Masuzawa, Chen and Tokura proved [16]. Inoue et al. present an algorithm that solves the lattice agreement problem. Then, the reduction by Attiya, Herlihy and Rachman [8] provides a linear snapshot implementation with multi-writer registers. However, this reduction requires unbounded registers (and even unbounded memory). Hence, the F-snapshot limitations forbid using this implementation for the $Flags$ snapshot object in our algorithm. Therefore, the question if there is a linear F-snapshot implementation using multi-writer registers is not answered here, although there is a linear snapshot implementation that uses multi-writer registers.

The main idea behind our algorithm is, in some sense, orthogonal to the classical bounded timestamps problem. In a time-stamp system, the processes label their writes to an array of data values. These labels provide a linear-ordering that extends the actual partial-ordering between the writes to the array. In our algorithm, we use bounded data to label the ordering between scan events and not between write events. We are not aware of a formulation of this abstract problem. This is a possible direction for further research, influenced by ideas behind our algorithm.

By the essence of the F-snapshot problem, an interesting complexity measure is the size of the bounded registers that are accessed during an Fscan operation (named "flags"). For convenience, we assume that the range of F, $D = \{0, \ldots, |D| - 1\}$ and we note that $\Omega(\log |D|)$ is a trivial lower bound for the size of each flag. In our algorithm, each register writes to the snapshot object $Flags$ a data value of type $flag$. This data type consists of several fields when the largest are the sets: $winners$ and $losers$ that require $O(n)$ bits, and the field ans stores elements from D thus can be assumed to be consist of $\log |D|$ bits. However, the $Flags$ implementation require additional fields when the largest one stores a view: n-tuple of values of type $flag$. Therefore, the size of each flag in our algorithm is $O(n^2 + n \log |D|)$. We believe that this can be significantly improved.

In our algorithm, the Fscan procedure accesses only bounded registers due to the problem constrains and the update procedure accesses unbounded registers (otherwise the problem is unsolvable). The segments of the snapshot object V store elements from $Vals$ (which might be infinite), and counters that infinitely grow. Hence, if the function F has a finite domain, the update procedure will still access unbounded registers. Thus, in those cases, it is better to use some other implementation such as the bounded version of the algorithm in [9]. An interesting question that arises is whether there is an F-snapshot algorithm that satisfies both properties:

1. If F has a finite range, then the Fscan procedure accesses only bounded registers.
2. If F has a finite domain, then only bounded registers are accessed.

References

1. Afek, Y., Attiya, H., Dolev, D., Gafni, E., Merritt, M., Shavit, N.: Atomic snapshots of shared memory. J. ACM **40**(4), 873–890 (1993)
2. Aguilera, M.K., Gafni, E., Lamport, L.: The mailbox problem (extended abstract). In: Taubenfeld, G. (ed.) DISC 2008. LNCS, vol. 5218, pp. 1–15. Springer, Heidelberg (2008). doi:10.1007/978-3-540-87779-0_1
3. Aguilera, M.K., Gafni, E., Lamport, L.: The mailbox problem. Distrib. Comput. **23**(2), 113–134 (2010)
4. Alur, R., McMillan, K., Peled, D.: Model-checking of correctness conditions for concurrent objects. In: Logic in Computer Science (LICS), pp. 219–228 (1996)
5. Amram, G.: On the signaling problem. In: International Conference on Distributed Computing and Networking, pp. 44–65 (2014)
6. Anderson, J.: Composite registers. Distrib. Comput. **6**(3), 141–154 (1993)
7. Aspnes, J., Herlihy, M.: Wait-free data structures in the asynchronous PRAM model. In: Proceedings of the 2nd Annual ACM Symposium on Parallel Architectures and Algorithms, pp. 340–349 (1990)
8. Attiya, H., Herlihy, M., Rachman, O.: Atomic snapshots using lattice agreement. Distrib. Comput. **8**(3), 121–132 (1995)
9. Attiya, H., Rachman, O.: Atomic snapshots in $O(n \log n)$ operations. In: Proceedings of the 12th Annual ACM Symposium on Principles of Distributed Computing, pp. 29–40 (1993)
10. Dolev, D., Shavit, N.: Bounded concurrent time-stamp systems are constructible. In: Proceedings of 21st STOC, pp. 454–466 (1989)
11. Dolev, D., Shavit, N.: Bounded concurrent time-stamping. SIAM J. Comput. **26**(2), 418–455 (1997)
12. Dwork, C., Waarts, O.: Simple and efficient bounded concurrent timestamping or bounded concurrent timestamp systems are comprehensible! In: Proceedings of the Twenty-Fourth Annual ACM Symposium on Theory of Computing, pp. 655–666 (1992)
13. Guerraoui, R., Ruppert, E.: Linearizability is not always a safety property. In: Proceeding of the 2nd International Conference, Networked Systems, pp. 57–69 (2014)
14. Herlihy, M., Shavit, N.: The Art of Multiprocessor Programming. Morgan Kaufmann, New York (2008)
15. Herlihy, M., Wing, J.: Linearizability: a correctness condition for concurrent objects. ACM TOPLAS **12**(3), 463–492 (1990)
16. Inoue, M., Masuzawa, T., Chen, W., Tokura, N.: Linear-time snapshot using multi-writer multi-reader registers. In: Tel, G., Vitányi, P. (eds.) WDAG 1994. LNCS, vol. 857, pp. 130–140. Springer, Heidelberg (1994). doi:10.1007/BFb0020429
17. Israeli, A., Li, M.: Bounded time-stamps. Distrib. Comput. **6**(4), 205–209 (1993)
18. Israeli, A., Shaham, A., Shirazi, A.: Linear-time snapshot implementations in unbalanced systems. Math. Syst. Theor. **28**(5), 469–486 (1995)
19. Israeli, A., Shirazi, A.: The time complexity of updating snapshot memories. Inf. Process. Lett. **65**(1), 33–40 (1998)

20. Jayanti, P.: f-arrays: implementation and applications. In: Proceedings of the 21st Annual Symposium on Principles of Distributed Computing, pp. 270–279 (2002)
21. Jayanti, P., Tan, K., Toueg, S.: Time, space lower bounds for nonblocking implementations. SIAM J. Comput. **30**(2), 438–456 (2000)
22. Lamport, L.: Proving the correctness of multiprocess Programs. IEEE Trans. Softw. Eng. **3**(2), 125–145 (1977)

t-Resilient Immediate Snapshot Is Impossible

Carole Delporte[1], Hugues Fauconnier[1],
Sergio Rajsbaum[2], and Michel Raynal[3(✉)]

[1] IRIF/GANG, Université Paris Diderot, Paris, France
[2] Instituto de Matemáticas, UNAM, 04510 México D.F., Mexico
[3] IUF, IRISA (Université de Rennes), Rennes, France
raynal@irisa.fr

Abstract. An immediate snapshot object is a high level communication object, built on top of a read/write distributed system in which all except one processes may crash. It allows each process to write a value and obtains a set of pairs (process id, value) such that, despite process crashes and asynchrony, the sets obtained by the processes satisfy noteworthy inclusion properties.

Considering an n-process model in which up to t processes are allowed to crash (t-crash system model), this paper is on the construction of t-resilient immediate snapshot objects. In the t-crash system model, a process can obtain values from at least $(n - t)$ processes, and, consequently, t-immediate snapshot is assumed to have the properties of the basic $(n - 1)$-resilient immediate snapshot plus the additional property stating that each process obtains values from at least $(n - t)$ processes. The main result of the paper is the following. While there is a (deterministic) $(n-1)$-resilient algorithm implementing the basic $(n-1)$-immediate snapshot in an $(n-1)$-crash read/write system, there is no t-resilient algorithm in a t-crash read/write model when $t \in [1 \ldots (n - 2)]$. This means that, when $t < n - 1$, the notion of t-resilience is inoperative when one has to implement t-immediate snapshot for these values of t: the model assumption "at most $t < n - 1$ processes may crash" does not provide us with additional computational power allowing for the design of a genuine t-resilient algorithm (genuine meaning that such an algorithm would work in the t-crash model, but not in the $(t + 1)$-crash model). To show these results, the paper relies on well-known distributed computing agreement problems such as consensus and k-set agreement.

Keywords: Asynchronous system · Atomic read/write register · Consensus · Distributed computability · Immediate snapshot · Impossibility · Iterated model · k-Set Agreement · Linearizability · Process crash failure · Snapshot object · t-Resilience · Wait-freedom

1 Introduction

Immediate Snapshot Object and Iterated Immediate Snapshot Model. The *immediate snapshot* (IS) communication object was first introduced in [6,33], and

© Springer International Publishing AG 2016
J. Suomela (Ed.): SIROCCO 2016, LNCS 9988, pp. 177–191, 2016.
DOI: 10.1007/978-3-319-48314-6_12

then further investigated as an "object" in [5]. The associated *iterated immediate snapshot* (IIS) model was introduced in [7,20]. This distributed computing model consists of n asynchronous processes, among which any subset of up to $(n-1)$ processes may crash[1], which execute a sequence of asynchronous rounds. One and only one immediate snapshot (IS) object is associated with each round, which allows the processes to communicate during this round. More precisely, for any $x > 0$, a process accesses the x-th immediate snapshot only when it executes the x-th round, and it accesses it only once.

From an abstract point of view, an IS object $IMSP$, can be seen as an initially empty set, which can then contain at most n pairs (one per process), each made up of a process index and a value. This object provides the processes with a single operation denoted write_snapshot(), that each process may invoke only once. The invocation $IMSP$.write_snapshot(v) by a process p_i adds the pair $\langle i, v \rangle$ to $IMSP$ and returns a set of pairs belonging to $IMSP$ such that the sets returned to the processes that invoke write_snapshot() satisfy specific inclusion properties. It is important to notice that, in the IIS model, the processes access the sequence of IS objects one after the other, in the same order, and asynchronously.

The noteworthy feature of the IIS model is the following. It has been shown by Borowsky and Gafni in [7], that this model is equivalent to the usual read/write wait-free model ($(n-1)$-crash model) for task solvability with the wait-freedom progress condition (any non-faulty process obtains a result). Its advantage lies in the fact that its runs are more structured and easier to analyze than the runs in the basic read/write shared memory model [27]. It is also the basis of the combinatorial topology approach for distributed computing (e.g., [17]). Hence, IS objects constitute the algorithmic foundation of distributed iterated computing models.

It has been shown in [30] that trying to enrich the IIS model with (non trivial) failure detectors is inoperative. This means that, for example, enriching IIS with the failure detector Ω (which is the weakest failure detector that allows consensus to be solved in the basic read/write communication model [10,24]) does not allow to solve consensus in such an enriched IIS model. However, it has been shown in [29] that it is possible to capture the power of a failure detector (and other partially synchronous systems) in the IIS model by appropriately restricting its set of runs, giving rise to the *Iterated Restricted Immediate Snapshot* (IRIS) model. This approach has been further investigated in [32].

The IIS model has many interesting features among which the following two are noteworthy. The first is on the foundation side of distributed computing, namely IIS established a strong connection linking distributed computing and algebraic topology (see [6,17,19,21,33]). The second one lies on the algo-

---

[1] From a terminology point of view, we say *t-failure* model (in the present case *t-crash model*) if the model allows up to t processes to fail. We keep the term *t-resilience* for algorithms. The $(n-1)$-crash model is also called *wait-free* model [16]. Several progress conditions have been associated with *(n-1)-resilient* algorithms: wait-freedom [16], non-blocking [22], or obstruction-freedom [18]. (See a unified presentation in Chap. 5 of [31].).

rithmic and programming side, namely IIS allows for a recursive formulation of algorithms solving distributed computing problems. This direction, initiated in [5,15], has also been investigated in [28,31].

Another line of research is investigated in [14]. This paper considers models of distributed computations defined as subsets of the runs of the iterated immediate snapshot model. In such a context, it uses topological techniques to identify the tasks that are solvable in such a model.

t-Crash Model and t-Resilient Algorithms. The previous basic read/write model and IIS model consider that all but one process may crash. Differently, a *t*-crash model assumes that at most *t* processes may crash, i.e., by assumption, at least $(n - t)$ of them never crash. As already said, an algorithm designed for such a model is said to be *t*-resilient.

One of the most fundamental results of distributed computing is the impossibility to design a 1-resilient consensus algorithm in the 1-crash *n*-process model, be the communication medium an asynchronous message-passing system [13] or a read/write shared memory [25]. Differently, other problems, such as renaming (introduced in the context of *t*-resilient message-passing systems where $t < n/2$ [3]), can be solved by $(n - 1)$-resilient algorithms in the $(n - 1)$crash read/write shared memory model (such renaming algorithms are described in several textbooks, e.g. [4,31,34]).

Contribution of the Paper. When considering the *t*-crash *n*-process model where $t < n - 1$, and assuming that each correct process writes a value, a process may wait for values written by $(n - t)$ processes without risking being blocked forever. This naturally leads to the notion of a *t*-crash *n*-process iterated model, generalizing the IIS model to any value of *t*. To this end the paper introduces the notion of a *k*-immediate snapshot object, which generalizes the basic $(n - 1)$-immediate snapshot object. More precisely, when considering a *t*-immediate snapshot object in a *t*-crash *n*-process model, an invocation of write_snapshot() by a process returns a set including at least $(n - t)$ pairs (while it would return a set of *x* pairs with $1 \leq x \leq n$ if the object was an IS object). Hence, a *t*-immediate snapshot object allows processes to obtain as much information as possible from the other processes while guaranteeing progress.

The obvious question is then the implementability of a *t*-immediate snapshot object in the *t*-crash *n*-process model. This question is answered in this paper, which shows that it is impossible to implement a *t*-IS object in a *t*-crash *n*-process model when $0 < t < n-1$. More precisely we prove that implementing a *t*-IS object is equivalent[2] to implementing consensus when $t < n/2$ and enables to implement $(2t - n + 2)$-set agreement when $n/2 \leq t < n - 1$.

At first glance, this impossibility result may seem surprising. An IS object is a snapshot object (a) whose operations write() and snapshot() are glued together in a single operation write_snapshot(), and (b) satisfying an additional property linking the sets of pairs returned by concurrent invocations (called *Immediacy*

---

[2] A is equivalent to B if A can be (computationally) reduced to B and reciprocally.

property, Sect. 2.2). Then, as already indicated, a t-IS object is an IS object such that the sets returned by write_snapshot() contain at least $(n - t)$ pairs (*Output size* property, Sect. 2.4). The same Output size property on the sets returned by a snapshot object can be trivially implemented in a t-crash n-process model. Let us call t-snapshot such a constrained snapshot object. Hence, while a t-snapshot object can be implemented in the t-crash n-process model, a t-IS object cannot when $0 < t < n - 1$.

Roadmap. As previously indicated, the paper is on the computability power of t-IS objects in the t-crash computing model, for $t < n - 1$. Made up of Sect. 7 sections, it has the following content.

- Section 2 introduces the basic crash-prone read/write system model, immediate snapshot, a k-set agreement, and k-immediate snapshot (k-IS). It also proves a theorem which captures the additional computational power of k-immediate snapshot with respect to the basic $(n - 1)$-immediate snapshot.
- Assuming a majority of processes never crash, i.e. a t-crash read/write model in which $t < n/2$, Sect. 3 shows that it is impossible to implement t-immediate snapshot in such a model. The proof is a reduction of the consensus problem to t-immediate snapshot.
- Assuming $t \leq n - 1$, Sect. 4 presents a reduction of t-immediate snapshot to consensus in a t-crash read/write model. When combined with the result of Sect. 3, this shows that t-immediate snapshot and consensus have the same computational power in any t-crash model where $t < n/2$.
- Assuming a t-crash read/write model in which $n/2 \leq t < n - 1$, Sect. 5 shows that it is impossible to implement t-immediate snapshot in such a model. The proof is a reduction of the $(2t - n + 2)$-set agreement problem to t-immediate snapshot.
- By a simulation argument, Sect. 6 shows that consensus is not solvable with t-immediate snapshot when $n/2 \leq t < n$ proving that the computational power of t-immediate snapshot when $0 < t < n/2$ is strictly stronger than the computational power of t-immediate snapshot when $n/2 \leq t < n$.

Finally, Sect. 7 concludes the paper.

2   Immediate Snapshot, k-Set Agreement, and k-Immediate Snapshot

2.1   Basic Read/Write System Model

Processes. The computing model is composed of a set of $n \geq 3$ sequential processes denoted $p_1, ..., p_n$. Each process is asynchronous which means that it proceeds at its own speed, which can be arbitrary and remains always unknown to the other processes.

A process may halt prematurely (crash failure), but executes correctly its local algorithm until it possibly crashes. The model parameter t denotes the maximal number of processes that may crash in a run. A process that crashes in a run is said to be *faulty*. Otherwise, it is *correct* or *non-faulty*. Let us notice that, as a faulty process behaves correctly until it crashes, no process knows if it is correct or faulty. Moreover, due to process asynchrony, no process can know if another process crashed or is only very slow.

It is assumed that (a) $0 < t < n$ (at least one process may crash and at least one process does not crash), and (b) any process, until it possibly crashes, executes the algorithm assigned to it.

Communication Layer. The processes cooperate by reading and writing Single-Writer Multi-Reader (SWMR) atomic read/write registers [23]. This means that the shared memory can be seen as a set of arrays $A[1 \ldots n]$ where, while $A[i]$ can be read by all processes, it can be written only by p_i.

Notation. The previous model is denoted $\mathcal{CARW}_{n,t}[\emptyset]$ (which stands for "Crash Asynchronous Read/Write with n processes, among which up to t may crash"). A model constrained by a predicate on t (e.g. $t < x$) is denoted $\mathcal{CARW}_{n,t}[t < x]$. Hence, as we assume at least one process does not crash, $\mathcal{CARW}_{n,t}[t < n]$ is a synonym of $\mathcal{CARW}_{n,t}[\emptyset]$, which (as always indicated) is called *wait-free* model. When considering t-crash models, $\mathcal{CARW}_{n,t}[t \leq \alpha]$ is less constrained than $\mathcal{CARW}_{n,t}[t < \alpha - 1]$.

Shared objects are denoted with capital letters. The local variables of a process p_i are denoted with lower case letters, sometimes suffixed by the process index i.

2.2   One-Shot Immediate Snapshot Object

The immediate snapshot (IS) object was informally presented in the introduction. It can be seen as a variant of the snapshot object introduced in [1,2]. While a snapshot object provides the processes with two operations (write() and snapshot()) which can be invoked separately by a process (usually write() before snapshot()), a immediate snapshot provides the processes with a single operation write_snapshot(). One-shot means that a process may invoke write_snapshot() at most once.

Definition. An IS object *IMSP* is a set, initially empty, that will contain pairs made up of a process index and a value. Let us consider a process p_i that invokes *IMSP*.write_snapshot(v). This invocation adds the pair $\langle i, v \rangle$ to *IMSP* (contribution of p_i to *IMSP*), and returns to p_i a set, called view and denoted $view_i$, such that the sets returned to the processes collectively satisfy the following properties.

- Termination. The invocation of write_snapshot() by a correct process terminates.
- Self-inclusion. $\forall i : \langle i, v \rangle \in view_i$.
- Validity. $\forall i : (\langle j, v \rangle \in view_i) \Rightarrow p_j$ invoked write_snapshot(v).

– Containment. $\forall\, i, j :\ (view_i \subseteq view_j) \lor (view_j \subseteq view_i)$.
– Immediacy. $\forall\, i, j :\ (\langle i, v \rangle \in view_j) \Rightarrow (view_i \subseteq view_j)$.

It is relatively easy to show that the Immediacy property can be re-stated as follows: $\forall\, i, j :\ \big((\langle i, - \rangle \in view_j) \land (\langle j, - \rangle \in view_i)\big) \Rightarrow (view_i = view_j)$.

Implementation. Implementations of an IS object in $\mathcal{CARW}_{n,t}[0 < t < n]$ (classical read/write wait-free model) are described in [5,15,28,31]. While both a one-shot snapshot object and an IS object satisfy the Self-inclusion, Validity and Containment properties, only an IS object satisfies the Immediacy property. This additional property creates an important difference, from which follows that, while a snapshot object is atomic (operations on a snapshot object can be linearized [22]), an IS object is not atomic (its operations cannot always be linearized). However, an IS object is set-linearizable (set-linearizability allows several operations to be linearized at the same point of the time line [9,26]).

The Iterated Immediate Snapshot (IIS) Model. In this model (introduced in [7]), the shared memory is composed of a (possibly infinite) sequence of IS objects: $IMSP[1]$, $IMSP[2]$, ... These objects are accessed sequentially and asynchronously by the processes according to the following round-based pattern executed by each process p_i. The variable r_i is local to p_i; it denotes its current round number.

> $r_i \leftarrow 0$; $\ell s_i \leftarrow$ initial local state of p_i (including its input, if any);
> **repeat forever** % asynchronous IS-based rounds
> $\quad r_i \leftarrow r_i + 1$;
> $\quad view_i \leftarrow IMSP[r_i].\mathsf{write\_snapshot}(\ell s_i)$;
> $\quad$ computation of a new local state ℓs_i (which contains $view_i$)
> **end repeat.**

As indicated in the Introduction, when considering distributed tasks (as formally defined in [8,21]), the IIS model and $\mathcal{CARW}_{n,t}[0 < t < n]$ have the same computational power [7].

2.3 k-Set Agreement

k-Set agreement was introduced by S. Chaudhuri [11] to investigate the relation linking the number of different values that can be decided in an agreement problem, and the maximal number of faulty processes. It generalizes consensus which corresponds to the case $k = 1$.

A k-set agreement object is a one-shot object that provides the processes with a single operation denoted $\mathsf{propose}_k()$. This operation allows the invoking process p_i to propose a value it passes as an input parameter (called *proposed* value), and obtain a value (called *decided* value). The object is defined by the following set of properties.

– Termination. The invocation of $propose_k()$ by a correct process terminates.
– Validity. A decided value is a proposed value.
– Agreement. No more than k different values are decided.

It is shown in $[6, 21, 33]$ that the problem is impossible to solve in $CARW_{n,t}[k \leq t]$.

2.4 k-Immediate Snapshot

A k-immediate snapshot object (denoted k-IS) is an immediate snapshot object with the following additional property.

– Output size. The set *view* obtained by a process is such that $|view| \geq n - k$.

Theorem 1. *A k-IS object cannot be implemented in $CARW_{n,t}[k < t]$.*

Proof. To satisfy the output size property, the view obtained by a process p_i must contain pairs from $(n - k)$ different processes. If t processes crash (e.g. initially), a process can obtain at most $(n - t)$ pairs. If $t > k$, we have $n - t < n - k$. It follows that, after it has obtained pairs from $(n - t)$ processes, a process can remain blocked forever waiting for the $(t - k)$ missing pairs.   □

Considering the system model $CARW_{n,t}[0 < t < n - 1]$, the next theorem characterizes the power of a t-IS object in term of the Containment property.

Theorem 2. *Considering the system model $CARW_{n,t}[0 < t < n - 1]$, and a t-IS object, let us assume that all correct processes invoke write_snapshot(). No process obtains a view with less than $(n - t)$ pairs. Moreover, if the size of the smallest view obtained by a process is ℓ $(\ell \geq n - t)$, there is a set S of processes such that $|S| = \ell \geq n - t$ and each process of S obtains the smallest view or crashes during its invocation of write_snapshot().*

Proof. It follows from the Output size property of the t-IS object that no view contains less than $(n - t)$ pairs. Let *view* be the smallest view returned by a process, and let $\ell = |view|$. We have $\ell \geq n - t$. Moreover, due to (a) the Immediacy property (namely $(\langle i, - \rangle \in view) \Rightarrow (view_i \subseteq view)$) and (b) the minimality of *view*, it follows that $view_i = view$. As this is true for each process whose pair participates in *view*, and $\ell = |view|$, it follows that there is a set S of processes such that $|S| = \ell \geq n - t$ and each of its processes obtains the view *view*, or crashed during its invocation of write_snapshot(). Due to the Containment property, the others processes crash or obtain views which strictly include *view*.   □

3 t-Immediate Snapshot Is Impossible in $CARW_{n,t}[0 < t < N/2]$

This section shows that it is impossible to implement a t-IS object when $0 < t < n/2$.

From t-IS to Consensus in $\mathcal{CARW}_{n,t}[0 < t < n/2]$. Algorithm 1 reduces consensus to t-IS in the system model $\mathcal{CARW}_{n,t}[0 < t < n/2]$. As at most $t < n/2$ process may crash, at least $n-t > n/2t$ processes invoke the consensus operation $\mathsf{propose}_1()$.

---

operation $\mathsf{propose}_1(v)$ **is**
(1) $view_i \leftarrow IMSP.\mathsf{write\_snapshot}(v); VIEW[i] \leftarrow view_i;$
(2) $\mathsf{wait}(|\{\ j \text{ such that } VIEW[j] \neq \bot\}| = t + 1);$
(3)   **let** $view$ **be** the smallest of the previous $(t + 1)$ views;
(4) $\mathsf{return}(\text{smallest proposed value in } view)$
end operation.

Algorithm 1: Solving consensus in $\mathcal{CARW}_{n,t}[0 < t < n/2, t\text{-IS}]$ (code for p_i)

In addition to a t-IS object denoted $IMSP$, the processes access an array $VIEW[1 \ldots n]$ of SWMR atomic registers, initialized to $[\bot, \cdots, \bot]$. The aim of $VIEW[i]$ is to store the view obtained by p_i from the t-IS object $IMSP$.

When it calls $\mathsf{propose}_1(v)$, a process p_i invokes first the t-IS object, in which it deposits the pair $\langle i, v \rangle$, and obtains a view from it, that it writes in $VIEW[i]$ to make it publicly known (line 1). Then, it waits (line 2) until it sees the views of at least $(t+1)$ processes (as $n - t \geq t+1$, p_i cannot block forever and at least one of these views is from a correct process). Process p_i extracts then of these views the one with the smallest cardinality (line 3), and finally returns proposed value contained in this smallest view (line 4).

Theorem 3. *Algorithm 1 reduces consensus to t-IS in* $\mathcal{CARW}_{n,t}[0 < t < n/2]$.

Proof. Let us first prove the consensus Termination property. As $n-t \geq t+1$, and there are at least $(n-t)$ correct processes, it follows that at least $(n-t)$ entries of $VIEW[1 \ldots n]$ are eventually different from $\bot$. Hence, no correct process can remain blocked forever at line 2, which proves consensus Termination.

Let us now consider the consensus Agreement property. It follows from Theorem 2 that there is a set of at least $\ell \geq n-t$ processes, that obtained the same view min_view (or crashed before returning from $\mathsf{write\_snapshot}()$), and this view is the smallest view obtained by a process and its size is $|min\_view| = \ell$. As $\ell \geq n - t$ and $(n - t) + (t + 1) > n$, it follows from the waiting predicate of line 2, that, any process that executes line 3, obtains a copy of min_view, and consequently we have $view = min\_view$ at line 3. It follows that no two processes can decide different values.

Finally, the consensus Validity property follows from the fact that any pair contained in a view is composed of a process index and the value proposed by the corresponding process. $\qquad\Box$

Corollary 1. *Implementing a t-IS object in* $\mathcal{CARW}_{n,t}[0 < t < n/2]$ *is impossible.*

Proof. The proof is an immediate consequence of Lemma 3, and the fact that consensus cannot be solved in $\mathcal{CARW}_{n,t}[0 < t < n/2]$ [25].   □

4   From Consensus to *t*-IS in $\mathcal{CARW}_{n,t}[0 < t \leq N - 1]$

Algorithm 2 describes a reduction of *t*-IS to consensus in $\mathcal{CARW}_{n,t}[0 < t \leq n-1]$. This algorithm uses two shared data structures. The first is an array $REG[1 \ldots n]$ of SWMR atomic registers (where $REG[i]$ is associated with p_i). The second is an array of $(t + 1)$ consensus objects denoted $CONS[(n - t) \ldots n]$.

---

operation write_snapshot(v_i) **is**
(1) $REG[i] \leftarrow v_i$; $view_i \leftarrow \emptyset$; $dec_i \leftarrow \emptyset$; $\ell \leftarrow -1$; launch the tasks $T1$ and $T2$.

(2)   **task** $T1$ **is**
(3)   **repeat** $\ell \leftarrow \ell + 1$;
(4)   wait$\big(\exists$ a set aux_i: $(dec_i \subset aux_i) \wedge (|aux_i| = n - t + \ell)$
                    $\wedge \ (aux_i \subseteq \{\langle j, REG[j]\rangle$ such that $REG[j] \neq \perp\})\big)$;
(5) $dec_i \leftarrow CONS[n - t + \ell].\text{propose}_1(aux_i)$;
(6)   **if** $(\langle i, v_i \rangle \in dec_i) \wedge (view_i = \emptyset)$ **then** $view_i \leftarrow dec_i$ **end if**
(7)   **until** $(\ell = t)$ **end repeat**
(8)   **end task** $T1$.

(9)   **task** $T2$ **is** wait($view_i \neq \emptyset$); return($view_i$) **end task** $T2$.
end operation.

Algorithm 2: Implementing *t*-IS in $\mathcal{CARW}_{n,t}[0 < t < n, \text{CONS}]$ (code for p_i)

The invocation of write_snapshot(v_i) by a process p_i deposits v_i in $REG[i]$, and launches two underlying tasks $T1$ and $T2$. The task $T2$ is a simple waiting task, which will return a view to the calling process p_i. The return() statement at line 9 terminates the write_snapshot() operation invoked by p_i. The termination of $T2$ does not kill the task $T1$ which may continue executing.

Task $T1$ (lines 2–8) has two aims: provide p_i with a view $view_i$ (line 6), and prevent processes from deadlocking, thereby allowing them to terminate. It consists in a loop that is executed $(t + 1)$ times. The aim of the ℓ-th iteration (starting at $\ell = 0$) is to allow processes to obtain a view including $(n - t + \ell)$ pairs. More precisely, we have the following.

– When it enters the ℓ-th iteration, a process p_i first waits until it obtains a set of pairs, denoted aux_i, which (a) contains $(n - t + \ell)$ pairs, (b) contains the set of pairs dec_i decided during the previous iteration, and (c) contains only pairs extracted from the array $REG[1 \ldots n]$. This is captured by the predicate of line 4.
– Then, p_i proposes the set aux_i to the consensus object $CONS[n - t + \ell]$ associated with the current iteration step (line 5). The set decided is stored in dec_i.

– Finally, if its pair $\langle i, v_i \rangle$ belongs to dec_i and p_i has not yet decided (i.e., no set has yet been assigned to $view_i$), it does it by writing dec_i in $view_i$. Let us notice that this ensures the Self-inclusion property of the t-IS object. Moreover, a process decides no more than once.

Whether a process decides or not during the current iteration step, it systematically proceeds to the next iteration step. Hence, a process that obtains its view during an iteration step x can help other processes to obtain a view during later iteration steps $y > x$.

Theorem 4. *Algorithm 2 reduces t-IS to consensus in $\mathcal{CARW}_{n,t}[0 < t \leq n-1]$.*

Proof. The Self-inclusion property follows directly from the predicate $\langle i, v_i \rangle \in dec_i$ used before assigning dec_i to $view_i$ at line 6.

The Validity property follows from (a) the fact that a process p_i assigns the value it wants to deposit in the t-IS object in $REG[i]$, (b) this atomic variable is written at most once (line 1), and (c) the predicate $REG[j] \neq \perp$ is used at line 4 to extract values from $REG[1 \ldots n]$.

The Output size property follows from the predicate of line 4, which requires that any set aux_i (and consequently any set dec_i output by a consensus object) contains at least $(n - t)$ pairs.

To prove the Immediacy property, let us consider any two processes p_i and p_j such that $\langle j, v_j \rangle \in view_i$ and $\langle i, v_i \rangle \in view_j$. Let $dec_x[\ell]$ denote the local variable dec_x after p_x assigned it a value at line 5 during iteration step ℓ.

Let ℓ_i be the iteration step at which p_i assigns dec_i to $view_i$ (due to the predicate $view_i = \emptyset$ used at line 5, such an assignment is done only once). It follows from the first predicate of line 6, that $\langle i, v_i \rangle \in dec_i[\ell_i] = view_i$ (otherwise, $view_i$ would not be assigned dec_i); ℓ_j, dec_j, and $view_j$ being defined similarly, we also have $\langle j, v_j \rangle \in dec_j[\ell_j] = view_j$. As by assumption we have $\langle j, v_j \rangle \in view_i$ and $\langle i, v_i \rangle \in view_j$, we also have $\{\langle i, v_i \rangle, \langle j, v_j \rangle\} \subseteq dec_i[\ell_i] = view_i$ and $\{\langle i, v_i \rangle, \langle j, v_j \rangle\} \subseteq dec_j[\ell_j] = view_j$. Due to the Agreement property of the consensus objects, we have $dec_i[\ell_i] = dec_j[\ell_i]$, and $dec_i[\ell_j] = dec_j[\ell_j]$.

Let us assume that $\ell_i < \ell_j$. This is not possible because, on the one side, $\langle j, v_j \rangle \in dec_i[\ell_i] = dec_j[\ell_i]$, and, on the other side, ℓ_j is the only iteration step at which we have $\langle j, v_j \rangle \in dec_j \wedge view_j = \emptyset$ (and consequently $view_j$ is assigned the value in $dec_j[\ell_j]$). For the same reason, we cannot have $\ell_i > \ell_j$. It follows that $\ell_i = \ell_j$. Hence, as $dec_i[\ell_i] = dec_j[\ell_j]$, p_i and p_j obtain the very same view (and this occurs during the same iteration step).

As far as the Containment property is concerned, we have the following. Considering the iteration number ℓ, let us first observe that, due to the predicate $|aux_i| = n - t + \ell$ (line 4), the set output by $CONS[n - t + \ell]$ contains $n - t + \ell$ pairs. Hence, the sequence of consensus outputs sets whose size is increased by 1 at each instance. Let us now observe that, due to the predicate $dec_i \subset aux_i$ (line 4), the set output by $CONS[n - t + \ell + 1]$ is a superset of the set output by the previous consensus instance $CONS[n - t + \ell]$. It follows that the sequence of pairs output by the consensus instances is such that each set of pairs includes the previous set plus one new element, from which the Containment property follows.

As far as the Termination property is concerned, let p be the number of processes that have deposited a value in $REG[1\dots n]$. We have $n - t \le p \le n$. It follows from the predicate in the wait statement (line 4), that no process can block forever at this line for $\ell \in [0\dots p-n+t]$. As there are at least $(n-t)$ correct processes, and none of them can be blocked forever at line 4, it follows that each of them invokes $CONS[n-t+\ell].\mathsf{propose}_1()$ (line 5), for each $\ell \in [0\dots p-n+t]$. Hence, the only reason for a correct process not to obtain a view (and terminate), is to never execute the assignment $view_i \leftarrow dec_i$ at line 7.

The sequence of consensus instances outputs a sequence of sets of pairs whose successive sizes are $(n - t)$, $(n - t + 1)$, ..., p, which means that the identity of every of the p processes that wrote in $REG[1\dots n]$ appears at least once in the sequence of consensus outputs. Hence, for each correct process p_i, there is a consensus instance whose output dec is such that, while $view_i = \emptyset$, we have $\langle i, v_i \rangle \in dec$, which concludes the proof of the Termination property. $\square$

Corollary 2. *Consensus and t-IS are equivalent in* $\mathcal{CARW}_{n,t}[0 < t < n/2]$.

Proof. The proof follows from Theorem 3 (Algorithm 1) and Theorem 4 (Algorithm 2). $\square$

5 t-Immediate Snapshot Is Impossible in $\mathcal{CARW}_{n,t}[n/2 \le T < n - 1]$

This section shows that it is impossible to implement a t-IS object in the system model $\mathcal{CARW}_{n,t}[n/2 \le t < n - 1]$. To this end, it presents a reduction of k-set agreement (in short k-SA) to t-IS for $k = 2t - n + 2$ (e.g., a reduction of $(n - 2)$-SA agreement to $(n - 2)$-IS in $\mathcal{CARW}_{n,t}[t = n - 2]$).

From t-IS to (2t-k+2)-set Agreement in $\mathcal{CARW}_{n,t}[n/2 \le t < n - 1, t - IS]$. Algorithm 3 reduces $(2t - n + 2)$-set agreement to t-IS in $\mathcal{CARW}_{n,t}[n/2 \le t < n - 1]$. As at most t process may crash, at least $(n - t)$ processes invoke the k-SA operation $\mathsf{propose}_k()$. This algorithm is very close to Algorithm 1. Its main difference lies in the replacement of $(t + 1)$ by $(n - t)$ at line 2.

---

operation $\mathsf{propose}_{2t-n+2}(v)$ **is**
(1) $view_i \leftarrow IMSP.\mathsf{write\_snapshot}(v);\ VIEW[i] \leftarrow view_i;$
(2) $\mathsf{wait}(|\{\ j\ \text{such that}\ VIEW[j] \ne \bot\}| = n - t);$
(3)   **let** $view$ **be** the smallest of the previous $(n - t)$ views;
(4)   **return**(smallest proposed value in $view$)
end operation.

Algorithm 3: Solving $(2t-n+2)$-set agreement in $\mathcal{CARW}_{n,t}[n/2 \le t < n-1, t\text{-IS}]$ (code for p_i)

Theorem 5. *Algorithm 3 reduces* $(2t - n + 2)$-*set agreement to* t-*IS in* $\mathcal{CARW}_{n,t}[n/2 \le t < n - 1]$.

Proof. Let $k = 2t - n + 2$.

Let us first consider the k-SA Termination property. There are at least $(n-t)$ correct processes, and each of them first invokes $IMSP$.write_snapshot() and then writes the view it obtained in the shared array $VIEW$ (line 1). Hence, at least $(n-t)$ entries of $VIEW$ are eventually different from $\bot$, from which follows that no process can block forever at line 2.

Let us now consider the k-SA Validity property. It follows from the Containment property of the t-IS object that any set of views deposited in $VIEW$ is not empty. Therefore, the view selected by a process at line 3 is not empty. As a view can only contain pairs, each including a proposed value (line 1), the k-SA Validity property follows.

Let us finally consider the k-SA Agreement property. Let us first observe that, due to the t-IS Containment property and Theorem 2, at most $n-(n-t)+1 = t+1$ different views can be written in the array $VIEW[1 \ldots n]$. Let $V(1)$ the smallest of these views (which contains $\ell \ge n - t$ pairs), $V(2)$ the second smallest, etc., until $V(t + 1)$ the greatest one. There are two cases according to the $(n - t)$ non-$\bot$ views obtained by a process p_i at line 2. Let us remind that, as $n \le 2t$, we have $n - t \le t$.

- Case 1. The view $V(1)$ belongs to the $(n - t)$ views obtained by p_i. In this case, p_i selects $V(1)$ at line 3 and decides at line 4 the smallest proposed value contained in $V(1)$.
- Case 2. The view $V(1)$ does not belong to the $(n - t)$ views obtained by p_i. Hence, the $(n - t)$ views obtained by any process of Case 2 belong to $\{V(2), \cdots, V(t+1)\}$.
  It follows that the $m = (n - t) - 1$ biggest views in $\{V(2), \cdots, V(t+1)\}$ will never be selected be the processes that are in Case 2, and consequently the set of these processes obtain at most $t - m = t - ((n - t) - 1) = 2t - n + 1$ different smallest views. Hence, these processes may decide at most $2t - n + 1$ different values at line 4.

When combining the two cases, at most $k = 2t - n + 2$ different values can be decided, which concludes the proof of the theorem. $\square$

Corollary 3. *Implementing a* t-*IS object in* $\mathcal{CARW}_{n,t}[n/2 \le t < n - 1]$ *is impossible.*

Proof. As $t \le n-2$, we have $2t-n+2 \le t$. The proof is an immediate consequence of Theorem 5, and the fact that $(2t - n + 2)$-set agreement cannot be solved in $\mathcal{CARW}_{n,t}[n/2 \le t < n - 1]$ [5,21,33]. $\square$

6 t-Immediate Snapshot and Consensus in $\mathcal{CARW}_{n,t}[n/2 \le T < n - 1]$

Theorem 6. *There is no* t-*resilient consensus algorithm using* t-*immediate snapshot in* $\mathcal{CARW}_{n,t}[n/2 \le t < n - 1]$.

Table 1. Summary of results presented in the paper

$1 \le t < n/2$	$n/2 \le t < n-1$
t-IS implements *t*-CONS (Theorem 3)	*t*-IS implements $(2t - n + 2)$-SA (Theorem 5)
	t-IS does not implement *t*-CONS (Theorem 6)
t-CONS implements *t*-IS (Theorem 4)	*t*-CONS implements *t*-IS (Theorem 4)

The proof the theorem is by contradiction. It assume that there is a *t*-resilient consensus algorithm $\mathcal{A}$ for a set of processes $\{p_1, \cdots, p_n\}$, which uses a *t*-immediate snapshot object in a system where $n = 2t$ (the cases for the other values of *t* can easily be reduced to this case).

The contradiction is obtained by simulating $\mathcal{A}$ with two processes Q_0 and Q_1, such that Q_0 and Q_1 solve consensus despite the possible crash of one of them. As there is no wait-free consensus algorithm for 2 processes, it follows that such a consensus algorithm $\mathcal{A}$ based on *t*-immediate snapshot objects does not exist. The proof can be found in [12].

7 Conclusion

This paper addressed the design of *t*-tolerant algorithms building a *t*-immediate snapshot (*t*-IS) object. Such an object in an immediate snapshot object (defined by Termination, Self-inclusion, Containment, and Immediacy properties), in a *t*-crash asynchronous system. Hence, it is required that each set returned to a process contains at least $(n - t)$ pairs. Immediate snapshot corresponds to $(n - 1)$-immediate snapshot.

The paper has shown that, while it is possible to build an $(n - 1)$-IS object in the asynchronous read/write $(n - 1)$-crash model, it is impossible to build a *t*-IS object in an asynchronous read/write *t*-crash model when $0 < t < n - 1$. It follows that the notion of an IIS distributed model seems inoperative for these values of *t*. The results of the paper are summarized in Table 1 where *t*-CONS denotes the consensus in the presence of up to *t* process crashes, and SA stands for "Set Agreement".

Interestingly, this study shows that there are two contrasting impossibility results in asynchronous read/write *t*-crash *n*-process systems, each one lying at an end of the *t*-resilience spectrum. Consensus is impossible as soon as $t > 0$, while *t*-immediate snapshot is impossible as soon as $t < n - 1$.

As a final remark, some computability problems remain open. As an example, is it possible to implement a *t*-IS object from $(2t - n + 2)$-Set agreement?

Acknowledgments. The authors want to thank the referees for their constructive comments. This work was been partially supported by the French ANR project DIS-PLEXITY devoted to the study of Computability and Complexity in distributed computing, and the UNAM-PAPIIT project IN107714.

References

1. Afek, Y., Attiya, H., Dolev, D., Gafni, E., Merritt, M., Shavit, N.: Atomic snapshots of shared memory. J. ACM **40**(4), 873–890 (1993)
2. Anderson, J.: Multi-writer composite registers. Distrib. Comput. **7**(4), 175–195 (1994)
3. Attiya, H., Bar-Noy, A., Dolev, D., Peleg, D., Reischuk, R.: Renaming in an asynchronous environment. J. ACM **37**(3), 524–548 (1990)
4. Attiya, H., Welch, J.: Distributed Computing: Fundamentals, Simulations and Advanced Topics, 2nd edn. Wiley-Interscience, New York (2004). 414 pages
5. Borowsky E., Gafni E.: Immediate atomic snapshots and fast renaming. In: Proceedings of the 12th ACM Symposium on Principles of Distributed Computing (PODC 1993), pp. 41–50 (1993)
6. Borowsky E. and Gafni E., Generalized FLP impossibility results for t-resilient asynchronous computations. In: Proceedings of the 25th ACM Symposium on Theory of Computation (STOC 1993), California, USA, pp. 91–100 (1993)
7. Borowsky E., Gafni E.: A simple algorithmically reasoned characterization of wait-free computations. In: Proceedings of the 16th ACM Symposium on Principles of Distributed Computing (PODC 1997), pp. 189–198. ACM Press (1997)
8. Borowsky, E., Gafni, E., Lynch, N., Rajsbaum, S.: The BG distributed simulation algorithm. Distrib. Comput. **14**, 127–146 (2001)
9. Castañeda, A., Rajsbaum, S., Raynal, M.: Specifying concurrent problems: beyond linearizability and up to tasks. In: Moses, Y. (ed.) DISC 2015. LNCS, vol. 9363, pp. 420–435. Springer, Heidelberg (2015). doi:10.1007/978-3-662-48653-5_28
10. Chandra, T., Hadzilacos, V., Toueg, S.: The weakest failure detector for solving consensus. J. ACM **43**(4), 685–722 (1996)
11. Chaudhuri, S.: More choices allow more faults: set consensus problems in totally asynchronous systems. Inf. Comput. **105**(1), 132–158 (1993)
12. Delporte, C., Fauconnier, H., Rajsbaum, S., Raynal, M.: t-resilient immediate snapshot is impossible. Technical report 2036, IRISA, Université de Rennes (F): http://hal.inria.fr/hal-01313556
13. Fischer, M.J., Lynch, N.A., Paterson, M.S.: Impossibility of distributed consensus with one faulty process. J. ACM **32**(2), 374–382 (1985)
14. Gafni E., Kuznetsov P., and Manolescu C., A generalized asynchronous computability theorem. In: Proceedings of the 33th ACM Symposium on Principles of Distributed Computing (PODC 1994), pp. 222–231. ACM Press (2014)
15. Gafni, E., Rajsbaum, S.: Recursion in distributed computing. In: Dolev, S., Cobb, J., Fischer, M., Yung, M. (eds.) SSS 2010. LNCS, vol. 6366, pp. 362–376. Springer, Heidelberg (2010). doi:10.1007/978-3-642-16023-3_30
16. Herlihy, M.P.: Wait-free synchronization. ACM Trans. Program. Lang. Syst. **13**(1), 124–149 (1991)
17. Herlihy, M.P., Kozlov, D., Rajsbaum, S.: Distributed Computing Through Combinatorial Topology. Morgan Kaufmann/Elsevier, New York (2014). 336 pages. ISBN 9780124045781
18. Herlihy, M.P., Luchangco, V., Moir, M.: Obstruction-free synchronization: double-ended queues as an example. In: Proceedings of the 23th International IEEE Conference on Distributed Computing Systems (ICDCS 2003), pp. 522–529. IEEE Press (2003)
19. Herlihy, M., Rajsbaum, S., Raynal, M.: Power and limits of distributed computing shared memory models. Theor. Comput. Sci. **509**, 3–24 (2013)

20. Herlihy, M.P., Shavit, N.: A simple constructive computability theorem for wait-free computation. In: Proceedings of the 26th ACM Symposium on Theory of Computing (STOC 1994), pp. 243–252. ACM Press (1994)
21. Herlihy, M.P., Shavit, N.: The topological structure of asynchronous computability. J. ACM **46**(6), 858–923 (1999)
22. Herlihy, M.P., Wing, J.M.: Linearizability: a correctness condition for concurrent objects. ACM Trans. Programm. Lang. Syst. **12**(3), 463–492 (1990)
23. Lamport, L.: On interprocess communication, Part I: basic formalism. Distrib. Comput. **1**(2), 77–85 (1986)
24. Lo, W.-K., Hadzilacos, V.: Using failure detectors to solve consensus in asynchronous shared-memory systems. In: Tel, G., Vitányi, P. (eds.) WDAG 1994. LNCS, vol. 857, pp. 280–295. Springer, Heidelberg (1994). doi:10.1007/BFb0020440
25. Loui, M., Abu-Amara, H.: Memory requirements for agreement among unreliable asynchronous processes. Adv. Comput. Res. **4**, 163–183 (1987)
26. Neiger G., Set-linearizability. In: Brief Announcement in Proceedings of the 13th ACM Symposium on Principles of Distributed Computing (PODC 1994), p. 396. ACM Press (1994)
27. Rajsbaum, S.: Iterated shared memory models. In: López-Ortiz, A. (ed.) LATIN 2010. LNCS, vol. 6034, pp. 407–416. Springer, Heidelberg (2010). doi:10.1007/978-3-642-12200-2_36
28. Rajsbaum, S., Raynal, M.: An introductory tutorial to concurrency-related distributed recursion. Bull. Eur. Assoc. TCS **111**, 57–75 (2013)
29. Rajsbaum, S., Raynal, M., Travers, C.: The iterated restricted immediate snapshot model. In: Hu, X., Wang, J. (eds.) COCOON 2008. LNCS, vol. 5092, pp. 487–497. Springer, Heidelberg (2008). doi:10.1007/978-3-540-69733-6_48
30. Rajsbaum, S., Raynal, M., Travers, C.: An impossibility about failure detectors in the iterated immediate snapshot model. Inf. Process. Lett. **108**(3), 160–164 (2008)
31. Raynal, M.: Concurrent Programming: Algorithms, Principles and Foundations. Springer, Heidelberg (2013). 515 pages. ISBN 978-3-642-32026-2
32. Raynal, M., Stainer, J.: Increasing the power of the iterated immediate snapshot model with failure detectors. In: Even, G., Halldórsson, M.M. (eds.) SIROCCO 2012. LNCS, vol. 7355, pp. 231–242. Springer, Heidelberg (2012). doi:10.1007/978-3-642-31104-8_20
33. Saks, M., Zaharoglou, F.: Wait-free *k*-set agreement is impossible: the topology of public knowledge. SIAM J. Comput. **29**(5), 1449–1483 (2000)
34. Taubenfeld, G.: Synchronization Algorithms and Concurrent Programming. Pearson Prentice-Hall, Upper Saddle River (2006). 423 pages. ISBN 0-131-97259-6

Mobile Agents

Linear Search by a Pair
of Distinct-Speed Robots

Evangelos Bampas[1], Jurek Czyzowicz[2], Leszek Gąsieniec[3], David Ilcinkas[4],
Ralf Klasing[4], Tomasz Kociumaka[5(✉)], and Dominik Pająk[6]

[1] LIF, CNRS, Aix-Marseille University, Marseille, France
evangelos.bampas@lif.univ-mrs.fr
[2] Département d'informatique,
Université du Québec en Outaouais, Gatineau, Canada
jurek.czyzowicz@uqo.ca
[3] Department of Computer Science, University of Liverpool, Liverpool, UK
L.A.Gasieniec@liverpool.ac.uk
[4] LaBRI, CNRS, University of Bordeaux, Talence, France
{david.ilcinkas,ralf.klasing}@labri.fr
[5] Institute of Informatics, University of Warsaw, Warsaw, Poland
kociumaka@mimuw.edu.pl
[6] Institute of Informatics, Wrocław University of Technology, Wrocław, Poland
dominik.pajak@pwr.edu.pl

Abstract. Two mobile robots are initially placed at the same point on
an infinite line. Each robot may move on the line in either direction not
exceeding its maximal speed. The robots need to find a stationary target
placed at an unknown location on the line. The search is completed when
both robots arrive at the target point. The target is discovered at the
moment when either robot arrives at its position. The robot knowing
the placement of the target may communicate it to the other robot. We
look for the algorithm with the shortest possible search time (i.e. the
worst-case time at which both robots meet at the target) measured as a
function of the target distance from the origin (i.e. the time required to
travel directly from the starting point to the target at unit velocity).

We consider two standard models of communication between the
robots, namely *wireless communication* and *communication by meeting*.
In the case of communication by meeting, a robot learns about the target
while sharing the same location with the robot possessing this knowledge.

J. Czyzowicz—Partially funded by NSERC. Part of this work was done while Jurek
Czyzowicz was visiting the LaBRI as a guest professor of the University of Bordeaux.
L. Gąsieniec—Sponsored in part by the University of Liverpool initiative Networks
Systems and Technologies NeST.
D. Ilcinkas—Partially funded by the ANR project MACARON (ANR-13-JS02-002).
This study has been carried out in the frame of the "Investments for the future"
Programme IdEx Bordeaux – CPU (ANR-10-IDEX-03-02).
R. Klasing—Partially funded by the ANR project DISPLEXITY (ANR-11-BS02-014).
D. Pająk—Partially funded by the National Science Centre, Poland - grant number
2015/17/B/ST6/01897.

© Springer International Publishing AG 2016
J. Suomela (Ed.): SIROCCO 2016, LNCS 9988, pp. 195–211, 2016.
DOI: 10.1007/978-3-319-48314-6_13

We propose here an optimal search strategy for two robots including the respective lower bound argument, for the full spectrum of their maximal speeds. This extends the main result of Chrobak et al. (SOFSEM 2015) referring to the exact complexity of the problem for the case when the speed of the slower robot is at least one third of the faster one. In addition, we consider also the wireless communication model, in which a message sent by one robot is instantly received by the other robot, regardless of their current positions on the line. In this model, we design an optimal strategy whenever the faster robot is at most 6 times faster than the slower one.

1   Introduction

Searching is a well-studied problem in which mobile robots need to find a specific target placed at some a priori unknown location. In some cases, a team of robots is involved, trying to coordinate their efforts in order to minimize the time. The complexity of the multi-robot searching is usually defined as the time when the first searcher arrives at the target position whose location is controlled by an adversary.

In distributed computing, one of the central problems is *rendezvous* when two mobile robots collaborate in order to meet in the smallest possible time. The efficiency of the rendezvous strategy is expressed as the time when the last involved robot reaches the meeting point, and the meeting point is arbitrary, i.e., the robots may choose the most convenient one.

In the *linear search* problem studied in the present paper, a pair of robots has to meet at an unknown fixed target point of the environment and the time complexity of the process is determined by the arrival of the second robot. More specifically we consider two mobile robots placed at the origin of an infinite line. Each robot has its maximal speed that it cannot exceed while moving in either direction along the line. There is a stationary target, placed at an unknown point of the line, that a robot discovers when arriving at its placement. The robot which possesses the knowledge of the target position may communicate it to the other robot. We consider two communication models of the robots: *communication by meeting* when the robots can exchange information only while being located at the same position, and *wireless communication* when the robot finding the target may instantaneously inform the other robot of its position. We want to schedule the movement of both robots so that eventually each of them arrives at the target location. The cost of the schedule is the first time when both robots are present at the target position. We express it as a function of the distance between the target and the origin.

1.1   Related Work

Numerous papers have been written on the searching problem, studying diverse models involving stationary or mobile targets, graph or geometric terrain, known or unknown environment, one or many searchers, etc. (cf. [1,3,17,21]). Depending on the setting, the problem is known under the name of treasure hunting, pursuit-evasion, cops and robbers, fugitive search games, etc. Sometimes the searching robot is not looking for an individual target point, attempting rather to evacuate being lost in an unknown environment or determine its position within a known map (e.g. [12,15]). Several of these research papers offer exciting challenges of combinatorial or algorithmic nature (see [17]). In most papers studying algorithmic issues, the objective is either to determine the feasibility of the search, (i.e., whether the search will succeed under all adversarial choices) or to minimize its cost represented by the search time, assuming some given speeds of searchers (and perhaps evaders).

Most of the time searching is considered for a single robot. As one robot usually cannot map the graph being explored (unless e.g., leaving pebbles at some nodes; see [8]), the second searcher makes the task feasible (cf. [9]). However, optimization of the search by the use of multiple robots often involves coordination issues, where the searchers need to communicate in order to synchronize their efforts and adequately split the entire task into portions assigned to individual robots (cf. [11,14,16,18]). As this objective is often not easy to achieve, some multi-robot search problems turn out to be NP-hard (e.g., see [18]).

Several papers on searching consider online algorithms (cf. [19]), where the information about the environment is acquired as the search progresses. The performance of an online algorithm is measured by its *competitive ratio*, i.e., the worst-case ratio of its cost with respect to the offline cost, which is the search time of the optimal algorithm with full *a priori* knowledge of the environment and the target placement. Many search problems, especially for geometric environments, are analyzed from this perspective, in particular when the cost of the offline solution is just the distance to the target; see [3,11,16,19].

The *linear search* problem for a single robot was introduced by Beck [6] and Bellman [7]. They proposed an optimal on-line algorithm with search time $9d$, where d is the distance from the origin to the target. This question was extended to the *cow-path problem* in [2], in which the searcher has more than two directions to follow, to searching in the plane [3], and numerous other variations. Bose et al. [10] recently studied a variant of these problems where upper and lower bounds on the distance to the target are given. On a line, without this information the time $9d$ cannot be improved even if the search is performed by a team of same-speed robots communicating by meeting if all robots have to reach the target [11]; see also [4]. Surprisingly, time $9d$ can still be achieved by distinct-speed robots if the slowest robot is at most 3 times slower than the fastest one.

The *rendezvous problem* has been central to distributed computing for many years. It was studied in various settings (cf. [22]), but even for environments as simple as a line or a ring, optimal solutions are not always known. Feasibility of the rendezvous problem is often determined by a symmetry breaking process, which must prevent the robots from falling into an infinite pattern avoiding the meeting. Searching and rendezvous may be viewed as problems with opposite objectives. Searching is a game between a searcher, who tries to find the target as fast as possible and the adversary, who knows the searching strategy and attempts to maximize the search time by its choice of the environment parameters, target placement (or its escape route), etc. Hence in searching, the two players have contradictory goals. In rendezvous the two players collaborate, trying to quickly find one another (see [1]). Contrary to the searching problem, the rendezvous destination is not given in advance but it may be decided by the robots.

Equivalent to our setting are *evacuation problems*, where a collection of mobile robots need to find an unknown exit in the environment and the exit must be reached by all involved robots. In previous research usually robots travelling at the same speed were considered (cf. [11,12]). For other problems considering robots with distinct speeds (e.g., the patrolling problem studied in [13,20]), only partial results were obtained. Optimal patrolling using more than two robots on a ring [13], or more than three robots on a segment [20], is unknown in general and all intuitive solutions have been proved sub-optimal for some configurations of the speeds of the robots. Another example is the long-standing *lonely runner* conjecture [23], concerning k entities moving with constant speeds around a circular track of unit-length. If the speeds are pairwise different, the conjecture states that at some moment all runners are located equidistantly on the cycle. The conjecture is open in general, having been verified for up to 7 runners [5].

1.2   Our Results

In this paper, we consider the linear search problem for two robots equipped with distinct maximal speeds. For the convenience of presentation we scale their speeds so that the speed of the faster robot is 1 and the slower one is $0 < v \leq 1$.

In the model with communication by meeting, we propose an optimal strategy for any value of v. And in particular our strategy works in time $\frac{1+3v}{v-v^2}d$, for any $v \leq \frac{1}{3}$ for the target being placed at unknown distance d from the origin. The remaining part of the spectrum has been covered in [11] where the authors provide: an implicit (in the limit) argument for the lower bound $9d$ when the robots share the maximal speed 1; and they show that this bound can be met from above when the slower robot's maximal speed is at least $\frac{1}{3}$.

In the model with wireless communication, we design a strategy achieving search time $\frac{2+v+\sqrt{v^2+8v}}{2v}d$. We show that this is optimal for any $v \geq \frac{1}{6}$. Note that for $v > \sqrt{17}-4 \approx 0.123$ our strategy for wireless communication outperforms the optimal strategy for communication by meeting, which shows that the feature of wireless communication is useful. On the other hand, one can observe that this feature becomes less significant as v decreases. For $v = 1$, the optimal algorithm for wireless communication is 3 times faster than the optimal algorithm for communication by meeting whereas for $v = \frac{1}{6}$, it is only 1.08 times faster.

2   Preliminaries

For any algorithm $\mathcal{A}$, we denote by $t(\mathcal{A}, p)$ the search time of algorithm $\mathcal{A}$ if the target is located at point p. We define $\tau(\mathcal{A}) = \limsup_{|p| \to \infty} \frac{t(\mathcal{A},p)}{|p|}$ as the main efficiency measure of the algorithms. Whereas all the lower bounds we derive hold for the efficiency measure $\tau(\mathcal{A})$, all the algorithms $\mathcal{A}$ we design actually satisfy the stronger property $t(\mathcal{A}, p) \leq \tau(\mathcal{A})|p|$ for every point $p \in \mathbb{R}$ (sometimes by making infinitesimal moves as the time approaches 0). In consequence, our bounds are in particular directly adaptable to a setting where the target placement must lie at a distance at least x from the origin, where x is a fixed constant, and one measures performance of the algorithms using $\sup\{\frac{t(\mathcal{A},p)}{|p|} : |p| > x\}$.

Having fixed an algorithm $\mathcal{A}$ for a set $\mathcal{R}$ of robots, each robot $\Gamma \in \mathcal{R}$ follows a fixed trajectory while it is unaware of the location of the target. We use $\Gamma(t)$ to denote the position of robot Γ at time t provided that the target location is not known to the robot. Our lower bounds rely on the analysis of the *progress speeds* $\limsup_{t\to\infty} \frac{|\Gamma(t)|}{t}$. The largest of these values over $\Gamma \in \mathcal{R}$ is called the *overall progress speed*. For each point p, the time $T(p) = \min\{t : \exists_{\Gamma \in \mathcal{R}} \Gamma(t) = p\}$ is called the *discovery time* of p (it is the first moment when any robot visits p) and $\phi(p)$ denotes the set of robots which visit p at time $T(p)$. To simplify notation, we will not make explicit the dependence of $\Gamma(t)$, $T(p)$, and $\phi(p)$ on the algorithm $\mathcal{A}$. Our results are primarily designed for a set $\mathcal{R}$ of two of robots, denoted R and r. Their speed limits are 1 and v ($v \leq 1$), respectively.

3   Communication by Meeting

In this model, once a robot finds the target, it must walk to meet the other robot, and then the robots travel to the target. Naturally, the schedule consists of three phases: *exploration phase* while the target is unknown, *pursuit phase* where the informed robot chases after the other one in order to tell it about the target, and *target phase* when both robots walk to the target location. Recall that for robots with equal speeds, one of the possible optimal solutions consists in both robots following together a cow-path trajectory [4,11], thus the pursuit and target phases may be nonexistent.

3.1   The Upper Bound

A robot following a standard cow-path trajectory visits, in order of increasing k, the points $p_k := (-2)^k$, $k \in \mathbb{Z}$, on alternating sides of the origin, travelling at full speed between consecutive points p_k.[1] In this strategy, the robot discovers new locations after it passes p_k on the way from p_{k+1} to p_{k+2}. This happens from

---

[1] Note that the sequence $(p_k)_{k\in\mathbb{Z}}$ is understood as prescribing infinitesimally small moves for the robot in the two directions around the origin at the beginning of the execution (when time is in the neighborhood of 0, i.e., at the beginning of the execution, the robot visits points p_k for k in the neighborhood of $-\infty$, hence it makes infinitesimal moves). Algorithm $\mathcal{A}^*$, described below, has similar behavior. In order to avoid this, we could start the sequence p_k from any finite k (instead of $-\infty$). This would result in small constant additive terms appearing throughout the calculations, but the asymptotic behavior of the algorithm and in particular the efficiency measure $\tau(\mathcal{A})$ would be unaffected.

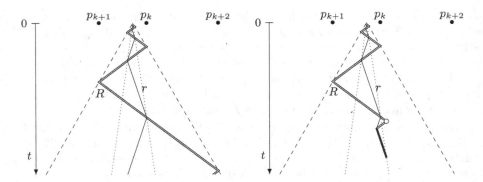

Fig. 1. Illustration of algorithm $\mathcal{A}^*$ before target detection (left), and when the target has been located (right). The horizontal axis represents the line searched and the vertical axis represents the time. The empty circle denotes the target discovery. Double and single solid lines represent the trajectories of the faster and the slower robot, respectively. Dashed lines correspond to the overall progress speed and dotted lines to the search time.

time $t_k := |p_k| + 2 \sum_{j=-\infty}^{k+1} |p_j| = 9 \cdot 2^k = 9|p_k|$ to $t'_{k+2} := |p_{k+2}| + 2 \sum_{j=-\infty}^{k+1} |p_j| = 12 \cdot 2^k = 3|p_{k+2}|$. Consequently, the search time is bounded from above by $9|p|$.

As observed by Chrobak et al. [11], this strategy generalizes to a collection of two robots with speed limits 1 and $\frac{1}{3}$. Both robots follow the cow-path trajectory at their maximal speed, which means that they meet in p_k at time $t_k = 3t'_k$. When the faster robot R discovers the target at a point p between p_k and p_{k+2}, it pursues the slower robot r and brings it to the target, which turns out to be feasible within time $9|p|$; see Fig. 1.

Algorithm $\mathcal{A}^*$ [for two robots with communication by meeting]

---

1. Until the target is located, both robots visit, in order of increasing k, the points $p_k = (-c)^k$ for all $k \in \mathbb{Z}$, where $c = \frac{1+\tilde{v}}{2\tilde{v}}$ and $\tilde{v} = \min(v, \frac{1}{3})$. Robot R moves with speed 1 between consecutive points, and robot r with speed $\tilde{v}$.
2. When R finds the target, it moves with speed 1 to meet and notify r.
3. After the meeting, robots move together to the target at speed $\tilde{v}$.

We extend this strategy to allow $v < \frac{1}{3}$ as the speed limit of the slower robot r. We insist on the two robots meeting in points p_k at times t_k for adjusted values p_k and t_k. The smaller speed v of r allows R to travel further before going back to p_k. More formally, we increase the ratio $|p_{k+1}|/|p_k|$ and instead of taking $p_k = (-2)^k$, we set $p_k = (-c)^k$ for some $c > 2$. We still make both robots visit consecutive points p_k at their full speeds, and we choose c so that they meet in p_k while r is there for the first time and R for the second time. A condition inductively forcing the meeting at p_k to be followed by a meeting in p_{k+1} can be expressed as $\frac{1}{v}|p_{k+1} - p_k| = t_{k+1} - t_k = |p_{k+1} - p_{k+2}| + |p_{k+2} - p_k|$, i.e., $\frac{1}{v}(c+1) = 2c^2 + c - 1$. This gives $c = \frac{1+v}{2v}$, which we use for our algorithm $\mathcal{A}^*$.

The following theorem bounds the search time by robots using this strategy.

Theorem 1. *For the algorithm $\mathcal{A}^*$ and every point $p \in \mathbb{R}$, we have:*

$$t(\mathcal{A}^*, p) = \frac{1+3v}{v-v^2}|p| \qquad\qquad \text{if } v \leq \tfrac{1}{3}, \qquad\qquad (1)$$

$$t(\mathcal{A}^*, p) = 9|p| \qquad\qquad \text{if } \tfrac{1}{3} < v \leq 1. \qquad\qquad (2)$$

Proof. First, let us show (1). Let us choose k so that the target p is located between p_k and p_{k+2}. The meeting time in p_k is

$$t_k = \frac{1}{v}\left(|p_k| + 2\sum_{j\leq k-1}|p_j|\right) \leq \frac{1}{v}c^k\left(1 + \frac{2}{c-1}\right) = \frac{1}{v}c^k\frac{c+1}{c-1} = c^k\frac{1+3v}{v-v^2}.$$

Suppose that $|p - p_k| = \delta$. After meeting r in p_k, robot R needs time δ to discover the target. At that time, the distance between the robots is $\delta(1 + v)$ since they were going in opposite directions with their maximal speeds until time $t_k + \delta$. Then, the faster robot pursues the slower one. With the speed difference of $1 - v$ this takes $\frac{\delta(1+v)}{1-v}$ units of time. Next, the robots go back to the target at speed v which requires time $\frac{\delta(1+v)}{v-v^2}$, i.e., $\frac{1}{v}$ times more than the pursuit. In total, the time between t_k and the moment when both robots reach the target is

$$\delta + \frac{\delta(1+v)}{1-v} + \frac{\delta(1+v)}{v-v^2} = \delta\frac{v-v^2+v+v^2+1+v}{v-v^2} = \delta\frac{1+3v}{v-v^2}.$$

Since $t_k = |p_k|\frac{1+3v}{v-v^2}$, the total search time is $t(\mathcal{A}, p) = (|p_k| + \delta)\frac{1+3v}{v-v^2} = |p|\frac{1+3v}{v-v^2}$, as claimed.

To show (2), we simply observe that, for $v = \frac{1}{3}$, we have $\frac{1+3v}{v-v^2} = 9$. Note that for $v > \frac{1}{3}$, the searcher moving at velocity $\frac{1}{3}$ could increase its speed to v, but no additional gain in efficiency is possible (see the lower bounds in [4, 11] and in Sect. 3.2). $\qquad\square$

3.2 The Lower Bound

We show that the strategy from Sect. 3.1 is optimal, achieving the best possible bound on the search time. In fact, some results of this section are presented in order to work for collections $\mathcal{R}$ of any number of robots. Consequently, in this section v denotes the slowest maximal speed among all the robots in $\mathcal{R}$, and r denotes some robot with maximal speed v. We also define $\tau^* = \frac{1+3v}{v-v^2}$ and, for any fixed algorithm $\mathcal{A}$, the overall progress speed $w = \max_{\Gamma \in \mathcal{R}} \limsup_{t\to\infty} \frac{|\Gamma(t)|}{t}$. Note that the functions Γ and w depend on $\mathcal{A}$, but we do not make this relation explicit in our notation.

Before we proceed with the actual lower bound, let us prove a lemma relating the search time and the overall progress speed for any collection $\mathcal{R}$ of robots.

Lemma 2. *For any algorithm $\mathcal{A}$ and any collection $\mathcal{R}$ of robots with speeds not exceeding 1, we have $\tau(\mathcal{A}) \geq \frac{1+3w}{w-w^2}$, when $w \in (0,1)$. If $w = 0$ or $w = 1$, then $\tau(\mathcal{A})$ cannot be bounded from above by any finite number.*

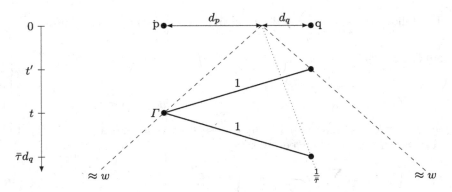

Fig. 2. Illustration of notions used in the proof of Lemma 2. Rays starting from the origin as well as thick lines representing constraints are all annotated with the corresponding speeds. Here, robot Γ, while in p at time t, must know that the target is not in q, or it must be able to reach q before the deadline.

Proof. We proceed with a proof by contradiction. That is, we suppose that $\tau(\mathcal{A})$ can be bounded from above if $w \in \{0, 1\}$, and that $\tau(\mathcal{A}) < \frac{1+3w}{w-w^2}$ if $w \in (0, 1)$. In both cases, the assumption implies the existence of a finite $\bar{\tau}$ such that $\tau(\mathcal{A}) < \bar{\tau}$ and $(w - w^2)\bar{\tau} < 1 + 3w$. The former condition yields that there exists d_0 such that $t(\mathcal{A}, p) < \bar{\tau}|p|$ for $|p| \geq d_0$. We will obtain a contradiction with respect to the latter condition.

Let us fix $\varepsilon > 0$. Note that there exists t_0 such that $\frac{|\Gamma(t)|}{t} \leq w + \varepsilon$ for every $t \geq t_0$ and every robot $\Gamma \in \mathcal{R}$. Also, there exists a robot Γ and arbitrarily large time values t such that $\frac{|\Gamma(t)|}{t} \geq w - \varepsilon$. We fix such a robot Γ and time t, which satisfies $t \geq (\bar{\tau} - 1) \max(d_0, t_0)$.

Let $p = \Gamma(t)$ and $d_p = |p|$. Also, consider a point q at distance $d_q = \frac{t+d_p}{\bar{\tau}-1}$ from the origin on the opposite side of p; see Fig. 2. Note that $d_q \geq d_0$, so $t(\mathcal{A}, q) < \bar{\tau}d_q$.

Suppose that robot Γ at time t cannot exclude the possibility that the target is located at q. Then, it must be able to reach q by the deadline, at $t(\mathcal{A}, q) < \bar{\tau}d_q$, starting at time t from point p. The robot cannot exceed the speed limit of 1, so we conclude $\bar{\tau}d_q - t > d_p + d_q$. However, the distance d_q is defined so that $\bar{\tau}d_q - t = d_p + d_q$, a contradiction.

Consequently, robot Γ must already know at time t that the target is not at point q. Since robots can only communicate by meeting and their speeds are limited by 1, this information needs $d_q + d_p$ time to travel from q to p. In other words, some robot Γ' must have visited q at time $t' \leq t - d_p - d_q$.

On the other hand, the speed limit of Γ' is at most 1, so we have $t' \geq d_q \geq t_0$. Hence, we can use a stronger bound using progress speed: $d_q = \Gamma'(t') \leq t'(w+\varepsilon)$. Consequently, we obtain $d_q \leq (w+\varepsilon)(t-d_p-d_q)$. Plugging in the definition of d_q, after some term rearrangements, we get $(1+w+\varepsilon)(t+d_p) \leq (t-d_p)(w+\varepsilon)(\bar{\tau}-1)$. Equivalently, $d_p \leq t\frac{(w+\varepsilon)(\bar{\tau}-2)-1}{(w+\varepsilon)\bar{\tau}+1}$. However, recall that time t was chosen so that

$d_p \geq (w-\varepsilon)t$. Therefore, $w-\varepsilon \leq \frac{(w+\varepsilon)(\bar{\tau}-2)-1}{(w+\varepsilon)\bar{\tau}+1}$. As $\varepsilon > 0$ can be chosen arbitrarily close to 0, we conclude that $w \leq \frac{w(\bar{\tau}-2)-1}{w\bar{\tau}+1}$, that is $(w - w^2)\bar{\tau} \geq 1 + 3w$. This contradicts the definition of $\bar{\tau}$.   □

The following immediate corollary gives an alternative proof of the optimality of $\mathcal{A}^*$ for $v \geq \frac{1}{3}$. (Recall the lower bound of 9 in [11]; see also [4].)

Corollary 3. *For any algorithm $\mathcal{A}$ and any collection $\mathcal{R}$ of robots with speeds not exceeding 1, we have $\tau(\mathcal{A}) \geq 9$.*

Proof. It suffices to observe that $\frac{1+3w}{w-w^2} \geq 9$ for any $w \in (0,1)$.   □

We continue the analysis assuming that $v < \frac{1}{3}$ and $w \in (0,1)$. We provide a series of lemmas, each imposing certain constraints on hypothetical algorithms $\mathcal{A}$ satisfying $\tau(\mathcal{A}) < \tau^*$. Eventually, we deduce that some of these constraints exclude each other. Due to space restrictions, in this version of the paper some proofs are only sketched, with rigorous arguments deferred to the full version.

Lemma 4. *If $v < \frac{1}{3}$ and $\tau(\mathcal{A}) < \tau^*$, then $w < \frac{1-v}{1+3v}$.*

Proof. Suppose that $w \geq \frac{1-v}{1+3v}$. Note that $w \geq \frac{1-v}{1+3v} > \frac{1-\frac{1}{3}}{1+1} = \frac{1}{3}$ (because $v < \frac{1}{3}$) and the function $f(x) = \frac{1+3x}{x-x^2}$ is increasing on $(\frac{1}{3},1)$. Thus $\frac{1+3v}{v-v^2} = f(\frac{1-v}{1+3v}) \leq f(w) = \frac{1+3w}{w-w^2}$. Consequently, Lemma 2 implies $\tau(\mathcal{A}) \geq f(w) \geq \frac{1+3v}{v-v^2} = \tau^*$.   □

Lemma 5. *If $v < \frac{1}{3}$ and $\tau(\mathcal{A}) < \tau^*$, then $\limsup_{t\to\infty} \frac{|r(t)|}{t} < \frac{vw\tau^* - wv - v - w}{vw\tau^* + 1}$.*

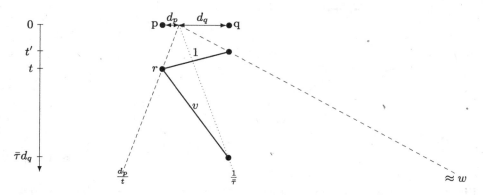

Fig. 3. Illustration of notions used in the proof of Lemma 5. The slowest robot r, while in p at time t, must know that the target is not in q or it must be able to reach q before the deadline.

Proof (sketch). We choose an arbitrarily large time t. Let $p = r(t)$ and $d_p = |p|$. We also consider a point q at distance $d_q = \frac{vt+d_p}{v\bar{\tau}-1}$ from the origin on the opposite side of p. Here, $\bar{\tau}$ is an arbitrary value such that $\tau(\mathcal{A}) < \bar{\tau} < \tau^*$ and $\frac{1}{v} < \bar{\tau} < \tau^* = \frac{1}{v} + \frac{4}{1-v}$. We may assume $t(\mathcal{A}, q) < \bar{\tau}d_q$ if t is sufficiently large.

The distance d_q is defined so that $\frac{1}{v}(d_q + d_p) + t = \bar{\tau}d_q$. Hence, it is impossible for the slower robot r to reach point q before $\bar{\tau}d_q > t(\mathcal{A}, q)$, starting from p at time t. Consequently r already knows at time t that the target is not located at q. Hence, some robot must have visited q at time $t' \leq t - d_p - d_q$, where the inequality is due to the fact that information cannot travel faster than at speed 1. On the other hand, the progress speed w gives an upper bound on $\frac{d_q}{t'}$ as t' approaches infinity. We combine these two inequalities to bound $\frac{d_p}{t}$ from above and derive the claimed result.   □

Corollary 6. *If $v < \frac{1}{3}$ and $\tau(\mathcal{A}) < \tau^*$, then $\limsup_{t \to \infty} \frac{|r(t)|}{t} < \frac{1}{\tau^*}$. In particular, the set $\{p : r \in \phi(p)\}$ of points discovered by r is bounded.*

Proof (sketch). By Lemma 4, we may assume $w < \frac{1-v}{1+3v}$. Upon substituting this inequality into the upper bound of Lemma 5, this implies $\frac{vw\tau^* - wv - v - w}{vw\tau^* + 1} < \frac{1}{\tau^*}$. Thus, the slowest robot visits sufficiently far points only after the deadline. To arrive at some location earlier, it must be notified by some other robot about the target location.   □

While Lemmas 4 and 5 and Corollary 6 hold for arbitrary collections of robots, this is not the case for the following lemma.

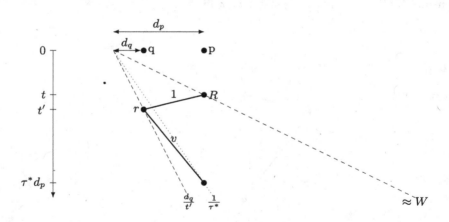

Fig. 4. Illustration of notions used in the proof of Lemma 7. The faster robot R, having discovered at time t the target located at p, must be able to catch the slower robot r and bring it to the target before the deadline.

Lemma 7. *If $v < \frac{1}{3}$ and $\tau(\mathcal{A}) < \tau^*$, then $\limsup_{t \to \infty} \frac{|r(t)|}{t} \geq \frac{vW+v+W-vW\tau^*}{vW\tau^*+1}$ where $W = \frac{w-w^2}{1+3w}$.*

Proof (sketch). By Corollary 6, we may assume that the faster robot discovers all sufficiently far locations. Thus, its own progress speed is equal to the overall progress speed w. Moreover, the trajectory of R can be interpreted as a search algorithm for a collection $\mathcal{R} = \{R\}$ consisting of the faster robot R only. The search time of this algorithm is $T(p)$, and therefore Lemma 2 lets us conclude that $\limsup_{|p| \to \infty} \frac{T(p)}{|p|} \geq \frac{1+3w}{w-w^2} = \frac{1}{W}$.

We choose a sufficiently far point p such that $\frac{T(p)}{|p|}$ is arbitrarily close to $\frac{1}{W}$ and set $d_p = |p|$. By Corollary 6, we may assume that the slower robot does not reach p on its own before the deadline. Thus, once the faster robot discovers the target located at p, its optimal strategy is to pursue the slower robot (moving at speed 1) and then bring it to the target (moving at speed v). We define t' and $q = r(t')$ as the time and location where R catches r. The search deadline is earlier than $\tau^* d_p$, which lets us derive a lower bound on $\frac{|q|}{t'}$ and consequently bound the progress speed of the slower robot from below. $\square$

Lemma 8. *If $v < \frac{1}{3}$ and $\tau(\mathcal{A}) < \tau^*$, then $w \geq \frac{1-v}{1+3v}$.*

Proof (sketch). We obtain $\frac{vw\tau^* - wv - v - w}{vw\tau^* + 1} > \frac{vW + v + W - vW\tau^*}{vW\tau^* + 1}$ using Lemmas 5 and 7. Since τ^* and W are defined using v and w only, this is an inequality on these two variables. It yields $w \geq \frac{1-v}{1+3v}$ after elementary calculations. $\square$

Lemmas 4 and 8 give conflicting constraints for any algorithm $\mathcal{A}$ such that $\tau(\mathcal{A}) < \tau^*$, which implies the following theorem.

Theorem 9. *For any line search algorithm $\mathcal{A}$, if $v < \frac{1}{3}$, then $\tau(\mathcal{A}) \geq \tau^*$.*

4 Wireless Communication

In this model, we have only the *exploration phase* and the *target phase*. We show that, for robots travelling at speeds with low relative difference (i.e., if $v \geq \frac{1}{6}$), in order to achieve the optimal search time, both robots need to participate in the exploration.

4.1 The Upper Bound

The optimal strategy for two robots travelling at the same speed [4] is very simple: Both robots explore in opposite directions at full speeds. When a robot learns that the other robot has found the target, it changes its direction towards the target.

Let us analyze the performance of this strategy for robots with distinct speeds. The total search time is a sum of three terms: the time for a robot to discover the target, the time for the other robot to go back to the origin and the time for that robot to reach the target. We consider two cases. First, suppose that the faster robot R discovers the target at distance d from the origin. Then the total search time is $d + d + \frac{1}{v}d = (2 + \frac{1}{v})d$. On the other hand, if the slower robot r discovers the target, the search time is worse: $\frac{1}{v}d + \frac{1}{v}d + d = (\frac{2}{v} + 1)d$.

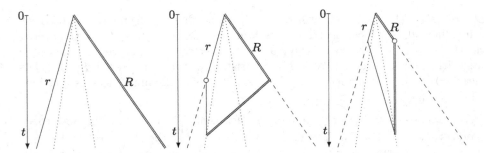

Fig. 5. Illustration of algorithm $\mathcal{B}^*$ before target discovery (left), when the target is discovered by r (middle), and by R (right). The horizontal axis represents the line searched and the vertical axis represents the time. The empty circle denotes the target discovery. Double and single solid lines represent the trajectories of the faster and the slower robot, respectively. Dashed lines correspond to the progress speeds of the two robots and dotted lines to the search time.

Intuitively, the faster robot explores too fast and it thus spends too much time going back to the origin. Hence, we limit the exploration speed of R to $v' < 1$. When it already knows the target, the faster robot is still allowed to use its full speed equal to 1. Now, the total search times are $\frac{1}{v'}d + \frac{1}{v'}d + \frac{1}{v}d = (\frac{2}{v'} + \frac{1}{v})d$ and $\frac{1}{v}d + \frac{v'}{v}d + d = \frac{1+v'+v}{v}d$, respectively. We choose v' to minimize the maximal of these two quantities. As they are, respectively, a decreasing and an increasing function of v', for the optimal value v' these terms are equal to each other, i.e., v' satisfies $\frac{1+v'+v}{v} = \frac{2}{v'} + \frac{1}{v}$ or, equivalently, $v'^2 + v'v = 2v$.

Algorithm $\mathcal{B}^*$ [for two robots with wireless communication]

1. Until the target is discovered, the two robots move in opposite directions. Robot r moves with its maximal speed v and robot R with speed $v' = \frac{1}{2}(\sqrt{v^2 + 8v} - v)$.
2. When either robot finds the target, it notifies the other one using wireless communication and the other robot moves to the target using its maximal speed.

The following fact, with a simple yet technical proof deferred to the full version of the paper, gives the right values of v' and of the search time τ^*. This lets us complete the description of the algorithm $\mathcal{B}^*$ (see Fig. 5), whose analysis follows immediately from the discussion above.

Fact 10. *For any speed $v \in (0, 1]$, define $\tau^* = \frac{2+v+\sqrt{v^2+8v}}{2v}$ and $v' = \frac{\sqrt{v^2+8v}-v}{2}$.*

$\quad$ (a) $\quad \tau^* = \frac{1+v+v'}{v}, \qquad$ (b) $\quad \tau^* = \frac{1}{v} + \frac{2}{v'}, \qquad$ and $\quad$ (c) $\quad v'^2 + v'v = 2v$.

Moreover, if $v \geq \frac{1}{6}$, then $3v \geq v' \geq \frac{1}{2}$.

Theorem 11. *For the algorithm $\mathcal{B}^*$ we have $t(\mathcal{B}^*, p) = \tau^*|p|$ for every $p \in \mathbb{R}$.*

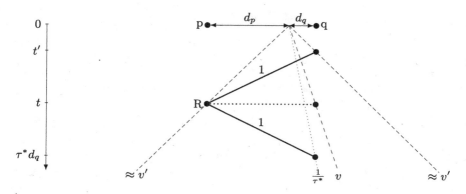

Fig. 6. Illustration of notions used in the proof of Lemma 12. The faster robot R, while in p at time t, must know that the target is not in q or it must be able to reach q before the deadline. For the former, either the slower robot r must have visited q prior to t, or R must have visited q on its own and traveled all the way to p.

4.2   The Lower Bound

By Theorem 11, for all points p we have $t(\mathcal{B}^*, p) = \tau^*|p|$ and thus $\tau(\mathcal{B}^*) = \tau^*$. We will show that for $v \geq \frac{1}{6}$ no algorithm $\mathcal{B}$ admits a smaller value of $\tau(\mathcal{B})$. As in Sect. 3.2, we impose some constraints on the hypothetical algorithms, two of which are going to be inconsistent. Due to space restrictions, here we present proof sketches only; full arguments are deferred to the full version.

Lemma 12. *If $v \geq \frac{1}{6}$ and $\tau(\mathcal{B}) < \tau^*$, then $\limsup_{t \to \infty} \frac{|R(t)|}{t} \leq v'$.*

Proof (sketch). For a proof by contradiction we suppose that the progress speed of R exceeds v'. Then, we may choose arbitrarily large time t such that $\frac{|R(t)|}{t} > v'$. Let $p = R(t)$ and $d_p = |p|$. We also consider a point q at distance $d_q = tv$ from the origin on the opposite side of p; see Fig. 6. If the time t is chosen sufficiently large, we may assume that $t(\mathcal{B}, q) < \tau^* d_q$.

The distance d_q is defined so that the robot R is unable to reach q prior to the deadline starting from p at time t. Thus, some robot must visit point q at time $t' < t$. The speed restriction for the slower robot is too strong for it to arrive at q early enough. Therefore, it must be the faster robot R which discovers q. Consequently, t' must be small enough for R to travel from q to p during time $t - t'$. On the other hand, the progress speed gives a lower bound on t' as t approaches infinity. We combine these two bounds to derive a contradiction. □

Lemma 13. *If $v \geq \frac{1}{6}$ and $\tau(\mathcal{B}) < \tau^*$, then the set $\{p : r \in \phi(p)\}$ of points discovered by the slower robot r is bounded.*

Proof (sketch). For a proof by contradiction, we suppose that there are arbitrarily far points discovered by robot r. We choose such a point p at distance $d_p = |p|$

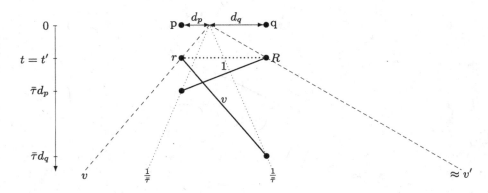

Fig. 7. Illustration of notions used in the proof of Lemma 13. The slower robot r, while discovering p at time t, must know that the target is not in q or it must be able to reach q before the deadline. For the former, the robot discovering q at t' prior to t, must be able to reach p before the deadline.

from the origin. We also consider a point q at distance $d_q = \frac{2d_p}{\bar{\tau}v-1}$ from the origin on the opposite side of p. Let $\bar{\tau}$ be an arbitrary value such that $\tau(\mathcal{B}) < \bar{\tau} < \tau^*$ and $\frac{1}{v} < \bar{\tau} < \tau^* = \frac{1}{v} + \frac{2}{v'}$. We may assume $t(\mathcal{B}, p) < \bar{\tau}d_p$ and $t(\mathcal{B}, q) < \bar{\tau}d_q$ if p is chosen sufficiently far.

We analyze the discovery times $t = T(p)$ and $t' = T(q)$, and distinguish two cases depending on which is smaller. If $t \leq t'$, then robot r, while in p at time t, must be able to reach q before the deadline. The distance d_q is defined so that it is unable to do so if $t \geq \frac{d_p}{v}$, and the latter inequality easily follows from the speed limit of the slower robot r.

On the other hand, if $t' \leq t$, then the robot which visits q at time t' must be able to reach p before $\bar{\tau}d_p$. This gives an upper bound on $t' \leq \bar{\tau}d_q - d_q - d_p$ due to the speed limits. Combined with the bound of Lemma 12 on the progress speed, this yields a contradiction if the initial point p is chosen sufficiently far.   □

Lemma 14. *If $v \geq \frac{1}{6}$ and $\tau(\mathcal{B}) < \tau^*$, then the set $\{p : r \in \phi(p)\}$ is unbounded.*

Proof (sketch). For a proof by contradiction, we suppose that the set is bounded. Then, the faster robot cannot go to infinity in one direction only, and it must pass the origin at arbitrarily large moments of time. Let us fix a sufficiently large t such that $R(t) = 0$. Consider two points p_l and p_r at distance $d = \frac{tv}{\bar{\tau}v-1-v}$ on each side of the origin. Let $\bar{\tau}$ be an arbitrary value such that $\tau(\mathcal{B}) < \bar{\tau} < \tau^*$ and $1 + \frac{1}{v} < \bar{\tau} < \tau^* = 1 + \frac{1}{v} + \frac{v'}{v}$. If t is chosen large enough, we may assume that $t(\mathcal{B}, p_l) < \bar{\tau}d$ and $t(\mathcal{B}, p_r) < \bar{\tau}d$ and that both points must be discovered by R.

Since $R(t) = 0$, points p_l and p_r can only be discovered no later than at time $t - d$ or no sooner than at time $t + d$. The distance d is defined so that the slower robot r starting from any position at time $t + d$ is either unable to reach p_l or unable to reach p_r. Hence, one of these points must be discovered (by R) at time

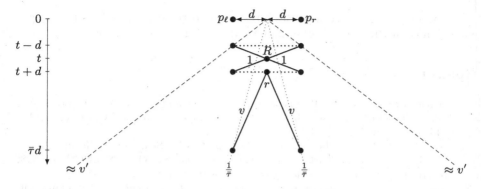

Fig. 8. Illustration of notions used in the proof of Lemma 14. The slower robot r at time $t + d$ must either be able to reach both p_ℓ and p_r before the deadline, or the faster robot R must have visited one of these points prior to time $t + d$, which actually could only happen prior to time $t - d$.

$t' \leq t - d$. For sufficiently large t, this contradicts the bound of Lemma 12 on the progress speed. □

When combined, Lemmas 13 and 14 exclude any algorithm with $\tau(\mathcal{B}) < \tau^*$.

Theorem 15. *For any line search algorithm $\mathcal{B}$, if $\frac{1}{6} \leq v \leq 1$, then $\tau(\mathcal{B}) \geq \tau^*$.*

5   Conclusions and Open Questions

Clearly, any search strategy for the communication by meeting model may also be used in the wireless communication model. The bound of $\frac{1+3v}{v-v^2}$ obtained in Sect. 3.1 outperforms $\frac{2+v+\sqrt{v^2+8v}}{2v}$ from Sect. 4.1 for small values of v. More precisely, an interested reader may observe that for $v \leq \sqrt{17} - 4 \approx 0.123$ we have $\frac{1+3v}{v-v^2} \leq \frac{2+v+\sqrt{v^2+8v}}{2v}$. This immediately shows that our strategy from Sect. 4.1 is not optimal in general. We conjecture that for $v \leq \sqrt{17} - 4$ some variation of the strategy from Sect. 3.1, when the faster robot is the only one responsible for exploration, will be optimal also for the wireless communication model. (Note that for general target points, it is not possible to improve the performance of the algorithm from Sect. 3.1 for wireless communication just by making the slower robot change direction immediately once the target is discovered by the faster robot.) As both strategies are fundamentally different, it would also be interesting to see what happens for the speeds $\sqrt{17} - 4 < v < \frac{1}{6}$.

The above fact may be viewed from another, perhaps more interesting perspective. Two unit-speed robots perform linear search in $9d$ time when communicating by meeting and in $3d$ time for the less restrictive wireless communication. Is it true that, for significantly different robot speeds, the wireless communication model loses its advantage over the communication by meeting model, and the linear search takes the same time in both models?

Another possible area of research is to extend the considerations to a larger collection of distinct-speed robots for both communication models.

References

1. Alpern, S., Gal, S.: The Theory of Search Games and Rendezvous. International Series in Operations Research & Management Science, vol. 55. Kluwer Academic Publishers, New York (2002)
2. Baeza-Yates, R.A., Culberson, J.C., Rawlins, G.J.E.: Searching with uncertainty extended abstract. In: Karlsson, R., Lingas, A. (eds.) SWAT 1988. LNCS, vol. 318, pp. 176–189. Springer, Heidelberg (1988). doi:10.1007/3-540-19487-8_20
3. Baeza-Yates, R.A., Culberson, J.C., Rawlins, G.J.E.: Searching in the plane. Inf. Comput. **106**(2), 234–252 (1993)
4. Baeza-Yates, R.A., Schott, R.: Parallel searching in the plane. Comput. Geom. **5**, 143–154 (1995)
5. Barajas, J., Serra, O.: The lonely runner with seven runners. Electr. J. Comb. **15**(1), 1–18 (2008). http://www.combinatorics.org/Volume_15/Abstracts/v15i1r48.html
6. Beck, A.: On the linear search problem. Israel J. Math. **2**(4), 221–228 (1964)
7. Bellman, R.: An optimal search. SIAM Rev. **5**(3), 274 (1963)
8. Bender, M.A., Fernández, A., Ron, D., Sahai, A., Vadhan, S.P.: The power of a pebble: exploring and mapping directed graphs. Inf. Comput. **176**(1), 1–21 (2002)
9. Bender, M.A., Slonim, D.K.: The power of team exploration: two robots can learn unlabeled directed graphs. In: 35th IEEE Annual Symposium on Foundations of Computer Science, FOCS 1994, pp. 75–85. IEEE Computer Society (1994)
10. Bose, P., Carufel, J.D., Durocher, S.: Searching on a line: a complete characterization of the optimal solution. Theor. Comput. Sci. **569**, 24–42 (2015)
11. Chrobak, M., Gąsieniec, L., Gorry, T., Martin, R.: Group search on the line. In: Italiano, G.F., Margaria-Steffen, T., Pokorný, J., Quisquater, J.-J., Wattenhofer, R. (eds.) SOFSEM 2015-Testing. LNCS, vol. 8939, pp. 164–176. Springer, Heidelberg (2015)
12. Czyzowicz, J., Gąsieniec, L., Gorry, T., Kranakis, E., Martin, R., Pajak, D.: Evacuating robots via unknown exit in a disk. In: Kuhn, F. (ed.) DISC 2014. LNCS, vol. 8784, pp. 122–136. Springer, Heidelberg (2014)
13. Czyzowicz, J., Gąsieniec, L., Kosowski, A., Kranakis, E.: Boundary patrolling by mobile agents with distinct maximal speeds. In: Demetrescu, C., Halldórsson, M.M. (eds.) ESA 2011. LNCS, vol. 6942, pp. 701–712. Springer, Heidelberg (2011)
14. Dereniowski, D., Disser, Y., Kosowski, A., Pająk, D., Uznański, P.: Fast collaborative graph exploration. Inf. Comput. **243**, 37–49 (2015)
15. Dudek, G., Romanik, K., Whitesides, S.: Localizing a robot with minimum travel. SIAM J. Comput. **27**(2), 583–604 (1998)
16. Feinerman, O., Korman, A., Lotker, Z., Sereni, J.: Collaborative search on the plane without communication. In: Kowalski, D., Panconesi, A. (eds.) ACM Symposium on Principles of Distributed Computing, PODC 2012, pp. 77–86. ACM (2012)
17. Fomin, F.V., Thilikos, D.M.: An annotated bibliography on guaranteed graph searching. Theor. Comput. Sci. **399**(3), 236–245 (2008)
18. Fraigniaud, P., Gąsieniec, L., Kowalski, D.R., Pelc, A.: Collective tree exploration. Networks **48**(3), 166–177 (2006)
19. Jaillet, P., Stafford, M.: Online searching. Oper. Res. **49**(4), 501–515 (2001)

20. Kawamura, A., Kobayashi, Y.: Fence patrolling by mobile agents with distinct speeds. Distrib. Comput. **28**(2), 147–154 (2015)

21. Nahin, P.J.: Chases and Escapes: The Mathematics of Pursuit and Evasion. Princeton Puzzlers, Princeton University Press, Princeton (2012)

22. Pelc, A.: Deterministic rendezvous in networks: a comprehensive survey. Networks **59**(3), 331–347 (2012)

23. Wills, J.M.: Zwei Sätze über inhomogene diophantische Approximation von Irrationalzahlen. Monatsh. Math. **71**(3), 263–269 (1967)

Deterministic Meeting of Sniffing Agents in the Plane

Samir Elouasbi[(✉)] and Andrzej Pelc

Département d'informatique, Université du Québec en Outaouais,
Gatineau, Québec J8X 3X7, Canada
{elos02,pelc}@uqo.ca

Abstract. Two mobile agents, starting at arbitrary, possibly different times from arbitrary locations in the plane, have to meet. Agents are modeled as discs of diameter 1, and meeting occurs when these discs touch. Agents have different labels which are integers from the set $\{0, \ldots, L-1\}$. Each agent knows L and knows its own label, but not the label of the other agent. Agents are equipped with compasses and have synchronized clocks. They make a series of moves. Each move specifies the direction and the duration of moving. This includes a *null* move which consists in staying inert for some time, or forever. In a non-null move agents travel at the same constant speed, normalized to 1.

Agents have sensors enabling them to estimate the distance from the other agent, but not the direction towards it. We consider two models of estimation. In both models an agent reads its sensor at the moment of its appearance in the plane and then at the end of each move. This reading (together with the previous ones) determines the decision concerning the next move. In both models the reading of the sensor tells the agent if the other agent is already present. Moreover, in the *monotone model*, each agent can find out, for any two readings in moments t_1 and t_2, whether the distance from the other agent at time t_1 was smaller, equal or larger than at time t_2. In the weaker *binary model*, each agent can find out, at any reading, whether it is at distance less than ρ or at distance at least ρ from the other agent, for some real $\rho > 1$ unknown to them. Such distance estimation mechanism can be implemented, e.g., using chemical sensors. Each agent emits some chemical substance (scent), and the sensor of the other agent detects it, i.e., *sniffs*. The intensity of the scent decreases with the distance. In the monotone model it is assumed that the sensor is ideally accurate and can measure any change of intensity. In the binary model it is only assumed that the sensor can detect the scent below some distance (without being able to measure intensity) above which the scent is too weak to be detected.

We show the impact of the two ways of sensing on the time of meeting, measured from the start of the later agent. For the monotone model we show an algorithm achieving meeting in time $O(D)$, where D is the initial distance between the agents. This complexity is optimal. For the binary model we show that, if agents start at distance smaller than ρ

A. Pelc—Supported in part by NSERC discovery grant 8136 – 2013 and by the Research Chair in Distributed Computing of the Université du Québec en Outaouais.

© Springer International Publishing AG 2016
J. Suomela (Ed.): SIROCCO 2016, LNCS 9988, pp. 212–227, 2016.
DOI: 10.1007/978-3-319-48314-6_14

(i.e., when they sense each other initially) then meeting can be guaranteed within time $O(\rho \log L)$, and that this time cannot be improved in general. Finally we observe that, if agents start at distance $\alpha\rho$, for some constant $\alpha > 1$ in the binary model, then sniffing does not help, i.e., the worst-case optimal meeting time is of the same order of magnitude as without any sniffing ability.

Keywords: Algorithm · Rendezvous · Mobile agent · Synchronous · Deterministic · Plane · Distance

1   Introduction

The background and the problem. Two mobile agents, starting at arbitrary, possibly different times from arbitrary locations in the plane, have to meet. Agents are modeled as discs of diameter 1, and meeting occurs when these discs touch (i.e., the centers of the agents get at distance 1). This way of formulating the meeting problem in the plane is equivalent to the problem of *approach* [11], where agents are modeled as points moving in the plane, and the approach is defined as these points getting at distance at most 1 from each other. This is one of the versions of the well-known rendezvous problem in which two or more agents have to meet in some environment. This problem has been studied in many variations: in the plane and in networks, in the synchronous vs. asynchronous setting, and using deterministic vs. randomized algorithms. In applications, mobile agents may be rescuers trying to find a lost tourist in the mountains, animals trying to find a mate, or mobile robots that have to meet in order to compare previously collected data.

We are interested in deterministic algorithms for the task of meeting. If agents were anonymous (identical), then, if started simultaneously, they would trace identical trajectories and hence could never meet. In order to break the symmetry, we assume that agents have different labels which are integers from the set $\{0, \ldots, L-1\}$. Each agent knows L and knows its own label which it can use as a parameter in the algorithm that they both execute, but it does not know the label of the other agent.

Agents are equipped with compasses showing the cardinal directions, and have synchronized clocks. The adversary places each agent at some point of the plane at possibly different times. The clock of the agent starts at the moment of its appearance in the plane. Each agent makes a series of moves. A move specifies the direction and the duration of moving. This includes a *null* move which consists in staying inert for some time, or forever. In a non-null move agents travel at the same constant speed, normalized to 1.

We assume that agents have sensors enabling them to estimate the distance from the other agent (defined as the distance between centers of discs), but not the direction towards it. We consider two models of estimation. In both models an agent reads its sensor at the moment of its appearance in the plane and then at the end of each move. This reading (together with the previous ones) determines the decision concerning the next move. In both models the reading of the sensor tells the agent if the other agent is already present. Moreover, in the *monotone model* each agent can find out, for any two readings in moments t_1 and t_2, whether the distance from the other agent at time t_1 was smaller, equal or larger than at time t_2. In the weaker *binary model* each agent can find out, at any reading, whether it is at distance less than ρ or at distance at least ρ from the other agent, for some real $\rho > 1$ unknown to them. (We assume that $\rho > 1$ because agents are always at distance larger than 1 before touching, hence $\rho \leq 1$ would be useless for sensing.) Such distance estimation mechanism can be implemented, e.g., using chemical sensors. Each agent emits some chemical substance (scent), and the sensor of the other agent detects it, i.e., *sniffs*. If at some time the agent is still alone in the plane, the reading of its sensor is 0. Otherwise, the reading of the sensor is positive in the monotone model, and in the binary model it is 1 if the distance between the agents is less than ρ and 0 if it is at least ρ. The intensity of the scent decreases with the distance. In the monotone model it is assumed that the sensor is ideally accurate: it can measure any change of scent intensity and hence can compare distances at any two readings. (The name *monotone* comes from the fact that the intensity of the scent accurately sensed by the agent is a strictly decreasing function of the distance. We do not assume anything else about this function: an agent cannot learn, e.g., the value of its distance to the other agent.) In the binary model it is only assumed that the sensor can detect the scent below some distance (without being able to measure its intensity) above which the scent is to weak to be detected.

Note that the monotone model is similar to the model of distance-aware agents from [9]. The differences are the environment (networks in the case of [9] vs. the plane with the use of compasses in our case) and sensing after each round in [9] vs. sensing at times decided by the agent in our case. These differences are important and will lead to a more efficient algorithm in our setting.

Our results. We show the impact of the two ways of sensing on the time of meeting, measured from the start of the later agent. For the monotone model we show an algorithm achieving meeting in time $O(D)$, where D is the initial distance between the agents. This complexity is optimal. For the binary model we show that, if agents start at distance smaller than ρ (i.e., when they sense each other initially) then meeting can be guaranteed within time $O(\rho \log L)$, and that this time cannot be improved in general. Indeed we show that, for some initial distance less than ρ, and for some labels of the agents, time $\Omega(\rho \log L)$ is needed to meet in the binary model. Finally we observe that if agents start at distance $\alpha\rho$, for some constant $\alpha > 1$ in the binary model, then sniffing does not help, i.e., the worst-case optimal meeting time is of the same order of magnitude as without any sniffing ability.

Our results show a separation between the two models of sensing accuracy. Suppose that agents start at distance $\rho/3$. Then in the monotone model the optimal meeting time is $\Theta(\rho)$, while in the binary model it is $\Theta(\rho \log L)$.

Due to lack of space, several proofs are omitted.

Related work. The large literature on rendezvous can be classified according to the mode in which agents move (deterministic or randomized) and the environment where they move (a network modeled as a graph or a terrain in the plane). An extensive survey of randomized rendezvous in various scenarios can be found in [1], cf. also [2,16].

Deterministic rendezvous in networks was surveyed in [19]. In this setting a lot of effort has been dedicated to the study of the feasibility of rendezvous, and to the time required to achieve this task, when feasible, under the synchronous scenario. For instance, deterministic rendezvous with agents equipped with tokens used to mark nodes was considered, e.g., in [17]. Time of deterministic rendezvous of agents equipped with unique labels was discussed in [10,20]. Memory required by the agents to achieve deterministic rendezvous was studied in [3,15] for trees and in [8] for general graphs. Fault-tolerant rendezvous was studied, e.g., in [5]. In [18] the authors studied tradeoffs between the time of rendezvous and the total number of edge traversals by both agents until the meeting. In [9] the authors considered distance-aware agents operating in networks. As mentioned above, this is a model similar to our monotone model. They showed a rendezvous algorithm polynomial in local parameters of the problem (initial distance between agents, maximum degree and length of the shorter label). They also established a lower bound for this time which exceeds $\Theta(D)$. This shows a separation between our setting and theirs.

Other works were devoted to asynchronous rendezvous in networks, cf. e.g., [4,12], when the agent chooses the edge which it decides to traverse but the adversary controls the speed. Under this assumption rendezvous in a node cannot be guaranteed even in very simple graphs, and hence the rendezvous requirement is relaxed to permit the agents to meet inside an edge.

Rendezvous of two or more agents in the plane was mainly considered in two settings. In one of them, cf. e.g., [6,7,13,14], agents can see the positions of other agents, and make the decisions based on these observations, usually in an asynchronous way. Another scenario does not allow agents to make any observations. In [11] the authors proposed an algorithm for the asynchronous version of the problem of approach in the plane (equivalent to our meeting), with cost polynomial in the initial distance and in the length of the smaller label. The results from [4] are for asynchronous rendezvous in the grid but imply solutions for the approach problem in the plane as well. They use a strong assumption of knowledge of initial positions of the agents in some global system of coordinates, but achieve approach at cost $O(D^2 polylog(D))$, where D is the initial distance.

2   Terminology and Preliminaries

The direction North – South is called the *vertical* direction, and the direction East – West is called the *horizontal* direction. We say that agent a is North of agent b, if the horizontal line containing the center of agent a is North of the horizontal line containing the center of agent b. The three other expressions ("South of", "East of" and "West of") have analogous meaning. We say that agent a is at distance x from b in the vertical direction, if the horizontal lines containing the centers of the agents are at distance x. The distance in the horizontal direction is defined similarly.

Let $(a_1 \ldots a_m)$ be the binary representation of the label ℓ of an agent. The *transformed label* $T(\ell)$ is obtained by padding the string $(a_1 \ldots a_m)$ by a prefix of $\lambda - m$ zeroes, where $\lambda = \lceil \log L \rceil$. Hence every transformed label has length λ. Notice that if the labels are different, then there exists an index for which the corresponding bits of their transformed labels differ (this is not necessarily the case for the binary representations of the original labels, since one of them might be a prefix of the other). Moreover, if $\ell_1 < \ell_2$, then $T(\ell_1)$ is lexicographically smaller than $T(\ell_2)$.

3   The Monotone Model

In this section we present an algorithm that accomplishes the meeting in the monotone model in time $O(D)$, where D is the initial distance between the agents. This is of course optimal, as time $(D-1)/2$ is a lower bound, because the speed of the agents is 1.

There are two elementary instructions in the monotone model: $read(C)$ and $move(card, x)$. The instruction $read(C)$ results in reading the current value of the scent intensity sensor into the variable C. Recall that the value of the sensor is 0 if the agent is alone in the plane, and it is some positive real otherwise. The instruction $move(card, x)$, where $card$ is one of the cardinal directions (N, E, S, W) and x is a positive real, results in moving the agent in the direction $card$ during time x. Since the speed of the agent is 1, this means that the agent travels distance x in direction $card$.

At a high level, the idea of the algorithm is the following. In the beginning the agent reads its sensor. If its value is 0, this means that the other agent is not yet in the plane, which enables the agent to break symmetry. The agent stays inert forever and will be eventually found by the other agent. The other agent must realize that the first agent is inert and find it by first getting at distance at most 1 in the vertical direction and then getting at distance at most 1 in the horizontal direction, which results in meeting.

If the initial readings of sensors of both agents are positive, this means that both of them are placed in the plane simultaneously. In this case, the only way to break symmetry is using the (transformed) labels of the agents, which are different by assumption. The agents must realize that they are in this more difficult situation, and then approach, first in the vertical and then in the horizontal direction. This is done by moving North and South for the vertical approach (and East and West for the horizontal approach) according to the bits of the transformed label of the agent. At the first bit where their transformed labels differ, the symmetry between the agents will be broken, they will realize it by reading their sensors, and then accomplish the approach. In order to keep the time $O(D)$ (and not $O(D + \log L)$) the moves of the agents corresponding to consecutive bits shrink by a factor of 2 at each consecutive bit.

We now proceed to a detailed description of the algorithm. We will use the following elementary procedures. The first of them compares the previous reading of the sensor with the current one, and assigns the result of the comparison to the variable *compare*.

Procedure Test

$P \leftarrow C$
$read(C)$
if $P < C$ **then** *compare* $\leftarrow$ *larger*
if $P = C$ **then** *compare* $\leftarrow$ *equal*
if $P > C$ **then** *compare* $\leftarrow$ *smaller*

The aim of the next procedure is to approach the other agent either in the vertical or in the horizontal direction by walking in steps of prescribed length x. It is used when the agent already realized that the other agent is North (resp. South) of it, or it is East (resp. West) of it. We formulate the procedure for the parameter *card* which can be equal to N, E, S, or W.

Procedure GetCloser $(card, x)$

while *compare* $=$ *smaller* **do**
    $move(card, x)$; **Test**

Our next procedure is called in the case when agents appear simultaneously in the plane, and its aim is to break symmetry between them in this case. This is done by having the value of the variable *compare* in Procedure **Test** become different from *equal*. This occurs when agents process the bit of their transformed labels in which they differ. Note the two attempts at getting the value of the variable *compare* different from *equal*. This is necessary in the special case when transformed labels of the agents differ in only one bit, say with index j, the distance in the vertical direction between the agents is $\frac{1}{2^j}$, and the agent whose jth bit is 1 is South of the agent whose jth bit is 0. In this special case, a single attempt at breaking symmetry would go unnoticed: agents would switch positions in the vertical direction and the value of variable *compare* would remain *equal*. The procedure is executed by an agent with label ℓ.

Procedure Dance

$T(\ell) \leftarrow (c_1 \ldots c_\lambda)$
$i \leftarrow 1$
while *compare = equal* **do**
      if $c_i = 1$ **then**
           $move(N, \frac{1}{2^i})$; Test
           if *compare = equal* **then**
               $move(N, \frac{1}{2^i})$; Test
      else
           $move(S, \frac{1}{2^i})$; Test
           if *compare = equal* **then**
               $move(S, \frac{1}{2^i})$; Test
        $i \leftarrow i + 1$

We now describe two main procedures of our algorithm which will be called one after another. The aim of the first of them is to get the agents at distance at most 1 in the vertical direction and the aim of the second is to get the agents at distance at most 1 in the horizontal direction. In Procedure VerticalApproach, the later agent, or both agents if they start simultaneously, first realize which of these two cases occurs. This is done as follows. The agent, call it a, moves North by 1. If it decreased the distance from the other agent, it learns that the other agent is inert, and then approaches it by steps of length $1/2$, getting at distance at most 1 in the vertical direction. If it increased the distance, it also learns that the other agent is inert, goes back (i.e. South by 1), and then approaches the other agent by steps of length $1/2$, getting at distance at most 1 in the vertical direction. If the distance did not change, the situation is still unclear: the start may have been simultaneous and the other agent also moved North by 1, or the other agent may be inert and was at vertical distance $1/2$ North before the move of agent a. Agent a clarifies this by moving North by 1 again. Agent a could not decrease its distance from b after this move. If it increased the distance, it goes back and approaches agent b as before. On the other hand, if the distance did not change again, agent a learns that it is in the simultaneous start situation. It then performs Procedure Dance, at the end of which the value of the variable *compare* is different from *equal*. The j-th bit whose processing caused this change is the first bit where the transformed labels of the agents differ. This breaks symmetry. There are two cases. If, at the end of Dance, *compare = smaller*, then the agent whose j-th bit is 1 was South of the other agent *before* the last move. The agents backtrack by a distance of $\frac{1}{2^j}$ and then approach each other: the agent whose j-th bit is 1 going North, and the agent whose j-th bit is 0 going South. If, at the end of Dance, *compare = larger*, then the agent whose j-th bit is 1 is North of the other agent *after* the last move. Now there is no need of backtracking: the agents simply approach each other: the agent whose j-th bit is 1 going South, and the agent whose j-th bit is 0 going North. In both cases, the steps during approach (Procedure GetCloser) are now of length $1/4$ instead of $1/2$, because agents perform approach simultaneously.

Procedure VerticalApproach

$sim \leftarrow false$; $move(N, 1)$; Test
if $compare = smaller$ **then** GetCloser $(N, \frac{1}{2})$
else

    if $compare = larger$ **then**
        $move(S, 1)$; GetCloser $(S, \frac{1}{2})$
    else

        $move(N, 1)$; Test
        if $compare = larger$ **then**
            $move(S, 1)$; GetCloser $(S, \frac{1}{2})$
        else (*$compare = equal$*)
            $sim \leftarrow true$; Dance; $j \leftarrow i - 1$
            if $compare = smaller$ **then**
                if $c_j = 1$ **then**
                    $move(S, \frac{1}{2^j})$; GetCloser $(N, \frac{1}{4})$
                else
                    $move(N, \frac{1}{2^j})$; GetCloser $(S, \frac{1}{4})$
            else (* $compare = larger$ *)
                if $c_j = 1$ **then**
                    GetCloser $(S, \frac{1}{4})$
                else
                    GetCloser $(N, \frac{1}{4})$

The following Procedure HorizontalApproach will be called after Procedure VerticalApproach, and uses global variables sim and j whose values are set in the above procedure. Procedure HorizontalApproach is simpler because, when it is called, the agent has two important pieces of information. First, it already knows that its vertical distance from the other agent is at most 1, and hence agents cannot pass each other horizontally without meeting. Second, the agent knows if the start was simultaneous, or if the other agent started first and hence is inert. This information is coded in the boolean variable sim, which is set to $true$ in Procedure VerticalApproach if and only if the start was simultaneous. Moreover, if the start was simultaneous, the agent knows already the first bit in which its transformed label differs from that of the other agent: the index of this bit is j. Hence, if $sim = false$, the agent approaches the other (inert) agent similarly as before, and if $sim = true$, then agents approach each other horizontally by going either East or West, depending on the value of the jth bit of their transformed label, until meeting.

```
Procedure HorizontalApproach

if sim = false then
        move(E, 1)
        Test
        if compare = smaller then GetCloser (E, 1)
        else
                move(W, 1)
                GetCloser (W, 1)
else
        if c_j = 1 then
                move(E, 1)
                Test
                if compare = smaller then GetCloser (E, 1)
                else
                        move(W, 1)
                        GetCloser (W, 1)
        else
                move(W, 1)
                Test
                if compare = smaller then GetCloser (W, 1)
                else
                        move(E, 1)
                        GetCloser (E, 1)
```

Now our main algorithm for the monotone model can be formulated succinctly as follows. It is executed by an agent with label ℓ (that intervenes in Procedure Dance), with the understanding that when agents touch (i.e., get at distance 1), the algorithm is interrupted, as meeting is then accomplished. We will prove that this will always occur by the end of the execution of the algorithm.

```
Algorithm MeetingWithPreciseSensor

read(C)
if C = 0 then stay inert forever
else
        VerticalApproach
        HorizontalApproach
```

Theorem 1. *The meeting of two agents that are arbitrarily placed in the plane and execute Algorithm* MeetingWithPreciseSensor *in the monotone model, occurs by the end of the execution of this algorithm. If agents are at initial distance D, then the meeting occurs in time $O(D)$ after the appearance of the later agent.*

4   The Binary Model

In this section we present an algorithm that accomplishes the meeting in the binary model in time $O(\rho \log L)$, assuming that agents are initially at distance smaller than ρ, i.e., that they can initially sense each other, when both agents are already in the plane. We also show a matching lower bound on the time of the meeting, for some initial positions at distance smaller than ρ and for some labels of the agents.

There are two elementary instructions in the binary model: $check(C)$ and $move(card, x)$. The instruction $check(C)$ results in reading a single bit into the variable C. This bit is 1, if the other agent is at distance less than ρ, and it is 0 otherwise (i.e., if the other agent is either still absent from the plane, or if it is at distance at least ρ). The instruction $move(card, x)$, where $card$ is one of the cardinal directions (N, E, S, W) and x is a positive real, is identical as in the monotone model: it results in moving in the direction $card$ during time x.

At a high level, the idea of the algorithm is the following. In the beginning the agent reads its sensor. If its value is 0, this means that the other agent is not yet in the plane, which enables the agent to break symmetry as in the monotone model. The agent stays inert forever and will be eventually found by the other agent. As explained below, this will be done in a much different way than in the monotone model, as sensing of the other agent is less precise.

If the initial readings of sensors of both agents are 1, this means that both of them are placed in the plane simultaneously. In this case, they break symmetry using their transformed labels, which are different by assumption. Initially, an agent has no way of deciding if it was placed in the plane later, or if both agents were placed in the plane simultaneously. Hence, when the bit initially read in $check(C)$ is 1, the agent performs actions that will enable it to *lose contact* with the other agent (i.e., to get the reading 0 in $check(C)$), regardless of which of these situations occurs. Losing contact will always happen when one agent moves and the other stays. This will enable the agents to break symmetry: the agent during whose move the contact was lost will find the other agent that becomes inert (in the case of non-simultaneous start, the agent that will perform the finding is the later agent, and the earlier agent is inert from the start). Once symmetry is broken, the moving agent accomplishes the meeting using the binary readings of its sensor and geometric properties of the plane.

We now give a detailed description of the algorithm. The aim of the first procedure is losing contact between the agents, i.e., having both of them (in the case of simultaneous start), or the later agent (in the case of non-simultaneous start) get the reading 0 in $check(C)$). This is done by going North or staying inert, for increasing periods of time, according to the bits of the transformed label: when the bit is 1, the agent moves, when it is 0, it stays inert. It will be proved that agents eventually get at distance at least ρ, which they will realize by reading their sensors. Losing contact occurs when one agent moves North and the other agent is inert. This happens when agents, in the case of simultaneous start (or only the later agent, otherwise) process the jth bit of their transformed label. When contact is lost, the agent during whose move this occurred, i.e., the agent for which $c_j = 1$, goes back South, this time by small steps, until contact is regained (the reading of C is 1 again). This is done to determine the point when contact has been lost, with sufficient precision. The procedure is executed by an agent with label ℓ, and called when the value of C is 1.

Procedure LoseContact (C)

$T(\ell) \leftarrow (c_1 \ldots c_\lambda)$
$d \leftarrow 1$
leading $\leftarrow$ *false*
while $C = 1$ **do**
    $i \leftarrow 1$
    while $(C = 1$ **and** $i < \lambda)$ **do**
        if $c_i = 1$ **then** $move(N, d)$
        else stay inert for time d
        $i \leftarrow i + 1$
        $check(C)$
    $d \leftarrow 2d$
$j \leftarrow i - 1$
if $c_j = 1$ **then**
    leading $\leftarrow$ *true*
    while $C = 0$ **do**
        $move(S, \frac{1}{2})$
        $check(C)$

The second procedure is based on the following observation (cf. Fig. 1).

Consider two points in the plane, X and A, where A is North of X (in the sense defined in Sect. 2). Suppose that the distance between A and X is ρ. Let p be the vertical line containing point A, and let B be the other point of line p at distance ρ from X. (This includes the case when A and B coincide). Let Z be the midpoint of the segment AB. Then Z is on the same horizontal line as X because the triangle XAB is isoceles.

The above observation can be used to construct the next procedure, called after executing Procedure LoseContact, when (the center of) one of the agents, call it a, approximates a point A North of the inert agent, with center X, such that the distance between A and X is ρ. Agent a starts the procedure with the reading 1 of its sensor. It goes South with steps of length $1/2$ (counted in the counter t), until the reading of its sensor becomes 0. At this time its center approximates the other point B at distance ρ from X, on the vertical line along which it travelled. Then the agent goes back (i.e., North) on this vertical line, at distance $\lceil t/2 \rceil \cdot \frac{1}{2}$. Upon completing this move it reaches a point which approximates the midpoint Z of the segment AB. Now it goes horizontally, trying directions East and West in increasing leaps whose lengths double at each time, until meeting the inert agent. Since steps used by the agent, when it moved vertically, were sufficiently small, the approximations are good enough for the meeting to eventually occur.

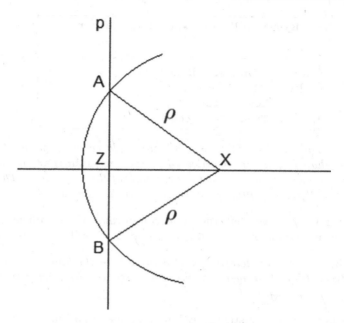

Fig. 1. Geometric setting for procedure `TriangleSearch`

Procedure `TriangleSearch` (C)

if $leading = true$ **then**
    $t \leftarrow 0$
    while $C = 1$ **do**
        $move(S, \frac{1}{2})$; $check(C)$; $t \leftarrow t + 1$
    $move(N, \lceil t/2 \rceil \cdot \frac{1}{2})$
    $d \leftarrow 1$
    repeat until meeting
        $move(E, d)$; $move(W, 2d)$; $move(E, d)$
        $d \leftarrow 2d$

Now our main algorithm for the binary model can be formulated succinctly as follows. It is executed by an agent with label ℓ (that intervenes in Procedure `LoseContact`), with the understanding that when agents touch (i.e., get at distance 1), the algorithm is interrupted, as meeting is then accomplished. We will prove that this will always occur by the end of the execution of the algorithm.

Algorithm MeetingWithBinarySensor

$check(C)$
if $C = 0$ **then** stay inert forever
else
      LoseContact (C)
      TriangleSearch (C)

Theorem 2. *The meeting of two agents that are placed in the plane at an initial distance less than ρ and execute Algorithm* MeetingWithBinarySensor *in the binary model, occurs by the end of the execution of this algorithm. The meeting occurs in time $O(\rho \log L)$ after the appearance of the later agent.*

Our last result shows that the time complexity of the above Algorithm MeetingWithBinarySensor cannot be improved in general, in the binary model.

Theorem 3. *There is no deterministic algorithm for the binary model, guaranteeing meeting of any two agents starting at distance less than ρ in the plane, and working in time $o(\rho \log L)$.*

Proof. Consider a deterministic algorithm $\mathcal{A}$ for the binary model. Assume that $\rho > 30$. Define a tiling of the plane into pairwise disjoint squares of size $\rho/5$. Each tile includes its North and East edges. For a fixed tile 0, enumerate the 8 neighboring tiles by numbers $1, \ldots, 8$, starting with the tile North of tile 0 and going clockwise. Let A and B be two tiles on the same horizontal line, separated by one tile. Call such tiles *conjugate*. Place one agent in any point of A, the other agent in any point of B, and start them simultaneously at time 0.

Suppose that algorithm $\mathcal{A}$ works in time at most $c\rho\lambda$, where $\lambda = \lceil \log L \rceil$ and $c < \frac{1}{24 \log 9}$ is a constant. Divide the time into consecutive segments of length $\rho/6$. If the agent is in some tile at the beginning of a time segment, then at the end of this segment it is either in the same or in one of the neighboring tiles. The *behavior pattern* of an agent is the sequence $(a_1, \ldots, a_k)$ with terms $0, 1, \ldots, 8$, defined as follows. If at the beginning of the ith segment the agent is in some tile 0, then $a_i = j$ if and only if the agent is in tile j at the end of the segment. Since the algorithm works in time at most $c\rho\lambda$, the length of the behavior pattern is at most $k = \lceil 6c\lambda \rceil$. Since $c < \frac{1}{24 \log 9}$, we have $k < \frac{\log L}{\log 9}$, for sufficiently large L, and hence $9^k < L$.

As long as the agents are at distance less than ρ, and hence the reading of their sensors is always 1, the actions of each agent depend only on its label. Agents start in conjugate tiles. Consequently they start at distance at most $\frac{\sqrt{10}\rho}{5}$. Since time segments are of length $\rho/6$, the distance between the agents during the first time segment is always at most $\frac{\sqrt{10}\rho}{5} + \frac{2\rho}{6} < \rho$. Hence the actions of each agent during the first time segment depend only on its label. There is a number in the set $\{0, 1, \ldots, 8\}$, such that for a set S_1 of at least $L/9$ labels, if agents with distinct labels from this set start respectively in conjugate tiles that are given number 0, then the tile numbers in which agents finish the first segment are the same. Hence agents with such labels finish the first time segment in conjugate tiles as well. Consequently, the distance between the agents during the second time segment is always at most $\frac{\sqrt{10}\rho}{5} + \frac{2\rho}{6} < \rho$. Hence the behavior of each agent during the second time segment depends only on its label. There exists a set $S_2 \subseteq S_1$ of size at least $|S_1|/9$, such that agents with labels in this set finish the second time segment in conjugate tiles as well.

Since $9^k < L$, by induction on the index of the time segment, there are at least two labels such that agents starting in tiles A and B with these labels have the same behavior pattern. Assign these labels to agents starting in tiles A and B. It follows that these agents are in conjugate tiles at the end of each time segment. Since time segments are of length $\rho/6$, we show that agents are always at distance at least $\frac{\ell}{5} - \frac{\ell}{6}$. Indeed, suppose that during a time segment agents get at distance smaller than $\frac{\ell}{5} - \frac{\ell}{6}$. This could occur only when both agents are inside the tile separating them at the beginning of the segment. However, during this time segment, each agent can penetrate this separating tile only at distance at most $\rho/12$, because at the end of the segment they have to be again separated by this tile. Since at the beginning of the time segment agents were at distance at least $\frac{\ell}{5}$, this distance could not decrease by more than $\frac{\ell}{6}$ during the time segment, which results in the distance at least $\frac{\ell}{5} - \frac{\ell}{6} = \frac{\ell}{30} > 1$ at all times. Hence agents cannot meet. This contradiction implies that the time of the algorithm must be larger than $c\rho \log L$, for sufficiently large ρ and L, which concludes the proof.

We conclude this section with the observation that if the agents start at distance $\alpha\rho$, for some constant $\alpha > 1$, in the binary model, then sniffing does not help.

Proposition 1. *If the agents start at distance $\alpha\rho$, for some constant $\alpha > 1$, in the binary model, then the worst-case optimal meeting time is of the same order of magnitude as without any sniffing ability.*

5   Conclusion

We provided optimal algorithms for the task of meeting of two agents equipped with sniffing sensors, in two models of accuracy of these sensors. In the monotone model it is assumed that sensors are perfectly accurate, i.e., they can notice any change of distance between the agents, although they cannot measure the distance itself, nor recognize the direction in which the other agent is located. In the binary model, the sensor can only tell the agent if the other agent is close or far, for some threshold of closeness. We showed a separation between the two models: while in the monotone model meeting is guaranteed in time proportional to the initial distance between the agents, in the binary model we showed that optimal meeting time is $\Theta(\rho \log L)$, where ρ is the sensing threshold and L is the size of the label space.

In both models we assume that both agents travel at the same speed 1, and this assumption is heavily used in our algorithms and their analysis. It is an interesting open problem how our results would change if agents moved in an asynchronous way, or at least were allowed the same degree of asynchrony as in [11], i.e., each agent moved at a constant speed, but possibly different from the other agent.

References

1. Alpern, S., Gal, S.: The Theory of Search Games and Rendezvous. International Series in Operations Research and Management Science. Kluwer Academic Publisher, Dordrecht (2002)
2. Anderson, E., Fekete, S.: Two-dimensional rendezvous search. Oper. Res. **49**, 107–118 (2001)
3. Baba, D., Izumi, T., Ooshita, F., Kakugawa, H., Masuzawa, T.: Space-optimal rendezvous of mobile agents in asynchronous trees. In: Patt-Shamir, B., Ekim, T. (eds.) SIROCCO 2010. LNCS, vol. 6058, pp. 86–100. Springer, Heidelberg (2010). doi:10.1007/978-3-642-13284-1_8
4. Bampas, E., Czyzowicz, J., Gąsieniec, L., Ilcinkas, D., Labourel, A.: Almost optimal asynchronous rendezvous in infinite multidimensional grids. In: Lynch, N.A., Shvartsman, A.A. (eds.) DISC 2010. LNCS, vol. 6343, pp. 297–311. Springer, Heidelberg (2010). doi:10.1007/978-3-642-15763-9_28
5. Chalopin, J., Dieudonné, Y., Labourel, A., Pelc, A.: Fault-tolerant rendezvous in networks. In: Esparza, J., Fraigniaud, P., Husfeldt, T., Koutsoupias, E. (eds.) ICALP 2014. LNCS, vol. 8573, pp. 411–422. Springer, Heidelberg (2014). doi:10.1007/978-3-662-43951-7_35
6. Cieliebak, M., Flocchini, P., Prencipe, G., Santoro, N.: Distributed computing by mobile robots: gathering. SIAM J. Comput. **41**, 829–879 (2012)
7. Czyzowicz, J., Gasieniec, L., Pelc, A.: Gathering few fat mobile robots in the plane. Theor. Comput. Sci. **410**, 481–499 (2009)
8. Czyzowicz, J., Kosowski, A., Pelc, A.: How to meet when you forget: log-space rendezvous in arbitrary graphs. Distrib. Comput. **25**, 165–178 (2012)
9. Das, S., Dereniowski, D., Kosowski, A., Uznański, P.: Rendezvous of distance-aware mobile agents in unknown graphs. In: Halldórsson, M.M. (ed.) SIROCCO 2014. LNCS, vol. 8576, pp. 295–310. Springer, Heidelberg (2014). doi:10.1007/978-3-319-09620-9_23
10. Dessmark, A., Fraigniaud, P., Kowalski, D., Pelc, A.: Deterministic rendezvous in graphs. Algorithmica **46**, 69–96 (2006)
11. Dieudonné, Y., Pelc, A.: Deterministic polynomial approach in the plane. Distrib. Comput. **28**, 111–129 (2015)
12. Dieudonné, Y., Pelc, A., Villain, V.: How to meet asynchronously at polynomial cost. SIAM J. Comput. **44**, 844–867 (2015)
13. Flocchini, P., Prencipe, G., Santoro, N., Widmayer, P.: Gathering of asynchronous robots with limited visibility. Theor. Comput. Sci. **337**, 147–168 (2005)
14. Flocchini, P., Santoro, N., Viglietta, G., Yamashita, M.: Rendezvous of two robots with constant memory. In: Moscibroda, T., Rescigno, A.A. (eds.) SIROCCO 2013. LNCS, vol. 8179, pp. 189–200. Springer, Heidelberg (2013). doi:10.1007/978-3-319-03578-9_16
15. Fraigniaud, P., Pelc, A.: Delays induce an exponential memory gap for rendezvous in trees. ACM Trans. Algorithms 9 (2013). Article 17

16. Kranakis, E., Krizanc, D., Morin, P.: Randomized rendez-vous with limited memory. In: Laber, E.S., Bornstein, C., Nogueira, L.T., Faria, L. (eds.) LATIN 2008. LNCS, vol. 4957, pp. 605–616. Springer, Heidelberg (2008). doi:10.1007/978-3-540-78773-0_52

17. Kranakis, E., Krizanc, D., Santoro, N., Sawchuk, C.: Mobile agent rendezvous in a ring. In: Proceedings 23rd International Conference on Distributed Computing Systems (ICDCS 2003), pp. 592–599

18. Miller, A., Pelc, A.: Time versus cost tradeoffs for deterministic rendezvous in networks. In: Proceedings 33rd Annual ACM Symposium on Principles of Distributed Computing (PODC 2014), pp. 282–290

19. Pelc, A.: Deterministic rendezvous in networks: a comprehensive survey. Networks **59**, 331–347 (2012)

20. Ta-Shma, A., Zwick, U.: Deterministic rendezvous, treasure hunts and strongly universal exploration sequences. In: Proceedings 18th ACM-SIAM Symposium on Discrete Algorithms (SODA 2007), pp. 599–608

Distributed Evacuation in Graphs
with Multiple Exits

Piotr Borowiecki[1], Shantanu Das[2], Dariusz Dereniowski[1],
and Łukasz Kuszner[1(✉)]

[1] Faculty of Electronics, Telecommunications and Informatics,
Gdańsk University of Technology, Gdańsk, Poland
{pborowie,deren,kuszner}@eti.pg.gda.pl
[2] Aix Marseille Univ., CNRS, LIF, Marseille, France
shantanu.das@lif.univ-mrs.fr

Abstract. We consider the problem of efficient evacuation using multiple exits. We formulate this problem as a discrete problem on graphs where mobile agents located in distinct nodes of a given graph must quickly reach one of multiple possible exit nodes, while avoiding congestion and bottlenecks. Each node of the graph has the capacity of holding at most one agent at each time step. Thus, the agents must choose their movements strategy based on locations of other agents in the graph, in order to minimize the total time needed for evacuation. We consider two scenarios: (i) the centralized (or offline) setting where the agents have full knowledge of initial positions of other agents, and (ii) the distributed (or online) setting where the agents do not have prior knowledge of the location of other agents but they can communicate locally with nearby agents and they must modify their strategy in an online fashion while they move and obtain more information. In the former case we present an offline polynomial time solution to compute the optimal strategy for evacuation of all agents. In the online case, we present a constant competitive algorithm when agents can communicate at distance two in the graph. We also show that when the agents are heterogeneous and each agent has access to only a subgraph of the original graph then computing the optimal strategy is NP-hard even with full global knowledge. This result holds even if there are only two types of agents.

Keywords: Discrete evacuation · Distributed algorithm · Mobile agents · Network flow

1 Introduction

Coordinated action of multiple autonomous agents is a subject of study in many contexts. Frequently, the communication capabilities of agents are limited, and

Research partially supported by the Polish National Science Center grant DEC-2011/02/A/ST6/00201 and by the ANR (France) project MACARON (anr-13-js02-0002).

© Springer International Publishing AG 2016
J. Suomela (Ed.): SIROCCO 2016, LNCS 9988, pp. 228–241, 2016.
DOI: 10.1007/978-3-319-48314-6_15

the exchange of information between agents is only possible when they are located close to each other, which creates challenges for coordination problems. We consider here the *evacuation* problem which requires multiple mobile entities to reach designated safe area, in a coordinated manner. This can correspond to evacuating a building with a crowd of people inside, who are required to leave the building through emergency exits due to flood, fire, bomb attack, gas leak or other dangers. One could also imagine other scenarios, e.g., a swarm of mobile robots that are required to gather in selected places (exits in the context of evacuation terminology).

In practical situations it can be desired to calculate the evacuation strategy 'on the fly' – just in the time, when the evacuation process is about to start. In fact it is possible to compute a customized solution adapted to the current situation, depending on the number of agents and their locations. Centralized computation of the evacuation strategy has an important drawback as the security of the system relies strongly on the central computing unit and the communication between central unit and agents. Hence, for the safety reasons, a distributed approach could be a better solution. Indeed, recently some attention has been paid to the study of non-centralized evacuation control systems for example assuming a usage of handheld devices (like smartphones) by evacuees [16].

In this paper we try to answer the question of how the lack of knowledge about the positions of other agents may influence the quality of a solution, e.g., by creating bottlenecks when too many agents try to go to the same exit. We compare the distributed (online) solutions to the problem with centralized (offline) solutions. Another important issue for evacuation is when the mobile agents have intrinsic characteristics preventing some of them from visiting some areas of the environment. For example, disabled people might not be able to go through steep stairs inside a building. To address such situations we consider a reachability function defined for each of the agents separately and investigate the coordinated evacuation of such heterogenous agents.

1.1 Related Work

Evacuation models can be classified as macro- and microscopic [13], where macroscopic models are based on optimization approaches of dynamic network flows and do not consider individual characteristics of evacuees while microscopic models are based on simulations in which physical abilities of evacuees are considered.

Indeed, the evacuation problem has been widely described as an application of flows over time (dynamic flows) in time expanded graphs [10,11] (see the PhD thesis by Jan-Philipp Kappmeier [15] for the recent survey and the plethora of literature references).

Graphs based models are the first choice in the street network modelling [14] but also such environments like buildings [2,4] or caves [3] can be modelled by graphs. In this case, additional properties of graphs can be considered to reflect properties of the real network such as the transit times, connection capacities, node capacities etc. The motivation for the graph model considered in this paper is lattice based, discretization graph where the Euclidean space is divided into small cells in the shape of squares or hexagons. Such an approach has been mostly considered for the sake of microscopic simulations [15].

Another type of investigations has been done in the context of distributed algorithms for collaborative agents. Chrobak et al. [5] investigate evacuation from the line and Czyżowicz et al. [6,7] study the problem of evacuation through an unknown exit from a disc shaped continuous space. The goal in these papers is to minimize the time when the last agent reaches an exit, which is in contrast to a collaborative search problem studied earlier [1], where the goal is to find an exit by the first robot as fast as possible. Many of these results attempt to provide the best algorithms in terms of evacuation time, compared to the obvious optimal solution when the exit is known to the agents. In the above investigations, the number of agents is small and the issue of congestion or collisions does not appear at all.

1.2 Our Results

In this paper, we consider the evacuation problem in the (discrete) graph model where the agents start from distinct nodes and they know the environment (i.e., the graph and the designated exits) but may not know the initial position of other agents. We first consider the centralized or offline version of our problem which we formulate in Sect. 2 while in Sect. 3 we present polynomial time algorithms for computing the evacuation strategy in general graphs using the time expansion technique. Section 4 gives a formal statement of the distributed version of the problem when the agents can only communicate locally and thus the strategy must be computed in online fashion. As a first step towards a distributed solution, we consider tree networks and we present and analyze distributed strategies for evacuation from tree networks in Sect. 5. In particular, in Sect. 5.1 we prove that there does not exist any distributed algorithm for evacuating agents in less than 2 times the offline optimal t_{opt} steps even in tree networks (Theorem 2). On the other hand, in Sect. 5.3 we give a distributed algorithm for trees proving its correctness and bounding the evacuation time by $72 \cdot t_{opt}$ steps (Theorem 4). Finally, in Sect. 6, we consider the evacuation of heterogenous agents having additional restrictions, namely every agent has access to a predefined subset of edges in the graph (see the similar concept applied to rendezvous problem in [9]). We show that computing the optimal evacuation strategy is NP-hard in this case even if there are only two types of agents and even if the evacuation time is a small constant.

2 Problem Formulation

In this section we formulate the discrete evacuation problem for the offline setting. Additional assumptions required in the distributed setting will be given in Sect. 4. In a basic version of the problem we are given a simple graph $G = (V, E)$ with node set V, edge set E, and size $n = |V(G)|$, and the set $\mathcal{A}$ of k agents initially placed on preselected nodes of the graph G, called *homebases* such that no two agents occupy the same homebase. In what follows the set of all home-bases is denoted by H. We also distinguish a subset X of nodes called *exits*. Time is divided into *steps* of unit duration. In each step the following actions are performed by each agent:

(A1) an agent either changes its location from the currently occupied node v to one of its neighbors, or remains at v,

(A2) an agent that occupies some node in X, *evacuates* from the graph.

An agent that evacuates is removed from the graph. If in some step an agent is not evacuated, then we say that the agent is *present* in the graph. The node occupied by an agent i at the end of action (A1) in step s is denoted by $\nu_i(s)$, while $\nu_i(0)$ denotes the initial position (the homebase) of the agent. A sequence $(\nu_1, \ldots, \nu_k)$ of the above functions is called an *evacuation strategy* if for each agent $i \in \mathcal{A}$ there is a step s_i such that $\nu_i(s_i) \in X$, and for each pair of distinct agents $i, j \in \mathcal{A}$ in each step $s \in \{1, \ldots, \min\{s_i, s_j\}\}$ it holds $\nu_i(s) \neq \nu_j(s)$. The *length* of an evacuation strategy is defined as $\max\{s_1, \ldots, s_k\}$. Note that according to (A1) we allow any pair of agents i, j located in neighbouring nodes to move in a single step s in such a way that $\nu_i(s) = \nu_j(s - 1)$ provided that $\nu_j(s - 1) \neq \nu_j(s)$. The decision version of discrete evacuation problem EVAC is defined as follows:

Problem EVAC

Input: a graph G, an integer l, a set X of exits and a set H of homebases keeping k agents.

Question: does there exist an evacuation strategy of length at most l?

3   The Complexity of EVAC

In this section we argue that there exists a polynomial-time algorithm for the problem EVAC. The solution is obtained using classical results for the maximum flow problem. More precisely, for an input (G, l, X, H) of EVAC we construct an *evacuation digraph* $Q = (U, A)$ for which there exists a flow of size k if and only if the answer to EVAC is YES.

Construction 1. Let $V = \{v_1, \ldots, v_n\}$ be the set of nodes of a given graph G and let $Q_0, Q_1, \ldots, Q_l$ be disjoint digraphs such that for each digraph $Q_j = (V_j^{\text{in}} \cup V_j^{\text{out}}, A_j)$ with $V_j^{\text{in}} = \{v_{j,1}^{\text{in}}, \ldots, v_{j,n}^{\text{in}}\}$ and $V_j^{\text{out}} = \{v_{j,1}^{\text{out}}, \ldots, v_{j,n}^{\text{out}}\}$, the arc set A_j is defined as follows $A_j = \{(v_{j,p}^{\text{in}}, v_{j,p}^{\text{out}}) \mid p \in \{1, \ldots, n\}\}$. The node set U of digraph Q consists of the nodes of all digraphs Q_j and the two additional nodes s and t. More formally $U = \{s, t\} \cup \bigcup_{j \in \{0,\ldots,l\}} (V_j^{\text{in}} \cup V_j^{\text{out}})$. To obtain the arc set A of Q we take all arcs of digraphs $Q_0, Q_1, \ldots, Q_l$, and for each $j \in \{0, \ldots, l-1\}$ between the nodes of Q_j and Q_{j+1} we add the arcs: $(v_{j,p}^{\text{out}}, v_{j+1,p}^{\text{in}})$ for each $p \in \{1, \ldots, n\}$, and $(v_{j,p}^{\text{out}}, v_{j+1,q}^{\text{in}})$ if for the corresponding nodes v_p, v_q of the graph G it holds $\{v_p, v_q\} \in E(G)$. We also add to A the following arcs: $(s, v_{0,p}^{\text{in}})$ if the corresponding vertex v_p of G belongs to H, and $(v_{j,p}^{\text{out}}, t)$ whenever

v_p belongs to X and $j \in \{0, \ldots, l\}$. Less formally, we add arcs outgoing from s to the nodes in V_0^{in} corresponding to all homebases of agents, and we add arcs incoming to t from all nodes in $V_0^{\text{out}} \cup \cdots \cup V_l^{\text{out}}$ corresponding to the exits. All arcs in Q are assumed to have unit capacities. Clearly, for every (G, l, X, H) the corresponding evacuation digraph can be constructed in polynomial time.   □

Lemma 1. *There exists an s-t flow of size k in evacuation digraph Q if and only if the answer to* EVAC *is* YES *for the input* (G, l, X, H).

Proof. Suppose that we have an s-t flow of size k in the digraph Q. Clearly, such a flow is made of k s-t paths that have only their endpoints in common. Each such a path P encodes the moves of a corresponding agent in an evacuation strategy. Indeed, for each $j \in \{0, \ldots, l-1\}$, the path P contains exactly one node $v_{j,p}^{\text{out}}$ in V_j^{out} and exactly one node $v_{j+1,q}^{\text{in}}$ in V_{j+1}^{in}, and $(u, v) \in A$. Then, the evacuation strategy dictated by P moves the corresponding agent from v_p to v_q in G in step $j + 1$. Since the internal nodes of the paths that constitute the flow have no internal nodes in common, the evacuation strategy obtained in this way is guaranteed to have no 'collisions' between agents. Also, since the only arcs incoming to t are outgoing from nodes in V_j^{out} that correspond to the exit set X, it is ensured that each agent reaches an exit within the time limit l.   □

Since the maximum flow problem is polynomial [8] we have the following.

Theorem 1. *There exists a polynomial-time algorithm for solving the problem* EVAC.   □

4   Distributed Evacuation Model

We now consider the distributed version of the evacuation problem where each agent must autonomously compute its strategy to move based on local information and communication with the agents it encounters. As before the model is synchronous and during each time step, each agent first communicates with other agents and performs local computations to decide on the action to be performed in this step; then the agent performs the action (i.e. it moves, stays or exits). Recall that the two possible actions are (A1) and (A2) defined in Sect. 2. Since each node can hold at most one agent at a given time, the actions selected by the agents in each time step must satisfy this restriction. However it is possible for two agents in adjacent nodes to swap their positions in a given step (there are no capacity restrictions on edges of the graph in our model).

Our next assumption is that each agent knows in advance the network and its own homebase. This includes the knowledge of the location of all exits. The nodes of the network have unique identifiers (thus the agents also have unique identifiers as each agent may 'inherit' the identifier of its homebase). Each agent has the information necessary for navigating in the graph i.e. the agent knows which edge to follow to reach the node selected for its next location. However the locations and the number of other agents are initially unknown to the agents.

The communication model is as follows. Each agent can directly communicate with any other agent that is within a distance of at most 2 in the graph. Note that the exchange of messages within distance of two (not just one) is a necessary assumption that follows from the necessity of avoiding collisions between agents. Otherwise two agents at distance two, being unaware of each other may decide to move to the same node. We assume that communication is instantaneous (or it takes negligible time compared to the duration of a time step). Thus, at the beginning of each step, any two agents can exchange any number of messages if the distance between them is not larger than 2. Agents can use message passing to communicate indirectly with other agents via intermediate agents. For example, see Fig. 1 where the agents form two 'groups' of communicating agents such that within each group any two agents i_1 and i_j can communicate either directly (if they are within the distance of two) or indirectly if i_1 and i_j are at distance greater than two but there exist agents $i_2, \ldots, i_{j-1}$ such that for each $p \in \{1, \ldots, j-1\}$ the agents i_p and i_{p+1} are at distance at most 2. In this example, agents i_1 and i_4 can communicate by passing messages through agents i_2 and i_3. Since we do not impose any restrictions on the amount of messages exchanged between any two agents at the beginning of each step, we may assume without loss of generality that each agent begins each step by identifying positions of all agents with whom it may communicate directly or indirectly. This can be achieved by performing a broadcast algorithm by the agents and since this part is straightforward, we omit the details and assume also in our algorithms that each step begins with collecting information about all agents that can be reached directly or indirectly by an agent in that step.

5   Evacuation in Tree Networks

In this section we investigate evacuation from tree networks in the distributed setting as formulated above. The performance measure we use for a distributed algorithm is the *competitive ratio*, defined as the worst case ratio of the evacuation time achieved by the algorithm over t_{opt} the optimal evacuation time in the

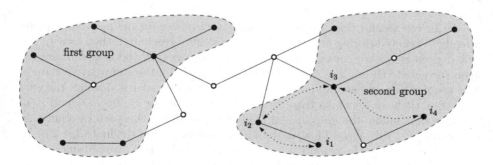

Fig. 1. Example graph with two groups of communicating agents. Black nodes contains agents and white nodes are unoccupied.

offline setting for the same instance (we consider the worst case scenario over all tree networks and all possible locations of homebases). We start by showing a lower bound on the competitive ratio of any distributed strategy, in Sect. 5.1. Then, in Sect. 5.2, we introduce some notations used in Sect. 5.3 which provides a distributed evacuation algorithm for trees.

5.1   A Lower Bound

Theorem 2. *The competitive ratio of any distributed evacuation algorithm is at least 2 even for trees.*

Proof. Given two integers $k, p \geq 1$, let us consider a tree with $3kp + 1$ nodes: a node v_0 and $3kp$ nodes $v_{i,j}$, where $i \in \{1, 2, \ldots, k\}$ and $j \in \{1, 2, \ldots, 3p\}$. There are $k + 1$ exits, $X = \{v_0, v_{1,3p}, v_{2,3p}, \ldots, v_{k,3p}\}$. Node v_0 is adjacent to k nodes: $v_{1,1}, v_{2,1}, \ldots, v_{k,1}$ and for every $i \in \{1, 2, \ldots, k\}$ and $j \in \{1, \ldots, 3p - 1\}$ there is an edge $\{v_{i,j}, v_{i,j+1}\}$. There are k agents $a_1, \ldots, a_k$, where $\nu_{a_i}(0) = v_{i,p}$.

Informally speaking, there are k paths of length $3p$ joined by an exit node v_0, with k agents, one in each path, placed p steps away from the exit v_0 and $2p$ steps from the other exit $v_{i,3p}$ at the end of the path. We take $k = 4p$. Note that the optimum evacuation time for this instance is $t_{\text{opt}} = 2p$.

Let A be any distributed algorithm for EVAC. Let us consider the step $2p - 2$ of the execution of A. There are two cases:

(i) There exists $i \in \{1, 2, \ldots, k\}$ such that there is an agent a on the path connecting $v_{i,p+1}$ and $v_{i,3p-2}$. Clearly $a = a_i$. Note that the agent a_i was unable to reach the node $v_{i,1}$ adjacent to the exit v_0 and thus this agent performed no communication with another agent within the first $2p - 2$ steps. If so, let us consider a different instance of the problem in which the graph is the same as in the former one but there exists only one agent, namely a_i with the same homebase $v_{i,p}$. Thus, the input to the algorithm executed by the agent a_i is the same as in the previous scenario. Therefore, due to the lack of communication as argued above, the behaviour of the agent a_i is exactly the same in the latter scenario as in the former one. Thus, the agent a_i will not evacuate before step $2p$. The evacuation time in the latter scenario with single agent present is clearly p and hence in this case the competitive ratio is 2.

(ii) In the second case, every agent a_i is somewhere on the path $v_{i,1}, v_{i,p}$ or already evacuated during the first $2p - 2$ steps. If a_i evacuated, then clearly the exit it used is v_0. We consider two additional subcases. In the first subcase, all agents evacuate through v_0. Since $k = 4p$ and $t_{\text{opt}} = 2p$, the competitive ratio is 2 as required. In the second subcase, some agent a_i evacuates through the exit $v_{i,3p}$. But then, a_i needs to traverse the path connecting $\nu_{a_i}(2p - 2)$ and $v_{i,3p}$ in steps that follow the step $2p - 2$. This path contains the path of length $2p$ connecting $v_{i,p}$ with $v_{i,3p}$. Thus, a_i does not evacuate within the first $(2p-2)+2p$ steps. Therefore, in the second subcase the competitive ratio we obtain is $\frac{4p-2}{2p}$ which can be made arbitrarily close to 2 by taking p large enough.   □

5.2 Additional Notations

The distance between a pair of nodes u and v denoted by $d(u, v)$ is the length of the path (i.e., the number of edges) connecting these nodes. By the distance between two agents i and j in step s we mean the distance between the nodes in which the agents are located at the end of step s, i.e., $d(\nu_s(i), \nu_s(j))$.

The exit *associated with node* v is the exit $x \in X$ with the smallest identifier among the exits at minimum distance from v. By the *primary exit* of agent i, denoted by $\mathrm{pe}(i)$, we mean the exit associated with $\nu_i(0)$, the homebase of the agent. Clearly, there may be many agents having the same node as the primary exit but for each agent the primary exit is unique. Also note that since the primary exit depends only on the homebase of the agent, it remains the same even if agent changes its position during evacuation. As we will see later, in the early stage of our algorithm each agent attempts to evacuate through its primary exit. This leads to the formation of groups of agents having the same primary exits and allows for computation of group rather than individual strategies. We address this issue in more detail during analysis of our algorithm.

For the description of the algorithm we also need a specific partition $\mathcal{V} = (V_{x_1}, \dots, V_{x_\eta})$ of the node set of a given tree T with the set of exits $X = \{x_1, \dots, x_\eta\}$. Namely, for each exit $x \in X$ the set V_x of $\mathcal{V}$ is defined as a set consisting of all nodes having x as the associated exit. Naturally, $\bigcup_{x \in X} V_x = V(T)$ and since primary exits are uniquely determined, for any pair of distinct exits x_p, x_q it holds $V_{x_p} \cap V_{x_q} = \emptyset$. It is also not hard to see that for each $x \in X$ the subgraph induced by V_x is connected. Let T_x denote the tree induced by V_x.

At the beginning of its computation each agent roots the tree T at the node with the smallest identifier. This ensures that all agents select the same root. Without loss of generality assume that agents rooted T in node v_r that belongs to the tree T_{x_r} which corresponds to exit x_r (see in Fig. 2(a) where $r = 2$). According to the above assumption we construct a tree $\widetilde{T}$ with the node set X, root x_r, and edge between each pair of nodes x_p, x_q for which T contains an edge $\{v, u\}$ such that $v \in V_{x_p}$ and $u \in V_{x_q}$, $p, q \in \{1, \dots, \eta\}$ (see Fig. 2(b) with x_1, x_4 corresponding to v, u in Fig. 2(a)). Following "away from the root" natural orientation of the edges of $\widetilde{T}$ one can easily relate subtrees T_x of T. Namely, for any two exits x_p, x_q we say that T_{x_p} is a *child* of T_{x_q} if x_q is the parent of x_p under the above orientation of $\widetilde{T}$.

The moves of agents in the early stage of our algorithm result in grouping the agents at exits. To address this more accurately let $x \in X$ and let s be a step. A *group of agents at x in step s*, denoted by $\mathcal{G}_{x,s}$ is a maximal set of agents for which $\nu_i(s) \in V(T_x)$ and each node of the path connecting $\nu_i(s)$ and x is occupied by an agent form $\mathcal{G}_{x,s}$. We say that an agent i *joins the group* $\mathcal{G}_{x,s}$ *in step s* if $i \notin \mathcal{G}_{x,s-1}$ but $i \in \mathcal{G}_{x,s}$. An agent i *joins exit x in step s* if in step s it joins the group $\mathcal{G}_{x,s}$ (we allow that the group is empty before step s).

Finally, let $\mathcal{A}_x$ denote the set of agents with their homebases in V_x, i.e., $\mathcal{A}_x = \{i \in \mathcal{A} \mid \mathrm{pe}(i) = x\}$. Note that $\mathcal{A}_x$ is not necessarily a group at x.

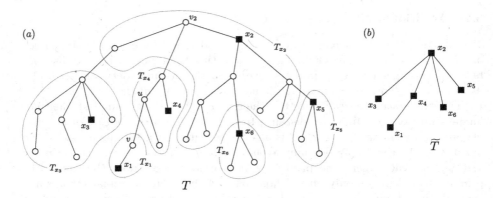

Fig. 2. Tree T with exits $x_1, \ldots, x_6$, subtrees $T_{x_1}, \ldots, T_{x_6}$, and the corresponding tree $\widetilde{T}$.

5.3 The Algorithm

Our algorithmic result obtained in this section is achieved in two steps. First, we give an evacuation algorithm that receives as an input an upper bound B on the optimum time required for the evacuation t_{opt} and we prove that this algorithm has a constant competitive ratio. Then, we provide our main procedure that uses the 'doubling technique' (the first algorithm mentioned above is called several times with exponentially increasing values on possible upper bounds) to disregard the assumption that an upper bound on evacuation time is known. As we prove, the competitive ratio of the second algorithm increases by a constant with respect to the first one. Both algorithms are formulated for an executing agent i.

Our first algorithm called BoundedDTE (*Bounded Distributed Tree Evacuation*). We start with an informal description providing the key ideas. The algorithm can be seen as having two phases. In the first phase, the agents located in the subtree composed of nodes in V_x want to communicate with each other to establish the best strategy for them. This strategy that they aim to establish ignores possible agents located outside of V_x. However, the homebases in V_x are possibly spread in such a way that the communication between the agents is not possible initially. Thus, all agents have to meet and they do so by going to x and forming a group at x. It can be estimated when all agents joined the group at x thanks to the fact that an upper bound B on t_{opt} is given as an input. Once all agents joined the group (we point out that at this point some agents possibly evacuated through x) they compute a strategy in which some of the agents remain to evacuate through x while some agents will follow to exit x' such that $T_{x'}$ is a child of T_x. Then the second phase starts in which each agent goes to the assigned exit.

In the pseudocode we will use the phrase *if possible* to indicate that the executing agent i verifies if the specified move can be performed. In particular, if the agent i wants to move to a node u, then communication with all agents located at u and its neighbors is required. This communication is needed to determine all agents that want to move to u in this given round and if there is more than one such an agent, then the agents decide which one (as there can be no more than one) will perform the move. In our algorithm such ties can be resolved arbitrarily.

Procedure 1. BoundedDTE(T, B) for agent i

Input: the underlying tree T, the time bound B.
1: **for** $k = 1$ **to** B **do**
2:    If possible, then move to the adjacent node that is closer to pe(i).
3: **end for**
4: Compute the optimum evacuation strategy for the group $\mathcal{G}_{\mathrm{pe}(i),B}$ taking into account pe(i) and exits in child trees of $T_{\mathrm{pe}(i)}$. Let *secondary exit*, se(i), be the exit for agent i computed in this strategy.
5: **for** $k = 1$ **to** $8 \cdot B$ **do**
6:    If possible, then move to the adjacent node that is closer to se(i).
7: **end for**

Theorem 3. *If B is an upper bound on the evacuation time t_{opt}, then Procedure* BoundedDTE *evacuates agent i in $9 \cdot B$ steps.*[1]

We now proceed to the description of our main algorithm DTE (*Distributed Tree Evacuation*) in which we disregard the previous assumption that an upper bound on t_{opt} is known to agents, and again we start with an informal description. The algorithm, as indicated earlier, proceeds by guessing the possible upper bounds B on t_{opt}. For each choice of B, DTE makes a call to BoundedDTE with input B. If the value of B is indeed at least t_{opt}, then all agents will successfully evacuate. Otherwise, the agents who still did not evacuate reverse their movements to return to their homebases.

Theorem 4. *Procedure* DTE *evacuates all agents in $\Theta(t_{\mathrm{opt}})$ steps.*

Proof. Once the variable B achieves value B' which is larger than t_{opt} ($t_{\mathrm{opt}} < B' \leq 2 \cdot t_{\mathrm{opt}}$) then Procedure DTE eventually evacuates all remaining agents. Let us estimate t_2, the number of steps required for the execution of Procedure DTE:

$$t_2 \leq 18 \cdot (2 + 4 + \cdots + B') \leq 36 \cdot B' \leq 72 \cdot t_{\mathrm{opt}}.$$

Now it is required to justify that agents are able to realize the described strategy. The analysis is very similar to that of Procedure DTE but there are two possibilities of additional traffic jams that must be considered.

---

[1] Due to the space constraint, proofs of Theorems 3 and 6 have been omitted. They can be found in the appendix of the pre-proceedings version of the paper, available online at the conference site: http://sirocco2016.hiit.fi.

Procedure 2. DTE(T) for agent i

Input: the underlying tree T.

1: $B \leftarrow 1$
2: **while** true **do**
3:      $B \leftarrow 2 \cdot B$
4:      **if** $d(\nu_i(0), \text{pe}(i)) > B$ **then**
5:          stay immobilized for next $18 \cdot B$ steps.
6:      **else**
7:          Apply Procedure BoundedDTE for $9 \cdot B$ steps or until the evacuation of i.
8:          reverse $9 \cdot B$ steps made according to Procedure BoundedDTE reaching its home-base.
9:      **end if**
10: **end while**

First, it can happen that agents from the group $\mathcal{G}_{\text{pe}(i),B}$ executing statements described in lines: 7, 8 of Procedure DTE encounter immobilized agents from the rest of $\mathcal{A}_{\text{pe}(i)}$. In this case a swap of agents will be applied. Suppose that traveling agent a at node u is about to swap with immobilized agent j at node v at the path uvw. Then, a swaps with j and they move: a moves from v to w and j moves from u back to v. Observe here also, that during the reverse phase (line 8) of Procedure DTE immobilized agents are still immobilized in the same place, thus making the reverse phase possible.

An interaction in between agents from $\mathcal{G}_{\text{pe}(i),B}$ and agents from the parent subtree of $T_{\text{pe}(i)}$ is covered by the correctness proof of Procedure DTE.   □

6   Generalizations of EVAC

In this section we introduce the problem RESTEVAC, a generalization of EVAC in which the moves of agents are restricted so that each agent has access to a preselected subset of edges.

6.1   Evacuation with Restricted Access to Edges

We start by defining an evacuation strategy in this scenario. All restrictions of EVAC carry over, except that additionally for each agent $i \in \mathcal{A}$, a set $R_i \subseteq E(G)$ of *permitted edges* is given as a part of the input, while action (A1) defined in Sect. 2 is replaced by:

(RA1) each agent either changes its location from the currently occupied node v to one of its neighbors u provided that $\{u,v\} \in R_i$, or the agent remains at v.

A sequence of functions $(\nu_1, \ldots, \nu_k)$ that satisfy (RA1) and (A2) is called*restricted evacuation strategy*. The decision version of our new problem, called RESTEVAC can be defined as follows:

Problem RestEvac

> Input: a graph G, an integer l, a set X of exits, a set H of homebases keeping k agents, and sets $R_1, \ldots, R_k$ specifying permitted edges.
> Question: does there exist a restricted evacuation strategy of length at most l?

6.2   The Complexity of RestEvac – Restricted Length

We now show that the RestEvac problem is NP-hard by reducing the NP-complete 3-dimensional matching problem, denoted by 3DM, to RestEvac. We first recall the former problem. The input to 3DM consists of three pairwise disjoint m-element sets A, B, C and a set of triples $M \subseteq A \times B \times C$. The answer to 3DM is YES if and only if there exists $M' \subseteq M$ such that $|M'| = m$ and for every two distinct triples (a, b, c) and (a', b', c') in M' it holds $a \neq a'$, $b \neq b'$ and $c \neq c'$.

Theorem 5. *The problem* RestEvac *is NP-complete even if the input parameter l equals 2.*

Proof. We describe our reduction by constructing the input to RestEvac. Suppose that (A, B, C, m, M) is the input to 3DM. For any triple $(a, b, c) \in M$ define for brevity $\xi((a, b, c)) = \{\{a, b\}, \{b, c\}\}$. Let G be defined as follows:

$$G = \left(A \cup B \cup C, \bigcup_{Z \in M} \xi(Z) \right).$$

Set $l = 2$, $X = C$, $k = m$ and let the k agents be initially placed on the nodes in A, i.e., set $H = A$. For each agent $i \in A$ having its homebase $a_i \in A$, let

$$R_i = \bigcup_{Z \in N} \xi(Z), \text{ where } N = M \cap (\{a_i\} \times B \times C).$$

This completes the construction and it remains to prove its correctness. We argue that the answer to 3DM is YES if and only if the answer to RestEvac is YES.

First, suppose that the answer to 3DM is YES and let $M' \subseteq M$ be the corresponding solution. We construct a restricted evacuation strategy $(\nu_1, \ldots, \nu_k)$ as follows. For each $i \in A$ there exists $(a_i, b_i, c_i) \in M'$ and we define:

$$\nu_i(0) = a_i, \quad \nu_i(1) = b_i, \quad \nu_i(2) = c_i.$$

It follows from the construction of the input to 3DM and from the definition of X that $\nu_i(0) \neq \nu_j(0)$ for $i \neq j$. Since M' is a solution to 3DM, we obtain that $\nu_i(s) \neq \nu_j(s)$ for $i \neq j$ and $s \in \{1, 2\}$. Consider an edge $e = \{\nu_i(s), \nu_i(s+1)\}$ traversed by agent i in step $s \in \{0, 1\}$. We have that $e \in \xi((a_i, b_i, c_i))$ and hence $e \in R_i$. Finally, $\nu_i(2) \in X$ by construction for each $i \in A$, which proves that $(\nu_1, \ldots, \nu_k)$ is indeed a restricted evacuation strategy of length 2.

Next, suppose that the answer to RestEvac is YES and let $(\nu_1, \ldots, \nu_k)$ be a restricted evacuation strategy of length 2. Since the homebases of agents are in A and $X = C$, we have that for each agent $i \in \mathcal{A}$ it holds $\nu_i(0) \in A$, $\nu_i(1) \in B$ and $\nu_i(2) \in C$. Define $M' = \{(\nu_i(0), \nu_i(1), \nu_i(2)) \mid i \in \mathcal{A}\}$. We have that $\{\nu_i(0), \nu_i(1)\} \in R_i$ and $\{\nu_i(1), \nu_i(2)\} \in R_i$, which for each agent $i \in \mathcal{A}$ implies $(\nu_i(0), \nu_i(1), \nu_i(2)) \in M$. Thus M' is a solution to 3DM. $\square$

6.3 The Complexity of RESTEVAC – Two Types of Agents

We prove that RESTEVAC is computationally hard even if there are only two types of agents, i.e., when there exist two subsets $E_1, E_2 \subseteq E(G)$ of permitted edges such that for each agent $i \in \mathcal{A}$ it holds $R_i \in \{E_1, E_2\}$. Our proof is by a reduction from the problem $(3,4)$-SAT, in which a logical formula $F = C_1 \wedge C_2 \wedge \cdots \wedge C_m$ over variables $\mathbf{x}_1, \ldots, \mathbf{x}_p$ is a part of the input, where each clause C_i is a disjunction of exactly three literals and each variable occurs at most four times in the logical formula. The question in $(3,4)$-SAT is whether there exists a Boolean assignment to variables that satisfies F. The problem $(3,4)$-SAT is known to be NP-complete [17]. The main result of this section is that RESTEVAC is computationally hard even for fixed length evacuation strategies and only two types of agents whose respective sets of permitted edges are disjoint.

Theorem 6. *The problem* RESTEVAC *is* NP-*complete even if the input parameter l equals 5 and for each agent $i \in \mathcal{A}$ it holds $R_i \in \{E_1, E_2\}$, where $E_1 \cap E_2 = \emptyset$.*

7 Conclusions and Open Problems

The goal of this work was to introduce a natural distributed model for the discrete evacuation problem and to analyze its basic properties by looking at its complexity and solvability in the mobile agent setting. One assumption that we made in this model, which greatly affects the algorithmic approach, is that local computations of agents and passing of messages take negligible time with respect to the movements of agents. However, one can consider scenarios in which the amount of communication performed in each synchronous step is somewhat more restricted. For example, one could analyze the number of messages exchanged by agents besides the evacuation time.

Another research direction could lead towards dropping the assumption that the network is known to agents. In such scenario, an algorithmic approach should adopt some concepts from the well studied exploration problems for unknown networks. As an intermediate scenario, one could consider providing the agents only partial information about the network by performing, e.g., a quantitative analysis of the amount of input information (we refer here to the advice complexity introduced in [12]).

In terms of the competitive ratio achievable by any online algorithm, we studied tree networks, and it is interesting to see how this parameter would behave in other network topologies. For example, are there networks in which the competitive ratio is not constant, e.g., a function of the number of agents or some other input parameter? Another open question worth investigating is what happens when the agents cannot communicate but have only local visibility.

References

1. Alspach, B.: Searching and sweeping graphs: a brief survey. Le Matematiche (Catania) **59**, 5–37 (2004)
2. Berlin, G.: The use of directed routes for assessing escape potential. Fire Technol. **14**(2), 126–135 (1978)
3. Breisch, R.: An intuitive approach to speleotopology. Southwestern Cavers **6**, 72–78 (1967)
4. Chalmet, L., Francis, R., Saunders, P.: Network models for building evacuation. Fire Technol. **18**(1), 90–113 (1982)
5. Chrobak, M., Gasieniec, L., Gorry, T., Martin, R.: Group search on the line. In: Italiano, G.F., Margaria-Steffen, T., Pokorný, J., Quisquater, J.-J., Wattenhofer, R. (eds.) SOFSEM 2015. LNCS, vol. 8939, pp. 164–176. Springer, Heidelberg (2015). doi:10.1007/978-3-662-46078-8_14
6. Czyzowicz, J., Gasieniec, L., Gorry, T., Kranakis, E., Martin, R., Pajak, D.: Evacuating robots via unknown exit in a disk. In: Kuhn, F. (ed.) DISC 2014. LNCS, vol. 8784, pp. 122–136. Springer, Heidelberg (2014). doi:10.1007/978-3-662-45174-8_9
7. Czyzowicz, J., Georgiou, K., Kranakis, E., Narayanan, L., Opatrny, J., Vogtenhuber, B.: Evacuating robots from a disk using face-to-face communication (Extended Abstract). In: Paschos, V.T., Widmayer, P. (eds.) CIAC 2015. LNCS, vol. 9079, pp. 140–152. Springer, Heidelberg (2015). doi:10.1007/978-3-319-18173-8_10
8. Edmonds, J., Karp, R.: Theoretical improvements in algorithmic efficiency for network flow problems. J. ACM **19**(2), 248–264 (1972)
9. Farrugia, A., Gasieniec, L., Kuszner, L., Pacheco, E.: Deterministic rendezvous in restricted graphs. In: Italiano, G.F., Margaria-Steffen, T., Pokorný, J., Quisquater, J.-J., Wattenhofer, R. (eds.) SOFSEM 2015. LNCS, vol. 8939, pp. 189–200. Springer, Heidelberg (2015). doi:10.1007/978-3-662-46078-8_16
10. Ford, L., Fulkerson, D.: Constructing maximal dynamic flows from static flows. Oper. Res. **6**(3), 419–433 (1958)
11. Ford, L., Fulkerson, D.: Flows in Networks. Princeton University Press, Princeton (1962)
12. Fraigniaud, P., Ilcinkas, D., Pelc, A.: Oracle size: a new measure of difficulty for communication tasks. In: PODC 2006, pp. 179–187. ACM (2006)
13. Hamacher, H., Tjandra, S.: Mathematical modelling of evacuation problems - a state of the art. In: Pedestrian and Evacuation Dynamics, pp. 227–266. Springer, Heidelberg (2002)
14. Higashikawa, Y., Golin, M., Katoh, N.: Multiple sink location problems in dynamic path networks. Theor. Comput. Sci. **607**, 2–15 (2015)
15. Kappmeier, J.P.: Generalizations of flows over time with applications in evacuation optimization. Ph.D. thesis, Technische Universität Berlin (2015)
16. del Moral, A.A., Takimoto, M., Kambayashi, Y.: Distributed evacuation route planning using mobile agents. Trans. Comput. Collective Intell. **17**, 128–144 (2014)
17. Tovey, C.: A simplified NP-complete satisfiability problem. Discr. Appl. Math. **8**, 85–89 (1984)

Universal Systems of Oblivious Mobile Robots

Paola Flocchini[1], Nicola Santoro[2], Giovanni Viglietta[1(✉)],
and Masafumi Yamashita[3]

[1] University of Ottawa, Ottawa, Canada
{paola.flocchini,gviglieet}@uottawa.ca
[2] Carleton University, Ottawa, Canada
santoro@scs.carleton.ca
[3] Kyushu University, Fukuoka, Japan
mak@csce.kyushu-u.ac.jp

Abstract. An oblivious mobile robot is a stateless computational entity located in a spatial universe, capable of moving in that universe. When activated, the robot observes the universe and the location of the other robots, chooses a destination, and moves there. The computation of the destination is made by executing an algorithm, the same for all robots, whose sole input is the current observation. No memory of all these actions is retained after the move. When the spatial universe is a graph, distributed computations by oblivious mobile robots have been intensively studied focusing on the conditions for feasibility of basic problems (e.g., gathering, exploration) in specific classes of graphs under different schedulers. In this paper, we embark on a different, more general, type of investigation.

With their movements from vertices to neighboring vertices, the robots make the system transition from one configuration to another. Thus the execution of an algorithm from a given configuration defines in a natural way the computation of a discrete function by the system. Our research interest is to understand which functions are computed by which systems. In this paper we focus on identifying sets of systems that are *universal*, in the sense that they can collectively compute all finite functions. We are able to identify several such classes of fully synchronous systems. In particular, among other results, we prove the universality of the set of all graphs with at least one robot, of any set of graphs with at least two robots whose quotient graphs contain arbitrarily long paths, and of any set of graphs with at least three robots and arbitrarily large finite girths. We then focus on the minimum size that a network must have for the robots to be able to compute all functions on a given finite set. We are able to approximate the minimum size of such a network up to a factor that tends to 2 as n goes to infinity.

The main technique we use in our investigation is the *simulation* between algorithms, which in turn defines *domination* between systems.

This work has been supported in part by the Natural Sciences and Engineering Research Council of Canada through the Discovery Grant program; by Prof. Flocchini's University Research Chair; and by the Scientific Grant in Aid by the Ministry of Education, Culture, Sports, Science, and Technology of Japan.

© Springer International Publishing AG 2016
J. Suomela (Ed.): SIROCCO 2016, LNCS 9988, pp. 242–257, 2016.
DOI: 10.1007/978-3-319-48314-6_16

If a system dominates another system, then it can compute at least as many functions. The other ingredient is constituted by *path* and *ring* networks, of which we give a thorough analysis. Indeed, in terms of implicit function computations, they are revealed to be fundamental topologies with important properties. Understanding these properties enables us to extend our results to larger classes of graphs, via simulation.

1   Introduction

Consider a network, represented as a finite graph G, where the vertices are unlabeled, and edge labels are possibly not unique. In G operate k *oblivious mobile robots* (or simply "robots"), that is, indistinguishable computational entities with no memory, located at the vertices of the network, and capable of moving from vertex to neighboring vertex of G. Robots are activated by an adversarial *scheduler S*. Whenever activated, a robot observes the location of the other robots in the graph (the current *configuration*); it computes a destination (a neighboring vertex or the current location); and it moves there. The computation of the destination is made by executing an algorithm, the same for all robots, whose sole input is the current configuration. The current activity terminates after the move, and no memory of the computation is retained; in other words, the entities are *stateless*. The overall system is represented by the triplet (G, k, S). Notice that, even if the algorithm A the robots execute is deterministic, its executions may still be non-deterministic. Indeed, since the network's port numbers may not be unique, it may be impossible for an algorithm to unambiguously indicate where each robot has to move. This model, introduced by Klasing, Markou, and Pelc [23] as an extension of the model of oblivious robots in continuous spaces (e.g., [14]), has been extensively employed and investigated, focusing on basic problems in specific classes of graphs under different schedulers: *gathering* and *scattering* (e.g., [4–7,10,17,18,20,22,23,26,27]), and *exploration* and *traversal* (e.g., [1–3,8,9,11–13,24,25]). Note that, with the exception of [3], the literature assumes unlabelled edges. In this paper, we consider both labelled and unlabelled edges, and focus on the fully synchronous scheduler $\mathcal{F}$, which simply activates every robot at every turn. We then embark on a different, more general, type of investigation.

Consider the system $(G, k, \mathcal{F})$. Whenever the robots move in the graph according to algorithm A, the system transitions from the current configuration to a (possibly) different one. The obliviousness of the robots implies that always the same (or equivalent) transition occurs from the same given configuration. Consider now the *configuration graph* where there is a directed edge from one configuration to another if some algorithm dictates such a transition. Then the execution of A in $(G, k, \mathcal{F})$ from a given configuration is just a walk in this graph from that configuration. The execution can be viewed in a natural way as the computation of a discrete function f by the system, where f maps a configuration C into the configuration $f(C)$ reached by executing (one step of) A from C, defining a subgraph of the configuration graph, called function graph.

The concept of function computation and function graph are formally defined in Sect. 2.

We seek to understand which functions are computed by which systems. Knowing the structure of such functions gives us information on the robots' behavior as they execute an algorithm, and what tasks the robots can and cannot perform in a network. For instance, if an algorithm computes a function whose graph has no cycles, it means that the robots will eventually be stationary regardless of their initial position; if the function has a unique fixed point, it means that the algorithm solves a *pattern formation* problem. On the other hand, if the function's graph has only cycles of length $p > 1$ (possibly with some "branches" attached), the robots are collectively implementing a *self-stabilizing clock* of period p. If such graphs can be embedded in the configuration graph, then we know that such algorithms exist, and that the corresponding problems are solvable in the system.

In this paper we focus on identifying sets of systems that are *universal*, in the sense that they compute all finite functions. In Sect. 4, we identify several classes of universal fully synchronous systems. In particular, among other results, we prove that

Theorem. The following families of systems are universal:

(a) $\{(G, 1, \mathcal{F}) \mid G$ is an unlabeled network$\}$,
(b) $\{(G_n, 2, \mathcal{F}) \mid$ the quotient graph of G_n contains a sub-path of length at least $n\}$,
(c) $\{(G_n, 3, \mathcal{F}) \mid$ the girth of G_n is at least n and finite$\}$.

In Sect. 5, we focus on computing discrete functions using the smallest possible networks, perhaps at the cost of employing a large numbers of robots. In particular, for a given finite set X, we study the minimum size that a network must have for the robots to be able to compute all functions from X to X. We are able to approximate the minimum size of such a network up to a factor that tends to 2 as n goes to infinity.

The main tool we use in our investigation is the *simulation* between algorithms, which in turn defines *domination* between systems. If system Ψ dominates system Ψ', then Ψ computes at least all the functions computed by Ψ'. The other tool is constituted by the *path* and *ring* graphs (Sect. 3). These are the main ingredients of all our stronger results, because rings and paths are fundamental topologies with important properties that can be extended to other graphs via simulation.

Full proofs can be found in the extended version of this paper [15].

2   Definitions

In this section we introduce the models of mobile robots that we are going to study. Informally, we consider networks with port numbers, which are represented as graphs where each vertex has a label on each outgoing edge.

Port numbers are not required to be unique, which allows us to model anonymous networks with unlabeled edges, as well.

On a network we may place any number of robots, which are indistinguishable mobile entities with no memory. At all times, each robot must be located at a vertex of the network, and any number of robots may occupy the same vertex. All robots follow the same algorithm, which takes as input the network and the robots' positions, and tells each robot to which adjacent vertex it has to move next (or it may tell it to stay still). Time is discretized, and we assume that robots can move to adjacent vertices instantaneously.

Even if algorithms are deterministic, their executions may still be non-deterministic. This is partly because the network's port numbers may not be unique, and therefore it may be impossible for an algorithm to unambiguously indicate where each robot has to move. Another potential source of non-determinism is the scheduler, which is an adversary that decides which robots are going to be activated next. In this paper we will focus on the fully synchronous scheduler, which simply activates every robot at every turn. We will also briefly discuss the semi-synchronous scheduler in Sect. 6.

Labeled Graphs. A *labeled graph* is a triplet $G = (V, E, \ell)$, where (V, E) is an undirected graph called the *base graph*, and ℓ is a function that maps each ordered pair (u, v), such that $\{u, v\} \in E$, to a non-negative integer called *label*. A labeled graph is also referred to as a *network*. A network is *unlabeled* if all its labels are equal.

An *automorphism* of a labeled graph $G = (V, E, \ell)$ is a permutation α of V preserving adjacencies and labels, i.e., for all $u, v \in V$, if $\{u, v\} \in E$, then $\{\alpha(u), \alpha(v)\} \in E$ and $\ell(u, v) = \ell(\alpha(u), \alpha(v))$. If there exists an automorphism that maps a vertex u to a vertex v, then u and v are *equivalent vertices* in G. The *quotient graph* G^* is the labeled graph G obtained by identifying equivalent vertices, and preserving adjacencies and labels.

Configuration Spaces. Let $\mathbb{N}_n = \{0, 1, \cdots, n - 1\}$, for every $n \geq 1$. An *arrangement* of k robots on a network $G = (V, E, \ell)$ is a mapping from $\mathbb{N}_k$ to V. An arrangement specifies the locations of k *distinguishable* robots on a network whose vertices are all *distinguishable*. However, we ultimately intend to model *identical* robots, which *cannot distinguish* between equivalent vertices of the network, unless such vertices are occupied by different amounts of robots. The following definition serves this purpose: two arrangements $a_1, a_2 \colon \mathbb{N}_k \to V$, are *equivalent* if there exist an automorphism $\alpha \colon V \to V$ and a permutation π of $\mathbb{N}_k$ such that $\alpha \circ a_1 = a_2 \circ \pi$.

The *configuration space* $\mathcal{C}(G, k)$, where G is a network and k is a positive integer, is the quotient of the set of arrangements of k robots on G under the above equivalence relation between arrangements. The elements of the configuration space are called *configurations*.

Say that an arrangement a is equivalent to itself under an automorphism α and a permutation π, as defined above. Then, whenever $\alpha(v) = v'$ and $\pi(r) = r'$, we say that v and v' are *equivalent vertices* in a, and r and r' are *equivalent robots* in a.

A class of *indistinguishable* vertices U (respectively, a class of *indistinguishable* robots R) of a configuration $C \in \mathcal{C}(G, k)$ is a mapping from each arrangement $a \in C$ to an equivalence class of vertices U_a (respectively, an equivalence class of robots R_a) of a such that, for all $a_1, a_2 \in C$ and all automorphisms α and permutations π under which a_1 and a_2 are equivalent, $\alpha(U_{a_1}) = U_{a_2}$ (respectively, $\pi(R_{a_1}) = R_{a_2}$).

Configuration Graphs. While the configuration space contains all the configurations that are distinguishable, either by the base graph's topology, or by the labels, or by the robots' positions, the *configuration graph* specifies which configurations can reach which other configurations "in one step". Of course, this depends on a notion of algorithm, and on a notion of scheduler.

An *algorithm* for k robots on a network G is a function that maps a pair (C, U) into a set U', where $C \in \mathcal{C}(G, k)$ (describing the network's configuration at the moment the algorithm is executed), and U and U' are classes of indistinguishable vertices of C (indicating the executing robot's location and its destination, respectively) such that, for every arrangement $a \in C$ and every vertex $u \in U(a)$, there exists a vertex $u' \in U'(a)$ such that either $u = u'$ or u' is adjacent to u. According to this definition, a robot can only specify its destination as a class of indistinguishable vertices, representing either a null movement or a movement to some adjacent vertex.

An *execution* for k robots in a network G is a sequence of configurations of $\mathcal{C}(G, k)$. A *scheduler* for k robots in a network G is a binary relation between algorithms and executions. The *possible* executions of an algorithm under some scheduler are the executions that correspond to the algorithm under the relation specified by such a scheduler. A *system of oblivious mobile robots* is a triplet $\Psi = (G, k, S)$, where G is a labeled graph, $k \geq 1$, and S is a scheduler for k robots in G.

The *configuration graph* $\mathcal{G}(\Psi) = (\mathcal{C}(G, k), \mathcal{E}(\Psi))$, where $\Psi = (G, k, S)$ is a system of oblivious mobile robots, is a directed graph on the configuration space $\mathcal{C}(G, k)$, where $(C, C') \in \mathcal{E}(G, k)$ if there is an algorithm A and a possible execution $E = (C_i)_{i \geq 0}$ of A under S, such that there exists an index i satisfying $C = C_i$ and $C' = C_{i+1}$.

The *deterministic configuration graph* $\mathcal{G}'(\Psi) = (\mathcal{C}(G, k), \mathcal{E}'(\Psi))$, where $\Psi = (G, k, S)$ is a system of oblivious mobile robots, is a directed graph on the configuration space $\mathcal{C}(G, k)$, where $(C, C') \in \mathcal{E}'(G, k)$ if there is an algorithm A such that, for all possible executions $E = (C_i)_{i \geq 0}$ of A under S, and for every index i satisfying $C = C_i$, we have $C' = C_{i+1}$.

Intuitively, $\mathcal{G}'(\Psi)$ is a subgraph of $\mathcal{G}(\Psi)$ whose edges represent moves that can be deterministically done by the robots, i.e., on which all the scheduler's choices yield the same result. If $\mathcal{G}(\Psi) = \mathcal{G}'(\Psi)$, then Ψ is said to be a *deterministic* system.

Fully Synchronous Scheduler. Given an algorithm A for k robots on a network, we say that a configuration C' *yields* from a configuration C under algorithm A if, for every arrangement $a \in C$ there is an arrangement $a' \in C'$ such that, for every $r \in \mathbb{N}_k$, either $a(r) = a'(r)$ or $a(r)$ is adjacent to $a'(r)$ and, if U is the class of indistinguishable vertices of C such that $a(r) \in U(a)$, then $a'(r) \in U'(a)$, where $U' = A(C, U)$. The *fully synchronous scheduler* $\mathcal{F}$ is defined as follows: $(A, E = (C_i)_{i \geq 0}) \in \mathcal{F}$ if, for every $i \geq 0$, C_{i+1} yields from C_i. In the rest of the paper, we will write $\mathcal{F}(G, k)$ instead of $(G, k, \mathcal{F})$.

In other words, the fully synchronous scheduler lets every robot move at every turn to the destination it computes. However, if a robot's destination consists of several indistinguishable vertices, the scheduler may arbitrarily decide to move the robot to any of those vertices, provided that it can be reached in at most one hop. All these choices are made by the scheduler at each turn and for each robot, independently.

Simulating Algorithms. To define the concept of simulation, we preliminarily define a relation on executions. Given an execution $E = (C_i)_{i \geq 0}$ for k robots on a network G, an execution $E' = (C'_i)_{i \geq 0}$ for k' robots on a network G', and a surjective partial function $\varphi \colon \mathcal{C}(G, k) \to \mathcal{C}(G', k')$, we say that E is *compliant* with E' under φ if either φ is undefined on C_0, or there exists a weakly increasing surjective function $\sigma \colon \mathbb{N} \to \mathbb{N}$ such that, for every $i \in \mathbb{N}$, φ is defined on C_i, and $\varphi(C_i) = C'_{\sigma(i)}$.

An algorithm A under system Ψ *simulates* an algorithm A' under system Ψ' if there is a surjective partial function $\varphi \colon \mathcal{C}(G, k) \to \mathcal{C}(G', k')$ such that each execution of A under Ψ is compliant under φ with at least one execution of A' under Ψ'.

In this definition, φ "interprets" some configurations of the simulating system Ψ as configurations of the simulated system Ψ', in such a way that every configuration of Ψ' is represented by at least one configuration of Ψ. Moreover, the definition of compliance ensures that the simulating algorithm A makes configurations transition in a way that agrees with A' under φ.

Computing Functions. We define the implicit computation of a function as the simulation of a system consisting in a single robot on a network whose shape is given by the function itself.

The network *induced* by a function $f \colon X \to X$ is defined as $\Gamma_f = (X, f, \ell)$, where $\ell \colon (u, v) \mapsto v$. Hence the base graph of Γ_f has edges of the form $(x, f(x))$, and the labeling ℓ makes all vertices of Γ_f distinguishable from each other. The algorithm A_f *associated* to the function f is the algorithm for one robot on Γ_f that always makes the robot move from any vertex $x \in X$ to the vertex $f(x)$.

We say that an algorithm A *computes* a function $f \colon X \to X$ under system Ψ if it simulates the algorithm A_f under $\mathcal{F}(\Gamma_f, 1)$.

What this definition intuitively means is that each element of X is represented by a set of robot configurations; an algorithm computes f if any execution from a configuration representing $x \in X$ eventually yields a configuration representing

$f(x)$ without passing through configurations that represent other elements of X (or that represent no element of X).

If an algorithm under system Ψ computes a function f (respectively, simulates an algorithm A'), then we say that Ψ computes f (respectively, simulates A'). Moreover, a system Ψ *dominates* Ψ' if every algorithm under Ψ' is simulated by some algorithm under Ψ.

We use the notation $X \preceq Y$ to indicate all the concepts defined above: X may be a function computed by an algorithm Y (under some system), or it can be an algorithm simulated by Y, or a system dominated by a system Y, etc.

3   Basic Results

Proposition 1. *The relation $\preceq$ is transitive.*   □

Corollary 1. *If a system Ψ dominates a system Ψ', then all functions computed by Ψ' are also computed by Ψ.*

Proof. Suppose that $\Psi' \preceq \Psi$. Then, for any function f such that $f \preceq \Psi'$, the transitivity of $\preceq$ implies that $f \preceq \Psi$.   □

3.1   General Graphs

Proposition 2. *For every network G, the system $\mathcal{F}(G, 1)$ is deterministic, and its configuration graph is isomorphic to the graph obtained from the quotient graph G^* by replacing each unoriented edge $\{u, v\}$ with the two oriented edges (u, v) and (v, u), and adding a self-loop (v, v) to each vertex v.*   □

A fundamental question is whether adding robots to a network allows to compute more functions. We can at least prove that adding robots does not reduce the set of computable functions, provided that the network is not pathologically small.

Theorem 1. *For all networks G with at least three vertices and all $k \geq 1$, $\mathcal{F}(G, k+1) \not\preceq \mathcal{F}(G, k)$.*

Proof. It suffices to show that $|\mathcal{C}(G, k+1)| > |\mathcal{C}(G, k)|$. For each configuration in $\mathcal{C}(G, k)$, choose a vertex that contains the largest number of robots, and add one robot to it. This way we obtain $|\mathcal{C}(G, k)|$ distinct configurations of $\mathcal{C}(G, k+1)$. We can generate yet another configuration by placing $\lfloor (k + 1)/2 \rfloor$ robots on a vertex, $\lfloor (k + 1)/2 \rfloor$ robots on another vertex, and the remainder on a third vertex.   □

We can also show that a single robot does not compute more functions than $k \geq 1$ robots, in any network G.

Theorem 2. *For all networks G and all $k \geq 1$, $\mathcal{F}(G, 1) \preceq \mathcal{F}(G, k)$.*

Proof. In the simulation we use only the configurations of $\overset{*}{\mathcal{F}}(G, k)$ in which all robots lie in equivalent vertices of G. Then each robot pretends to be the only robot in the network, and makes the move that the unique robot of $\mathcal{F}(G, 1)$ would make. This is a well-defined simulation even if $\mathcal{F}(G, k)$ is not deterministic, due to Proposition 2. □

We can extend this idea to show that $\mathcal{F}(G, k) \preceq \mathcal{F}(G, 2k)$, provided that $\mathcal{F}(G, 2k)$ is deterministic.

Theorem 3. $\mathcal{F}(G, k) \preceq \mathcal{F}(G, k')$, *provided that* $\mathcal{F}(G, k')$ *is deterministic and* $k' \geq 2k$.

Proof. The configurations of $\mathcal{F}(G, k')$ that we use in our simulation are only those in which there is a (unique) vertex v occupied by at least $k' - k + 1$ robots. Each of these configurations is mapped to the configuration of $\mathcal{F}(G, k)$ that is obtained by removing $k' - k$ robots from v. This mapping is surjective. The simulation can be carried out because $\mathcal{F}(G, k')$ is deterministic, and therefore robots occupying the same vertex can never be separated, implying that there is always going to be a vertex with at least $k' - k + 1$ robots. □

Finally, we conjecture that adding robots increases a system's computational capabilities.

Conjecture 1. For all networks G and all $k \geq 1$, $\mathcal{F}(G, k) \preceq \mathcal{F}(G, k + 1)$.

3.2   Path Graphs

The first special type of network we consider is the one whose base graph consists of a single *path*. This fundamental configuration will turn out to be of great importance in Sects. 4 and 5, when studying universal classes of systems. In terms of labeling, we focus on two extreme cases: a labeling that gives a consistent orientation to the whole network (i.e., each vertex in the path has port labels indicating which neighbor is on the "left" and which one is on the "right"), and the anonymous unlabeled network. In the first case we have an *oriented path*, and in the second case we have an *unoriented path*. By $\overrightarrow{\mathcal{P}}_n^k$ and $\mathcal{P}_n^k$ we denote, respectively, the oriented and the unoriented path with n vertices and k robots, under the fully synchronous scheduler.

Oriented Paths. Let us study the configuration graph of $\overrightarrow{\mathcal{P}}_n^k$. Since the path has an orientation, no two vertices are equivalent. Therefore, by Proposition 2, $\mathcal{G}\left(\overrightarrow{\mathcal{P}}_n^1\right)$ consists of a path of length n with bidirectional edges and a self-loop on each vertex. In general, the configuration space of $\overrightarrow{\mathcal{P}}_n^k$ is in bijection with the set of weakly increasing k-tuples of integers in $\mathbb{N}_n$. Hence, for a fixed k, the size of the configuration space is

$$\binom{n + k - 1}{k} \sim \frac{n^k}{k!}.$$

If these k-tuples are thought of as points of $\mathbb{R}^k$, they constitute the set of lattice points in the k-dimensional simplex whose $k+1$ vertices have the form $(0, 0, \cdots, 0, 1, 1, \cdots, 1)$. This simplex has $k+1$ facets, two of which correspond to configurations in which the first or the last vertex of the network is occupied by a robot, while the other $k-1$ facets correspond to configurations in which exactly $k-1$ vertices are occupied (i.e., exactly two robots share a vertex).

The edges of the configuration graph (that are not self-loops) connect bidirectionally all pairs of points whose Chebyshev distance is at most 1, with the exception of the points that lie on the aforementioned $k-1$ facets. Indeed, since no algorithm can separate two robots that occupy the same vertex, it follows that those facets (as well as all their intersections) can never be left once they are reached. Figure 1(a) shows the configuration graph of $\overrightarrow{\mathcal{P}}_5^2$.

Property 1. For all $n \geq 1$, $\mathcal{G}\left(\overrightarrow{\mathcal{P}}_{2n}^2\right)$ contains an $n \times n$ grid with bidirectional edges and self-loops.

Since an oriented path gives the robots a sense of direction, an algorithm can unambiguously indicate to which neighbor each robot is supposed to move. Therefore $\mathcal{G}\left(\overrightarrow{\mathcal{P}}_n^k\right) = \mathcal{G}'\left(\overrightarrow{\mathcal{P}}_n^k\right)$.

Property 2. For all $n, k \geq 1$, the system $\overrightarrow{\mathcal{P}}_n^k$ is deterministic.

Unoriented Paths. Let us study the configuration graph of $\mathcal{P}_n^k$. Since the network in this system is unlabeled, if two vertices are symmetric with respect to the center of the path, they are equivalent. So, the configuration space is in bijection with the set of weakly increasing k-tuples of integers in $\mathbb{N}_n$, where each k-tuple $(a_1, \cdots, a_k)$ is identified with its "symmetric" one, $(n - a_k - 1, \cdots, n - a_1 - 1)$. Elementary computations reveal that, if k is fixed, these k-tuples are

$$\frac{1}{2} \cdot \binom{n+k-1}{k} + O\left(n^{\lfloor k/2 \rfloor}\right) \sim \frac{n^k}{2 \cdot k!}.$$

Geometrically, the configuration space of $\mathcal{P}_n^k$ can be represented as the set of lattice points in a truncated k-dimensional simplex, which is obtained by cutting the simplex of $\overrightarrow{\mathcal{P}}_n^k$ roughly in half, along a suitable hyperplane. Figures 1(b) and (c) show the configuration graphs of $\mathcal{P}_5^2$ and $\mathcal{P}_6^2$.

If n is even, we have $\mathcal{G}\left(\mathcal{P}_n^k\right) = \mathcal{G}'\left(\mathcal{P}_n^k\right)$, because each robot has a unique closest endpoint of the path, which it can use to specify unambiguously in which direction it intends to move. However, if $n > 1$ is odd and $k \geq 2$, the two graphs differ. Indeed, if the configuration is symmetric and the central vertex is occupied by more than one robot, then it is impossible to guarantee that all the central robots will move in the same direction: the adversary will decide how many of these robots go left, and how many go right. For instance, $\mathcal{G}'\left(\mathcal{P}_5^2\right)$ differs from $\mathcal{G}\left(\mathcal{P}_5^2\right)$ in that the vertex in $(2, 2)$ has no outgoing edges in $\mathcal{G}'\left(\mathcal{P}_5^2\right)$, because these correspond to non-deterministic moves.

Property 3. For all $n, k \geq 1$, the system $\mathcal{P}_{2n}^k$ is deterministic.

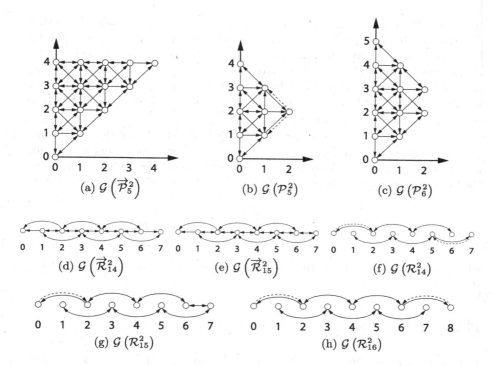

Fig. 1. Configuration graphs of some oriented and unoriented paths and rings. Dashed arrows represent non-deterministic moves. For clarity, self-loops have been omitted from all vertices.

3.3 Ring Graphs

Now we consider *ring* networks, which are networks whose base graph is a single cycle. This is another fundamental class of networks, which will have a great importance in Sects. 4 and 5. Like a path, a ring can be *oriented* if its labeling gives a consistent sense of direction to the robots in the network (i.e., each vertex has port labels indicating which neighbor lies in the "clockwise" direction, and which one lies in the "counterclockwise" direction), and *unoriented* if the network is unlabeled. Therefore we have the two systems $\vec{\mathcal{R}}_n^k$ and $\mathcal{R}_n^k$, denoting, respectively, the oriented and the unoriented ring with n vertices and k robots, under the fully synchronous scheduler.

Oriented Rings. Let us study the structure of $\mathcal{G}\left(\vec{\mathcal{R}}_n^k\right)$. Note that, in a ring network, all vertices are equivalent. Therefore, by Proposition 2, $\mathcal{G}\left(\vec{\mathcal{R}}_n^1\right)$ consists of a single vertex with a self-loop.

In the case of $k = 2$ robots, a configuration is uniquely identified by the distance d of the two robots on the ring, which may be any integer between 0 and $\lfloor n/2 \rfloor$. If $d = 0$, the robots are bound to remain on the same vertex.

If $d = n/2$ (hence n is even), the robots are located on indistinguishable vertices, and they are bound to remain on indistinguishable vertices. In all other cases, it is possible to distinguish the two robots and move them independently, thanks to the orientation of the ring. Therefore, if $0 < d < n/2$, an algorithm may move the two robots independently in any direction, thus adding any integer between -2 and $+2$ to d (subject to the $0 \leq d \leq \lfloor n/2 \rfloor$ constraint). Figures 1(d) and (e) show the configuration graphs of $\vec{\mathcal{R}}_{14}^2$ and $\vec{\mathcal{R}}_{15}^2$.

In general, the configuration space of $\vec{\mathcal{R}}_n^k$ is in bijection with the set of binary *necklaces* of length n and density k, i.e., the binary strings having k zeros and $n - k$ ones, taken modulo rotations. To count them, the Pólya enumeration theorem can be applied, as in [16]. If k is fixed, the configuration space has size

$$\frac{1}{n+k} \cdot \sum_{d \mid \gcd(k,n)} \phi(d) \cdot \binom{(n+k)/d}{k/d} \sim \frac{n^{k-1}}{k!},$$

where ϕ is Euler's totient function. Since the ring is oriented, $\mathcal{G}\left(\vec{\mathcal{R}}_n^k\right) = \mathcal{G}'\left(\vec{\mathcal{R}}_n^k\right)$.

Property 4. For all $n, k \geq 1$, the system $\vec{\mathcal{R}}_n^k$ is deterministic.

Unoriented Rings. The structure of $\mathcal{G}\left(\mathcal{R}_n^k\right)$ is similar to that of $\mathcal{G}\left(\vec{\mathcal{R}}_n^k\right)$, except that now two configurations are indistinguishable also if they are "reflections" of each other. The case $k = 1$ is again trivial and yields a single configuration, but the case $k = 2$ is more interesting. As before, a configuration of $\mathcal{R}_n^2$ is identified by an integer d with $0 \leq d \leq \lfloor n/2 \rfloor$, but the two robots are now indistinguishable, and therefore they must always make symmetric moves. Hence d may only change by -2 or $+2$; the only exception is when n is odd and $d = \lfloor n/2 \rfloor$, which can change to $d = \lfloor n/2 \rfloor - 1$ (as well as to $d = \lfloor n/2 \rfloor - 2$), and vice versa. So, if n is odd, $\mathcal{G}\left(\mathcal{R}_n^2\right)$ is isomorphic to $\mathcal{G}\left(\vec{\mathcal{P}}_{\lceil n/2 \rceil}^1\right)$. If n is even, $\mathcal{G}\left(\mathcal{R}_n^2\right)$ consists of two connected components: the one corresponding to even d's is isomorphic to $\mathcal{G}\left(\vec{\mathcal{P}}_{\lceil (n+2)/4 \rceil}^1\right)$, and the one corresponding to odd $d's$ is isomorphic to $\mathcal{G}\left(\vec{\mathcal{P}}_{\lfloor (n+2)/4 \rfloor}^1\right)$. Figures 1(f), (g), and (h) show the configuration graphs of $\mathcal{R}_{14}^2$, $\mathcal{R}_{15}^2$, and $\mathcal{R}_{16}^2$.

Property 5. For all $n, k \geq 1$, $\mathcal{G}\left(\mathcal{R}_n^k\right)$ consists of either a single path or two disjoint paths.

In general, instead of representing the configurations of $\mathcal{R}_n^k$ with necklaces as before, we use *bracelets* of length n and density k, i.e., binary strings having k zeros and $n - k$ ones, taken modulo rotations and reflections. The size of the configuration space can be computed again with the Pólya enumeration theorem, this time using the dihedral group instead of the cyclic group. For fixed k, its size is

$$\frac{1}{2(n+k)} \cdot \sum_{d \mid \gcd(k,n)} \phi(d) \cdot \binom{(n+k)/d}{k/d} + O\left(n^{\lfloor k/2 \rfloor}\right) \sim \frac{n^{k-1}}{2 \cdot k!}.$$

With unoriented rings, the deterministic configuration graph is slightly different. If $d = 0$ or $d = n/2$, the adversary may choose to keep d unvaried, by making both robots always move in the same direction. Therefore, the configurations corresponding to $d = 0$ and $d = n/2$ have no outgoing edges in $\mathcal{G}'\left(\mathcal{R}_n^k\right)$. Other than that, the two graphs are the same.

3.4 Domination Relations Between Paths and Rings

Next we describe how different systems of paths and rings dominate each other.

Theorem 4. *For all* $n, k \geq 1$ *and* $k' \geq 2k$, $\overrightarrow{\mathcal{P}}_n^k \preceq \overrightarrow{\mathcal{P}}_n^{k'}$, $\mathcal{P}_{2n}^k \preceq \mathcal{P}_{2n}^{k'}$, *and* $\overrightarrow{\mathcal{R}}_n^k \preceq \overrightarrow{\mathcal{R}}_n^{k'}$.

Proof. By Properties 2, 3, and 4, the systems $\overrightarrow{\mathcal{P}}_n^{k'}$, $\mathcal{P}_{2n}^{k'}$, and $\overrightarrow{\mathcal{R}}_n^{k'}$ are deterministic. Thus all relations follow from Theorem 3. □

Theorem 5. *For all* $n, k \geq 1$, $\overrightarrow{\mathcal{P}}_n^k \preceq \mathcal{P}_{2n}^k$.

Proof. We place the k robots of $\mathcal{P}_{2n}^k$ on the first n vertices of the path, leaving the other half empty. This gives an implicit orientation to the path, and now the simulation is trivial. □

Theorem 6. *For all* $n \geq 1$ *and* $k \geq 2$, $\overrightarrow{\mathcal{P}}_n^k \preceq \overrightarrow{\mathcal{R}}_{2n}^{k+1}$.

Proof. We use only the configurations of $\overrightarrow{\mathcal{R}}_{2n}^{k+1}$ having a vertex occupied by a single robot, followed in clockwise order by $n - 1$ empty vertices. The simulation is performed by the other k robots on the remaining n vertices, which are always unambiguously identified because $k \geq 2$. □

Theorem 7. *For all* $n \geq 1$ *and* $k \geq 2$, $\mathcal{P}_n^k \preceq \mathcal{R}_{3n-1}^{k+1}$.

Proof. We use only the configurations of $\mathcal{R}_{3n-1}^{k+1}$ having a vertex occupied by one robot, surrounded on both sides by sequences of $n - 1$ empty vertices. The simulation is performed by the other k robots on the remaining n vertices, which are always unambiguously identified because $k \geq 2$. □

4 Universality

A system Ψ is *universal for* $\mathbb{N}_n$ if it computes every function on $\mathbb{N}_n$. In this case, we write $\mathbb{N}_n \preceq \Psi$. Note that this extension of the relation $\preceq$ preserves its transitivity. A set of systems Υ is *universal* if, for every $n \geq 1$ and every function $f \colon \mathbb{N}_n \to \mathbb{N}_n$, there is a system $\Psi \in \Upsilon$ that computes f.

Theorem 8. $\{\mathcal{F}(G,1) \mid G$ *is an unlabeled network*$\}$ *is universal.*

Proof. Given a function $f \colon \mathbb{N}_n \to \mathbb{N}_n$, take the complete graph K_n and attach i dangling vertices to its i-th vertex, for all $0 \le i < n$. To compute f, instruct the robot to move from the vertex with i dangling vertices to the one with $f(i)$ dangling vertices (which is always distinguishable). □

However, complete graphs are very demanding networks. If no vertex in the network has more than two neighbors, universality requires more robots: two for paths and oriented rings, and three for unoriented rings.

Theorem 9. $\left\{\overrightarrow{\mathcal{P}}_n^1 \mid n \ge 1\right\}$, $\{\mathcal{P}_n^1 \mid n \ge 1\}$, $\left\{\overrightarrow{\mathcal{R}}_n^1 \mid n \ge 1\right\}$, *and* $\{\mathcal{R}_n^2 \mid n \ge 1\}$ *are not universal.* □

Next we show that two robots are indeed sufficient to compute every function on arbitrarily long oriented and unoriented paths and oriented rings, and that three robots are sufficient for unoriented rings. First we introduce a definition: an algorithm is *sequential* if it never instructs two robots to move at the same time.

Lemma 1. *For all* $n \ge 1$, $\mathbb{N}_n \preceq \overrightarrow{\mathcal{P}}_{2n}^2$. *The algorithms used in the computations are sequential.* □

Theorem 10. $\left\{\overrightarrow{\mathcal{P}}_n^2 \mid n \ge 1\right\}$, $\{\mathcal{P}_n^2 \mid n \ge 1\}$, $\left\{\overrightarrow{\mathcal{R}}_n^2 \mid n \ge 1\right\}$, *and* $\{\mathcal{R}_n^3 \mid n \ge 1\}$ *are universal.* □

We can generalize this result in two directions. First, to systems of two robots on networks whose quotient graphs contain arbitrarily long paths.

Theorem 11. $\{\mathcal{F}(G_n, 2) \mid G_n^*$ *contains a sub-path of length at least* $n\}$ *is universal.*

Proof. We prove that $\overrightarrow{\mathcal{P}}_n^2 \preceq \mathcal{F}(G_n, 2)$, and the universality follows from Lemma 1. By assumption, the robots can agree on an oriented path P of length n in G_n^*, since all its vertices are distinguishable, due to the definition of quotient graph. Then, a robot located at some vertex of G_n is interpreted as lying on the corresponding vertex of G_n^* and follows whatever algorithm it is simulating, remaining on vertices of G_n corresponding to P; this way, the two robots can simulate $\overrightarrow{\mathcal{P}}_n^2$ on G_n. □

Theorem 10 can also be generalized to systems of three robots on networks with arbitrarily long *girths* (the girth of a graph being the length of its shortest cycle, or infinity if there are no cycles).

Theorem 12. $\{\mathcal{F}(G_n, 3) \mid$ *the girth of* G_n *is at least* n *and finite*$\}$ *is universal.* □

Our conjecture is that the above results characterize the universal classes of systems with at least three robots on unlabeled networks.

Conjecture 2. The set $\{\mathcal{F}(G_i, k) \mid i \geq 0\}$, where $k \geq 3$ and every G_i is an unlabeled network, is universal if and only if either the quotient graphs G_i^* have unboundedly long sub-paths, or the graphs G_i have unboundedly long shortest cycles.

5   Optimizing Network Sizes

In this section, our goal is to compute all the functions on $\mathbb{N}_n$ under the fully synchronous scheduler, using the smallest possible network, and perhaps a large number of robots. We are able to approximate the minimum size of such a network up to a factor that tends to 2 as n goes to infinity, using very short oriented paths. Nonetheless, thanks to the simulation tools developed in Sect. 3.4, we could as well use unoriented paths or rings, again achieving the optimum size up to factors that tend to small constants.

Lemma 2. *For all* $n, k \geq 1$, $\overrightarrow{\mathcal{P}}_{n!}^{k} \preceq \overrightarrow{\mathcal{P}}_{kn}^{kn(n-1)/2}$.

Proof. We divide the base graph of $\overrightarrow{\mathcal{P}}_{kn}^{kn(n-1)/2}$ into k sub-paths of length n. In every sub-path, we place a different amount of robots on each vertex, from 0 to $n-1$ robots. The possible placements of such robots within a sub-path correspond to the $n!$ permutations of n distinct objects. It is well known that the set of permutations can be ordered in such a way that each permutation is obtained from the previous one by swapping only two adjacent objects [19]. If we let the i-th permutation under this ordering encode the i-th vertex of $\overrightarrow{\mathcal{P}}_{n!}^{k}$, we can simulate a move of i-th robot of $\overrightarrow{\mathcal{P}}_{n!}^{k}$ by simply swapping the robots occupying two adjacent vertices of the i-th sub-path of $\overrightarrow{\mathcal{P}}_{kn}^{kn(n-1)/2}$. $\square$

Theorem 13. *For all* $n \geq 1$, $\mathbb{N}_n \preceq \overrightarrow{\mathcal{P}}_{2m}^{m(m-1)}$, *with* $(m-1)! < 2n \leq m!$.

Proof. Immediate from Lemma 1, Lemma 2 with $k = 2$, and the transitivity of $\preceq$. $\square$

This tells us that, on a network with $|V|$ vertices, all the functions on a set of size $2(|V|/2)!$ can be computed, provided that enough robots are available. We can also show that, on the same network, it is impossible to compute all functions on a set of size $|V|! + 1$.

Theorem 14. *For all networks* $G = (V, E, \ell)$ *and all* $n, k \geq 1$, *if* $\mathbb{N}_n \preceq \mathcal{F}(G, k)$, *then* $|V|! \geq n$.

Proof. Let A be an algorithm that computes the cycle function $\lambda_n = \{(i, i+1) \mid i \in \mathbb{N}_{n-1}\} \cup \{(n-1, 0)\}$ under $\mathcal{F}(G, k)$, according to the surjective partial function by $\varphi \colon \mathcal{C}(G, k) \to \mathbb{N}_n$. For any execution $E = (C_i)_{i \geq 0}$ of A such that $C_0 \in \varphi^{-1}(0)$, the sequence $(\varphi(C_i))_{i \geq 0}$ must span the whole range $\mathbb{N}_n$ infinitely often. Moreover, since $\mathcal{C}(G, k)$ is finite, there is a configuration C that occurs infinitely many times in E. Hence there are two such occurrences, say $C_j = C_{j'} = C$, between which E spans at least $d \geq n$ different configurations. During this fragment of the execution, no two separate sets of robots may end up on the same vertex at the same time, or they become impossible to separate deterministically, contradicting the fact that C must be reached again. In particular, the number of vertices that are occupied by some robots remains the same, q. It follows that d cannot be larger than $|V|!/(|V| - q)!$. Thus $n \leq d \leq |V|!/(|V| - q)! \leq |V|!$, as desired.   □

6   Further Work

In addition to the fully synchronous scheduler, also a semi-synchronous and an asynchronous one can be defined. Both schedulers may activate any subset of the robots at each turn, keeping the others quiescent. The asynchronous scheduler may even delay the robots, making them move based on obsolete observations of the network.

Our universality results can be extended to both these schedulers, by observing that all the algorithms we used in our simulations are sequential. We can also prove a weaker version of Theorem 13 for these schedulers: $\mathbb{N}_n \preceq \overrightarrow{\mathcal{P}}_{2m}^m$, with $m = O(\log n)$; we have a matching lower bound for both schedulers, as well. These results indicate that the semi-synchronous and asynchronous schedulers, albeit not drastically reducing the robots' computing powers, make them somewhat less efficient.

We leave two open problems: Conjectures 1 and 2.

References

1. Blin, L., Milani, A., Potop-Butucaru, M., Tixeuil, S.: Exclusive perpetual ring exploration without chirality. In: Lynch, N.A., Shvartsman, A.A. (eds.) DISC 2010. LNCS, vol. 6343, pp. 312–327. Springer, Heidelberg (2010). doi:10.1007/978-3-642-15763-9_29
2. Bonnet, F., Milani, A., Potop-Butucaru, M., Tixeuil, S.: Asynchronous exclusive perpetual grid exploration without sense of direction. In: Fernàndez Anta, A., Lipari, G., Roy, M. (eds.) OPODIS 2011. LNCS, vol. 7109, pp. 251–265. Springer, Heidelberg (2011). doi:10.1007/978-3-642-25873-2_18
3. Chalopin, J., Flocchini, P., Mans, B., Santoro, N.: Network exploration by silent and oblivious robots. In: Thilikos, D.M. (ed.) WG 2010. LNCS, vol. 6410, pp. 208–219. Springer, Heidelberg (2010). doi:10.1007/978-3-642-16926-7_20
4. D'Angelo, G., Di Stefano, G., Klasing, R., Navarra, A.: Gathering of robots on anonymous grids without multiplicity detection. Theoret. Comput. Sci. **610**, 158–168 (2016)
5. D'Angelo, G., Di Stefano, G., Navarra, A.: Gathering six oblivious robots on anonymous symmetric rings. J. Discrete Algorithms **26**, 16–27 (2014)
6. D'Angelo, G., Di Stefano, G., Navarra, A.: Gathering on rings under the look-compute-move model. Distrib. Comput. **27**(4), 255–285 (2014)
7. D'Angelo, G., Di Stefano, G., Navarra, A., Nisse, N., Suchan, K.: Computing on rings by oblivious robots: a unified approach for different tasks. Algorithmica **72**(4), 1055–1096 (2015)

8. Devismes, S., Lamani, A., Petit, F., Tixeuil, S.: Optimal torus exploration by oblivious mobile robots. Inria Technical Report HAL-00926573 (2014)
9. Devismes, S., Petit, F., Tixeuil, S.: Optimal probabilistic ring exploration by semi-synchronous oblivious robots. Theoret. Comput. Sci. **498**, 10–27 (2013)
10. Elor, Y., Bruckstein, A.M.: Uniform multi-agent deployment on a ring. Theoret. Comput. Sci. **412**, 783–795 (2011)
11. Flocchini, P., Ilcinkas, D., Pelc, A., Santoro, N.: Remembering without memory: tree exploration by asynchronous oblivious robots. Theoret. Comput. Sci. **411**(14–15), 1583–1598 (2010)
12. Flocchini, P., Ilcinkas, D., Pelc, A., Santoro, N.: How many oblivious robots can explore a line. Inf. Process. Lett. **111**(20), 1027–1031 (2011)
13. Flocchini, P., Ilcinkas, D., Pelc, A., Santoro, N.: Ring exploration by asynchronous oblivious robots. Algorithmica **65**(3), 562–583 (2013)
14. Flocchini, P., Prencipe, G., Santoro, N.: Distributed Computing by Oblivious Mobile Robots. Morgan & Claypool, San Rafeal (2012)
15. Flocchini, P., Santoro, N., Viglietta, G., Yamashita, M.: Universal systems of oblivious mobile robots [cs.DC]. arXiv:1602.04881 (2016)
16. Gilbert, E., Riordan, J.: Symmetry types of periodic sequences. Ill. J. Math. **5**(4), 657–665 (1961)
17. Guilbault, S., Pelc, A.: Gathering asynchronous oblivious agents with local vision in regular bipartite graphs. Theoret. Comput. Sci. **509**, 86–96 (2013)
18. Izumi, T., Izumi, T., Kamei, S., Ooshita, F.: Mobile robots gathering algorithm with local weak multiplicity in rings. In: Patt-Shamir, B., Ekim, T. (eds.) SIROCCO 2010. LNCS, vol. 6058, pp. 101–113. Springer, Heidelberg (2010). doi:10.1007/978-3-642-13284-1_9
19. Johnson, S.: Generation of permutations by adjacent transposition. Math. Comput. **17**, 282–285 (1963)
20. Kamei, S., Lamani, A., Ooshita, F., Tixeuil, S.: Gathering an even number of robots in an odd ring without global multiplicity detection. In: Rovan, B., Sassone, V., Widmayer, P. (eds.) MFCS 2012. LNCS, vol. 7464, pp. 542–553. Springer, Heidelberg (2012). doi:10.1007/978-3-642-32589-2_48
21. Kant, G.: Drawing planar graphs using the canonical ordering. Algorithmica **16**(1), 4–32 (1996)
22. Klasing, R., Kosowski, A., Navarra, A.: Taking advantage of symmetries: gathering of many asynchronous oblivious robots on a ring. Theoret. Comput. Sci. **411**, 3235–3246 (2010)
23. Klasing, R., Markou, E., Pelc, A.: Gathering asynchronous oblivious mobile robots in a ring. Theoret. Comput. Sci. **390**, 27–39 (2008)
24. Kosowski, A., Navarra, A.: Graph decomposition for improving memoryless periodic exploration. In: Královič, R., Niwiński, D. (eds.) MFCS 2009. LNCS, vol. 5734, pp. 501–512. Springer, Heidelberg (2009). doi:10.1007/978-3-642-03816-7_43
25. Lamani, A., Potop-Butucaru, M.G., Tixeuil, S.: Optimal deterministic ring exploration with oblivious asynchronous robots. In: Patt-Shamir, B., Ekim, T. (eds.) SIROCCO 2010. LNCS, vol. 6058, pp. 183–196. Springer, Heidelberg (2010). doi:10.1007/978-3-642-13284-1_15
26. Millet, L., Potop-Butucaru, M., Sznajder, N., Tixeuil, S.: On the synthesis of mobile robots algorithms: the case of ring gathering. In: Felber, P., Garg, V. (eds.) SSS 2014. LNCS, vol. 8756, pp. 237–251. Springer, Heidelberg (2014). doi:10.1007/978-3-319-11764-5_17
27. Ooshita, F., Tixeuil, S.: On the self-stabilization of mobile oblivious robots in uniform rings. Theoret. Comput. Sci. **568**, 84–96 (2015)

Collaborative Delivery with Energy-Constrained Mobile Robots

Andreas Bärtschi[1(✉)], Jérémie Chalopin[2], Shantanu Das[2], Yann Disser[3], Barbara Geissmann[1], Daniel Graf[1], Arnaud Labourel[2], and Matúš Mihalák[4]

[1] ETH Zürich, Zürich, Switzerland
{andreas.baertschi,barbara.geissmann,daniel.graf}@inf.ethz.ch
[2] LIF, CNRS and Aix-Marseille Université, Marseille, France
{jeremie.chalopin,shantanu.das,arnaud.labourel}@lif.univ-mrs.fr
[3] TU Berlin, Berlin, Germany
disser@math.tu-berlin.de
[4] Maastricht University, Maastricht, The Netherlands
matus.mihalak@maastrichtuniversity.nl

Abstract. We consider the problem of collectively delivering some message from a specified source to a designated target location in a graph, using multiple mobile agents. Each agent has a limited energy which constrains the distance it can move. Hence multiple agents need to collaborate to move the message, each agent handing over the message to the next agent to carry it forward. Given the positions of the agents in the graph and their respective budgets, the problem of finding a feasible movement schedule for the agents can be challenging. We consider two variants of the problem: in *non-returning* delivery, the agents can stop anywhere; whereas in *returning* delivery, each agent needs to return to its starting location, a variant which has not been studied before. We first provide a polynomial-time algorithm for returning delivery on trees, which is in contrast to the known (weak) NP-hardness of the non-returning version. In addition, we give resource-augmented algorithms for returning delivery in general graphs. Finally, we give tight lower bounds on the required resource augmentation for both variants of the problem. In this sense, our results close the gap left by previous research.

1 Introduction

We consider a team of mobile robots which are assigned a task that they need to perform collaboratively. Even simple tasks such as collecting information and delivering it to a target location can become challenging when it involves the cooperation of several agents. The difficulty of collaboration can be due to several limitations of the agents, such as limited communication, restricted vision or the lack of persistent memory, and this has been the subject of extensive research (see [19] for a recent survey). When considering agents that move physically (such as mobile robots or automated vehicles), a major limitation of the agents are their energy resources, which restricts the travel distance of the agent. This is particularly true for small battery operated robots or drones, for which the

J. Suomela (Ed.): SIROCCO 2016, LNCS 9988, pp. 258–274, 2016.
DOI: 10.1007/978-3-319-48314-6_17

energy limitation is the real bottleneck. We consider a set of mobile agents where each agent i has a budget B_i on the distance it can move, as in [2,9]. We model their environment as an undirected edge-weighted graph G, with each agent starting on some vertex of G and traveling along edges of G, until it runs out of energy and stops forever. In this model, the agents are obliged to collaborate as no single agent can usually perform the required task on its own.

The problem we consider is that of moving some information from a given source location to a target location in the graph G using a subset of the agents. Although the problem sounds simple, finding a valid schedule for the agents to deliver the message, is computationally hard even if we are given full information on the graph and the location of the agents. Given a graph G with designated source and target vertices, and k agents with given starting locations and energy budgets, the decision problem of whether the agents can collectively deliver a single message from the source to the target node in G is called BUDGETEDDELIVERY. Chalopin et al. [9,10] showed that *non-Returning* BUDGETEDDELIVERY is weakly NP-hard on paths and strongly NP-hard on general graphs.

Unlike previous papers, we also consider a version of the problem where each agent needs to return to its starting location after completing its task. This is a natural assumption, e.g. for robots that need to return to their docking station for maintenance or recharging. We call this variant *Returning* BUDGETEDDELIVERY. Surprisingly, this variant of the problem is easier to solve when the graph is a tree (unlike the original version of the problem), but we show it to be strongly NP-hard even for planar graphs. We present a polynomial time algorithm for solving *Returning* BUDGETEDDELIVERY on trees.

For arbitrary graphs, we are interested in resource-augmented algorithms. Since finding a feasible schedule for BUDGETEDDELIVERY is computationally hard when the agents have just enough energy to make delivery possible, we consider augmenting the energy of each robot by a constant factor γ, to enable a polynomial-time solution to the problem. Given an instance of BUDGETEDDELIVERY and some $\gamma > 1$, we have a γ-resource-augmented algorithm, if the algorithm, running in polynomial time, either (correctly) answers that there is no feasible schedule, or finds a feasible schedule for the modified instance with augmented budgets $\hat{B}_i = \gamma \cdot B_i$ for each agent i.

Our Model. We consider an undirected edge-weighted graph $G = (V, E)$ with $n = |V|$ vertices and $m = |E|$ edges. The weight $w(e)$ of an edge $e \in E$ defines the energy required to cross the edge in either direction. We have k mobile agents which are initially placed on arbitrary nodes $p_1, \ldots, p_k$ of G, called starting positions. Each agent i has an initially assigned budget $B_i \in \mathbb{R}_{\geq 0}$ and can move along the edges of the graph, for a total distance of at most B_i (if an agent travels only on a part of an edge, its travelled distance is downscaled proportionally to the part travelled). The agents are required to move a message from a given source node s to a target node t. An agent can pick up the message from its current location, carry it to another location (a vertex or a point inside an edge), and drop it there. Agents have global knowledge of the graph and may communicate freely.

Given a graph G with vertices $s \neq t \in V(G)$ and the starting nodes and budgets for the k agents, we define BUDGETEDDELIVERY as the decision problem of whether the agents can collectively deliver the message without exceeding their individual budgets. In *Returning* BUDGETEDDELIVERY each agent needs to return to its respective starting position before using up its energy budget; in the *non-returning* version we do not place such a restriction on the agents and an agent may terminate at any location in the graph.

A solution to BUDGETEDDELIVERY is given in the form of a *schedule* which prescribes for each agent whether it moves and if so, the two locations in which it has to pick up and drop off the message. A schedule is *feasible* if the message can be delivered from s to t.

Related Work. Delivery problems in the graph have been usually studied for a single agent moving in the graph. For example, the well known *Travelling salesman problem* (TSP) or and the *Chinese postman problem* (CPP) require an agent to deliver packets to multiple destinations located in the nodes of the graph or the edges of the graph. The optimization problem of minimizing the total distance traveled is known to be NP-hard [3] for TSP, but can be solved in polynomial time for the CPP [18].

When the graph is not known in advance, the problem of exploring a graph by a single agent has been studied with the objective of minimizing the number of edges traversed (see e.g. [1,23]). Exploration by a team of two agents that can communicate at a distance, has been studied by Bender and Slonim [6] for digraphs without node identifiers. The model of energy-constrained robot was introduced by Betke et al. [7] for single agent exploration of grid graphs. Later Awerbuch et al. [4] studied the same problem for general graphs. In both these papers, the agent could return to its starting node to refuel and between two visits to the starting node, the agent could traverse at most B edges. Duncan et al. [15] studied a similar model where the agent is tied with a rope of length B to the starting location and they optimized the exploration time, giving an $\mathcal{O}(m)$ time algorithm.

For energy-constrained agents without the option of refuelling, multiple agents may be needed to explore even graphs of restricted diameter. Given a graph G and k agents starting from the same location, each having an energy constraint of B, deciding whether G can be explored by the agents is NP-hard, even if graph G is a tree [20]. Dynia et al. studied the online version of the problem [16,17]. They presented algorithms for exploration of trees by k agents when the energy of each agent is augmented by a constant factor over the minimum energy B required per agent in the offline solution. Das et al. [12] presented online algorithms that optimize the number of agents used for tree exploration when each agent has a fixed energy bound B. On the other hand, Dereniowski et al. [14] gave an optimal time algorithm for exploring general graphs using a large number of agents. Ortolf et al. [22] showed bounds on the competitive ratio of online exploration of grid graphs with obstacles, using k agents.

When multiple agents start from arbitrary locations in a graph, optimizing the total energy consumption of the agents is computationally hard for several formation problems which require the agents to place themselves in desired configurations (e.g. connected or independent configurations) in a graph. Demaine et al. [13] studied such optimization problems and provided approximation algorithms and inapproximability results. Similar problems have been studied for agents moving in the visibility graphs of simple polygons and optimizing either the total energy consumed or the maximum energy consumed per agent can be hard to approximate even in this setting, as shown by Bilo et al. [8].

Anaya et al. [2] studied centralized and distributed algorithms for the information exchange by energy-constrained agents, in particular the problem of transferring information from one agent to all others (*Broadcast*) and from all agents to one agent (*Convergecast*). For both problems, they provided hardness results for trees and approximation algorithms for arbitrary graphs. The budgeted delivery problem was studied by Chalopin et al. [9] who presented hardness results for general graphs as well as resource-augmented algorithms. For the simpler case of lines, [10] proved that the problem is weakly NP-hard and presented a quasi-pseudo-polynomial time algorithm. Czyzowicz et al. [11] recently showed that the problems of budgeted delivery, broadcast and convergecast remain NP-hard for general graphs even if the agents are allowed to exchange energy when they meet.

Our Contribution. This is the first paper to study the *Returning* version of BUDGETEDDELIVERY. We first show that this problem can be solved in $\mathcal{O}(n + k \log k)$ time for lines and trees (Sect. 2). This is in sharp contrast to the *non-Returning* version which was shown to be weakly NP-hard [10] even on lines. In Sect. 4, we prove that *Returning* BUDGETEDDELIVERY is NP-hard even for planar graphs. For arbitrary graphs with arbitrary values of agent budgets, we present a 2-resource-augmented algorithm and we prove that this is the best possible, as there exists no $(2 - \epsilon)$-resource-augmented algorithm unless P = NP (Sect. 5). We show that this bound can be broken when the agents have the same energy budget and we present a $(2 - 2/k)$-resource-augmented algorithm for this case.

For the *non-Returning* version of the BUDGETEDDELIVERY, we close the gaps left open by previous research [9,10]. In particular we prove that this variant of the problem is also strongly NP-hard on planar graphs, while it was known to be strongly NP-hard for general graphs and weakly NP-hard on trees. We also show tightness of the 3-resource-augmented algorithm for the problem, presented in [9]. Finally, in Sect. 6, we investigate the source of hardness for BUDGETEDDELIVERY and show that the problem becomes easy when the order in which the agents pick up the message is known in advance.

2   Returning BudgetedDelivery on the Tree

We study the *Returning* BUDGETEDDELIVERY on a tree and show that it can be solved in polynomial time. We immediately observe that this problem is reducible to the *Returning* BUDGETEDDELIVERY on a path: There is a unique s-t path on a tree and we can move each agent from her starting position to the nearest node on this s-t path while subtracting from her budget twice the distance traveled. The path problem now has an equivalent geometric representation on the line: the source node s, the target node t, and the starting positions of the agents p_i are coordinates of the real line. We assume $s < t$, i.e., the message needs to be delivered from left to right.

Without loss of generality, we consider schedules in which every agent i that moves uses all its budget B_i. Because every agent needs to return to its starting position, an agent i can carry the message on any interval of size $B_i/2$ that contains the starting position p_i. For every agent i, let $l_i = p_i - B_i/2$ denote the leftmost point where she can pick a message, and let $r_i = p_i + B_i/2$ be the rightmost point to where she can deliver the message. The *Returning* BUDGETEDDELIVERY on a line now becomes the following *covering problem*: Can we choose, for every i, an interval I_i of size $B_i/2$ that lies completely within the *region* $[l_i, r_i]$ such that the segment $[s, t]$ is covered by the chosen intervals, i.e., such that $[s, t] \subseteq \cup_i I_i$?

The following *greedy* algorithm solves the covering problem. The algorithm works iteratively in rounds $r = 1, 2, \ldots$. We initially set $s_1 = s$. We stop the algorithm whenever $s_r \geq t$, and return true. In round r, we pick i^* having the smallest r_{i^*} among all agents i with $l_i \leq s_r < r_i$, and set $s_{r+1} = \min\{r_{i^*}, s_r + B_{i^*}/2\}$ and $I_{i^*} = (s_{r+1} - B_{i^*}/2, s_{r+1})$, and continue with the next round $r + 1$. If we cannot choose i^*, we stop the algorithm and return false.

Theorem 1. *There is an $\mathcal{O}(n + k \log k)$-time algorithm for* Returning BUD-GETEDDELIVERY *on a tree.*

Proof. The reduction from a tree to a path takes $\mathcal{O}(n)$ time using breadth-first search from s and the algorithm *greedy* can be implemented in time $\mathcal{O}(k \log k)$ using a priority queue.

For the correctness, we now show that greedy returns a solution to the covering problem if and only if there exists one. Greedy can be seen as advancing the cover of $[s, t]$ from left to right by adding intervals I_i. Whenever it decides upon I_i, it will set s_r to the respective endpoint of I_i, and never ever consider i again or change the placement of I_i within the boundaries $[l_i, r_i]$. Thus, whenever $s_r \geq t$, the intervals I_i form a cover of $[s, t]$.

We now show that if a cover exists, greedy finds one. Observe first that a cover can be given by a subset of the agents $\{i_1, \ldots, i_t\}$, $t \leq k$, and by their ordering $(i_1, i_2, \ldots)$, according to the right endpoints of their intervals I_{i_j}, since we can reconstruct a covering by always placing the respective interval I_{i_j} at the rightmost possible position.

Suppose, for contradiction, that greedy fails. Let $(i_1^*, i_2^*, \ldots)$ be a minimal cover of $[s, t]$ that agrees with the greedy schedule $(i_1, i_2, \ldots)$ in the maximum

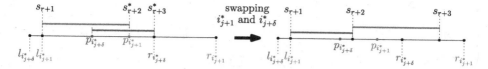

Fig. 1. Changing the order of agents i_{j+1}^* and $i_{j+\delta}^*$ in the schedule.

number of first agents $i_1, \ldots, i_j$. Hence, $j + 1$ is the first position such that $i_{j+1}^* \neq i_{j+1}$. The left endpoints of I_{j+1}^* and I_{j+1} correspond to s_{r+1} in our algorithm. If agent i_{j+1} does not appear in the solution $(i_1^*, i_2^*, \ldots)$, adding i_{j+1} to that solution and deleting some of the subsequent ones results in a minimal cover that agrees on the first $j + 1$ agents, a contradiction. If agent i_{j+1} appears in the solution $(i_1^*, i_2^*, \ldots)$, say, as agent $i_{j+\delta}^*$, then we modify this cover by swapping i_{j+1}^* with $i_{j+\delta}^*$. We claim that the new solution still covers $[s, t]$. This follows immediately by observing that greedy chose i_{j+1} to have smallest r_i among all agents that can extend the covering beyond s_{r+1}. Since every agent i covers at least half of its region $[l_i, r_i]$, we know that i_{j+1}^* and $i_{j+\delta}^*$ together cover the region $[s_{r+1}, r_{i_{j+\delta}^*}]$, and therefore by minimality $i_{j+\delta}^* = i_{j+2}^*$. Finally, if we change the order of the two agents, they will still cover the region $[s_{r+1}, r_{i_{j+\delta}^*}]$ (see Fig. 1). □

3   Resource Augmentation Algorithms

We now look at general graphs $G = (V, E)$. As we will see in the next section, BUDGETEDDELIVERY is NP-hard, hence we augment the budget of each agent by a factor $\gamma > 1$ to allow for polynomial-time solutions. For non-returning agents, a $\min\{3, 1 + \max \frac{B_i}{B_j}\}$-resource-augmented algorithm was given by Chalopin et al. [9]. We first provide a 2-resource-augmented algorithm for *returning* BUDGETEDDELIVERY. This is tight as there is no polynomial-time $(2 - \varepsilon)$-resource-augmented algorithm, unless P = NP (Sect. 5). If, however, the budgets of the agents are similar, we can go below the 2-barrier: In this case, we present a $(1 + \frac{k-2}{k} \max \frac{B_i}{B_j})$-resource-augmented algorithm. Throughout this section, we assume that there is no feasible schedule with a single agent, which we can easily verify.

Preliminaries. We denote by $d(u, v)$ the distance of two points $u, v \in G$. Assume an agent i with budget B_i starts in u and moves first to v. Which locations in the graph (vertices and positions on the edges) are then still reachable by i so that he has suffcient energy left to move back to u? We define the ellipsoid $\mathcal{E}(u, v, B_i) = \{p \in G \mid d(u, v) + d(v, p) + d(p, u) \leq B_i\}$ and the ball $\mathcal{B}(u, \frac{B_i}{2}) = \mathcal{E}(u, u, B_i)$. It is easy to see that $\mathcal{E}(u, v, B_i)$ can be (i) computed in polynomial time by running Dijkstra's shortest path algorithm from both u and v and (ii) represented in linear space: We store all vertices $p \in V$ with

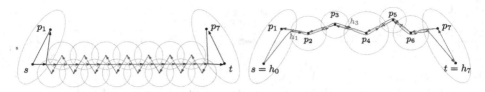

Fig. 2. (left) Feasible schedule. (right) $\left(1 + \frac{5}{7}\max\frac{B_j}{B_i}\right)$-resource-augmented schedule.

$p \in \mathcal{E}(u, v, B_i)$, and for each edge $(p, q) \in E$ with $p \in \mathcal{E}(u, v, B_i), q \notin \mathcal{E}(u, v, B_i)$ we store the furthest point of (p, q) still reachable by i.

Theorem 2. *There is a polynomial-time 2-resource-augmented algorithm for Returning* BUDGETEDDELIVERY.

Proof. Denote by p_i the starting position of agent i. We consider the balls $\mathcal{B}_i := \mathcal{B}(p_i, \frac{B_i}{2})$ around all agents, as well as the balls $\mathcal{B}(s, 0)$ and $\mathcal{B}(t, 0)$ of radius 0 around s and t. We compute the *intersection graph* G_I of the balls, which can be done in polynomial time. If there is a feasible schedule, then there must be a path from $\mathcal{B}(s, 0)$ to $\mathcal{B}(t, 0)$ in G_I (for example the path given by the balls around the agents in the feasible schedule).

If there is no path from $\mathcal{B}(s, 0)$ to $\mathcal{B}(t, 0)$, then the algorithm outputs that there is no feasible schedule with non-augmented budgets. Otherwise we can get a 2-resource-augmentation as follows: Pick a shortest path from $\mathcal{B}(s, 0)$ to $\mathcal{B}(t, 0)$ in G_I and denote by $\ell \le k$ the number of agents on this path, labeled without loss of generality $1, 2, \ldots, \ell$. For each edge on the shortest path, we specify a handover point $h_i \in \mathcal{B}_i \cap \mathcal{B}_{i+1}$ in G (where we set $h_0 = s$ and $h_\ell = t$). Then each agent i, $i = 1, \ldots, \ell$ walks from its starting position p_i to the handover point h_{i-1} to pick up the message, goes on to the handover point h_i to drop the message there, and returns home to p_i. Since $h_{i-1}, h_i \in \mathcal{B}(p_i, \frac{B_i}{2})$, the budget needed by agent i to do so is at most $d(p_i, h_{i-1}) + d(h_{i-1}, h_i) + d(h_i, p_i) \le \frac{B_i}{2} + 2 \cdot \frac{B_i}{2} + \frac{B_i}{2} = 2B_i$. □

Theorem 3. *There is a polynomial-time $(1 + \frac{k-2}{k}\max\frac{B_j}{B_i})$-resource-augmented algorithm for Returning* BUDGETEDDELIVERY.

Proof. We first "guess" the first agent a and the last agent b of the feasible schedule (by trying all $\binom{k}{2}$ pairs). In contrast to Theorem 2, we can in this way get a 2-resource-augmented solution in which a and b only need their original budgets. Intuitively, we can evenly redistribute the remaining part of $\hat{B}_a$ and $\hat{B}_b$ among all k agents, such that for each agent i we have $\hat{B}_i \le B_i + \frac{k-2}{k}\max B_j$. Without loss of generality, we assume that agent a walks from its starting position on a shortest path to s to pick up the message, and that agent b walks home directly after dropping the message at t. Hence consider the ellipsoids $\mathcal{B}_a := \mathcal{E}(p_a, s, B_a)$ and $\mathcal{B}_b := \mathcal{E}(p_b, s, B_b)$ as well as the balls $\mathcal{B}_i := \mathcal{B}(p_i, \frac{B_i}{2})$ around the starting positions of all other agents and compute their intersection graph G_I.

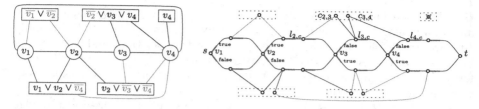

Fig. 3. (left) A plane embedding of a 3CNF F which is satisfied by $(v_1, v_2, v_3, v_4) =$ (true, false, false, true). (right) Its transformation to the corresponding delivery graph.

We denote by $i = 1, \ldots, \ell$ the agents on a shortest path from $\mathcal{B}_a$ to $\mathcal{B}_b$ in G_I (if any), where $a = 1$, $b = \ell \leq k$ and we specify the following points: $h_0 = s$, $h_i \in \mathcal{B}_i \cap \mathcal{B}_{i+1}$, and $h_\ell = t$. If the agents handover the message at the locations h_i, we get a 2-resource-augmentation where the agents 1 and ℓ use only their original budget. Instead we let them help their neighbours 2 and $\ell - 1$ by $\frac{\ell-2}{\ell} B_2$ and $\frac{\ell-2}{\ell} B_{\ell-1}$, respectively. Those agents further propagate the surplus towards the agent(s) in the middle, see Fig. 2 (right). We achieve a resource augmentation of $1 + \frac{\ell-2}{\ell} \max \frac{B_j}{B_i} \leq 1 + \frac{k-2}{k} \max \frac{B_j}{B_i}$. Details are given in the full version of the paper [5]. $\qquad \square$

4   Hardness for Planar Graphs

In this section, we show that BUDGETEDDELIVERY in a planar graph is strongly NP-hard, both for the *Returning* version and the *non-Returning* version. Both proofs are based on the same reduction from PLANAR3SAT.

Planar 3SAT. Let F be a conjunctive normal form 3CNF with a set of variables $V = \{v_1, \ldots, v_x\}$ and a set of clauses $C = \{c_1, \ldots, c_y\}$. Each clause is a disjunction of at most three literals $\ell(v_i) \vee \ell(v_j) \vee \ell(v_k)$, where $\ell(v_i) \in \{v_i, \overline{v_i}\}$. We can represent F by a graph $H(F) = (B \cup V, A_1 \cup A_2)$ which we build as follows: We start with a bipartite graph with the node set N consisting of all clauses and all variables and an edge set A_1 which contains an edge between each clause c and variable v if and only if v or $\overline{v}$ is contained in c, $A_1 = \{\{c_i, v_j\} \mid v_j \in c_i \text{ or } \overline{v_j} \in c_i\}$. To this graph we add a cycle A_2 consisting of edges between all pairs of consecutive variables, $A_2 = \{\{v_j, v_{j+1}\} \mid 1 \leq j < x\} \cup \{v_x, v_1\}$. We call F *planar* if there is a plane embedding of $H(F)$ which *at each variable node* has all paths representing positive literals on one side of the cycle A_2 and all paths representing negative literals on the other side of A_2. The decision problem PLANAR3SAT of finding whether a given planar 3CNF F is satisfiable or not is NP-complete, a result due to Lichtenstein [21]. We assume without loss of generality that every clause contains at most one literal per variable. For an example of such an embedding, see Fig. 3 (left).

Building the Delivery Graph. We first describe how to turn a plane embedding of a planar 3CNF graph $H(F)$ into a delivery graph $G(F)$, see Fig. 3. Only later we will define edge weights, the agents' starting positions and their energy budgets. We will focus on *Returning* BUDGETEDDELIVERY; the only difference for non-returning agents lie in their budgets, we provide adapted values for non-returning agents in footnotes.

We transform the graph in four sequential steps: First we dissolve the edge $\{v_x, v_1\}$ and replace it by an edge $\{v_x, v_{x+1}\}$. Secondly, denote by $\deg_{H(F),A_1}(v)$ the total number of positive literal edges and negative literal edges adjacent to v. Then we can "disconnect" and "reconnect" each variable node v_i ($1 \le i \le n$) from all of its adjacent clause nodes as follows: We delete all edges $\{\{v_i, c\}\} \subseteq A_1$ and split $\{v_i, v_{i+1}\}$ into two paths $p_{i,\text{true}}$ and $p_{i,\text{false}}$, on which we place a total of $\deg_{A_1}(v)$ internal *literal nodes* $l_{i,c}$: If v_i is contained in a clause c – and thus we previously deleted $\{v_i, c\}$ – we place $l_{i,c}$ on $p_{i,\text{false}}$ and "reconnect" the variable by adding an edge between $l_{i,c}$ and the clause node c. Else if $\overline{v}$ is contained in c we proceed similarly (putting the node $l_{i,c}$ on $p_{i,\text{true}}$ instead). As a third step, depending on the number of literals of each clause c, we may modify its node: If c contains only a single literal, we delete the c node. If c contains two literals $\ell(v_i), \ell(v_j)$, we rename the node to $c_{i,j}$. If c is a disjunction of three literals $\ell(v_i), \ell(v_j), \ell(v_k)$, we split it into two nodes $c_{i,j}$ (connected to $l_{i,c}, l_{j,c}$) and $c_{j,k}$ (connected to $l_{j,c}, l_{k,c}$). Finally, we place the message on the first variable node $s := v_1$ and set its destination to $t := v_{x+1}$.

We remark that all four steps can be implemented such that the resulting delivery graph $G(F)$ is still planar, as illustrated in Fig. 3 (in each path tuple $(p_{i,\text{true}}, p_{i,\text{false}})$ the order of the internal nodes follows the original circular order of adjacent edges of v_i, and for each clause $c = \ell(v_i) \vee \ell(v_j) \vee \ell(v_k)$ the nodes $c_{i,j}$ and $c_{j,k}$ are placed close to each other).

Reduction Idea. We show that the message can't be delivered via any of the clause nodes. Thus the message has to be routed in each path pair $(p_{i,\text{true}}, p_{i,\text{false}})$ through exactly one of the two paths. If the message is routed via the path $p_{i,\text{true}}$, we interpret this as setting $v_i = \text{true}$ and hence we can read from the message trajectory a satisfiable assignment for F.

Agent Placement and Budgets. We will use greek letters for weights (namely ζ and δ) when the weights depend on each other or on the input. We place three kinds of agents on G:

1. *Variable agents:* x agents which are assigned to the variable nodes $v_1, \ldots, v_x$. These agents will have to decide whether the message is delivered via $p_{i,\text{true}}$ or via $p_{i,\text{false}}$, thus setting the corresponding variable to true or to false. We give all of them a budget of 2ζ.[1]

---

[1] In the *nonreturning* version we want agents to have the same "range", hence we set their budget to ζ.

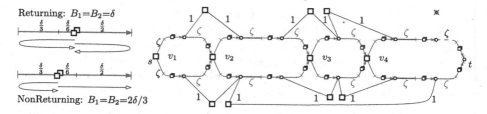

Fig. 4. (left) Two examples of δ-tubes for both versions of BUDGETEDDELIVERY. (right) Agent placement and edge weights on $G(F)$; agents are depicted by squares.

2. *Clause agents:* One agent per created clause *node*, e.g. a clause c containing three literals gets two agents, one in each of the two clause nodes. We think of these agents as follows: If in $c = \ell(v_i) \vee \ell(v_j) \vee \ell(v_k)$ the literal $\ell(v_j)$ is false, then clause c needs to send one of its agents down to the corresponding path node $l_{j,c}$ to help transporting the message over the adjacent "gap" of size ζ (depicted blue in Figs. 3 (right), 4). A 3CNF F will be satisfiable, if and only if no clause needs to spend more agents than are actually assigned to it respectively its node(s) in $G(F)$. We give all clause agents a budget of $2 \cdot (1 + \zeta)$.[2]

3. *Separating agents:* These will be placed in-between the agents defined above, to ensure that the variable and clause agents actually need to solve the task intended for them (they should not be able to deviate and help out somewhere else – not even their own kind). The separating agents will be placed in pairs inside δ-*tubes*, which we define next.

Remark 1. Strictly speaking, a reduction without variable agents works as well. In terms of clarity, we like to think of variable agents as the ones setting the variables to true or false.

δ-Tubes. We call a line segment a δ-*tube* if it satisfies the following four properties: (i) It has a length of δ. (ii) It contains exactly two agents which both have budget at most δ. (iii) Neither agent has enough energy to leave the line segment on the left or on the right by more than a distance of $\frac{\delta}{3}$. (iv) The agents can collectively transport a message through the line segment from left to right.

δ-tubes exist for both BUDGETEDDELIVERY versions, examples are given in Fig. 4 (left). The reader may think of these examples, whenever we talk about δ-tubes.

Edge Weights. We define edge weights on our graph $G(F)$ as follows: All edges between clause nodes and internal path nodes get weight 1 (in particular this means that if a clause agent walks to the path, it has a remaining range of ζ). Each path consists of alternating pieces of length ζ and of δ-tubes. We

---

[2] In the *nonreturning* version we assign a budget of $(1 + \zeta)$ to clause agents.

choose $\delta := \frac{4\zeta}{3} > \zeta$. This means that neither variable nor clause agents can cross a δ-tube (because their budget is not sufficiently large). Furthermore the distance any separating agent can move outside of its residential tube is at most $\frac{\delta}{3} = \frac{4\zeta}{9} < \frac{\zeta}{2}$. In particular separating agents are not able to collectively transport the message over a ζ-segment, since from both sides they are not able to reach the middle of the segment to handover the message. At last we set $\zeta := \frac{1}{8}$.

Lemma 1 (Returning BudgetedDelivery). *A planar 3CNF F is satisfiable if and only if it is possible to deliver a message from s to t in the corresponding delivery graph $G(F)$, such that all agents are still able to return to their starting points in the end.*

Proof (Sketch). "$\Rightarrow$" The schedule is straightforward: Each *variable agent* chooses, according to the assignment to v_i, the true-path $p_{i,\text{true}}$ or the false-path $p_{i,\text{false}}$. *Separating agents* and *clause agents* help wherever they are needed.

"$\Leftarrow$" One can show that the message cannot be delivered via any clause node. Hence we set $v_i = $ true if and only if the message moves through $p_{i,\text{true}}$. Now, each clause must have one satisfied literal, otherwise its agents could not have helped to bridge all ζ-segments.

We refer to the full version of the paper for further details: [5, Appendix A]. □

The same arguments work for *non-Returning* BudgetedDelivery as well. Recall that a delivery graph $G(F)$ created from a planar 3CNF F is planar. Furthermore the size of $G(F)$, as well as the number of agents we use, is polynomial in the number of clauses and variables. The agents' budgets and the edge weights are polynomial in ζ, δ and thus constant. Thus Lemma 1 shows NP-hardness of BudgetedDelivery on planar graphs. Finally, note that hardness also holds for a *uniform* budget B: One can simply add an edge of length $(B - B_i)/2$ to the starting location of each agent i and relocate i to the end of this edge.[3]

Theorem 4 (Hardness of BudgetedDelivery). *Both versions of* BUD-GETEDDELIVERY *are strongly NP-hard on planar graphs, even for uniform budgets.*

5   Hardness of Resource Augmentation

Main Ideas. We show that for all $\varepsilon > 0$, there is no polynomial-time $(2 - \varepsilon)$-resource-augmented algorithm for *Returning* BudgetedDelivery, unless P = NP. The same holds for $(3 - \varepsilon)$-resource-augmentation for the *non-Returning* version. Intuitively, an algorithm which finds out how to deliver the message with resource-augmented agents will at the same time solve 3SAT. We start by taking the reduction from PLANAR3SAT from Sect. 4. However, in addition to the previous delivery graph construction $G(F)$, we need to replace the δ-tubes and ζ-segments in order to take care of three potential pitfalls. We illustrate the modification into the new graph $G'(F)$ in Fig. 6:

---

[3] We relocate a non-returning agent by adding an edge of length $(B - B_i)$.

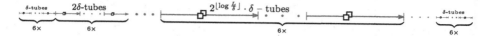

Fig. 5. L-δ-chains consist of blocks of 6 tubes of exponentially increasing and decreasing size.

1. In a resource-augmented setting, δ-tubes are no longer able to separate the clause and variable agents: These agents might be able to cross the δ-tube, or the separating agents residing inside the δ-tube can help out in the ζ-segments (there is no value for δ to prevent both). We will tackle this issue below by replacing δ-tubes by a chain of logarithmically many tubes with exponentially increasing and decreasing δ-values.

2. In the reduction for the original decision version of BUDGETEDDELIVERY, a clause c with three literals gave rise to two clause nodes $c_{i,j}$, $c_{j,k}$ that were adjacent to the same path node $l_{j,c}$. Hence the agent on $c_{i,j}$, now with resource-augmented budget, could pick up the message at $l_{j,c}$ and bring it close to the second resource-augmented agent stationed at $c_{j,k}$. This agent then might transport the message via its own clause node to the distant literal node $l_{k,c}$. To avoid this, we replace every ζ-segment adjacent to such a "doubly" reachable path node $l_{j,c}$ by two small parallel arcs. Both arcs contain exactly one ζ-segment, reachable from only one clause node (the message can then go over either arc), as well as a chain of tubes to provide the necessary separation.

3. A single clause agent stationed at $c_{i,j}$ might retrieve the message from the first literal node $l_{i,c}$, walk back to its origin and then on to the second literal $l_{j,c}$, thus transporting the message over a clause node. This can always be done by 2-resource-augmented agents; however for $(2 - \varepsilon)$-resource-augmentation we can prevent this by carefully tuning the weights of the ζ-segments, e.g. such that $(2 - \varepsilon) \cdot (1 + \zeta) \ll 2$.[4]

We now give a more formal description of the ideas mentioned above. Recall that a δ-tube had length δ and contained two agents with budget at most δ each. If these agents are now γ-resource-augmented, $\gamma < 3$, they can move strictly less than 3δ to the right or to the left of the δ-tube. In the following we want to uncouple the length of the line segment from the range the agents have left to move on the outside of the line segment.

L-δ-Chains. We call a line segment an L-δ-*chain* if it satisfies the following three properties: (i) Its length is at least L (a constant). (ii) No γ-resource-augmented agent ($1 \leq \gamma < 3$) contained in the chain has enough energy to leave the line segment by 3δ or more. (iii) The agents contained in the chain can collectively

---

[4] Non-returning clause agents can do this if they are 3-resource-augmented; and we can prevent it for $(3-\varepsilon)$-resource-augmentation by setting ζ such that $(3-\varepsilon) \cdot (1+\zeta) \ll 3$ (in fact the value of ζ will be the same as for *returning* BUDGETEDDELIVERY, but we will use different bounds in the proof).

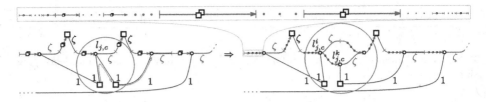

Fig. 6. (top-to-bottom) We replace δ-tubes in $G(F)$ by L-δ-chains in $G'(F)$. (left-to-right) We replace each ζ-segment connected to two clause agents by two parallel arcs.

transport a message through the line segment from left to right (already with their original budget).

We can create L-δ-chains for both BUDGETEDDELIVERY versions simply by using the respective δ-*tubes* as a blackbox: We start our line segment by adding a block of six δ-tubes next to each other, followed by a block of six 2δ-tubes, a block of six 4δ-tubes and so on until we get a block of length at least $6 \cdot 2^{\lfloor \log L/\delta \rfloor} \cdot \delta > L$. The same way we continue to add blocks of six tubes with lengths decreasing by powers of 2, see Fig. 5. Obviously properties (i) and (iii) are satisfied. To see (ii), note that any agent contained in the first or last block of δ-tubes cannot leave its tube (and thus the L-δ-chain) by 3δ or more. On the other hand, none of the inner blocks' agents is able to even cross the preceeding or the following block of six tubes, since their total length is larger than its budget.

Arc Replacement of ζ-Segments. Next we decouple any pair of clause agents (stationed at nodes $c_{i,j}$, $c_{j,k}$) that can directly go to the same literal node $l_{j,c}$ (so as not to allow them to transport the message via clause node with their augmented budgets, depicted in red in Fig. 6 (left)). We replace the adjacent ζ-segment by two small arcs which represent alternative ways over which the message can be transported. Each arc consists of one L-δ-chain and of one ζ-segment, see Fig. 6.

The *inner arc* begins with the ζ-segment – whose beginning $l_{j,c}^i$ can be reached through an edge of length 1 by the first clause agent (stationed at $c_{i,j}$) – and ends with the L-δ-chain. The *outer arc* first has the L-δ-chain and then the ζ-segment. The node in between these two parts, denoted by $l_{j,c}^k$, is connected via an edge of length 1 to the second clause agent's starting position $c_{j,k}$.

We conclude the replacement with three remarks: Firstly, it is easy to see that the described operation respects the planarity of the graph. Secondly, we are able to give values for L and δ in the next subparagraph such that a single clause agent is still both necessary and (together with agents inside the newly created adjacent L-δ-chain) sufficient to transport a message over one of the parallel arcs from left to right. Finally, the clause agent starting at $c_{i,j}$ is no longer able to meet the clause agent starting at $c_{j,k}$.

Budgets and Edge Weights. Recall that our agents have the following budgets: *separating agents* have a budget according to their position in the L-δ-chain, *variable agents* a budget of 2ζ and *clause agents* a budget of $2(1+\zeta)$.[5] Now these budgets are γ-resource-augmented, with $\gamma < 3$. We would like to prevent clause and variable agents from crossing L-δ-chains or even meeting inside of them, hence we set $L := 9$, which shall exceed the augmented budget of every agent by a factor of more than 2. Furthermore we don't want separating agents to help out too much outside of their residential chain, hence we set $\delta := \frac{\zeta}{9}$. A resource-augmented separating agent can thus walk only as far as $3\delta = \frac{\zeta}{3}$ to the outside of the tube. In particular, separating agents cannot transport the message over a ζ-segment.

Next we choose ζ such that an augmented clause agent stationed at a clause node $c_{i,j}$ is not able to transport the message from $l_{i,c}$ to $l_{j,c}$, not even in collaboration with the separating agents that can reach the two literal nodes. We set $\zeta := \frac{\varepsilon}{6-\varepsilon}$. The edges $\{c_{i,j}, l_{i,c}\}$, $\{c_{i,j}, l_{j,c}\}$ have length 1. In each edge, separating agents can help by at most $3\delta = \frac{\zeta}{3}$, leaving at least a distance of $1 - \frac{\zeta}{3}$ for the clause agent to cover. First note that for $0 < \varepsilon < 1$, we have $\zeta = \frac{\varepsilon}{6-\varepsilon} < \frac{\varepsilon}{5} < \frac{2\varepsilon}{3}$ and $(6-\varepsilon) > 3(2-\varepsilon)$. Hence a γ-resource-augmented clause agent has a budget of only $\gamma \cdot 2(1+\zeta) = 2(2-\varepsilon)(1+\zeta) = 2(2 - \varepsilon + \frac{(2-\varepsilon)\varepsilon}{6-\varepsilon}) < 2(2 - \frac{2\varepsilon}{3}) < 2(2-\zeta) < 4 \cdot (1 - \frac{\zeta}{3})$, and thus cannot transport the message via its clause node and return home in the end.[6]

Lemma 2 (Resource-augmented Returning BudgetedDelivery). *A planar 3CNF F is satisfiable if and only if it is possible to deliver a message with $(2-\varepsilon)$-resource-augmented agents from s to t in the corresponding delivery graph $G'(F)$, such that the agents are still able to reach their starting point in the end.*

Proof (Sketch) We follow the ideas of the proof of Lemma 1, and use the modifications to the graph structure and the weights presented in this section.
Details can be found in [5, Appendix B]. □

The same arguments work for *non-Returning* BUDGETEDDELIVERY as well, if we replace the inequalities for returning $(2 - \varepsilon)$-resource-augmented agents with the corresponding inequalities for non-returning $(3-\varepsilon)$-resource-augmented agents, given in (see Footnote 6).

Compare the new delivery graph $G'(F)$ with the original graph $G(F)$. The only topological changes we introduced with our replacements were the parallel arcs replacing the ζ-segments reachable by two clause nodes. We have already seen that this change respected the planarity of the delivery graph. Relevant changes to the edge weights and agent numbers, on the other hand, were added

---

[5] In the *nonreturning* version, variable agents have a budget of ζ and clause agents a budget of $1 + \zeta$.

[6] For non-returning agents we use (for $\varepsilon < 2$) the inequalities: $\zeta = \frac{\varepsilon}{6-\varepsilon} < \frac{\varepsilon}{4} < \frac{\varepsilon}{2}$ and $(6 - \varepsilon) > 2(3 - \varepsilon)$. Hence a non-returning γ-resource-augmented clause agent has a budget of $\gamma(1 + \zeta) = (3 - \varepsilon)(1 + \zeta) = 3 - \varepsilon + \frac{(3-\varepsilon)\varepsilon}{6-\varepsilon} < 3 - \frac{\varepsilon}{2} < 3 - \zeta = 3 \cdot (1 - \frac{\zeta}{3})$, and thus cannot transport the message via its clause node.

by replacing δ-tubes with L-δ-chains: Each chain consists of blocks of six δ-tubes of exponentially increasing size, hence we need a logarithmic number of tubes per chain, namely $\mathcal{O}\left(\log \frac{L}{\delta}\right)$ many. We have fixed the values of L and δ to $L = 9$ and $\delta = \frac{5}{9}$. With $\zeta^{-1} = \frac{9}{\varepsilon} - 1 \in \Theta(\varepsilon^{-1})$ we get $\mathcal{O}\left(\log \frac{L}{\delta}\right) = \mathcal{O}(\log(\zeta^{-1})) = \mathcal{O}(\log(\varepsilon^{-1}))$ many agents per chain. The number of chains is clearly polynomially bounded by the number of variables and clauses and the edge weights depend on ε only as well. Hence we conclude:

Theorem 5 (Inexistence of better resource augmentation for Budgeted Delivery). *There is no polynomial-time $(2-\varepsilon)$-resource-augmented algorithm for returning* BUDGETEDDELIVERY *and no $(3 - \varepsilon)$-resource-augmented algorithm for nonreturning* BUDGETEDDELIVERY, *unless* P = NP.

6   Conclusions

We gave a polynomial time algorithm for the returning variant of the problem on trees, as well as a best-possible resource-augmented algorithm for general graphs. On the other hand, we have shown that BUDGETEDDELIVERY is NP-hard, even on planar graphs and even if we allow resource augmentation. Our bounds on the required resource augmentation are tight and complement the previously known algorithm [9] for the non-returning case.

Our results show that BUDGETEDDELIVERY becomes hard when transitioning from trees to planar graphs. It is natural to investigate other causes for hardness. Chalopin et al. [9] gave a polynomial algorithm for the *non-Returning* version under the assumptions that (i) the order in which the agents move is fixed and (ii) the message can only be handed over at vertices. Using a dynamic program, we are able to drop assumption (ii), allowing handovers within edges [5]. Our result holds for both versions of BUDGETEDDELIVERY.

Theorem 6. BUDGETEDDELIVERY *is solvable in time $\mathcal{O}(k(n+m)(n \log n + m))$ if the agents are restricted to a fixed order in which they move.*

Corollary 1. *For a constant number of agents k,* BUDGETEDDELIVERY *is solvable in time $\mathrm{poly}(n, m)$ by brute forcing the order of the agents.*

An interesting open problem is to understand collaborative delivery of multiple messages at once. For example, the complexity of the problem on paths remains open. In this setting, it may be resonable to constrain the number of agents, the number of messages, or the ability of transporting multiple messages at once, in order to allow for efficient algorithms. Also, in general graphs, the problem may not become easy if the order in which agents move is fixed.

Acknowledgments. This work was partially supported by the project ANR-ANCOR (anr-14-CE36-0002-01) and the SNF (project 200021L_156620).

References

1. Albers, S., Henzinger, M.R.: Exploring unknown environments. SIAM J. Comput. **29**(4), 1164–1188 (2000)
2. Anaya, J., Chalopin, J., Czyzowicz, J., Labourel, A., Pelc, A., Vaxès, Y.: Convergecast and broadcast by power-aware mobile agents. Algorithmica **74**(1), 117–155 (2016)
3. Applegate, D.L., Bixby, R.E., Chvatal, V., Cook, W.J.: The Traveling Salesman Problem: A Computational Study (Princeton Series in Applied Mathematics). Princeton University Press, Princeton (2007)
4. Awerbuch, B., Betke, M., Rivest, R.L., Singh, M.: Piecemeal graph exploration by a mobile robot. Inf. Comput. **152**(2), 155–172 (1999)
5. Bärtschi, A., Chalopin, J., Das, S., Disser, Y., Geissmann, B., Graf, D., Labourel, A., Mihalák, M.: Collaborative Delivery with Energy-Constrained Mobile Robots, arXiv preprint. https://arxiv.org/abs/1606.05538 (2016)
6. Bender, M.A., Slonim, D.K.: The power of team exploration: two robots can learn unlabeled directed graphs. In: 35th Symposium on Foundations of Computer Science, FOCS 1994, pp. 75–85 (1994)
7. Betke, M., Rivest, R.L., Singh, M.: Piecemeal learning of an unknown environment. Mach. Learn. **18**(2), 231–254 (1995)
8. Biló, D., Disser, Y., Gu_alá, L., Mihal'ák, M., Proietti, G., Widmayer, P.: Polygon-constrained motion planning problems. In: Flocchini, P., Gao, J., Kranakis, E., der Heide, F.M. (eds.) ALGOSENSORS 2013. LNCS, vol. 8243, pp. 67–82. Springer, Heidelberg (2014)
9. Chalopin, J., Das, S., Mihalák, M., Penna, P., Widmayer, P.: Data delivery by energy-constrained mobile agents. In: Flocchini, P., Gao, J., Kranakis, E., der Heide, F.M. (eds.) ALGOSENSORS 2013. LNCS, vol. 8243, pp. 111–122. Springer, Heidelberg (2014)
10. Chalopin, J., Jacob, R., Mihalák, M., Widmayer, P.: Data delivery by energy-constrained mobile agents on a line. In: Esparza, J., Fraigniaud, P., Husfeldt, T., Koutsoupias, E. (eds.) ICALP 2014, Part II. LNCS, vol. 8573, pp. 423–434. Springer, Heidelberg (2014)
11. Czyzowicz, J., Diks, K., Moussi, J., Rytter, W.: Communication problems for mobile agents exchanging energy. In: 23rd International Colloquium on Structural Information and Communication Complexity SIROCCO 2016 (2016)
12. Das, S., Dereniowski, D., Karousatou, C.: Collaborative exploration by energy-constrained mobile robots. In: 22th International Colloquium on Structural Information and Communication Complexity SIROCCO 2015, pp. 357–369 (2015)
13. Demaine, E.D., Hajiaghayi, M., Mahini, H., Sayedi-Roshkhar, A.S., Oveisgharan, S., Zadimoghaddam, M.: Minimizing movement. ACM Trans. Algorithms **5**(3), 1–30 (2009)
14. Dereniowski, D., Disser, Y., Kosowski, A., Pajkak, D., Uznański, P.: Fast collaborative graph exploration. Inf. Comput. **243**, 37–49 (2015)
15. Duncan, C.A., Kobourov, S.G., Kumar, V.S.A.: Optimal constrained graph exploration. In: 12th ACM Symposium on Discrete Algorithms, SODA 2001, pp. 807–814 (2001)
16. Dynia, M., Korzeniowski, M., Schindelhauer, C.: Power-aware collective tree exploration. In: Grass, W., Sick, B., Waldschmidt, K. (eds.) ARCS 2006. LNCS, vol. 3894, pp. 341–351. Springer, Heidelberg (2006)

17. Dynia, M., Łopuszański, J., Schindelhauer, C.: Why robots need maps. In: Prencipe, G., Zaks, S. (eds.) SIROCCO 2007. LNCS, vol. 4474, pp. 41–50. Springer, Heidelberg (2007)
18. Edmonds, J., Johnson, E.L.: Matching, euler tours and the chinese postman. Math. Program. **5**(1), 88–124 (1973)
19. Flocchini, P., Prencipe, G., Santoro, N.: Distributed Computing by Oblivious Mobile Robots. Morgan & Claypool, San Rafeal (2012)
20. Fraigniaud, P., Gąsieniec, L., Kowalski, D.R., Pelc, A.: Collective tree exploration. Networks **48**(3), 166–177 (2006)
21. Lichtenstein, D.: Planar formulae and their uses. SIAM J. Comput. **11**(2), 329–343 (1982)
22. Ortolf, C., Schindelhauer, C.: Online multi-robot exploration of grid graphs with rectangular obstacles. In: 24th ACM Symposium on Parallelism in Algorithms and Architectures, SPAA 2012, pp. 27–36 (2012)
23. Panaite, P., Pelc, A.: Exploring unknown undirected graphs. J. Algorithms **33**(2), 281–295 (1999)

Communication Problems for Mobile Agents Exchanging Energy

Jurek Czyzowicz[1]($\boxtimes$), Krzysztof Diks[2], Jean Moussi[1], and Wojciech Rytter[2]

[1] Département d'informatique, Université du Québec en Outaouais,
Gatineau, Québec, Canada
{jurek,Jean.Moussi}@uqo.ca
[2] Faculty of Mathematics, Informatics and Mechanics,
University of Warsaw, Warsaw, Poland
{diks,rytter}@mimuw.edu.pl

Abstract. A set of mobile agents is deployed in the nodes of an edge-weighted network. Agents originally possess amounts of energy, possibly different for all agents. The agents travel in the network spending energy proportional to the distance traversed. Some nodes of the network may keep information which is acquired by the agents visiting them. The meeting agents may exchange currently possessed information, as well as any amount of energy. We consider communication problems when the information initially held by some network nodes have to be communicated to some other nodes and/or agents. The paper deals with two communication problems: data delivery and convergecast. These problems are posed for a centralized scheduler which has full knowledge of the instance. It is already known that, without energy exchange, both problems are NP-complete even if the network is a line. In this paper we show that, if the agents are allowed to exchange energy, both problems have linear-time solutions on trees. On the other hand for general undirected and directed graphs we show that these problems are NP-complete.

1 Introduction

A set of n agents is placed at nodes of an edge-weighted graph G. An edge weight represents its length, i.e., the distance between its endpoints along the edge. Initially, each agent has an amount of energy (possibly distinct for different agents).

Agents walk in a continuous way along the network edges using amount of energy proportional to the distance travelled. An agent may stop at any point of a network edge (i.e. at any distance from the edge endpoints, up to the edge weight). Initially, at the nodes of the graph is stored the information (different for each node), which may be collected by agents visiting such nodes. Each agent has memory in which it can store all collected information.

When two agents meet, one of them can transfer a portion of currently possessed energy to another one. Moreover, two meeting agents exchange their currently possessed information, so that after the meeting both agents keep in their memories the union of the information previously held by each of them.

J. Suomela (Ed.): SIROCCO 2016, LNCS 9988, pp. 275–288, 2016.
DOI: 10.1007/978-3-319-48314-6_18

We consider two problems:

1. *Data delivery* **problem:** Given two nodes s, t of G, is it possible to transfer the initial packet of information placed at node s to node t?
2. *Convergecast* **problem:** Is it possible to transfer the initial information of all nodes to the same node?

We will look for schedules of agent movements which will not only result in completing the desired task, but also attempt to maximize the unused energy. We call such schedules *optimal*. We conservatively suppose that, whenever two agents meet, they automatically exchange the entire information they hold. This information exchange procedure is never explicitly mentioned in our algorithms, supposing, by default, that it always takes place when a meeting occurs.

1.1   Our Results

We show that both communication have linear time algorithms on trees. On the other hand, for general undirected and directed graphs we show that these problems are NP-complete.

1.2   Related Work

Recent development in the network industry triggered the research interest in mobile agents computing. Several applications involve physical mobile devices like robots, motor vehicles or various wireless gadgets. Mobile agents are sometimes interpreted as software agents, i.e., programs migrating from host to host in a network, performing some specific tasks. Examples of agents also include living beings: humans (e.g. soldiers or disaster relief personnel) or animals. Most studied problems for mobile agents involve some sort of environment search or exploration (cf. [3,8,10–12]). In the case of a team of collaborating mobile agents, the challenge is to balance the workload among the agents in order to minimize the exploration time. However this task is often hard (cf. [13]), even in the case of two agents in a tree, [6]. The tree exploration by energy-constrained mobile robots has been considered in [10].

The task of convergecast is important when agents possess partial information about the network (e.g. when individual agents hold measurements performed by sensors located at their positions) and the aggregate information is needed to make some global decision based on all measurements. The convergecast problem is often considered as a version of the data aggregation question (e.g. [16,17]) and it has been investigated in the context of wireless and sensor networks, where the energy consumption is an important issue (cf. [4,15]).

The power awareness question has been studied in different contexts. Energy management of (not necessarily mobile) computational devices has been studied in [2]. To reduce energy consumption of computer systems the methods proposed include power-down strategies (see [2,5,14]) or speed scaling (cf. [18]). Most of research on energy efficiency considers optimization of overall power used. When the power assignments are made by the individual system components, similar to our setting, the optimization problem involved has a flavor of load balancing (cf. [7]).

The problem of communication by energy-constrained mobile agents has been investigated in [1]. The agents of [1] all have the same initial energy and they perform efficient convergecast and broadcast in line networks. However the same problem for tree networks is proven to be strongly NP-complete in [1].

The closely related problem of data delivery, when the information has to be transmitted between two given network nodes by a set of energy constrained agents has been studied in [9]. This problem is proven to be NP-complete in [9] already for line networks, if the initial energy values may be distinct for different agents. However, in the setting studied in [1,9], the agents do not exchange energy. In the present paper we show that the situation is quite different if the agents are allowed to transfer energy between one another.

2 The Line Environment

In this section we start with a line environment and suppose that we are given a collection of agents $\{1, 2, \ldots, n\}$ on the line. Each agent i is initially placed at position a_i on the line and has initial energy e_i.

2.1 Data Delivery on the Line

We start with the delivery problem from point s to t. Assume that $a_i < a_j$ for $i < j$ and $s < t$. Indeed, in this case w.l.o.g. we may replace many agents starting at the same point may by a single agent holding the sum of their energy amounts.

The problem can be immediately reduced to the situation $s \leq a_1$, $a_n \leq t$. Otherwise, the agents on the left-hand side of s (starting from the leftmost one) walk left-to-right collecting energy of the encountered other agents. If some energy can be brought this way to s, we obtain an extra agent which will start at s. Symmetrically, the agents on the right-hand side of t act in order to possibly bring the maximal amount of energy to point t. It is easy to see that this is the best use of agents placed outside the interval $[s,t]$. Consequently, we may assume $s \leq a_1$, $a_n \leq t$.

Our first algorithm is only a decision version. Its main purpose is to show how certain useful table can be computed; all subsequent algorithms are based on computing similar type of tables.

Consider the partial delivery problem $\mathcal{D}_i$, in which agents larger than i are removed, together with their energy, and the goal is to deliver the packet from point a_1 to point a_i. We say that the problem $\mathcal{D}_i$ is *solvable* iff such a delivery is possible.

We define the following table $\overrightarrow{\Delta}$:

- If $\mathcal{D}_i$ is not solvable then $\overrightarrow{\Delta}_i = -\delta$, where δ is the *minimal* energy which needs to be added to e_i (to the energy of i-th agent) to make $\mathcal{D}_i$ solvable.

– If $\mathcal{D}_i$ is solvable then $\overrightarrow{\Delta}_i$ is the *maximal* unused energy which can remain in point a_i after delivering the packet from a_1 to a_i. Note that it is possible that $\overrightarrow{\Delta}_i > e_i$ since during delivery the unused energy of some other agents can be moved to point a_i.

Assume that points s and t are the starting points $s = a_0$ and $t = a_{n+1}$ of virtual dummy agents 0 and $n+1$, respectively, each having zero energy. Therefore, we may assume that the original positions of the agents are $s = a_0 \leq a_1 < a_2 < a_3 < \ldots < a_n \leq t = a_{n+1}$.

We have the following decision algorithm.

ALGORITHM DELIVERY-TEST-ON-THE-LINE ;

1. $A := e_0 = 0;\ a_0 := s;\ a_{n+1} := t;\ e_{n+1} := 0;$

2.   **for** $i = 1$ **to** $n + 1$ **do**

3. $d := a_i - a_{i-1};$

4.   **if** $A \geq d$ **then** $A := A - d$

5.   **else if** $A \geq 0$ **then** $A := -2(d - A)$

6.   **else** $A := A - 2d;$

7. $A := A + e_i;\ \overrightarrow{\Delta}_i := A;$

8.   **return** $(A \geq 0)$;

Example 1. Assume

$$[a_0, a_1, \ldots, a_4] = [0, 10, 20, 30, 40, 50], \quad [e_0, e_1, \ldots e_4] = [0, 24, 10, 40, 0].$$

Then (assuming, by convention, $\overrightarrow{\Delta}_0 = 0$, see also Fig. 1) we have

$$\overrightarrow{\Delta} = [0,\ 4,\ -2,\ 18,\ 8].$$

Remark. The values of $\overrightarrow{\Delta}_i$ are not needed to solve the decision-only version. However they will be useful in creating the delivery schedule and also in the *convergecast* problem.

Lemma 2. *The algorithm* DELIVERY-TEST-ON-THE-LINE *correctly computes the table* $\overrightarrow{\Delta}$ *(thus it solves the decision version of the delivery problem) in linear time.*

Proof. We prove by induction on i, that the value of $\overrightarrow{\Delta}_i$ is correctly computed in line 7 of the algorithm.

Suppose first the case $i = 1$. In the case $a_0 = a_1$, as $A = e_0 = 0$, in lines 4 and 7 we compute the value of $A = \overrightarrow{\Delta}_1 = e_1$, which is correct as agent 1 does not need to use any energy to pick up the packet at point a_0. Otherwise, if $a_0 < a_1$ we have $A = 0$ and $A < d$, so lines 5 and 7 are executed, in which case we have $A = \overrightarrow{\Delta}_1 = e_1 - 2d$. As agent 1 needs to cover distance d in both directions to bring the packet to point a_1 this is correct, independently whether the computed value negative or not.

Suppose now, by inductive hypothesis, that the algorithm computed correctly $A = \overrightarrow{\Delta}_{i-1}$ in the previous iteration. There are three cases:

Case 1 (line 4 of the algorithm). The instance $\mathcal{D}_{i-1}$ was solvable and after moving the packet from a_1 to a_{i-1} the maximal remaining energy was $\overrightarrow{\Delta}_{i-1}$. As in this case we have $\overrightarrow{\Delta}_{i-1} = A \geq d$, the energy $\overrightarrow{\Delta}_{i-1}$ is sufficient to move the packet from a_{i-1} to a_i. Consequently, we spent d energy to travers the distance d in one direction and we have $\overrightarrow{\Delta}_i = \overrightarrow{\Delta}_{i-1} - d + e_i$ as correctly computed in lines 4 and 7.

Case 2 (line 5). The instance $\mathcal{D}_{i-1}$ was still solvable but after moving the packet from a_1 to a_{i-1} the remaining energy $\overrightarrow{\Delta}_{i-1}$ is not sufficient to reach a_i without *help* from agents to the right of a_{i-1}. Then the $(i-1)$-st agent moves only one-way by distance $\overrightarrow{\Delta}_{i-1}$. The remaining distance $d - \overrightarrow{\Delta}_{i-1}$ to point a_i should be covered both-ways from a_i. Hence we need to use the amount of $2(d - \overrightarrow{\Delta}_{i-1})$ energy, which is expressed by statement 5. The value of $\overrightarrow{\Delta}_i$ is computed correctly independently whether the addition of e_i makes it positive or not.

Case 3 (line 6). In this case the instance $\mathcal{D}_{i-1}$ was not solvable, i.e. the agents $1, 2, \ldots, i-1$ could not deliver the packet to point a_{i-1}. Consequently, the interval $[a_{i-1}, a_i]$ has to be traversed entirely in both direction and we obtain $\overrightarrow{\Delta}_i = \overrightarrow{\Delta}_{i-1} - 2d + e_i$, which is correctly computed in lines 6 and 7.

The cases correspond to the statements in the algorithm, and show its correctness. This completes the proof.

Once the values of $\overrightarrow{\Delta}_i$ are computed, the schedule describing the behaviour of each agent is implicitly obvious, but we give it below for reference. Note that the action of each agent a_i is started once the process involving lower-numbered agents has been completed. We are not interested in this paper in finding the shortest time to complete the schedule (allowing agents to work in parallel).

ALGORITHM DELIVERY-SCHEDULE-ON-THE-LINE;

{ Delivering packet from s to t }

$pos := s$;

for $i = 1$ to n **do**

    if $\overrightarrow{\Delta}_i \geq 0$ and $pos < a_i$ **then**

        1. The i-th agent walks left collecting energy of all encountered agents until arriving at the packet position. It picks up the packet.

        2. The i-th agent walks right collecting energy of all encountered agents until exhausting its energy or reaching t.

        3. The i-th agent leaves the packet at the actual position pos.

Delivery is successful iff $pos = t$;

Figure 1 illustrates the execution of the above algorithm for Example 1. We conclude with the following theorem.

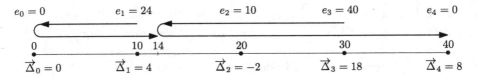

Fig. 1. Schedule of agent movements for a_i's and energies given in Example 1.

Theorem 3. *In linear time we can decide if the information of any agent can be delivered to any other agent and, if it is possible, we find the centralized scheduling algorithm which performs such a delivery.*

2.2   Convergecast on the Line

The convergecast consists in communication in which the union of initial information of all nodes arrives to the same node. In some other papers (e.g. [1]), the convergecast problem consists in determining whether the union of the entire information may be transferred to the same agent. However, for energy exchanging agents, this is not a problem: if convergecast is possible then any agent may be its target, as agents may swap freely when meeting.

We present below the algorithm finding if convergecast is possible. We will use algorithm DELIVERY-TEST-ON-THE-LINE to compute the values of $\overrightarrow{\Delta}_i$ as defined before, assuming that $s = a_1$ and $t = a_n$. Similarly we denote by $\overleftarrow{\Delta}_i$ the values of the energy potential at point a_i that the symmetric algorithm would compute while transferring the packet initially situated at the point a_n towards the target position at a_i. Therefore, $\overleftarrow{\Delta}_i$ equals the deficit or the surplus of energy during the transfer of information initially held by agent n to agent i using agents $i, i + 1, \cdots, n$.

ALGORITHM CONVERGECAST-ON-THE-LINE;

1.   For all $i = 1, 2, \cdots, n$ compute the values of $\overrightarrow{\Delta}_i$ and $\overleftarrow{\Delta}_i$ representing the energy potentials at a_i, for deliveries from a_1 to a_i and a_n to a_i, respectively

2.   **for** $i = 1$ **to** n **do**

3.   **if** $\overrightarrow{\Delta}_i \geq 0 \wedge \overleftarrow{\Delta}_{i+1} \geq 0 \wedge \overrightarrow{\Delta}_i + \overleftarrow{\Delta}_{i+1} - (a_{i+1} - a_i) \geq 0$ **then**

4.   **return** Convergecast possible;

5.   **return** Convergecast not possible;

We have the following theorem.

Theorem 4. *Algorithm* CONVERGECAST-ON-THE-LINE *in* $O(n)$ *time solves the convergecast problem.*

Proof. The convergecast is possible if and only if the information of agent a_1 and the information of agent a_n may be transferred to the same point of the line. This is equivalent to the existence of a pair of agents i and $i + 1$, such that transferring the information from point a_1 to a_i using agents $1, 2 \cdots, i$ results in a surplus of energy brought to point a_i, as well as that transferring the information from point a_n to a_{i+1} using agents $n, n-1 \cdots, i+1$ results in a surplus of energy brought to point a_{i+1}. Moreover, the sum of these two surpluses of energy must be sufficient to complete a walk along the entire segment $[a_i, a_{i+1}]$ permitting agents i and $i+1$ to meet. This is exactly what is verified at line 3 of algorithm CONVERGECAST-ON-THE-LINE.

An interested reader may observe, that the condition of the if clause from line 3 may be simplified to $\overrightarrow{\Delta}_i + \overleftarrow{\Delta}_{i+1} - (a_{i+1} - a_i) \geq 0$ as in such case the convergecast is also possible although the convergecast point may not be inside the interval $[a_i, a_{i+1}]$. However, the current condition at line 3 permits to identify all points of the environment to which the union of all node information may be transported. We call such points *convergecast points*. Indeed, if $\overrightarrow{\Delta}_i + \overleftarrow{\Delta}_{i+1} - (a_{i+1} - a_i) = 0$, then there exists a unique convergecast point inside the interval $[a_i, a_{i+1}]$. The surplus of energy permits to deliver the convergecast information to an interval of the line larger than a single point. We have the following Corollary.

Corollary 5. *If the condition in line 3 of algorithm* CONVERGECAST-ON-THE-LINE *is true, then the set of convergecast points of the line equals* $[a_{i+1} - \overleftarrow{\Delta}_{i+1}, a_i + \overrightarrow{\Delta}_i]$.

3   The Tree Environment

3.1   Data Delivery in the Tree

The technique developed for delivery in lines can be extended easily to delivery in undirected trees. In this case, the agents are placed at the nodes of the tree. Observe that from the original tree we can remove subtrees which do not contain s, t or any agents. Consequently, we obtain a connected tree whose every leaf either contains s or t or an initial position of some agent.

The delivery problem for a tree is easily reducible to the case of a line.

Theorem 6. *We can solve delivery problem and construct delivery-scenario on the tree in linear time.*

Proof. Consider the path π in the tree T connecting s with t. Suppose we remove from T all edges of path π. The tree splits into several subtrees *anchored* at nodes of π. For each such subtree we direct all edges towards the root, which is a node of π. The agents initially present at the leaves of such trees are walking up along the directed paths towards their roots accumulating energies at intermediate nodes. To avoid having two agents walking along the same edge it is sufficient to move agents present at leaves only and remove every such edge after the move is made. Agents having energy use it during their walk bringing the remainder to the intermediate nodes. Agents with zero energy are moved freely bringing no energy. The process terminates when the subtree is reduced to a single root belonging to path pi. This way we optimize the energy that can be brought to path pi. The problem of the delivery on the tree is now reduced to the delivery on the line π. Consequently, all steps of this construction may be computed in linear time. This completes the proof.

3.2   Convergecast on the Tree

In this section we extend to the case of trees the basic ideas developed for the problem of convergecast for the line environment. The tables $\overleftarrow{\Delta}$ and $\overrightarrow{\Delta}$ for lines were computed locally, looking only at neighboring nodes. Simiarly, the values of the corresponding table $\overrightarrow{\Delta}$ for a node in a tree is computed looking at the neighbors of this node. However, as the flow of the information passing through node v can be made in d_v directions, where d_v is the degree of v, for each node v we will compute d_v different values of Δ. For this purpose, though the input tree is undirected, we will consider direction of edges. For each undirected edge (u, v) we consider two directed edges $u \rightarrow v$, $v \rightarrow u$. We define the subtree $T_{v \rightarrow u}$ as the connected component containing v and resulting by removing from T the edge (v, u), see Fig. 2. Observe that at the moment of convergecast, there are two agents meeting at a point of some edge, that we call *convergecast point*, where these agents start possessing the initial information of all nodes.

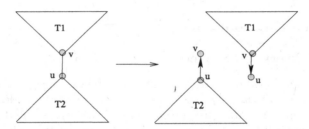

Fig. 2. Testing if there is a *convergecast point* on the undirected edge (u, v) is reduced to computation of the costs $\Delta_{u \rightarrow v}$ and $\Delta_{v \rightarrow u}$ of moving all packets in the trees $T2 = T_{u \rightarrow v}$ and $T1 = T_{v \rightarrow u}$.

In order to compute all needed values of Δ, for each directed edge $u \rightarrow v$ of the tree we define $\Delta_{u \rightarrow v}$ as the energy potential of moving all packets from the subtree $T_{u \rightarrow v}$ to its root u without interacting with any node outside $T_{u \rightarrow v}$. More exactly, if $\Delta_{u \rightarrow v} \geq 0$, then it represents the maximal amount of energy that can be delivered to u, together with all data packets originated at the nodes of $T_{u \rightarrow v}$. Observe that, if $T_{u \rightarrow v}$ initially does not contain any agents, then $\Delta_{u \rightarrow v}$ equals twice the sum of weights of all edges of $T_{u \rightarrow v}$. Indeed, in such case, the delivery must be performed by an agent starting at u and performing the DFS traversal of $T_{u \rightarrow v}$. If $T_{u \rightarrow v}$ initially contains some agents, the value of $\Delta_{u \rightarrow v}$ is smaller, but always equal at least the sum of weights of its edges. If $\Delta_{u \rightarrow v} < 0$ then $-\Delta_{u \rightarrow v}$ is the minimal amount of energy that we need to deliver to u by some agent, initially outside $T_{u \to v}$, so that this agent can bring to node u all data packets from the nodes of $T_{u \to v}$. In both cases, will be used all agents initially present inside $T_{u \to v}$ as well as their entire energy.

In order to correctly compute the values of Δ we define an order in which the consecutive directed edges of T will be treated by our algorithm. We denote $x \to y \prec y \to z$, when $x \neq z$, meaning that, for consecutive edges, an edge ending at a node *precedes* (according to relation $\prec$) an edge starting at this node.

Observation 1. *The relation $\prec$ in a tree is a partial order and it can be extended to a linear order X in $O(n)$ time.*

We propose the following algorithm.

ALGORITHM CONVERGECAST-ON-THE-TREE(T);

1. Compute a linear order X of directed edges of T according to relation $\prec$.
2. **for** each directed edge $u \to v$ taken in order X **do**
3.   COMPUTE $\Delta_{u \to v}$;
4. **for** each undirected edge (u, v) of T **do**
5.   **if** $(\Delta_{u \to v} \geq 0) \wedge (\Delta_{v \to u} \geq 0) \wedge (\Delta_{u \to v} + \Delta_{v \to u} \geq weight(u, v))$
6.   **then return** Convergecast is possible
7. **return** Convergecast is not possible

The values of $\Delta_{u \to v}$ are computed by the following procedure.

PROCEDURE COMPUTE $\Delta_{u \to v}$;

1. $\Delta_{u \to v} := e_u$; {initial energy of node u}
2. **for** each undirected edge $x \to u$, such that $x \neq v$ **do**
3.   **if** $\Delta_{x \to u} \geq weight(x, u)$
4.   **then** $\Delta_{u \to v} := \Delta_{u \to v} + \Delta_{x \to u} - weight(x, u)$
5.   **else if** $\Delta_{x \to u} > 0$
6.   **then** $\Delta_{u \to v} := \Delta_{u \to v} + 2 * (\Delta_{x \to u} - weight(x, u))$
7.   **else** $\Delta_{u \to v} := \Delta_{u \to v} + \Delta_{x \to u} - 2 * weight(x, u)$

We have the following theorem:

Theorem 7. *Algorithm* CONVERGECAST-ON-THE-TREE *in linear time solves the convergecast problem for trees.*

Proof. We show first the following claim:

Claim:, The for loop from line 2 of the algorithm correctly computes the value of $\Delta_{u \to v}$ for every directed edge $u \to v$.

The proof of the claim goes by induction on the consecutive iterations of the **for** loop from line 2. Consider the first directed edge $u \to v$ of X. As the tree $T_{u \to v}$ is then composed of a single node, we obtain correctly $\Delta_{u \to v} = e_u$, i.e. the initial energy of node u. The claim is also true for any other edge $u \to v$, treated later by the for loop of line 2, such that u is a terminal node.

Consider now the case when the for loop from line 2 takes an edge $u \to v$ for a non-terminal node u. Let $v_1, v_2, \cdots, v_p$ be all nodes adjacent to u, such that $v_i \neq v$, for $i = 1, \cdots, p$. Note that, at that moment, the values of $\Delta_{v_i \to u}$ for all $i = 1, \cdots, p$ have been already computed. Observe that, bringing packets from all $v_i \neq v$, $i = 1, \cdots, p$ to u needs to be done across the respective edges $v_i \to u$, sometimes bringing the unused energy to u and other times using some energy from $\Delta_{u \to v}$ to traverse twice edge (v_i, u), or its portion, by an agent coming from u.

Take any v_i and suppose first that $\Delta_{v_i \to u} \geq weight(v_i, u)$. Then by inductive assumption, the agents present at $T_{v_i \to u}$ can perform the convergecast to v_i bringing there the amount of $\Delta_{v_i \to u}$ extra energy. This energy is sufficient to transfer all packets of $T_{v_i \to u}$ through edge (v_i, u) and the remaining amount of $\Delta_{v_i \to u} - weight(v_i, u)$ energy is accumulated at $\Delta_{u \to v}$, which is correctly computed at line 4 of procedure COMPUTE $\Delta_{u \to v}$.

Suppose now, that $\Delta_{v_i \to u} \leq 0$. In order to bring all packets of $T_{v_i \to u}$ to u, an agent present at u must traverse the edge $u \to v_i$, bring the packets to node using $-\Delta_{v_i \to u}$ extra energy and then traverse the edge (u, v_i) in the opposite direction $v_i \to u$. For this purpose is needed the extra energy of $-\Delta_{x \to u} + 2 * weight(x, u)$, which is correctly suppressed from $\Delta_{u \to v}$ at line 7 of the procedure.

Consider now the remaining case when $0 < \Delta_{v_i \to u} < weight(v_i, u)$. In this case, all packets of $T_{v_i \to u}$ are brought to node v_i by some agent initially present within $T_{v_i \to u}$, but this agent does not have enough energy to traverse edge $v_i \to u$ by itself. Such agent will use its entire energy of $\Delta_{v_i \to u}$ to traverse a portion of edge $v_i \to u$ and some other agent need to come from u and to traverse the other portion in both directions in order to transfer the packets to u. The energy needed by the second agent equals $2 * (weight(v_i, u) - \Delta_{v_i \to u})$, which is correctly suppressed from $\Delta_{u \to v}$ at line 6 of the procedure. This completes the proof of the claim.

To complete the proof, consider the moment when in an optimal convergecast algorithm one agent obtains the union of the initial information of all nodes of the network. This happens while two agents meet on some edge (u, v), one of them carrying the union of information from the subtree $T_{u \to v}$, and the other one - from the subtree $T_{v \to u}$. These agents need to have enough positive energy to meet within the edge (u, v). This is equivalent to the condition tested in line 5 of the algorithm.

The condition from line 5 of algorithm CONVERGECAST-ON-THE-TREE permits to decide only if the convergecast is possible. However, similarly to the line case, an interested reader may observe that one can easily identify the set of all convergecast points. For this purpose we define the set of $D_{u,v}(d)$ containing a subset of points from the edges of T. Consider a point p and the simple path $\Pi(u, p)$ of T joining p with u. We define $p \notin D_{u,v}(d)$ if the path $\Pi(u, p)$ goes in the direction of edge (u, v) and its length exceeds d, i.e. $|\Pi(u, p)| > d$. All other points of T belong to $D_{u,v}(d)$. We have the following corollary.

Corollary 8. *If the condition in line 5 of algorithm* CONVERGECAST-ON-THE-TREE *is true, then the set of all convergecast points of the tree equals* $D_{u,v}(\Delta_{u \to v}) \cap D_{v,u}(\Delta_{v \to u})$.

4   NP-Completeness for digraphs and graphs

We use the following NP-complete problem:

Integer Set Partition (ISP): Given set $X = \{x_1, x_2, \ldots, x_n\}$ of positive integer values verify whether X can be partitioned into two subsets with equal total sums of values.

We have the following theorem.

Theorem 9. *The delivery and convergecast problems are NP-complete for general directed graphs.*

Proof. Denote $E = \sum_{i=1}^n x_i$. Given an instance of the ISP problem, we construct the following graph G_X (see Fig. 3). The set of $n+3$ nodes of G_X consists of three nodes s, t, a and the set of nodes $V = \{v_1, v_2, \ldots, v_n\}$. Each node v_i contains a single agent i having an initial energy $e = x_i$, for $i = 1, 2, \ldots, n$. The weights of edges outgoing from nodes of V are $w(v_i \to s) = x_i/3$ and $w(v_i \to a) = 0$ for $i = 1, 2, \ldots, n$. Moreover we have $w(s \to a) = E/3$ and $w(a \to t) = E/2$. Consider a delivery from s to t. W.l.o.g. we can suppose that this is done by some agent i, which must traverse the path $v_i \to s \to a \to t$. As no agent can do it using only its own energy (otherwise $x_i \geq 5E/6$ and the ISP trivially has no solution), some other agents of the collection must walk to s and some other ones must go directly to a, in order to deliver to agent i additional energy needed to complete its path $v_i \to s \to a \to t$.

Assume that X_1, X_2 are the sets of agents which directly move to s and a, respectively. Let

$$\alpha = \sum_{i \in X_1} x_i, \ \beta = \sum_{i \in X_2} x_i$$

Hence the energy delivered to s, unused by the agents X_1 incoming to s, is $\frac{2}{3}\alpha$. As this energy must be sufficient to traverse at least edge $s \to a$, we have

$$\frac{2}{3}\alpha \geq E/3 \tag{1}$$

Consider now the maximal energy, which may be available to agent i at point a. It is equal to the sum of energy β, which is brought to point a by agents X_2, and the energy unused by agent i, ending its traversal of edge $s \to a$, which equals $2\alpha/3 - E/3$. As the sum of these energies must suffice to traverse edge $a \to t$ of weight $E/2$ and $\alpha + \beta = E$ we have

$$\frac{E}{2} \leq \beta + \frac{2}{3}\alpha - E/3 = \frac{1}{3}\alpha + \frac{2}{3}\beta = \frac{E}{3} + \frac{1}{3}\beta \tag{2}$$

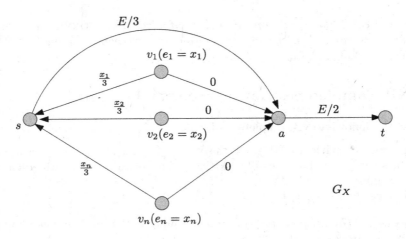

Fig. 3. Delivery from s to t is possible iff the set of weights x_i can be partitioned into two sub-sets of the same sum.

From (1) we have $\alpha \geq E/2$ and (2) leads to $\beta \geq E/2$, which implies $\alpha = \beta = E/2$.

Consequently, the delivery from s to t in graph G_X is possible if and only if the given instance of the integer partition problem is solvable. This implies NP-completeness of the delivery problem.

As t is the only node having paths incoming from all other nodes, the convergecast for G_X implies the delivery from s to t, hence the convergecast problem is also NP-complete.

Theorem 10. *The delivery and convergecast problems are NP-complete for general undirected graphs.*

Proof. Consider graph H_X - an undirected version of the graph from the previous proof (see Fig. 4). Increase the energy of every agent by E, i.e. agent i, initially placed at node v_i, now has energy $E + x_i$, for $i = 1, 2, \ldots, n$. Moreover increase by E the weight of each edge, which is incident to node v_i, i.e. $w(s, v_i) = E + x_i/3$ and $w(v_i, a) = E$, for $i = 1, 2, \ldots, n$.

Delivery. Consider delivery from s to t. Observe that no edge incident to v_i, for $i = 1, 2, \ldots, n$, can be used twice. Indeed, in order to transfer energies between agents they have to meet moving from their initial positions. However, at the moment of such meeting the sum of the remaining energies is smaller than E, which does not permit to traverse any edge incident to x_i for the second time. Clearly traversing directed edges $a \rightarrow s$ and $t \rightarrow a$ is also useless, hence the delivery from s to t in graph H_X is equivalent to the respective delivery in G_X.

Convergecast. If we consider t as the convergast node, the conergecast problem is equivalent to the delivery from s to t, which implies its NP-completeness.

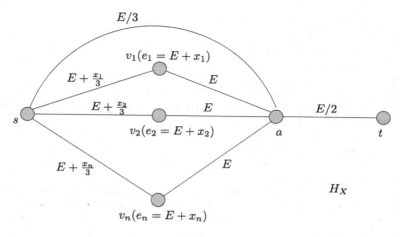

Fig. 4. The undirected version of the graph from Fig. 3. The weights of nodes v_i and lengths of edges incident to these nodes are increased by E.

5 Final Remarks

It is somewhat remarkable that, without energy exchange, even the simplest problem of data delivery is NP-complete in the simplest environment of the line, while, as we have shown in this paper, considered communication problems with energy exchange are solvable in linear time even for tree networks. On the other hand, it is not surprising that energy exchange in general graphs does not help and the problems are NP-complete.

There remain interesting open problems using energy-exchanging mobile agents for other communication protocols, like broadcast or gossiping. Observe that, in the case of data delivery and convergecast, each point of the tree is traversed at least once and at most twice (once in each direction) in the optimal solutions. However, in the case of broadcast, i.e. when the information of one node must be delivered to all other nodes of the tree, some tree edges need to be traversed by several agents. E.g., this is the case of weighted star with many agents starting at the same leaf. The problem of gossiping is even more involved.

An interested reader may try to extend the proposed solutions to the case when the data delivery needs to be performed from a given subset of nodes into another subset. Further possible extension is to realize one-to-one delivery using a configuration of energy-exchanging agents, i.e. when each of n source nodes must deliver to a specific target.

Acknowledgement. Research of the second, fourth and (in part) the first author is supported by the grant NCN2014/13/B/ST6/00770 of the Polish National Science Center. Research of the first and the third author is supported in part by NSERC Discovery grant.

References

1. Anaya, J., Chalopin, J., Czyzowicz, J., Labourel, A., Pelc, A., Vaxès, Y.: Collecting information by power-aware mobile agents. In: Aguilera, M.K. (ed.) DISC 2012. LNCS, vol. 7611, pp. 46–60. Springer, Heidelberg (2012)
2. Albers, S.: Energy-efficient algorithms. Comm. ACM **53**(5), 86–96 (2010)
3. Albers, S., Henzinger, M.R.: Exploring unknown environments. SIAM J. Comput. **29**(4), 1164–1188 (2000)
4. Annamalai, V., Gupta, S.K.S., Schwiebert, L.: On tree-based convergecasting in wireless sensor networks. IEEE Wirel. Commun. Netw. **3**, 1942–1947 (2003)
5. Augustine, J., Irani, S., Swamy, C.: Optimal powerdown strategies. SIAM J. Comput. **37**, 1499–1516 (2008)
6. Averbakh, I., Berman, O.: A heuristic with worst-case analysis for minimax routing of two traveling salesmen on a tree. Discr. Appl. Math. **68**, 17–32 (1996)
7. Azar, Y.: On-line load balancing. In: Fiat, A., Woeginger, G.J. (eds.) Online Algorithms. LNCS, vol. 1442, pp. 178–195. Springer, Heidelberg (1998). doi:10.1007/BFb0029569
8. Baeza-Yates, R.A., Schott, R.: Parallel searching in the plane. Comput. Geom. **5**, 143–154 (1995)
9. Chalopin, J., Jacob, R., Mihalák, M., Widmayer, P.: Data delivery by energy-constrained mobile agents on a line. In: Esparza, J., Fraigniaud, P., Husfeldt, T., Koutsoupias, E. (eds.) ICALP 2014, Part II. LNCS, vol. 8573, pp. 423–434. Springer, Heidelberg (2014)
10. Das, S., Dereniowski, D., Karousatou, C.: Collaborative exploration by energy-constrained mobile robots. In: Proceedings of SIROCCO, pp. 357–369 (2015)
11. Dynia, M., Korzeniowski, M., Schindelhauer, C.: Power-aware collective tree exploration. In: Grass, W., Sick, B., Waldschmidt, K. (eds.) ARCS 2006. LNCS, vol. 3894, pp. 341–351. Springer, Heidelberg (2006)
12. Fraigniaud, P., Gasieniec, L., Kowalski, D.R., Pelc, A.: Collective tree exploration. In: Farach-Colton, M. (ed.) LATIN 2004. LNCS, vol. 2976, pp. 141–151. Springer, Heidelberg (2004). doi:10.1007/978-3-540-24698-5_18
13. Frederickson, G., Hecht, M., Kim, C.: Approximation algorithms for some routing problems. SIAM J. Comput. **7**, 178–193 (1978)
14. Irani, S., Shukla, S.K., Gupta, R.: Algorithms for power savings. ACM Trans. Algorithms **3**(4), 41 (2007)
15. Kesselman, A., Kowalski, D.R.: Fast distributed algorithm for convergecast in ad hoc geometric radio networks. J. Parallel Distrib. Comput. **66**(4), 578–585 (2006)
16. Krishnamachari, L., Estrin, D., Wicker, S.: The impact of data aggregation in wireless sensor networks. In: ICDCS Workshops, pp. 575–578 (2002)
17. Rajagopalan, R., Varshney, P.K.: Data-aggregation techniques in sensor networks: a survey. Surv. Tutorials Comm. **8**(4), 48–63 (2006)
18. Yao, F.F., Demers, A.J., Shenker, S.: A scheduling model for reduced CPU energy. In Proceedings of 36th FOCS, pp. 374–382 (1995)

Data Dissemination and Routing

Asynchronous Broadcasting with Bivalent Beeps

Kokouvi Hounkanli and Andrzej Pelc[⊠]

Département d'informatique, Université du Québec en Outaouais,
Gatineau, Québec J8X 3X7, Canada
{houk06,pelc}@uqo.ca

Abstract. In broadcasting, one node of a network has a message that must be learned by all other nodes. We study deterministic algorithms for this fundamental communication task in a very weak model of wireless communication. The only signals sent by nodes are *beeps*. Moreover, they are delivered to neighbors of the beeping node in an asynchronous way: the time between sending and reception is finite but unpredictable. We first observe that under this scenario, no communication is possible, if beeps are all of the same strength. Hence we study broadcasting in the *bivalent beeping model*, where every beep can be either *soft* or *loud*. At the receiving end, if exactly one soft beep is received by a node in a round, it is heard as soft. Any other combination of beeps received in a round is heard as a loud beep. The cost of a broadcasting algorithm is the total number of beeps sent by all nodes.

We consider four levels of knowledge that nodes may have about the network: anonymity (no knowledge whatsoever), ad-hoc (all nodes have distinct labels and every node knows only its own label), neighborhood awareness (every node knows its label and labels of all neighbors), and full knowledge (every node knows the entire labeled map of the network and the identity of the source). We first show that in the anonymous case, broadcasting is impossible even for very simple networks. For each of the other three knowledge levels we provide upper and lower bounds on the minimum cost of a broadcasting algorithm. Our results show separations between all these scenarios. Perhaps surprisingly, the jump in broadcasting cost between the ad-hoc and neighborhood awareness levels is much larger than between the neighborhood awareness and full knowledge levels, although in the two former levels knowledge of nodes is local, and in the latter it is global.

Keywords: Algorithm · Asynchronous · Broadcasting · Deterministic · Graph · Network · Beep

1 Introduction

The background and the problem. Broadcasting is a fundamental communication task in networks. One node of a network, called the *source*, has a message

A. Pelc—Supported in part by NSERC discovery grant 8136 − 2013 and by the Research Chair in Distributed Computing of the Université du Québec en Outaouais.

© Springer International Publishing AG 2016
J. Suomela (Ed.): SIROCCO 2016, LNCS 9988, pp. 291–306, 2016.
DOI: 10.1007/978-3-319-48314-6_19

that must be learned by all other nodes. We study deterministic algorithms for this well-researched task in a very weak model of wireless communication. The only signals sent by nodes are *beeps*. Moreover, they are delivered to neighbors of the beeping node in an asynchronous way: the time between sending and reception is finite but unpredictable. Our aim is to study how the combination of two weaknesses of the communication model, very simple and short messages on the one hand, and the asynchronous way of delivery on the other hand, influences efficiency of communication. Each of these two model weaknesses separately has been studied before. Synchronous broadcasting and gossiping with beeps was studied in [7]. Asynchronous broadcasting in the radio model, where large messages can be sent in a round, was investigated in [3,4,20]. To the best of our knowledge, the combination of the two model weaknesses, i.e., very short messages and asynchronous delivery, has never been studied before.

We first observe that under this very harsh scenario, no communication is possible, if beeps are all of the same strength (see Sect. 2). Hence we study broadcasting in the *asynchronous bivalent beeping model*, where every beep can be either *soft* or *loud*, as this is, arguably, the weakest model under which asynchronous wireless broadcasting can be performed. At the receiving end, if exactly one soft beep is received by a node in a round, it is heard as soft. Any other combination of beeps received in a round is heard as a loud beep. The cost of a broadcasting algorithm is the total number of beeps sent by all nodes. This measures (the order of magnitude of) the energy consumption by the network, as the energy used to send a loud beep can be considered to be a constant multiple of that used to send a soft beep.

The model. Communication proceeds in rounds. In each round, a node can either listen, i.e., stay silent, or send a *soft beep*, or send a *loud beep*. For any beep sent by any node, an omniscient asynchronous adversary chooses a non-negative integer t, and delivers it to all neighbors of the sending node t rounds later. The delivery delay at all neighbors is the same for a given beep, but may be different for different beeps of the same node and for beeps of different nodes. The only rule that the adversary has to obey regarding delivery of different beeps sent by the same node, is that they must be delivered in the same order as they were sent, and cannot be collapsed in delivery, i.e. two beeps cannot be delivered as one beep. This type of asynchronous adversary was called the *node adversary* in [4] and the *strong adversary* in [3]. The motivation is similar as in [3,4]. Nodes execute the broadcasting protocol concurrently with other tasks. Beeps to be sent by a node are prepared for transmission (stored), and then each beep (soft or loud) is transmitted in order. The (unknown) delay between these actions is decided by the adversary. In our terminology, storing for transmission corresponds to sending and actual transmission corresponds to simultaneous delivery to all neighbors. We assume that, at short distances between nodes, the travel time of the beep is negligible. The delay between storing and transmitting (in our terminology, between sending and delivery) depends on how busy the node is with other concurrently performed computational tasks.

At the receiving end, a node can hear something only if it is silent in the delivery round. If exactly one soft beep is delivered to a node in a round, it is heard as soft. Any other combination of beeps delivered to a node from its neighbors in a round (a single strong beep, or more than one beep of any kind) is heard as a loud beep. This way of modeling reception corresponds to a threshold in the listening device: the strength of a single soft beep is below the threshold, and the strength of a loud beep, or the combined strength of more than one beep is above the threshold. The cost of a broadcasting algorithm is the total number of beeps sent by all nodes.

The network is modeled as an n-node simple connected undirected graph, referred to as *graph*. We use terms "network" and "graph" interchangeably. We consider four levels of knowledge that nodes may have about the network:

1. anonymous networks: nodes do not have any labels and know nothing about the network;

2. ad-hoc networks: all nodes have distinct labels and every node knows only its own label;

3. neighborhood-aware networks: all nodes have distinct labels, and every node knows its label and labels of all neighbors;

4. full-knowledge networks: all nodes have distinct labels, every node knows the entire labeled map of the network and the identity of the source.

The messages to be broadcast are from some set of size M, called the *message space*. Without loss of generality, let the message space be the set of integers $\{0, \ldots, M - 1\}$. Except for the anonymous networks, all nodes have different labels from the set of integers $\{0, \ldots, L - 1\}$, called the *label space*.

Our results. Our aim is to study how different levels of knowledge about the network influence feasibility and cost of broadcasting in the asynchronous bivalent beeping model. We first show that, in the anonymous case, broadcasting is impossible even for very simple networks. For each of the other three knowledge levels, broadcasting is feasible, and we provide upper and lower bounds on the minimum cost of a broadcasting algorithm, in terms of the sizes of the network, of the message space and of the label space. Showing an upper bound UB on the cost of broadcasting at a given knowledge level means showing an algorithm which accomplishes broadcasting at this cost, for any network with this knowledge level, and any message to be broadcast. Showing a lower bound LB means that, for any algorithm of lower cost, there is some network at this knowledge level, and some message for which the algorithm fails.

For ad-hoc networks we give an algorithm of cost $2^{O(L+M)^2}$[1]. Since this cost is very large, it is natural to ask if there are broadcasting algorithms of cost polynomial in L and M. The answer turns out to be negative: indeed, we prove a lower bound of $\Omega(2^L)$ on the cost of any broadcasting algorithm in ad-hoc networks. For neighborhood-aware networks we prove an upper bound of $O(n \log M + e \log L)$, where n is the number of nodes and e is the number of

---

[1] If one of the parameters, L or M, is known to the nodes, this complexity can be decreased to $2^{O(LM)}$ (see Sect. 4).

edges, and a lower bound of $\Omega(n \log M + n \log \log L)$. Finally, for full-knowledge networks, we provide matching upper and lower bounds of $\Theta(n \log M)$.

Note that the above bounds show separations, in terms of broadcasting cost, between all the knowledge levels, in the case often appearing in applications, when the message space is some predetermined dictionary independent of the network, i.e., its size M is $O(1)$. Indeed, since $L \geq n$, the lower bound $\Omega(2^L)$ for ad-hoc networks exceed the (worst-case) upper bound $O(n^2 \log L)$ for known-neighborhood networks, and the lower bound $\Omega(n \log \log L)$ for known-neighborhood networks exceed the tight bound $\Theta(n)$ for full-knowledge networks.

It is interesting to compare the sizes of the two broadcasting cost jumps: the jump between ad-hoc and known-neighborhood networks, and the jump between known-neighborhood and full-knowledge networks. We illustrate it for the commonly assumed case, when the size L of the label space is polynomial in the size n of the network (and the size of the message space is $O(1)$, as before). The first jump is at least from $\Omega(2^n)$ to $O(n^2 \log n)$, i.e., exponential in n. The second jump is at most from $O(n^2 \log n)$ to $\Theta(n)$, i.e., polynomial in n. This may seem slightly counterintuitive, because both in ad-hoc and in known-neighborhood networks, information available to nodes is local, while in full-knowledge networks it is global. So at first glance it would seem that the larger jump should occur between known-neighborhood and full-knowledge networks.

Related work. Broadcasting has been studied in various models for over four decades. Early work focused on the telephone model, where in each round communication proceeds between pairs of nodes forming a matching. Deterministic broadcasting in this model has been studied, e.g., in [21]. In [8] the authors studied randomized broadcasting. In the telephone model, studies focused on the time of the communication task and on the number of messages it uses. Early literature on communication in the telephone and related models is surveyed in [10,13]. In [2] the authors studied tradeoffs between the radius within which nodes know the network and broadcasting efficiency in the message passing model. Fault-tolerant aspects of broadcasting and gossiping are surveyed in [19].

More recently, broadcasting has been studied in the radio model. While radio networks are used to model wireless communication, similarly as the beeping model, in radio networks nodes send entire messages of some bounded, or even unbounded size in a single round, which makes communication drastically different from that in the beeping model. The focus in the literature on radio networks was usually on the time of communication. Deterministic broadcasting in the radio model was studied, e.g., in [5,16], and randomized broadcasting was studied in [17]. The book [15] is devoted to algorithmic aspects of communication in radio networks.

In all the above papers, radio communication was supposed synchronous, i.e., the message was delivered in the same round in which it was sent. Asynchronous broadcasting in radio networks was studied in [3,4,20]. It is important to stress a significant difference between the radio and the beeping models, in the context of asynchrony. Since in the radio model large messages can be sent and delivered in a single round, asynchrony cannot alter a message, it can only destroy it, by creating unwanted interference. In the beeping model, however, beeps from various senders can be simultaneously delivered by the adversary, thus altering the intended numbers and types of beeps, creating "new" messages.

The beeping model has been introduced in [6] for vertex coloring, and used in [1] to solve the MIS problem, and in [22] to construct a minimum connected dominating set. Randomized leader election in the radio and in the beeping model was studied in [11]. Deterministic leader election in the beeping model was investigated in [9]. In [14], the authors studied the tasks of global synchronization and consensus using beeps, in the presence of faults. In [12], the authors studied the quantity of computational resources needed to solve problems in complete networks using beeps. In [18], various distributed problems were investigated under several variations of the beeping model from [6], and randomized emulations between these models were shown. The time of synchronous broadcasting and gossiping with beeps was studied in [7].

2   Preliminaries

The following observation shows that asynchronous broadcasting with beeps of uniform strength is impossible even in very simple graphs. This is the reason why we use the bivalent beeping model.

Proposition 1. *Asynchronous broadcasting using beeps of uniform strength is impossible even in the two-node graph.*

In the rest of the paper we use the asynchronous bivalent beeping model, described in the introduction.

3   Anonymous Networks

In this section we show that, if nodes do not have labels, then broadcasting is impossible, even for very simple graphs, and even when nodes know the topology of the network.

Proposition 2. *Broadcasting for anonymous networks is impossible even in the cycle of size 4.*

Proof. Consider the anonymous cycle of size 4, and consider a hypothetical broadcasting algorithm A. For convenience, we label nodes a, b, c, d, in clockwise order. This is for the negative argument only: nodes do not have access to these labels. Suppose that node a is the source. Notice that, in any execution of algorithm A, nodes b and d send exactly the same beeps in the same rounds, as in each round they have the same history: indeed, they receive the same beeps in the same rounds, they are identical, and execute the same deterministic algorithm. Let m_1 and m_2 be two different messages that have to be broadcast by the source. Consider two executions of the algorithm A: execution E_1, in which the source broadcasts message m_1, and execution E_2, in which the source broadcasts message m_2. Let s_1 be the sequence of beeps (soft or loud) sent by b and d in execution E_1 and let s_2 be the sequence of beeps sent by b and d in execution E_2. Let k_1 be the length of s_1, and let k_2 be the length of s_2, where $k_1 \le k_2$, without loss of generality. In both executions, the adversary delivers consecutive beeps from b and from d in the same rounds. As a result, node c hears only loud beeps: k_1 of them in execution E_1, and k_2 of them in execution E_2. The choice of the rounds of delivery of bits from b and d is as follows. In execution E_1 these are consecutive rounds $r, r+1, \ldots, r+k_1-1$, starting from some round r. Suppose that s is the round in which node c correctly outputs message m_1. Then, in execution E_2, the adversary delivers the first k_1 beeps from b and d in rounds $r, r+1, \ldots, r+k_1-1$, and the remaining k_2-k_1 beeps in rounds $t+1, \ldots, t+k_2-k_1$, where $t = \max(s, r+k_1-1)$. In round s, node c has the same history in executions E_1 and E_2: it heard a loud beep in the same rounds, in both these executions. Hence, in execution E_2, it incorrectly outputs the message m_1 in round s.   □

4   Ad-hoc Networks

In this section we show that providing nodes with distinct labels makes broadcasting possible in arbitrary graphs, even if nodes do not have any initial knowledge about the network, except their own label. Let $\mathcal{N}$ denote the set of non-negative integers. Consider the function $\varphi : \mathcal{N} \times \mathcal{N} \longrightarrow \mathcal{N}$ given by the formula $\varphi(x, y) = x + (x+y)(x+y+1)/2$. This is a bijection with the property $\varphi(x, y) \in O((x+y)^2)$. Intuitively, this is the "snake function" arranging all couples of non-negative integers into one infinite sequence.

The following algorithm is executed by an active node with label ℓ. In the beginning, all nodes are active. The part *Receive* is executed by any node other than the source. Its result is outputting the source message. This part is skipped by the source, as it knows the message. The part *Send* is executed by the source at the beginning of the algorithm, and it is executed by every other node upon outputting the source message in the part *Receive*. After executing the part *Send*, the node becomes non-active.

Algorithm Ad-hoc

Part 1. *Receive*
Wait until the number of soft beeps received is at least $1/2$ of the number of loud beeps received.
Let t be the number of loud beeps received, and let z be the largest integer such that $8^z \le t$.
Compute the unique couple of non-negative integers (x, y), such that $\varphi(x, y) = z$.
Output y as the source message.

Part 2. *Send*
Compute $\varphi(\ell, y)$, where y is the source message.
Send $8^{\varphi(\ell, y)}$ loud beeps, followed by $8^{\varphi(\ell, y)}$ soft beeps.   ◇

The following result shows that Algorithm Ad-hoc is correct, and estimates its cost.

Theorem 1. *Upon completion of Algorithm* Ad-hoc *in an arbitrary graph, every node correctly outputs the source message. The cost of the algorithm is $2^{O((L+M)^2)}$.*

Proof. The proof of correctness is split into two parts. We first show that no node outputs the source message incorrectly, and then we prove that every node outputs the source message in finite time. Let m be the source message. The first part of the proof is by contradiction. Suppose that some node outputs the source message incorrectly, let r be the first round when this happens, and let u be a node with label ℓ, incorrectly outputting the source message in round r. Let $u_1, \ldots, u_k$ be the nodes adjacent to u whose at least one beep is delivered by round r, ordered in increasing order of their labels $\ell_1, \ldots, \ell_k$. Since u outputs the source message in round r, the set of nodes $\{u_1, \ldots, u_k\}$ is non-empty. Moreover, all nodes $u_1, \ldots, u_k$ must have outputted the source message before round r (because they already sent some beeps by round r), and hence they outputted it correctly. Let $1 \leq i \leq k$ be the largest integer j, such that at least one soft beep of node u_j was delivered by round r.

Suppose that t was the number of loud beeps heard by u by the round r. Since u outputted the source message incorrectly, the largest integer z', such that $8^{z'} \leq t$, cannot be equal to $z = \varphi(\ell_i, m)$. (If it were, node u would correctly compute the source message m because φ is a bijection.) The integer z' cannot be smaller than z because node u heard at least 8^z loud beeps sent by node u_i. Hence $z' \geq z + 1$. This implies that node u must have heard at least 8^{z+1} loud beeps by round r. How many soft beeps could it hear by round r? All these beeps could come only from nodes $u_1, \ldots, u_i$. The total number of soft beeps sent by these nodes is $\sum_{j=1}^{i} 8^{\varphi(\ell_j, m)}$. Since $\varphi(\ell_1, m) < \varphi(\ell_2, m) < \cdots < \varphi(\ell_i, m)$, we have $\sum_{j=1}^{i} 8^{\varphi(\ell_j, m)} < \frac{8}{7} \cdot 8^{\varphi(\ell_i, m)} = \frac{8}{7} \cdot 8^z$. On the other hand, the number of soft beeps heard by node u by round r must be at least $1/2$ of the number of loud beeps it heard by round r. This implies $\frac{8}{7} \cdot 8^z \geq \frac{1}{2} \cdot 8^{z+1}$, which is a contradiction. This completes the first part of the proof.

We now prove that every node outputs the source message in finite time. This part of the proof is also by contradiction. Suppose that some node never outputs the source message. Since the source itself knows the source message, and the graph is connected, there must exist adjacent nodes u and v, such that u outputs the source message in finite time and v does not. Let $v_1, \ldots, v_s$ be the nodes adjacent to v that ever send at least one beep, ordered in increasing order of their labels $\lambda_1, \ldots, \lambda_s$. The set of nodes $\{v_1, \ldots, v_s\}$ is non-empty. We show that, at some point, the number of soft beeps heard by v is at least $1/2$ of the number of loud beeps heard by v. Indeed, assume that this did not happen before all beeps of all nodes $v_1, \ldots, v_s$ are delivered. The number of all beeps sent by nodes $v_1, \ldots, v_{s-1}$ is $2 \cdot \sum_{j=1}^{s-1} 8^{\varphi(\lambda_j, m)}$. Since $\varphi(\lambda_1, m) < \varphi(\lambda_2, m) < \cdots < \varphi(\lambda_s, m)$, we have $2 \cdot \sum_{j=1}^{s-1} 8^{\varphi(\lambda_j, m)} < \frac{2}{7} \cdot 8^{\varphi(\lambda_s, m)}$. In the worst case, these beeps can be delivered by the adversary simultaneously with the same number (fewer than

$\frac{2}{7} \cdot 8^{\varphi(\lambda_s, m)}$) of soft beeps sent by node v_s, thus producing loud beeps heard by node v. This would decrease the number of soft beeps heard by v and increase the number of loud beeps heard by this node, but the change cannot be too big. Indeed, this gives fewer than $\frac{9}{7} \cdot 8^{\varphi(\lambda_s, m)}$ loud beeps heard by v. On the other hand, node v hears at least $\frac{5}{7} \cdot 4^{\varphi(\lambda_s, m)}$ soft beeps sent by v_s and left intact (not delivered simultaneously with other beeps) by the adversary. Hence the number of soft beeps heard by node v is at least $1/2$ of the number of loud beeps heard by it. It follows that node v outputs the source message, contrary to our assumption.

This completes the proof of correctness of Algorithm Ad-hoc. We now estimate its cost. A node with label ℓ sends $2 \cdot 8^{\varphi(\ell, m)}$ beeps, where m is the source message. Hence the cost of Algorithm Ad-hoc in an n-node network is at most $2n \cdot 8^{\varphi(L, M)}$. Since $\varphi(L, M) \in O((L + M)^2)$, and $\varphi(L, M) \geq L \geq n$, this gives the cost $2^{O((L+M)^2)}$. $\qquad\square$

Remark. Notice that, if nodes know one of the parameters, either L or M, then the bijection φ can be replaced by a more efficient one-to-one function from the product $\{0, \ldots, L-1\} \times \{0, \ldots, M-1\}$ to non-negative integers. For example, if L is known, then this function can be $\psi(\ell, m) = mL + \ell$, and if M is known, then this function can be $\psi'(\ell, m) = \ell M + m$. The values of these functions are in $O(LM)$, and hence, if we substitute one of them for φ, the cost of the algorithm becomes $2^{O(LM)}$.

As we have seen above, the cost of Algorithm Ad-hoc is very large: even with knowledge of L or M, it is exponential in the product of these parameters. Hence, it is natural to ask if there are broadcasting algorithms, for ad-hoc networks, with cost polynomial in L and M. Our next result shows that the answer is negative. Before proving it we recall a notion and a fact from [3].

A set S of positive integers is *dominated* if, for any finite subset T of S, there exists $t \in T$ such that t is larger than the sum of all $t' \neq t$ in T.

Lemma 1. *Let S be a finite dominated set and let k be its size. Then there exists $x \in S$ such that $x \geq 2^{k-1}$.*

Theorem 2. *For arbitrary integers $L \geq 4$, there exist L-node ad-hoc networks, for which the cost of every broadcasting algorithm is $\Omega(2^L)$.*

Proof. Let A be any broadcasting algorithm. For any set $S \subseteq \{1, \ldots, L-2\}$, of size at least 2, the graph G_S is defined as follows. G_S has $|S| + 2$ nodes with labels from the set $S \cup \{0, L-1\}$. Each of the nodes with labels in S is adjacent to each of the nodes with labels 0 and $L-1$, and there are no other edges in the graph. The node with label 0 is the source, and the node with label $L-1$ is called the *sink*.

We will consider executions of algorithm A in graphs G_S, in which the adversary obeys the following rules concerning the delivery of beeps sent by the source and the sink:

1. All beeps sent by the source after it heard some beep, are delivered after the round when the sink outputs the source message.

2. All beeps sent by the sink are delivered after the round in which the sink outputs the source message.

Since the considered networks are ad-hoc, i.e., a priori, every node knows only its own label, and the adversary obeys the above rules, the number of beeps sent by a node with a given label $\ell \in \{1, \ldots, L-2\}$ by the round in which the sink outputs the source message, depends only on this label and on the source message, and not on the graph G_S in which the algorithm is executed. Indeed, the history of a node with label $\ell \in \{1, \ldots, L-2\}$, by the round in which the sink outputs the source message, is the same in all graphs G_S, for a given source message m.

Consider the execution of algorithm A in the graph $G_{\{1,\ldots,L-2\}}$, for a fixed source message m. Let $B(\ell)$, for $\ell \in \{1, \ldots, L-2\}$, be the number of beeps of both kinds, that the node with label ℓ sends by the round in which the sink outputs the source message. If the set of integers $I = \{B(\ell) : \ell \in \{1, \ldots, L-2\}\}$ is dominated, then by Lemma 1, some integer in this set is at least 2^{L-3}, and we are done. Otherwise, there exists a subset $T \subseteq \{1, \ldots, L-2\}$, with the following property. If $t \in T$ is such that $B(t) \geq B(t')$, for all $t' \in T \setminus \{t\}$, then $B(t) \leq \sum_{t' \in T \setminus \{t\}} B(t')$. Consider the execution E of algorithm A in the graph G_T, for the same source message m. As observed above, the number of beeps of both kinds, that the node with label ℓ sends in this execution by the round in which the sink outputs the source message, is $B(\ell)$. The adversary delivers beeps sent by nodes with labels from T, in consecutive rounds, delivering simultaneously a beep sent by the node with label t with one or more beeps sent by nodes with labels $t' \in T \setminus \{t\}$, in such a way that in no round a single beep is delivered. This is possible due to the inequality $B(t) \leq \sum_{t' \in T \setminus \{t\}} B(t')$. Hence the sink hears only loud beeps.

Now, consider a different source message m'. The same argument as above shows that, if the cost of the algorithm A on the graph $G_{\{1,\ldots,L-2\}}$ is smaller than 2^{L-3}, then there exists some set $T' \subseteq \{1, \ldots, L-2\}$, such that, in the execution E' of the algorithm A on the graph $G_{T'}$, with the source message m', the sink hears only loud beeps.

Suppose that, by the time it outputs the source message, the sink hears k loud beeps in the execution E and hears k' loud beeps in the execution E'. Without loss of generality, assume that $k \leq k'$. The choice of rounds of delivery of these beeps by the adversary is the following.

In execution E, these are consecutive rounds $r, r+1, \ldots, r+k-1$, starting from some round r. Suppose that s is the round in which the sink correctly outputs message m. Then, in execution E', the adversary first delivers beeps in rounds $r, r+1, \ldots, r+k-1$, and the remaining $k'-k$ rounds of beep delivery are $z+1, \ldots, z+k'-k$, where $z = \max(s, , r+k-1)$. In round s, the sink has the same history in executions E and E': it heard only loud beeps, and this happened in the same rounds, in both these executions. Hence, in execution E', it incorrectly outputs the message m in round s.

The obtained contradiction comes from assuming that the cost of algorithm A on the graph $G_{\{1,\ldots,L-2\}}$ is smaller than 2^{L-3}, for all source messages. This completes the proof. $\square$

5   Neighborhood-Aware Networks

In this section we assume that all nodes have distinct labels, and that each of them knows its own label and the labels of all its neighbors. This seemingly small increase of knowledge, compared to ad-hoc networks (the knowledge of every node is still local) turns out to decrease the cost of broadcasting in a dramatic way. In order to guarantee a low cost of broadcasting, we have to encode messages by sequences of beeps very efficiently. The algorithm uses messages of two types: non-negative integers and triples of non-negative integers. These messages have to be encoded by strings of beeps of length logarithmic in the values of these integers, in such a way that the recipient knows when the string starts and ends, and can unambiguously decode the message from the string. However, as opposed to Algorithm Ad-hoc in which nodes sent exponentially many beeps, such efficient encoding is very vulnerable to possible actions of the adversary that can arbitrarily interleave delivered beeps coming from different neighbors of a node. In order to avoid this, we design our algorithm in such a way that beeps encoding a message sent by some node are delivered before any other node starts sending its own beeps. In this way, the danger of interleaving beeps is avoided.

Before presenting the algorithm, we define the encoding of integers and of their triples, announced above. We denote a loud beep by l, a soft beep by s, and we use the symbol $\cdot$ for the concatenation of sequences of beeps. Let k be a non-negative integer, and let $(c_1, \ldots, c_r)$ be its binary representation. Denote by $S(k)$ the sequence of $2r$ beeps resulting from $(c_1, \ldots, c_r)$ by replacing every bit $c_i = 0$ by (ls) and by replacing every bit $c_j = 1$ by (sl). The code of an integer k, denoted by $[k]$, is the sequence $(ll) \cdot S(k) \cdot (ll)$. The code of a triple (a, b, c) of integers, denoted by $[a, b, c]$, is the sequence $(ll) \cdot S(a) \cdot (ss) \cdot S(b) \cdot (ss) \cdot S(c) \cdot (ll)$. Note that a sequence of 2 loud beeps marks the beginning and end of a message, and all messages contain an even number of beeps, logarithmic in the integers transmitted. A node at the receiving end can determine the beginning of the message as a sequence σ of 2 consecutive loud beeps, and the end of the message as the first sequence σ' of 2 consecutive loud beeps starting after the end of σ at an odd position, where the first bit of the sequence σ is at position 1. In order to decode the content of the message $(ll) \cdot \alpha \cdot (ll)$, with the beginning and end already correctly identified, a node looks for separators (ss) starting at odd positions of α. There are either 0 or 2 such separators. In the first case, the transmitted message was an integer, and the node decodes its binary representation by replacing each couple (ls) by 0 and each couple (sl) by 1. In the second case, the node can unambiguously represent α as $\alpha_1 \cdot (ss) \cdot \alpha_2 \cdot (ss) \cdot \alpha_3$, where each α_i has even length, and decode $\alpha_1, \alpha_2, \alpha_3$ as above.

Using the above encoding, we are now able to describe our broadcasting algorithm. At a high level, it is organized as a depth-first traversal of the graph, starting from the source. We will use the instructions "send $[a]$" and "send $[a, b, c]$" that are procedures sending the above described sequences of beeps, in consecutive rounds. A message $[a]$, where $a \in \{0, 1, \ldots M - 1\}$, is always the source message to be broadcast. There are two kinds of messages of type "triple of integers": For $a, b \in \{0, 1, \ldots L-1\}$, a message of the form $[a, b, 0]$ corresponds to a forward DFS edge traversal from the node with label a to a node with label b, and a message of the form $[a, b, 1]$ corresponds to a backward DFS edge traversal from the node with label a to a node with label b.

The algorithm is executed by a node with label ℓ. The actions of the node alternate between executing "send" instructions and listening. The algorithm is organized in such a way that the following *disjointness property* is satisfied. Consider a node u executing some send instruction $I(u)$. Let $\sigma(u)$ be the segment of consecutive rounds between the sending of the first beep of instruction $I(u)$ and the delivery of the last bit of this instruction. Then, for any two nodes u and v, executing any send instructions $I(u)$ and $I(v)$, the segments of rounds $\sigma(u)$ and $\sigma(v)$ are disjoint. This property permits to identify circulating messages as distinct "packets", and use them to implement a DFS traversal.

When the node listens, it watches for the beginning and end of a message formed by the delivered beeps. When it detects a complete message, it reacts to it in one of two ways: it either keeps listening and watches for another complete message, or it reacts by executing some "send" instruction. More specifically, the actions of the node with label ℓ, other than the source, are as follows. After getting the source message and the first forward DFS message $[a, \ell, 0]$, addressed to it and coming from a node with label a, the node with label ℓ starts spreading the message to all its neighbors with labels a_i, except that with label a, by sending the decoded source message $[m]$ and sending forward DFS messages $[\ell, a_i, 0]$ addressed to them, in increasing order of labels. In order to transit from one neighbor to the next, the node ℓ waits for a backward message $[a_i, \ell, 1]$, addressed to it. In the meantime, node ℓ refuses all subsequent forward DFS messages $[b, \ell, 0]$, for $b \neq a$, addressed to it, responding by a backward DFS message $[\ell, b, 1]$. The actions of the source are similar.

The pseudocode of the algorithm follows.

Algorithm Neighborhood-aware

if the executing node is the source, and the source message is m **then**
    $message \leftarrow m$
    let $(a_1, a_2, \ldots, a_s)$ be labels of all the neighbors of the node,
    in increasing order
    Spread$(a_1, \ldots, a_s)$
    whenever a message $[b, \ell, 0]$, for some integer b, is decoded then
        send $[\ell, b, 1]$
else
    when a message $[m]$ is decoded for the first time, then
        $message \leftarrow m$
        output $message$ as the source message
    when a message $[a, \ell, 0]$ is decoded for the first time, then
        let $(a_1, a_2, \ldots, a_s)$ be labels of all the neighbors of the node,
        except a, in increasing order
        Spread$(a_1, \ldots, a_s)$
        send $[\ell, a, 1]$

whenever a message $[b, \ell, 0]$, for some $b \neq a$, is decoded then
$\qquad$ send $[\ell, b, 1]$ $\qquad\qquad$ ◇

The procedure Spread, used by the algorithm and executed by a node with label ℓ, is described as follows.

Procedure Spread($a_1, \ldots, a_s$)

send [*message*]
$i \leftarrow 1$
while $i \leq s$ **do**
$\qquad$ send $[\ell, a_i, 0]$
$\qquad$ when the message $[a_i, \ell, 1]$ is decoded then
$\qquad\qquad$ $i \leftarrow i + 1$ $\qquad\qquad$ ◇

Theorem 3. *Upon completion of Algorithm* Neighborhood-aware *in an arbitrary n-node graph with e edges, every node correctly decodes the source message. The cost of the algorithm is $O(n \log M + e \log L)$.*

Proof. In view of the disjointness property, all messages are correctly decoded by their addressees. Since the control messages $[a, b, 0]$ and $[a, b, 1]$ travel in a DFS fashion, and each message $[a, b, 0]$ is preceded by the source message $[m]$, all nodes get the source message and decode it correctly. This proves the correctness of the algorithm. To estimate its cost, note that each node sends the source message $[m]$ once, and, for any pair of adjacent nodes a and b, two control messages among $[a, b, 0]$, $[a, b, 1]$, $[b, a, 0]$, $[b, a, 1]$ are sent. Since the source message $[m]$ consists of $O(\log M)$ beeps, and each control message consists of $O(\log L)$ beeps, the total cost of the algorithm is $O(n \log M + e \log L)$. $\qquad\square$

In order to prove our lower bound on the cost of broadcasting algorithms in neighborhood-aware networks, we need the following two lemmas.

Lemma 2. *Every broadcasting algorithm has cost $\Omega(\log M)$ in the two-node graph.*

Lemma 3. *Every broadcasting algorithm has cost $\Omega(\log \log L)$ in some cycle of size 4.*

Proof. Consider a broadcasting algorithm A working for all neighborhood-aware cycles of size 4. Suppose that the cost of algorithm A in all such cycles is at most $\frac{1}{2} \log \log L$. Consider a cycle of size 4, and call its nodes a, b, c, d, in clockwise order. Suppose that node a is the source. Let 0 be the label of node a, and let $L-1$ be the label of node c. The adversary delivers all beeps possibly sent by node c, only after this node outputs the source message. Hence, before the decision by node c, nodes b and d hear only beeps from the source a. The adversary delivers all beeps sent by node a in consecutive rounds. Since node a can send at most $\frac{1}{2} \log \log L$ beeps, the set X of possible sequences of beeps heard by nodes b and d has size at most $\sqrt{\log L}$. Let $N = \{0, 1, \ldots, \lfloor \frac{1}{2} \log \log L \rfloor\}$. Since each of the nodes b and d can send at most $\frac{1}{2} \log \log L$ beeps, the number of beeps sent by each of these nodes must be an integer from the set N. For any label $\ell \in \{1, \ldots, L-2\}$, let $\Phi_\ell : X \longrightarrow N$ be the function defined as follows: $\Phi_\ell(x)$ is the number of beeps sent by the node b or d, if it has label ℓ, and if it obtained the sequence x of beeps. There are $|N|^{|X|} < L - 2$ such functions, for sufficiently large L. Hence there exist labels $\ell_1 \neq \ell_2$ from the set $\{1, \ldots, L-2\}$, for which $\Phi_{\ell_1} = \Phi_{\ell_2}$. Assign these labels to the nodes b and d. In the obtained cycle C, nodes b and d send the same number of beeps, regardless of the sequence of beeps obtained from a. In particular, this will happen in two executions, E_1 and E_2, of algorithm A on the cycle C, where execution E_1 corresponds to source message m_1, and execution E_2 corresponds to source message m_2, for $m_1 \neq m_2$.

In both executions, the adversary delivers consecutive beeps from b and from d in the same rounds. As a result, node c hears only loud beeps: k_1 of them in execution E_1, and k_2 of them in execution E_2. Without loss of generality, suppose that $k_1 \leq k_2$. The choice of the rounds of delivery of bits from b and d is as follows. In execution E_1 these are consecutive rounds $r, r+1, \ldots, r+k_1-1$, starting from some round r. Suppose that s is the round in which node c correctly outputs message m_1. Then, in execution E_2, the adversary delivers the first k_1 beeps from b and d in rounds $r, r+1, \ldots, r+k_1-1$, and the remaining $k_2 - k_1$ beeps in rounds $t+1, \ldots, t+k_2-k_1$, where $t = \max(s, r+k_1-1)$. In round s, node c has the same history in executions E_1 and E_2: it heard a loud beep in the same rounds, in both these executions. Hence, in execution E_2, it incorrectly outputs the message m_1. □

The following result gives a lower bound on the cost of any broadcasting algorithm in neighborhood-aware networks.

Theorem 4. *For arbitrarily large integers n, there exist n-node neighborhood-aware networks for which every broadcasting algorithm has cost $\Omega(n \log M + n \log \log L)$.*

Proof. For any positive integer k, consider the graph G_k defined as follows. Let P_k be a simple path of length k, with extremities a and b. Consider pairwise disjoint copies $C_1, \ldots, C_k$ of the cycle of size 4, whose all nodes are distinct from nodes of the path. Let a_i, b_i, c_i, d_i be the nodes of the ith copy in clockwise order. Join the node a_1 to the node b by an edge, and for every $1 \leq i < k$, join the node c_i to the node a_{i+1} by an edge. The obtained graph has $n \in \Theta(k)$ nodes. We now assign the labels to nodes of G_k as follows. Nodes b_i and d_i in cycles C_i, for $i = 1, \ldots, k$, are assigned distinct labels by induction. For any i, we consider the set of all labels that were not used previously and find among them two labels $\ell_1 \neq \ell_2$ for which $\Phi_{\ell_1} = \Phi_{\ell_2}$, where Φ_ℓ, for any label ℓ, was defined in the proof of Lemma 3. This can be done similarly as in the quoted proof, because the number of still available labels is $\Theta(L)$. Finally, nodes of the path and all nodes a_i and c_i are assigned consecutive distinct labels among the remaining labels.

Let the node a be the source, and consider any broadcasting algorithm on graph G_k. By Lemma 2, each node of the path, other than b, has to transmit $\Omega(\log M)$ beeps, for otherwise the next node cannot get the message. By Lemma 3, the total cost of the algorithm in each subgraph C_i, for $i < k$, must be $\Omega(\log \log L)$, for otherwise the nodes of the next copy cannot get the message. (Note that edges of the path P_k and edges joining consecutive copies of the cycle, are bridges in G_k.) Hence the total cost of the algorithm is $\Omega(k \log M + k \log \log L) = \Omega(n \log M + n \log \log L)$. $\square$

6   Full-Knowledge Networks

In this section we consider broadcasting in networks whose nodes have the entire labeled map of the network, and know the identity of the source. With this complete knowledge, all nodes can agree on the same spanning tree T of the network, rooted at the source. (All trees rooted at the source can be canonically coded by binary strings, and the tree T can be chosen as that with lexicographically smallest code.) Let S be a DFS traversal of the tree T in which children of every node are explored in increasing order of their labels. The Eulerian tour of the tree T corresponding to this traversal can be represented as a sequence $(a_1, \ldots, a_{2(n-1)})$ of length $2(n-1)$ of node labels with repetitions, where a_i corresponds to the ith edge traversal in the tour, from the node with label a_i to the node with label a_{i+1}.

The only message circulating in the network is the message $[m]$, where m is the source message, and $[m]$ is the encoding of this integer, described in Sect. 5. The instruction send $[m]$ is the procedure of sending beeps of the encoding $[m]$ in consecutive rounds. Similarly as in Algorithm Neighborhood-aware, the disjointness property is satisfied, and hence each message can be correctly decoded by adjacent nodes. The idea of the algorithm is the following. Every node knows to which terms of the sequence $(a_1, \ldots, a_{2(n-1)})$ its label corresponds. It sends the message $[m]$ when the turn of such a term of the sequence comes. (Many nodes send messages many times.) In order to know when this happens, the node computes how many previous messages it should get before from all adjacent nodes, and when this number of messages is received, it proceeds with the execution of the send $[m]$ instruction corresponding to the given term of the sequence.

The algorithm is executed by a node with label ℓ, when the source message is m. The pseudocode of the algorithm follows.

Algorithm Full-knowledge

if the executing node is not the source **then**
       when a message $[m]$ is decoded for the first time, then
          output *message* as the source message
identify all positions of label ℓ in the sequence $(a_1, \ldots, a_{2(n-1)})$
let $i_1, \ldots, i_r$ be these positions
let x_1 be the number of indices $1 \le j < a_1$, corresponding to labels a_j of nodes adjacent to the node with label ℓ
for $1 < i \le r$, let x_i be the number of indices $a_{i-1} < j < a_i$, corresponding to labels a_j of nodes adjacent to the node with label ℓ

for $1 < i \leq r$, let $y_i = \sum_{t=1}^{i} x_t$
for $k = 1$ to r do
　　when a total of y_k messages $[m]$ is received then send $[m]$　　◇

Theorem 5. *Upon completion of Algorithm* `Full-knowledge` *in an arbitrary n-node graph, every node correctly outputs the source message. The cost of the algorithm is $O(n \log M)$.*

Proof. The correctness of the algorithm follows from the fact that nodes send messages whenever their turn comes in the sequence $(a_1, \ldots, a_{2(n-1)})$ that corresponds to an Eulerian tour of a spanning tree T, and from the disjointness property guaranteeing that the source message is always correctly decoded. The total number of messages sent is $2(n-1)$. Since each message corresponds to $O(\log M)$ bits, the total cost of the algorithm is $O(n \log M)$.　□

The following proposition shows that the cost of Algorithm `Full-knowledge` is optimal in full-knowledge networks.

Proposition 3. *For arbitrary integers $n \geq 2$ there exist n-node graphs for which the cost of any broadcasting algorithm is $\Omega(n \log M)$.*

Proof. Consider the simple path P_n with n nodes, one of whose extremities is the source. Note that Lemma 2 holds for full-knowledge networks as well. By Lemma 2, each node of the path, other than the last node, has to transmit $\Omega(\log M)$ beeps, for otherwise the next node cannot get the message correctly. Hence the cost of any broadcasting algorithm is $\Omega(n \log M)$.　□

7　Conclusion

We considered the cost of asynchronous broadcasting in networks with four different levels of knowledge: anonymous, ad-hoc, neighborhood-aware, and full-knowledge. We proved that broadcasting in anonymous networks is impossible, and we showed upper and lower bounds on the cost of broadcasting for the other three levels of knowledge. Our results show cost separations between all of them. While the bounds for full-knowledge networks are asymptotically tight, the other bounds are not, and designing optimal-cost broadcasting algorithms for ad-hoc and for neighborhood-aware networks is a natural open problem.

References

1. Afek, Y., Alon, N., Bar-Joseph, Z., Cornejo, A., Haeupler, B., Kuhn, F.: Beeping a maximal independent set. In: Peleg, D. (ed.) Distributed Computing. LNCS, vol. 6950, pp. 32–50. Springer, Heidelberg (2011)
2. Awerbuch, B., Goldreich, O., Peleg, D., Vainish, R.: A trade-off between information and communication in broadcast protocols. J. ACM **37**, 238–256 (1990)

3. Calamoneri, T., Fusco, E.G., Pelc, A.: Impact of information on the complexity of asynchronous radio broadcasting. In: Baker, T.P., Bui, A., Tixeuil, S. (eds.) OPODIS 2008. LNCS, vol. 5401, pp. 311–330. Springer, Heidelberg (2008)
4. Chlebus, B.S., Rokicki, M.A.: Centralized asynchronous broadcast in radio networks. Theor. Comput. Sci. **383**, 5–22 (2007)
5. Chrobak, M., Gasieniec, L., Rytter, W.: Fast broadcasting and gossiping in radio networks. J. Alg. **43**, 177–189 (2002)
6. Cornejo, A., Kuhn, F.: Deploying wireless networks with beeps. In: Lynch, N.A., Shvartsman, A.A. (eds.) DISC 2010. LNCS, vol. 6343, pp. 148–162. Springer, Heidelberg (2010)
7. Czumaj, A., Davis, P.: Communicating with beeps [cs.DC] (2015). arXiv:1505.06107
8. Feige, U., Peleg, D., Raghavan, P., Upfal, E.: Randomized broadcast in networks. Random Struct. Alg. **1**, 447–460 (1990)
9. Förster, K.-T., Seidel, J., Wattenhofer, R.: Deterministic leader election in multi-hop beeping networks. In: Kuhn, F. (ed.) DISC 2014. LNCS, vol. 8784, pp. 212–226. Springer, Heidelberg (2014)
10. Fraigniaud, P., Lazard, E.: Methods and problems of communication in usual networks. Disc. Appl. Math. **53**, 79–133 (1994)
11. Ghaffari, M., Haeupler, B.: Near optimal leader election in multi-hop radio networks. In: Proceedings of 24th ACM-SIAM Symposium on Discrete Algorithms (SODA 2013), pp. 748–766
12. Gilbert, S., Newport, C.: The computational power of beeps. In: Moses, Y., et al. (eds.) DISC 2015. LNCS, vol. 9363, pp. 31–46. Springer, Heidelberg (2015). doi:10.1007/978-3-662-48653-5_3
13. Hedetniemi, S.M., Hedetniemi, S.T., Liestman, A.L.: A survey of gossiping and broadcasting in communication networks. Networks **18**, 319–349 (1988)
14. Hounkanli, K., Miller, A., Pelc, A.: Global Synchronization and Consensus Using Beeps in a Fault-Prone MAC. CoRR abs/1508.06583 (2015)
15. Kowalski, D.: Algorithmic Foundations of Communication in Radio Networks. Morgan & Claypool Publishers, California (2011)
16. Kowalski, D., Pelc, A.: Time of deterministic broadcasting in radio networks with local knowledge. SIAM J. Comput. **33**, 870–891 (2004)
17. Kushilevitz, E., Mansour, Y.: An Omega(D log (N/D)) lower bound for broadcast in radio networks. SIAM J. Comput. **27**, 702–712 (1998)
18. Métivier, Y., Robson, J.M., Zemmari, A.: On distributed computing with beeps, CoRR abs/1507.02721 (2015)
19. Pelc, A.: Fault-tolerant broadcasting and gossiping in communication networks. Networks **28**, 143–156 (1996)
20. Pelc, A.: Activating anonymous ad hoc radio networks. Distrib. Comput. **19**, 361–371 (2007)
21. Slater, P.J., Cockayne, E.J., Hedetniemi, S.T.: Information dissemination in trees. SIAM J. Comput. **10**, 692–701 (1981)
22. Yu, J., Jia, L., Yu, D., Li, G., Cheng, X.: Minimum connected dominating set construction in wireless networks under the beeping model. In: Proceedings of IEEE Conference on Computer Communications, (INFOCOM 2015), pp. 972–980 (2015)

Sparsifying Congested Cliques and Core-Periphery Networks

Alkida Balliu[1,2], Pierre Fraigniaud[1(✉)], Zvi Lotker[3], and Dennis Olivetti[1,2]

[1] CNRS, University Paris Diderot, Paris, France
Pierre.Fraigniaud@liafa.univ-paris-diderot.fr
[2] Gran Sasso Science Institute, L'Aquila, Italy
[3] Ben Gurion University, Beersheba, Israel

Abstract. The *core-periphery* network architecture proposed by Avin et al. [ICALP 2014] was shown to support fast computation for many distributed algorithms, while being much sparser than the *congested clique*. For being efficient, the core-periphery architecture is however bounded to satisfy three axioms, among which is the capability of the core to emulate the clique, i.e., to implement the all-to-all communication pattern, in $O(1)$ rounds in the CONGEST model. In this paper, we show that implementing all-to-all communication in k rounds can be done in n-node networks with roughly n^2/k edges, and this bound is tight. Hence, sparsifying the core beyond just saving a fraction of the edges requires to relax the constraint on the time to simulate the congested clique. We show that, for $p \gg \sqrt{\log n/n}$, a random graph in $\mathcal{G}_{n,p}$ can, w.h.p., perform the all-to-all communication pattern in $O(\min\{\frac{1}{p^2}, np\})$ rounds. Finally, we show that if the core can emulate the congested clique in t rounds, then there exists a distributed MST construction algorithm performing in $O(t \log n)$ rounds. Hence, for $t = O(1)$, our (deterministic) algorithm improves the best known (randomized) algorithm for constructing MST in core-periphery networks by a factor $\Theta(\log n)$.

1 Introduction

1.1 Context and Objectives

Inspired by social networks and complex systems, Avin, Borokhovicha, Lotker, and Peleg [1] proposed a novel network architecture for parallel and distributed computing, called *core-periphery*. Interestingly, the core-periphery architecture is not described explicitly (e.g., via the description of a specific graph family), but rather implicitly via three so-called *axioms*. Specifically, a core-periphery network $G = (V, E)$ has its node set partitioned into a *core* C and a *periphery* P, and the three properties to be satisfied are then the following:

P. Fraigniaud—Additional supports from ANR project DISPLEXITY, and Inria project GANG.
Z. Lotker—Additional supports from Foundation des Sciences Mathématiques de Paris (FSMP).

© Springer International Publishing AG 2016
J. Suomela (Ed.): SIROCCO 2016, LNCS 9988, pp. 307–322, 2016.
DOI: 10.1007/978-3-319-48314-6_20

1. **Core boundary:** For every node $v \in C$, $\deg_C(v) \simeq \deg_P(v)$, where, for $S \subseteq V$ and $v \in V$, $\deg_S(v)$ denotes the number of neighbors of v in S.
2. **Clique emulation:** The core can emulate the clique in a constant number of rounds in the CONGEST model. That is, there is a communication protocol running in a constant number of rounds in the CONGEST model such that, assuming that each node $v \in C$ has a message $M_{v,w}$ on $O(\log n)$ bits for every $w \in C$, then, after $O(1)$ rounds, every $w \in C$ has received all messages $M_{v,w}$, for all $v \in C$. In other words, the *all-to-all* communication pattern can be implemented in a constant number of rounds.
3. **Periphery-core convergecast:** There is a communication protocol running in a constant number of rounds in the CONGEST model such that, assuming that each node $v \in P$ has a message M_v on $O(\log n)$ bits, then, after $O(1)$ rounds, for every $v \in P$, at least one node in the core has received M_v.

Figure 1 provides an example of a core-periphery network, i.e., a graph satisfying the three axioms. It was proved in [1] that these three axioms alone enable to design efficient distributed algorithms in the CONGEST model for classical problems such as matrix multiplication and MST construction. Most of the proposed algorithms are optimal in a sense that there is an asymptotically matching lower bound on the number of rounds under the three axiomatic constraints. Moreover, it is shown that if only two out of three axioms were satisfied, then the round complexity of all the considered problems would increase quite significantly—typically, from $O(1)$ to $O(\text{poly}(n))$ in n-node networks. There was an exception though: while the best known lower bound in [1] for MST construction is $\Omega(1)$, the proposed (randomized) MST construction algorithm runs in $O(\log^2 n)$ rounds. (If only two out of three axioms were satisfied, then MST construction would require at least $\tilde{\Omega}(n^{\frac{1}{4}})$ rounds).

The core-periphery model provides an attractive alternative to the *congested clique* model [19]. Indeed, the latter assumes a complete network interconnecting the nodes, i.e., for every two (distinct) nodes u and v, there is an edge $\{u, v\}$ connecting these nodes. The n-node congested clique has therefore $\binom{n}{2}$ edges, and every node has degree $n - 1$. Instead, assuming a core with, e.g., $O(\sqrt{n})$ nodes, even connecting all nodes in the core as a clique would only result in $O(n)$ edges in the core, a number that is much more manageable in practice. On the other hand, it was proved in [1] that $\Omega(\sqrt{n})$ nodes is the limit of how small can be the core, and that the core C must be dense, with $\Theta(|C|^2)$ edges.

In this paper, our objective is twofold. First, we are aiming at establishing tradeoffs between the number of edges, and the capability of emulating the clique. More precisely, we consider the all-to-all communication pattern:

- **Input:** Every node v has a message $M_{v,w}$, for every node $w \neq v$;
- **Output:** Every node w has received the message $M_{v,w}$, for every node $v \neq w$.

In the CONGEST model, assuming all messages are on $O(\log n)$ bits, all-to-all can be performed in just a single round in the clique. Our first objective is to study the tradeoff between number of edges, and number of rounds for performing all-to-all in the CONGEST model.

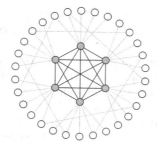

Fig. 1. Example of a core-periphery network, where the core (gray nodes) is a clique, and the periphery (white nodes) is a sparse graph.

Our second objective is to revisit one of the main problems left open in [1], namely the complexity of MST construction in the core-periphery model.

1.2   Our Results

We show that, in the CONGEST model, implementing all-to-all communication in k rounds can be done in n-node networks with roughly n^2/k edges, and this bound is essentially tight because every node must have degree at least $(n-1)/k$ to receive $n-1$ messages in at most k rounds. Hence, sparsifying the clique beyond just saving a fraction of the edges requires to relax the constraint on the time to simulate that clique.

Our first main result is about the ability of random graphs to emulate the clique. Let $\alpha = \sqrt{3e/(e-2)}$ where e is the basis of the natural logarithm. We show that, for $p \geq \alpha\sqrt{\ln n/n}$, a random graph in $\mathcal{G}_{n,p}$ can, w.h.p., perform all-to-all in $O(\min\{\frac{1}{p^2}, np\})$ rounds.

Our second main result is the design of a fast deterministic MST construction algorithm for core-periphery networks under the CONGEST model. Specifically, we show that if the core can emulate the clique in t rounds, then there exists a distributed MST construction algorithm performing in $O(t \log n)$ rounds. Hence, for $t = O(1)$, our deterministic algorithm performs in $O(\log n)$ rounds, improving the randomized algorithm in [1] by a factor $\Theta(\log n)$.

1.3   Related Work

The congested clique model has been widely studied in the literature. Lenzen [18] investigated the routing and sorting problems in the context of congested clique. He showed a deterministic algorithm that, if each node is the sender and receiver of at most n messages, allows to route all the messages in $O(1)$ rounds in a clique of size n using messages of size $O(\log n)$. He also showed an algorithm that allows to sort n^2 keys in constant time. Drucker et al. [5] proved that the congested clique is powerful enough to emulate certain classes of bounded depth circuits, which shows how difficult finding lower bounds for the congested clique is. In the case where each node can only broadcast, [5] gives upper and lower bounds for the problem of detecting some types of subgraphs. Hegeman et al. [15] investigated the metric facility location problem providing a $O(1)$ approximation algorithm that runs in expected $O(\log \log \log n)$ rounds. They also showed how to compute a 3-ruling set in the congested clique. In [14] it is shown that, under some restrictions, fast algorithms for the congested clique model can be translated into fast algorithms in the MapReduce framework. Censor-Hillel et al. [3] showed that matrix multiplication on congested clique can be computed in $O(n^{1-2/\omega})$ rounds, where $\omega < 2.3728639$ is the exponent of matrix multiplication. Also, they showed how to use matrix multiplication to solve a variety of graph related problems. In [19], Lotker et al. provided a deterministic MST construction algorithm that runs in $O(\log \log n)$ rounds in the congested clique. Then, Hegeman et al. [13] showed that in this context randomization can help, giving a randomized algorithm that requires $O(\log \log \log n)$ rounds. Recently, this complexity was even reduced further to $O(\log^* n)$ in [12].

In general, the MST construction problem has been widely studied. In the distributed asynchronous context, the most famous algorithm is the one of Gallager, Humblet and Spira [10] that runs in $O(n \log n)$. In the synchronous setting, the first sublinear algorithm was given by Garay et al. in [11] that runs in $O(D + n^{\frac{\ln 3}{\ln 6}} \log^* n)$, where D is the diameter of the graph. This complexity was later improved to $O(D + \sqrt{n} \log^* n)$ in [16]. Then, Peleg et al. [23] showed that this complexity is near optimal, giving a $\Omega(\frac{\sqrt{n}}{\log n})$ lower bound. This bound was later improved by Sarma et al. [24] to $\Omega(\sqrt{\frac{n}{\log n}})$ and then by Ookawa et al. [22] to $\Omega(\sqrt{n})$. All these lower bounds hold for graphs with diameter $\Omega(\log n)$. For constant diameter graphs, there is a bound $\widetilde{\Omega}(n^{1/3})$ rounds for diameter 4, a bound $\widetilde{\Omega}(n^{1/4})$ rounds for diameter 3, and a bound $O(\log n)$ rounds for diameter 2 (see [20]). Finally, Elkin [6] showed that if termination detection is not required, the diameter of the graph is not a lower bound, and that there exists an algorithm that requires $\widetilde{O}(\mu + \sqrt{n})$ rounds, where μ is the so-called MST-radius of the graph.

Feige et al. [7] studied the broadcast problem in random graphs, where a single node has a message that has to be received by all the nodes of the graph. They show that rumor spreading (which propagates the message to a randomly chosen neighbor at each step) is an efficient way to solve the broadcast problem for random graphs. Censor-Hillel et al. [4] studied the broadcast problem in the context where every node is the source of a message and it is limited to send the *same* message to each neighbor at each round. They give an efficient algorithm that solves the problem, also in case of failures.

Finally, it is worth mentioning that a problem related to our results, that is finding disjoint paths between pairs of nodes, has been largely investigated in expander graphs, which are sparse graphs that guarantee strong connectivity properties [2,8,9,17].

2   Deterministic Construction of Sparse Clique Emulators

In this section we provide a deterministic construction yielding a perfect tradeoff between number of edges and number of rounds in clique emulation.

Theorem 1. *Let $n \geq 1$, and $k \geq 3$. There is an n-node graph with at most $\lceil \frac{k-2}{(k-1)^2} n^2 \rceil$ edges that can emulate the n-node clique in k rounds. Also, there is an n-node graph with at most $\frac{1}{3}n^2$ edges that can emulate the n-node clique in 2 rounds.*

Proof. First, we show that there is an n-node graph with at most $\frac{1}{3}n^2$ edges that can emulate the n-node clique in 2 rounds. For this purpose, recall that the so-called Johnson graph $J(n, r)$ has vertex set composed of all the r-element subsets of the set $\{1, \ldots, n\}$, and two vertices are adjacent iff they meet in a $(r - 1)$-element set.

Fact 1. *There exists an independent set I of size at least $\lceil \frac{1}{n} \binom{n}{3} \rceil$ in the Johnson graph $J(n, 3)$.*

To establish this fact, for every k, $0 \leq k < n$, let us consider the set

$$I_k = \{\{x, y, z\} \in V(J(n, 3)) \mid x + y + z \equiv k \pmod{n}\}$$

Every set I_k is an independent set. Indeed, if two triples $\{x, y, z\}$ and $\{x, y, z'\}$ are both in I_k, then $x+y+z \equiv k \pmod{n}$ and $x+y+z' \equiv k \pmod{n}$. Therefore, $z \equiv z' \pmod{n}$, which implies $z = z'$, because $z, z' \in \{1, \ldots, n\}$. Observe that $\{I_0, \ldots, I_{n-1}\}$ is a partition of $V(J(n, 3))$. Therefore, one of them has size at least $\lceil \frac{1}{n} \binom{n}{3} \rceil$, which establishes Fact 1.

Let I as in Fact 1. Note that for any $\{a, b, c\} \in I$, none of the edges $\{a, b\}, \{a, c\}, \{b, c\}$ are appearing in any other triples of I. Thus, the edge $\{a, b\}$ of the complete graph can be emulated by the path $\{a, c\}, \{b, c\}$ without congestion resulting from the emulation of another edge $\{a', b'\}$. Moreover, the edge $\{a, b\}$ itself does not belong to any path used to emulate other edges. It follows that one can remove $|I|$ edges from K_n, one from each triple in the independent set I, and all removed edges can be emulated by edge-disjoint paths of length 2.

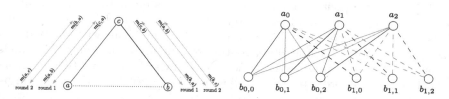

Fig. 2. (Left) Emulation of removed edge $\{a, b\}$ ($m(x, y)$ denotes the message from x to y). (Right) Emulating K_9 with $K_{3,6}$. The plain red path $(b_{0,1}, a_0, b_{0,2})$ is used at the 1st round for exchanging messages between $b_{0,1}$ and $b_{0,2}$ and, at the 2nd round, it is used for sending messages from $b_{0,1}$ to $b_{1,2}$, and from $b_{0,2}$ to $b_{1,1}$.

Figure 2(left) shows how to emulate the six communications $x \to y$ for every ordered pair (x, y), $x \in \{a, b, c\}$, $y \in \{a, b, c\}$, $x \neq y$, in just 2 rounds. It follows that there is an n-node graph with at most $\frac{n^2}{3}$ edges that can emulate the n-node clique in 2 rounds.

We now move on with the general case, that is, we show that there is an n-node graph with at most $\lceil \frac{n^2(k-2)}{(k-1)^2} \rceil$ edges that can emulate the n-node clique in k rounds.

Fact 2. *All-to-all communication between the nodes of the same part of the complete bipartite graph $K_{r,r}$ can be performed in 2 rounds.*

Indeed, let A and B be the two parts of $K_{r,r}$, where the nodes in A and B are labeled $a_0, \dots, a_{r-1}$ and $b_0, \dots, b_{r-1}$, respectively. Let us consider $a_i \in A$, and its message for node $a_j \in A$. This message is routed via node $b_k \in B$ where $i + j + k \equiv 0 \pmod{r}$. This guarantees that each edge is used at most once in each direction, at each round. Indeed, sender a_i chooses different intermediate nodes to route messages to the different receivers a_j, $j \neq i$. Similarly, for the same receiver j, different senders a_i, $i \neq j$, choose different intermediate nodes. This proves Fact 2.

By performing the above routing scheme in parallel, we directly get the following:

Fact 3. *Let A and B be the two parts of the complete bipartite graph $K_{r,kr}$, and let us partition the nodes of B into k groups $B_0, \dots, B_{k-1}$ of r nodes each. The k all-to-all communication patterns between the nodes of B_i can be performed in parallel for all $i \in \{0, \dots, k-1\}$, in 2 rounds, also in parallel to all-to-all communication between the nodes of A.*

We have now all the ingredients to establish the general case of Theorem 1. Let $k \geq 1$, and let $K_{r,kr}$ be the n-node complete bipartite graph with $r = \frac{n}{k+1}$ nodes in the first part A, and $kr = \frac{nk}{k+1}$ nodes in the other part B. Note that $K_{r,kr}$ has $kr^2 = \frac{n^2k}{(k+1)^2}$ edges. We show how to perform all-to-all in $K_{r,kr}$ in $k+2$ rounds. We divide the kr nodes of B into k groups $B_0, \dots, B_{k-1}$ of r nodes each. For $i \in \{0, \dots, k-1\}$, we set $B_i = \{b_{i,j}, 0 \leq j \leq r-1\}$—cf. Fig. 2(right). We describe a routing scheme that allows the kr nodes of B to perform all-to-all, by relaying their messages using the r nodes of A. Routing is achieved by repeating k times the all to all routing protocol in Fact 3, where, at each phase $s = 1, \dots, k$, nodes of B_i perform the communications with the nodes in $B_{j+s \bmod k}$. Importantly, the above routing scheme does not require $2k$ rounds but only $k+1$ rounds, because the kr nodes in B do not have to wait for receiving relayed messages in order to start sending new messages, and the phases can be pipelined. One more round is used to route the direct communication between every node in A and every node in B. Interestingly, during the $k+1$ rounds needed to perform all-to-all communications between the nodes in B, the edges are always used in both directions, except for the first and last round. We can use these two rounds to let the nodes in A perform their own all-to-all among them using the same routing pattern as in Fact 2. In total, in the $\frac{n^2 k}{(k+1)^2}$-edge graph $K_{r,kr}$, all-to-all is performed in $k + 2$ rounds.   □

We complete the section by showing that the bounds in Theorem 1 provide an essentially optimal tradeoff between the number of rounds k performed in the emulation, and the number of edges m of the emulator. A trivial lower bound $\frac{1}{2}\frac{n(n-1)}{k}$ can be obtained by noticing that every node must have degree at least $\frac{n-1}{k}$ for receiving $n - 1$ messages in k rounds. The following theorem improves this trivial bound by a factor 2, and matches with the bound in Theorem 1.

Property 1. Let $n \geq 1$, $k \in \{1, \ldots, n - 1\}$, and let G be an n-node graph that can emulate the n-node clique in k rounds. Then G has at least $\frac{n(n-1)}{k+1}$ edges.

Proof. Let m be the number of edges of G. There are $\binom{n}{2}$ pairs of nodes in K_n, communicating $n(n-1)$ messages in total. In G, only m pairs of nodes are directly connected. All the other $\binom{n}{2} - m$ pairs of nodes are not directly connected, and they are at least at distance 2 in G. Thus, the number of messages generated to route the messages corresponding to these pairs of nodes is at least $4(\binom{n}{2} - m)$. The total number of messages to be transferred is thus at least $2m + 4(\binom{n}{2} - m)$. Since one communication round in G can route at most $2m$ messages, it follows that any routing protocol requires at least $\frac{2m + 4\binom{n}{2} - 4m}{2m} = \frac{n(n-1)}{m} - 1$ rounds of communication. Thus, $k \geq \frac{n(n-1)}{m} - 1$, which implies $m \geq \frac{n(n-1)}{k+1}$.   □

3   Randomized Construction of Sparse Clique Emulators

In this section, we consider clique emulation by Erdős-Rényi random graphs $\mathcal{G}_{n,p}$. Our main result is the following.

Theorem 2. *Let $c \geq 0$, $n \geq 1$, $\alpha = \sqrt{(3 + c)e/(e - 2)}$ where e is the base of the natural logarithm, and $p \geq \alpha\sqrt{\ln n/n}$. For $G \in \mathcal{G}_{n,p}$,*

$$\Pr[G \text{ can emulate } K_n \text{ in } O(\min\{\tfrac{1}{p^2}, np\}) \text{ rounds}] \geq 1 - O(\tfrac{1}{n^{1+c}})$$

where the big-O notations hide the dependency in c.

Proof. Let $G \in \mathcal{G}_{n,p}$. The proof works as follows. For each missing edge in G between two nodes u and v, we route the messages between these nodes via an intermediate node w, i.e., along a path (u, w, v) of length 2. The intermediate node is picked at random among all nodes w such that $\{u, w\} \in E(G)$, and $\{w, v\} \in E(G)$. To analyze the load of the edges, we have to overcome two problems. First, the load of an edge is not necessarily independent from the load of another edge. Second, we are interested in the maximum, taken over all edges, of the load of the edges. As a consequence, an analysis based only on the expectation of the load of each edge may not yield accurate results. Instead, we base our analysis on a double application of a balls-into-bins protocol.

We aim at constructing a path for routing the messages between every pair of nodes that are not directly connected in G. As said before, the alternative paths used to replace missing edges are of length 2, and the probability expressed in the statement of the theorem reflects the probability that such paths exist, without too much congestion. More specifically, let us consider a missing edge $\{i, j\}$ in G. Let $S_{i,j}$ be the set of common neighbors to i and j in G. The message from i to j is aimed at being routed via some intermediate node $k \in S_{i,j}$. The first question to address is thus: how large is $S_{i,j}$? To answer this question, let $\mathcal{E}_{i,j}$ be the event "there are at least $\frac{np^2}{e}$ different paths of length 2 between i and j", and let $\mathcal{E} = \bigcap_{\{i,j\} \notin E(G)} \mathcal{E}_{i,j}$.

Fact 4. *Let $\alpha_c = \sqrt{(c+3)e/(e-2)}$, and $p \geq \alpha_c \sqrt{\ln n / n}$. Then*

$$\Pr[\mathcal{E}] \geq 1 - \frac{1}{n^{c+1}}.$$

To establish this fact, let $X_{i,j,k}$ be the Bernoulli random variable, for $\{i, j\} \notin E(G)$, such that $X_{i,j,k} = 1$ iff $k \in S_{i,j}$, i.e., $\{i, k\} \in E(G)$ and $\{k, j\} \in E(G)$. Then let $X_{i,j} = \sum_{k=1}^{n} X_{i,j,k}$. We have $\Pr[X_{i,j,k} = 1] = p^2$, and, for a fixed pair i, j, the variables $X_{i,j,k}$, $k = 1, \ldots, n$, are mutually independent. Thus, using Chernoff bounds, we get:

$$\Pr[X_{i,j} \leq \frac{np^2}{e}] \leq e^{(\frac{2}{e}-1)np^2}.$$

By union bound, it follows that

$$\Pr[\bigcup_{\{i,j\} \notin E(G)} \overline{\mathcal{E}_{i,j}}] \leq n^2 e^{(\frac{2}{e}-1)np^2} \leq \frac{1}{n^{c+1}}$$

as desired, where the last inequality holds because $p \geq \alpha_c \sqrt{\ln n / n}$.

In addition to Fact 4, we will also use the following known result:

Lemma 1 ([21]). *Let $X_1, \ldots, X_n$ be a sequence of random variables in an arbitrary domain, and let $Y_1, \ldots, Y_n$ be a sequence of binary random variables, with the property that Y_i is a function of the variables $X_1, \ldots, X_{i-1}$. If, for every $i = 1, \ldots, n$, we have $\Pr[Y_i = 1 | X_1, \ldots, X_{i-1}] \leq q$ then $\Pr[\sum_{i=1}^{n} Y_i \geq k] \leq \Pr[B(n, q) \geq k]$ where $B(n, q)$ denotes the binomial distribution of parameters n and q.*

Our path construction algorithm for every missing edge $\{i, j\} \notin E(G)$ is sequential, and proceeds as follows. For every $\{i, j\} \notin E(G)$, the path from i to j is not necessarily the same as the path from j to i. We process all ordered pairs of nodes (i, j) in n phases, where Phase i, $i = 1, \ldots, n$, constructs all paths (i, j) for $\{i, j\} \notin E(G)$, in increasing order of j. Assume already fixed a set of paths, corresponding to previously considered sender-receiver pairs, and consider now the pair (i, j) (of course corresponding to the missing edge $\{i, j\} \notin E(G)$). The previously constructed paths induce some load on each edge of G, corresponding to the number of paths using that edge. The choice of the path for (i, j) depends on this load, and is inspired from the power of two choices in balls-and-bins protocols. Precisely, for suitable parameters d and r, node i repeats r times the following: pick d incident edges $\{i, k\}$ uniformly at random, and select the least loaded one. Once this is done, node j picks the least loaded edge among the r edges selected by i.

Let $I_{i,j}$ be the node selected to route the message from sender i to receiver j. Messages from i to j will be routed along the path $P_{i,j} = (i, I_{i,j}, j)$. For $h \geq 0$, let $b_{i,h}(j)$ be the number of edges $\{i, k\}$ of load at least h after deciding the intermediate nodes $I_{i,1}, \ldots, I_{i,j}$ of the first j receivers for sender i. We define the following quantities:

$$x = \left\lceil \frac{e^{5+c}}{p^2} \right\rceil \text{ and } \beta = \frac{np^2}{e^{5+c}}.$$

Since $b_{i,x}(n) \leq n/x$, it follows from the above that $b_{i,x}(n) \leq \beta$. Now, let

$$\ell(j) = |\{j' \leq j : I_{i,j'} = I_{i,j}\}|.$$

We define the random variables $Z_{i,j}$ where

$$Z_{i,j} = \begin{cases} 1 \text{ if } \ell(j) \geq x + 1 \\ 0 \text{ otherwise.} \end{cases}$$

Hence $Z_{i,j} = 1$ is the bad event that the edge between node i and the intermediate node $I_{i,j}$ used to route from i to j is heavily loaded by i. Conditioned on the fact that $\mathcal{E}$ holds (cf. Fact 4), we get that

$$\Pr[Z_{i,j} = 1] \leq r \left(\frac{\beta}{np^2/e} \right)^d.$$

We let q be the right hand side of the above equation. Let us now consider $Z_i = \sum_{j=1}^n Z_{i,j}$. Observe that $Z_{i,j}$ is a function of $I_{i,1}, \ldots, I_{i,j-1}$. Therefore, by Lemma 1 we get that

$$\Pr[Z_i \geq k] \leq \Pr[B(n, q) \geq k].$$

So, in particular, $\Pr[Z_i \geq 1] \leq \Pr[B(n, q) \geq 1]$. We now set $d = \ln n$, and $r \leq n$ (a suitable r will be specified thereafter). Thanks to this choice of d and r, we have $q \leq \frac{1}{n^{3+c}}$, and therefore

$$\Pr[Z_i \geq 1] \leq \Pr[B(n, \frac{1}{n^{3+c}}) \geq 1] \leq \mathbf{E}[B(n, \frac{1}{n^{3+c}})] \leq \frac{1}{n^{2+c}}.$$

Let $Z = \sum_{i=1}^n Z_i$. By union bound, we get $\Pr[Z \geq 1] \leq \frac{1}{n^{1+c}}$.

Using a similar analysis, from the perspective of the receiver, and defining the corresponding random variables $Z'_{i,j}$ capturing the load of the edges incident to a receiver j, and $Z'_j = \sum_{i=1}^n Z'_{i,j}$, we get

$$\Pr[Z'_j \geq 1] \leq \Pr[B(n, q') \geq 1]$$

where

$$q' = \left(1 - \left(1 - \frac{e\beta}{np^2}\right)^d\right)^r.$$

We get $q' \leq \frac{1}{n^{3+c}}$ by setting $d = \ln n$ and $r = (c+3)\, n^\epsilon \, \ln n$ for $\epsilon = -\ln(1 - \frac{1}{e^{4+c}})$. By this setting of d and r, we get that

$$\Pr[Z'_j \geq 1] \leq \Pr[B(n, \frac{1}{n^{3+c}}) \geq 1] \leq \mathbf{E}[B(n, \frac{1}{n^{3+c}})] \leq \frac{1}{n^{2+c}}.$$

Let $Z' = \sum_{j=1}^n Z'_j$. By union bound, we get $\Pr[Z' \geq 1] \leq \frac{1}{n^{1+c}}$.
Therefore, altogether, we get that

$$\Pr[Z = 0 \text{ and } Z' = 0 \mid \mathcal{E}] \cdot \Pr[\mathcal{E}] \geq (1 - \frac{1}{n^{1+c}})^3 \geq 1 - \frac{3}{n^{1+c}}.$$

In other words, w.h.p., the load of all edges is no more than $x = O(1/p^2)$. On the other hand, with a similar argument as for proving that the degree is large, we have that, w.h.p., the degree of all nodes is at most enp, and therefore the load of an edge does not exceed enp.   □

4 MST Construction in Core-Periphery Networks

In [1], a randomized algorithm for Minimum Spanning Tree (MST) construction is presented. It runs in $O(\log^2 n)$ rounds with high probability. We improve this result by describing a deterministic algorithm for MST construction that runs in just $O(\log n)$ rounds. Recall that, for the MST construction task, every node is given as input the weight $w(e)$ of each of its incident edges e. These weights are supposed to be of values polynomial in the size n of the network, and thus each weight can be stored on $O(\log n)$ bits. The output of every node is a set of incident edges, such that the collection of all outputs forms an MST of the network. Without loss of generality, all weights are supposed to be different (since, otherwise, it is sufficient to add to each edge the identities of the extremities of that edge).

Theorem 3. *The MST construction task can be solved in $O(\log n)$ rounds in core-periphery networks under the* CONGEST *model.*

Proof. As usually in the distributed setting, the general idea of the algorithm is based on the sequential Borůvka's algorithm for MST construction, consisting in merging subtrees called *fragments*. Recall that, in Borůvka's algorithm, there are initially n fragments, where each node alone forms a fragment. Each fragment has an ID. Initially, the identity of each fragment is the ID of the single node in the fragment. Then the algorithm proceeds in at most $\lceil \log_2 n \rceil$ phases. At each phase, each fragment F computes the edge e_F of minimum weight incident to fragment F, and adds it to the MST. Fragments connected by such an edge merge, and a new phase begins. This procedure is repeated until there is only one fragment, which is the desired MST.

We first present a (deterministic) distributed algorithm running in $O(\log^2 n)$ rounds in core-periphery networks. This algorithm is composed of at most $\lceil \log_2 n \rceil$ phases, where each phase requires $O(\log n)$ rounds. Then, we show how to actually perform each phase in $O(1)$ rounds, obtaining the desired $O(\log n)$-round algorithm. Recall that a core-periphery network satisfies the three axioms listed in Sect. 1 where C and P denote the sets of nodes in the core and in the periphery, respectively.

The algorithm starts by an initialization phase, where each node in the periphery looks for a node in the core, which will be its *representative*. By Axiom 3 all nodes in the periphery can concurrently send messages to the core so that each message will be received by at least one node in the core after $O(1)$ rounds. So, each node in the periphery sends a request for a representative by sending its own ID to the core. Every node in the periphery then waits for an acknowledgment from nodes in the core that accepted its request. These acknowledgements follow the same route as the corresponding requests, backward. Hence, all acknowledgments are also received after $O(1)$ rounds. Every node takes as representative the core node whose acknowledgment reaches that node first. If a node receives several acknowledgments simultaneously, then it selects the one with the smallest ID. By Axiom 1, each node in the core can be the representative of at most $O(|C|)$ nodes in the periphery because its degree is at most $O(|C|)$, and thus it can receive at most $O(|C|)$ messages in $O(1)$ rounds. Every node in the core is its own representative.

We assume that the nodes in the core are sorted according to their IDs (this operation can be done in $O(1)$ rounds using all-to-all and Axiom 2). For every node in the core, we denote by $\text{succ}(u)$ and $\text{pred}(u)$ the successor and the predecessor of u in this order, respectively.

We heavily used the protocols in [18]. Note that the *routing* protocol in [18] requires that each node is the source and destination of at most n messages. However, it can be trivially adapted to be applied with $O(n)$ messages, still requiring $O(1)$ rounds. Similarly, the *sorting* protocol in [18] requires that each node receives at most n keys, but, again, it can be trivially modified for allowing each node to receive $O(n)$ keys, still requiring $O(1)$ rounds.

We now explain how every phase of Borůvka's algorithm is performed.

1. Every node sends the ID of its fragment to all its neighbors.
2. Let $r(v) \in C$ and $\text{id}(F)$ be the representative and the ID of the fragment F of node v, respectively. We denote by $e_F(v)$ the edge of minimum weight incident to v and connecting v to a node not in its fragment F. Each node v in the periphery sends $(e_F(v), w(e_F(v)), \text{id}(F), \text{id}(F'))$ to $r(v)$, where the tail of $e_F(v)$ belongs to F, and its head belongs to fragment $F' \neq F$. Observe that each node in the core receives $O(|C|)$ such messages.
3. Every node in the core, upon reception of 4-tuple $(e_F(v), w(e_F(v)), \text{id}(F), \text{id}(F'))$ from the nodes that it represents (including itself), selects the

ones with minimum weight for each fragment F. We denote by S_1 the set of the selected edges by all nodes in the core. Note that $|S_1| = O(|C|^2)$.

4. The algorithm assigns a *leader* to each fragment. The leaders are core nodes chosen in such a way that the fragments are equally distributed among leaders. Let

$$x = \lceil |S_1|/|C| \rceil.$$

Note that $x = O(|C|)$. Given a fragment F, its leader is

$$\ell(F) = 1 + \left\lfloor \frac{|\{(u,v) \in S_1 : \mathrm{id}(F_u) < \mathrm{id}(F)\}|}{x} \right\rfloor$$

where F_u is the fragment of u. Note that $1 \le \ell(F) \le |C|$. For each fragment F, all edges incident to F in S_1 are sent to $\ell(F)$ by its representative holding such edges—we shall explain hereafter how this is implemented in core-periphery networks. In this way each leader can select the edge e_F of minimum weight incident to fragment F. Let S_2 be the set of all edges e_F, where F is a fragment.

5. The algorithm then aims at merging the fragments. We call *merge tree* a tree whose nodes are fragments F, and whose edges are the edges e_F connecting these fragments. Note that, in a merge tree, there are two adjacent fragments F and F' connected by two possibly distinct edges e_F and $e_{F'}$. The fragment with smallest ID that is extremity of such an edge is the root of the merge tree. The algorithm proceeds so that each leader $\ell(F)$ of a fragment F in the merge tree becomes aware of the root of the tree. The ID of this root will become the ID of the fragment resulting from merging all the fragments in the merge tree. It is possible to find the root of a tree of height h in $O(\log h)$ steps using *pointer jumping*—we shall explain hereafter how this is precisely implemented in core-periphery networks.

6. By the previous step, for every fragment F, its leader $\ell(F)$ knows the ID of the merge tree it belongs to. Moreover, for each edge (u,v) that was received by a leader from the representative $r(u)$ in step 4, the leader saved $\mathrm{id}(r(u))$. This allows leaders to notify the right representatives of the ID of the root of the merge tree.

7. Finally, the ID of every merged fragment is sent to every node v of the periphery from its representative $r(v)$ in the core.

It remains to explain how steps 4 and 5 are actually performed.

Step 4 in More Details. First, observe that the parameter $x = \lceil |S_1|/|C| \rceil$ can be computed at each node of the core, as performing all-to-all communication in the core allows each core node to compute $|S_1|$. Now, we show how to distribute the fragments among the leaders such that leader $\ell(F)$ becomes aware of the edges $e_F(v) \in S_1$ incident to F.

The edges $(u,v) \in S_1$ are sorted according to the ID of the fragment F_u its tail belongs to, and are then split into groups of x edges. Again, this operation can be done in $O(1)$ rounds using the sorting protocol in [18] because $x = O(|C|)$. The kth group is assigned to the kth node of the core.

Let us consider a core node u, and let $\mathcal{F}(u)$ be the set of fragments F such that $\ell(F) = u$. Let us denote by $\mathrm{id}_{max}(u)$ (resp., $\mathrm{id}_{min}(u)$) the maximum ID (resp., minimum ID) of the fragments $F \in \mathcal{F}(u)$. Having sorted the set S_1 guaranties that the leader u receives all the edges assigned to it, except perhaps some edges starting from fragment $\mathrm{id}_{max}(u)$ that could have been delivered to $\mathrm{succ}(u)$. However, there are at most $x - 1$ such edges, since the representatives kept at most one edge per fragment. So, every core node u can send $\mathrm{id}_{max}(u)$ to $\mathrm{succ}(u)$, in order to let that node know that the leader of the fragment with ID equal to $\mathrm{id}_{max}(u)$ should be u, and not $\mathrm{succ}(u)$. Since each node u has then at most $x-1$ messages to transmit to $\mathrm{pred}(u)$, we can transmit these messages using the routing protocol in [18]. Now each leader u has all the outgoing edges of each fragment F with $\ell(F) = u$. Thus, u can compute e_F for each of these fragments. Finally, each node u in the core broadcasts the pair $(\mathrm{id}_{min}(u), \mathrm{id}_{max}(u))$ in the core so that every node in C learns the leader of each fragment.

Note that, while sorting and routing, every node keeps track of the ID of the representative nodes which originally received every edge that is manipulated by that node (this is needed in step 6).

Step 5 in More Details. We show how to perform the first step of pointer jumping. Recall that, for every fragment F, the leader $\ell(F)$ knows e_F. This latter edge is the one leading toward the root of the merge tree. Assume that $e_F = (u, v)$, with $u \in F$ and $v \in F'$. The objective for the leader $\ell(F)$ is to learn to which fragment F'' is pointing the edge $e_{F'} = (u', v')$ with $u \in F'$ and $v' \in F''$. In other words, if p denotes the parent relation in a merge tree, the leader $\ell(F)$ of fragment F wants to learn the ID of $p(p(F))$. The bad news is that $\ell(F)$ cannot directly ask $\mathrm{id}(p(p(F)))$ to $\ell(p(F))$ because this could create a bottleneck at $\ell(p(F))$. Nevertheless this issue can be overcame as follows.

First, the edges in S_2 are sorted according to the IDs of the fragment of their heads, and grouped into groups whose heads belong to the same fragment. In this way, only one request is sent for each group (to the leader of the corresponding fragment). Since $x = \lceil |S_1|/|C| \rceil$, we have $x = O(|C|)$, and thus the number of requests that each leader has to make is at most $O(|C|)$.

Second, every leader does not receive more than $O(|C|)$ requests. Indeed, let $q_{u,v}$ be the number of different fragments for which a node u in the core has to send a request to leader v. Let $F_{i_1}, F_{i_2}, \ldots, F_{i_{q_{u,v}}}$ be these fragments, with $\ell(F_{i_1}) = \ell(F_{i_2}) = \cdots = \ell(F_{i_{q_{u,v}}}) = v$, and $i_1 < i_2 < \cdots < i_{q_{u,v}}$. Recall that the edges in S_2 are sorted according to the IDs of the fragment of their heads. Thus, if $q_{u,v} > 1$ then the fragments $F_{i_2}, \ldots, F_{i_{q_{u,v}}}$ do not appear in any list of fragments assigned to nodes with identity smaller than $\mathrm{id}(u)$. Therefore, leader v receives at least $\sum_{u \in C}(q_{u,v} - 1)$ requests for different fragments. On the other hand, every core node v is the leader of at most x fragments. Therefore $\sum_{u \in C}(q_{u,v} - 1) \leq x$. Hence the number of requests received by v is $\sum_{u \in C} q_{u,v} = O(|C|)$.

These two facts, allow the routing protocol in [18] to be used, for sending the requests to the leaders, and for receiving back their answers. Once this is done, every node u sends $id(p(p(F)))$ to $\ell(F)$, for every $F \in \mathcal{F}(u)$ in a constant number of rounds, again using [18]. It follows that every leader u can learn the ID of $p(p(F))$ for every $F \in \mathcal{F}(u)$ in a constant number of rounds.

Time Analysis. The initialization phase can be performed in $O(1)$ rounds thanks to Axiom 3. Step 1 trivially requires $O(1)$ rounds. Step 2 also requires $O(1)$ rounds thanks to Axiom 3. Step 3 is executed locally by each node, thus it does not require communication. Step 4 can be executed in $O(1)$ rounds using the sorting protocol in [18] because $x = O(|C|)$. Step 6 can also be performed in $O(1)$ rounds using the routing protocol in [18] because each leader handles $O(|C|)$ edges (for which it has to send a fragment ID), and each representative has to receive $O(|C|)$ messages (one for each edge it has to receive a new fragment ID). The last step is the inverse of step 2, and thus can still be executed in $O(1)$ rounds. Step 5 however requires $O(\log n)$ rounds because the merge tree might be of height $\Omega(n^\epsilon)$ for some $\epsilon > 0$. Since the number of phases is also $O(\log n)$, the total number of rounds of this algorithm is $O(\log^2 n)$.

A Faster Algorithm. Now, we describe how to modify the above algorithm so that it uses only $O(1)$ rounds for each phase, hence $O(\log n)$ rounds in total. Since the only step that requires a non constant number of rounds is Step 5, we show how to perform that step in $O(1)$ rounds.

The idea is to use a technique introduced first in [20], and also used in Avin et al. [1], called *amortized pointer jumping*. The reduction of long chains of pointers is deferred to later phases of Borůvka's algorithm, and only a constant number of pointer jumps are performed at each phase. This technique exploits the fact that, if a chain is long, it must contain many fragments. As a consequence, when pointer jumping completes, the resulting fragment is quite large, and other nodes involved in small fragments may continue building the MST in parallel, without waiting for large fragments to be constructed.

We show how to do a constant number of pointer jumping steps, then freezing the procedure, and resuming it later in the next phase of Borůvka's algorithm. At each step of pointer jumping, every leader u can know, for every $F \in \mathcal{F}(u)$, if the root of the merge tree has been reached. Suppose that the root has not been reached by u after a constant number of pointer jumping (i.e., the leader does not know yet the new ID of the merged fragment), and that u is currently pointing at fragment F'. In the following, node u adds a flag in its messages, which specifies that the fragment has not been resolved yet, and that it stopped at F'. This flag will be propagated to all nodes that proposed edges that start from unresolved fragments. At the next phase of Borůvka's algorithm, these nodes will propose again the same edges, by specifying also F'. Fragment F' will be used as if it was the destination fragment of the edge. In this way, for every fragment F in a merge tree whose merging has not yet been performed, the same edge e_F as before will be chosen, and other steps of pointer jumping will be performed. This insures that nodes belonging to fragments in such merge trees do not propose new edges, thus emulating a full execution of pointer jumping.

After having reduced the number of rounds for performing step 5 from $O(\log n)$ to $O(1)$, amortized, we get that the resulting algorithm just requires $O(\log n)$ rounds to construct a MST. $\square$

5   Conclusion

We have shown how to emulate the clique by a random graph in $\mathcal{G}_{n,p}$ in time $O(\min\{\frac{1}{p^2}, np\})$ rounds, w.h.p. Hence, on dense random graphs (i.e., $p = \Omega(1)$), our simulation performs in just a multiplicative constant factor away from the optimal, and, on sparse graphs (i.e., $p \simeq \sqrt{\log n/n}$), it performs just a $\log n$ factor away from optimal. However, in general, whenever $p \gg \frac{1}{\sqrt[3]{n}}$, it performs in $O(\frac{1}{p^2})$ rounds, which is a factor $O(\frac{1}{p})$ away from the trivial lower bound $\Omega(\frac{1}{p})$. An intriguing question is whether the n-node clique can be simulated by $\mathcal{G}_{n,p}$ in just $O(\frac{1}{p})$ rounds.

Our deterministic MST algorithm for core-periphery networks performs in $O(\log n)$ rounds, improving the previously known (randomized) algorithm by a factor $\Theta(\log n)$. Recent advances in the congested clique model demonstrate that ultra fast MST algorithms exist for this later model, namely, a recent $O(\log^* n)$-round randomized algorithm [12], and a $O(\log \log n)$-round deterministic algorithm [19]. Another intriguing question is whether such ultra fast algorithms exist for core-periphery networks.

References

1. Avin, C., Borokhovich, M., Lotker, Z., Peleg, D.: Distributed computing on core-periphery networks: axiom-based design. In: Esparza, J., Fraigniaud, P., Husfeldt, T., Koutsoupias, E. (eds.) ICALP 2014. LNCS, vol. 8573, pp. 399–410. Springer, Heidelberg (2014). doi:10.1007/978-3-662-43951-7_34
2. Broder, A.Z., Frieze, A.M., Upfal, E.: Existence and construction of edge-disjoint paths on expander graphs. SIAM J. Comput. **23**(5), 976–989 (1994)
3. Censor-Hillel, K., Kaski, P., Korhonen, J.H., Lenzen, C., Paz, A., Suomela, J.: Algebraic methods in the congested clique. In: ACM Symposium on Principles of Distributed Computing (PODC), pp. 143–152 (2015)
4. Censor-Hillel, K., Toukan, T.: On fast and robust information spreading in the vertex-congest model. In: Scheideler, C. (ed.) Structural Information and Communication Complexity. LNCS, vol. 9439, pp. 270–284. Springer, Heidelberg (2015). doi:10.1007/978-3-319-25258-2_19
5. Drucker, A., Kuhn, F., Oshman, R.: On the power of the congested clique model. In: ACM Symposium on Principles of Distributed Computing (PODC), pp. 367–376 (2014)
6. Elkin, M.: A faster distributed protocol for constructing a minimum spanning tree. In: ACM-SIAM Symposium on Discrete Algorithms (SODA), pp. 359–368 (2004)
7. Feige, U., Peleg, D., Raghavan, P., Upfal, E.: Randomized broadcast in networks. Random Struct. Algorithms **1**(4), 447–460 (1990)
8. Frieze, A.M.: Disjoint paths in expander graphs via random walks: a short survey. In: Luby, M., Rolim, J.D.P., Serna, M. (eds.) RANDOM 1998. LNCS, vol. 1518, pp. 1–14. Springer, Heidelberg (1998). doi:10.1007/3-540-49543-6_1

9. Frieze, A.M.: Edge-disjoint paths in expander graphs. In: 11th ACM-SIAM Symposium on Discrete Algorithms (SODA), pp. 717–725 (2000)

10. Gallager, R.G., Humblet, P.A., Spira, P.M.: A distributed algorithm for minimum-weight spanning trees. ACM Trans. Program. Lang. Syst. 5(1), 66–77 (1983)

11. Garay, J.A., Kutten, S., Peleg, D.: A sublinear time distributed algorithm for minimum-weight spanning trees. SIAM J. Comput. 27(1), 302–316 (1998)

12. Ghaffari, M., Parter, M.: MST in log-star rounds of congested clique. In: 35th ACM Symposium on Principles of Distributed Computing (PODC) (2016)

13. Hegeman, J.W., Pandurangan, G., Pemmaraju, S.V., Sardeshmukh, V.B., Scquizzato, M.: Toward optimal bounds in the congested clique: graph connectivity and MST. In ACM Symposium on Principles of Distributed Computing (PODC), pp. 91–100 (2015)

14. Hegeman, J.W., Pemmaraju, S.V.: Lessons from the congested clique applied to MapReduce. Theor. Comput. Sci. 608, 268–281 (2015)

15. Hegeman, J.W., Pemmaraju, S.V., Sardeshmukh, V.B.: Near-constant-time distributed algorithms on a congested clique. In: Kuhn, F. (ed.) DISC 2014. LNCS, vol. 8784, pp. 514–530. Springer, Heidelberg (2014). doi:10.1007/978-3-662-45174-8_35

16. Kutten, S., Peleg, D.: Fast distributed construction of small k-dominating sets and applications. J. Algorithms 28(1), 40–66 (1998)

17. Leighton, T., Rao, S., Srinivasan, A.: Multicommodity flow and circuit switching. In: 31st Hawaii International Conference on System Sciences, pp. 459–465 (1998)

18. Christoph Lenzen. Optimal deterministic routing and sorting on the congested clique. In ACM Symposium on Principles of Distributed Computing (PODC), pp. 42–50, (2013)

19. Lotker, Z., Patt-Shamir, B., Pavlov, E., Peleg, D.: Minimum-weight spanning tree construction in O(log log n) communication rounds. SIAM J. Comput. 35(1), 120–131 (2005)

20. Lotker, Z., Patt-Shamir, B., Peleg, D.: Distributed MST for constant diameter graphs. In: 20th ACM Symposium on Principles of Distributed Computing (PODC), pp. 63–71 (2001)

21. Mitzenmacher, M., Upfal, E.: Probability and Computing - Randomized Algorithms and Probabilistic Analysis. Cambridge University Press, Cambridge (2005)

22. Ookawa, H., Izumi, T.: Filling logarithmic gaps in distributed complexity for global problems. In: Italiano, G.F., Margaria-Steffen, T., Pokorný, J., Quisquater, J.-J., Wattenhofer, R. (eds.) SOFSEM 2015. LNCS, vol. 8939, pp. 377–388. Springer, Heidelberg (2015). doi:10.1007/978-3-662-46078-8_31

23. Peleg, D., Rubinovich, V.: A near-tight lower bound on the time complexity of distributed minimum-weight spanning tree construction. SIAM J. Comput. 30(5), 1427–1442 (2000)

24. Sarma, A.D., Holzer, S., Kor, L., Korman, A., Nanongkai, D., Pandurangan, G, Peleg, D., Wattenhofer, R.: Distributed verification and hardness of distributed approximation. In: 43rd ACM Symposium on Theory of Computing (STOC), pp. 363–372 (2011)

Rumor Spreading with Bounded In-Degree

Sebastian Daum, Fabian Kuhn, and Yannic Maus[(✉)]

University of Freiburg, Freiburg im Breisgau, Germany
{daum,kuhn,yannic.maus}@cs.uni-freiburg.de

Abstract. In the gossip-based model of communication for disseminating information in a network, in each time unit, every node u can contact a single random neighbor v but can possibly be contacted by many nodes. In the present paper, we consider a restricted model where at each node only one incoming call can be answered in one time unit. We study the implied weaker version of the well-studied pull protocol, which we call *restricted pull*.

We prove an exponential separation of the rumor spreading time between two variants of the protocol (the answered call among a set of calls is chosen adversarial or uniformly at random). Further, we show that if the answered call is chosen randomly, the slowdown of restricted pull versus the classic pull protocol can w.h.p. be upper bounded by $O(\Delta/\delta \cdot \log n)$, where Δ and δ are the largest and smallest degree of the network.

Keywords: Rumor spreading · Gossiping · Pull · Push · Stochastic dominance · Coupling

1 Introduction

Gossip-based communication models have received a lot of attention as a simple, fault-tolerant, and in particular also scalable way to communicate and disseminate information in large networks. The classic application of gossip-based network protocols is the spreading of information in the network, specifically the problem of broadcasting a single piece of information to all nodes of a network, in this context also often known as *rumor spreading*, e.g., [2–7]. Gossip-based protocols have for example also been proposed for applications such as maintaining consistency in a distributed database [2], for data aggregation problems [8–10], or even to run arbitrary distributed computations [11].

The best studied gossip strategy is the *random phone call model*, which was first considered in [4]. We are given a network graph $G = (V, E)$ where initially a source node $s \in V$ knows some piece of information (*rumor*) and the objective is to disseminate the rumor to all nodes of G. Typically, time is divided into

A full version of this paper with all proofs is avalaible on https://arxiv.org/abs/1506.00828 [1].

Second and third author supported by ERC Grant No. 336495 (ACDC).

© Springer International Publishing AG 2016
J. Suomela (Ed.): SIROCCO 2016, LNCS 9988, pp. 323–339, 2016.
DOI: 10.1007/978-3-319-48314-6_21

synchronized rounds, where in each round, every node can contact a random neighbor and if u contacts v, an interaction between u and v is initiated for the current round. For spreading a rumor, two basic modes of operation are distinguished. Nodes that already know the rumor can PUSH the information to the randomly chosen neighbor [4] or nodes that do not yet know the rumor can PULL the information from the randomly chosen neighbor [2]. In much of the classic work, the network G is assumed to be a complete graph. In that case, it is not hard to see that PUSH and PULL both succeed in $O(\log n)$ rounds and that the total number of interactions of each node can also be bounded by $O(\log n)$. In [7], it is shown that when combining PUSH and PULL (in the following referred to as PUSH-PULL), the average number of interactions per node is only $\Theta(\log \log n)$.

Mostly in recent years, PUSH, PULL, and PUSH-PULL have also been studied for more general network topologies, e.g., [3,5,6,12–15], with [5,6] and [14,15] studying the time complexity as a function of the graph's conductance and vertex expansion, respectively. E.g., in [6], it is shown that with high probability (w.h.p.), the running time of PUSH-PULL can be upper bounded by $O((\log n)/\phi(G))$, where n is the number of nodes and $\phi(G)$ is the conductance of the network graph G.

While in gossip protocols, each node can initiate at most one interaction with some neighbor, even if each node contacts a uniformly random neighbor, the number of interactions a node needs to participate in each round can be quite large. In complete graphs and more generally in regular graphs, the total number of interactions per node and round can easily be upper bounded by $O(\log n)$. However in general topologies a single node might be contacted by up to $\Theta(n)$ neighboring nodes. As an extreme case, consider a star network where a single center node is connected to $n-1$ leaf nodes. Even if the rumor initially starts at a leaf node, PUSH-PULL manages to disseminate the rumor to all nodes in only 2 rounds. Clearly, in these 2 rounds, the center node has to interact with all $n-1$ leaf nodes. In fact, all recent papers which study the time complexity of the random PUSH-PULL protocol critically rely on the fact that a node can be contacted by many nodes in a single round, e.g., [6]. In some cases, this behavior might limit the implementability and thus the applicability of the proven results for this gossip protocol. In order to obtain scalable systems, ideally, we would like to not only limit the number of interactions each node initiates, but also the number of interactions each node participates in.

In the present paper, we therefore study a weaker variant of the described random gossip algorithms. In each round, every node can still initiate a connection to one uniformly random neighbor. However, if a single node receives several connection requests, only one of these connections is actually established. When disseminating a rumor by using the PUSH protocol, this restriction does not limit the progress of the algorithm. In a given round, a node v learns the rumor if and only if at least one PUSH request arrives at v. However, when using the PULL protocol, the restriction can have a drastic effect. If a node v receives several PULL requests from several nodes that still need to learn the rumor, only one of these nodes can actually learn the rumor in the current round. In our paper, we therefore concentrate on the PULL protocol and we define RPULL (*restricted* PULL) as the described weak variant of the PULL algorithm: In each RPULL round, every node that still needs to learn the rumor contacts a random neighbor. At every node that knows the rumor, one of the incoming requests (if there are any) is selected and the rumor is sent to the corresponding neighbor. By PUSH-RPULL we denote the combination of RPULL with a simultaneous execution of the classic PUSH protocol.

Contributions. We first consider two versions of the RPULL protocol which differ in the way how one of the incoming requests is selected. Assume that in a given round some informed node v receives RPULL requests from a set of neighbors R_v. In the *adversarial* RPULL protocol, an (adaptive) adversary picks some node $u \in R_v$ which will then learn the rumor. In the *random* RPULL protocol, we assume that a uniformly random node $u \in R_v$ learns the rumor (chosen independently for different nodes and rounds). While the choice of which neighbor a node (actively) contacts with a request is under the control of the protocol, it is not necessarily clear how one of the incoming requests in R_v is chosen. If the node can only answer one request per time unit and the requests do not arrive at exactly the same time, the first request might be served and all others dropped. Or even if requests arrive at the same time, it might be the underlying network infrastructure or operating system which picks one request and drops the others. If it is reasonable to assume that the incoming requests are served probabilistically and independently, we believe that random RPULL provides a good model. Otherwise, the adversarial assumption allows to study the worst-case behavior.

As a first result, we prove that the running times of the two RPULL variants are essentially the same on trees. Secondly, we show that there are instances for which there is an exponential gap between the running times of the two RPULL variants. We give an instance where for every source node the random RPULL protocol informs all nodes of the network in polylogarithmic time, w.h.p., whereas, for every source, the adversarial RPULL algorithm requires time $\Omega(\sqrt{n})$ to even succeed with a constant probability.

In the second part of the paper, we have a closer look at the performance of the random RPULL protocol. Consider a graph G and let δ and Δ denote the smallest and largest degree of G. In each round, in expectation, each informed node receives at most Δ/δ requests. Hence, if an uninformed node u sends an RPULL request to an informed node, u should receive the rumor with probability at least $\Omega(\delta/\Delta)$. Consequently, intuitively, the slowdown of using random RPULL instead of the usual PULL protocol should not be more than $\tilde{O}(\Delta/\delta)$.[1] We prove that this intuition is correct. For every given instance, we show that if the PULL algorithm informs all nodes in T rounds with probability p, for the same instance, the random RPULL algorithm manages to reach all nodes in time

---

[1] Here $\tilde{O}$ hides $\log(n)$ factors.

$O(\mathcal{T} \cdot \frac{\Delta}{\delta} \cdot \log n)$ with probability $(1 - o(1))p$.[2] While the statement might seem very intuitive, its formal proof turns out quite involved. Formally, we prove a stronger statement and show that a single round of the PULL protocol is w.h.p. *stochastically dominated* by $O(\frac{\Delta}{\delta} \cdot \log n)$ rounds of random RPULL in the following sense. We give a coupling between the random processes defined by PULL and random RPULL such that for every start configuration, w.h.p., the set of nodes informed after $O(\frac{\Delta}{\delta} \cdot \log n)$ rounds of random RPULL is a superset of the set of nodes informed in a single PULL round. The same holds for simulating one round of PUSH-PULL with PUSH-RPULL. A similar coupling between rumor spreading algorithms has been done in [16] where the authors couple $\log(n)$ rounds of asynchronous- with one round of synchronous PUSH-PULL. A coupling between PULL and RPULL in the classic sense, i.e., a coupling which does relinquish the w.h.p. term does not exist. We also show that for such a round-by-round analysis, our bound is tight. That is, there are configurations where $\Omega(\frac{\Delta}{\delta} \log n)$ random RPULL rounds are needed to dominate a single PULL round with high probability.

Notation and Preliminaries. Let $G = (V, E)$ be the n-node network graph. For a node $u \in V$, we use $N(u)$ to denote the set of neighbors of u and $d(u) := |N(u)|$ to denote its degree. Given a set of nodes $S \subseteq V$, we define $N_S(u) := N(u) \cap S$ to be the set of u's neighbors in S and $d_S(u) := |N_S(u)|$ for the number of neighbors of u in S. The smallest and largest degrees of G are denoted by δ and Δ, respectively. For a set $V' \subseteq V$ we denote with $G[V']$ the graph induced by V'. To indicate a disjoint union of two sets, i.e., $A \cup B$ with $A \cap B = \emptyset$, we write $A \dot\cup B$. For a set of natural numbers $\{1, \ldots, k\}$ we only write $[k]$.

When analyzing the progress of an algorithm ALG, the set S_t^{ALG} denotes the set of informed nodes after t rounds and U_t^{ALG} the set of uninformed nodes. When the algorithm is clear from the context we simply write S_t and U_t, furthermore we denote $S = S_0$ and $U = U_0$ for the initial configuration. Missing proofs appear in [1].

2   Separation of Adversarial and Random RPULL

We want to show that the adversarial RPULL can be exponentially slower than the randomized RPULL on general graphs. To show this, we first establish results on the run time of both algorithms on trees. These results might also be of independent interest.

In a tree network let $p_{v,u} = (v = v_0, v_1, v_2, \ldots, v_q = u)$ denote the unique path from v to u, though we use that notation also for the set of nodes on that path, i.e., $p_{v,u} = \{v, v_1, v_2, \ldots, v_{q-1}, u\}$. For a path p, let $M_p := \sum_{w \in p} d(w)$ be the sum of all degrees on the path.

The next lemma shows that on a tree any form of RPULL is asymptotically as fast as PULL plus an additive term in the order of the degree of the node that initially has the rumor.

---

[2] Actually, $\frac{\Delta}{\delta}$ can be replaced by $\max_{\{u,v\} \in E} d(u)/d(v)$ in all parts of the paper, where $d(u)$ and $d(v)$ denote the degrees of the nodes.

Lemma 1. *Let G be a tree network with $S_0 = \{r\}$ and let u be a node in U_0. Furthermore, let τ be the first round in which $u \in S_\tau$ holds, i.e., the number of rounds until u gets informed.*

(1) $\mathbf{E}[\tau] = \Theta(M_{p_{r,u}} - d(r))$ *for* PULL,
(2) $\mathbf{E}[\tau] = \Omega(M_{p_{r,u}} - d(r))$ *for every type of* RPULL,
(3) $\mathbf{E}[\tau] = O(M_{p_{r,u}})$ *for adversarial* RPULL.

Proof (Lemma 1). (1) We root the tree at the only informed node, r. Note that nodes are not aware of their own parent/child relationships. Consider some time t at which a node r' on the path $p_{r,u}$ is in $S_t \backslash S_{t-1}$, i.e., it just got informed. Thus its child $u' \in p_{r,u}$ on the path is not yet informed, i.e., $u' \in U_t$. In any round $t' \geq t$, in which u' is not informed yet, it requests its parent with probability $1/d(u')$. Thus each uninformed node $u' \in p_{r,u} \backslash \{r\}$ on the path needs $\Theta(d(u'))$ rounds in expectation before it can get informed. Linearity of expectation proves the claim for PULL.

(2) follows from the fact that RPULL is at most as fast as PULL.

(3) For adversarial RPULL divide all rounds $t' \geq t$ in which u' is not yet informed into two types: First rounds in which at least one *sibling* of u', i.e., the nodes in $N(r')\backslash\{u'\}$, requests from r' and secondly rounds in which no sibling of u' requests from r'. The first type of rounds is upper bounded by $d(r')$ because every neighbor of r' stops requesting after receiving the rumor. In expectation u' gets the rumor after $d(u')$ rounds of type two; thus in expectation u' is informed within $O(d(r')+d(u'))$ rounds. Applying this recursively to all uninformed nodes on the path $p_{r,u}$, we get the claimed result via linearity of expectation.

Lemma 2. *Let G be a tree network with $S_0 = \{r\}$. Then in both random and adversarial RPULL it takes $O\left(\max_{path\,p} M_p + \Delta \log n \right)$ rounds to fully inform all nodes in V, w.h.p..*

Proof (Sketch). Let u be a uninformed with informed parent v. In each round v either informs one child other than u or u has a probability of $1/d(u)$ to get the rumor. In expectation for each predecessor-successor pair (v, u) on a path p the time for u to get informed is at most $d(v)+d(u)$. A Chernoff bound for geometric random variables [1, Lemma B.1] provides the time bound of $O(M_p + \Delta \log n)$ for that path and a union bound over all paths concludes the proof.

Lemma 2 shows that random RPULL and adversarial RPULL are essentially the same on trees. This does not hold for general graphs.

Lemma 3. *There is a graph $G = (V, E)$ of size $\Theta(n)$ with a node $r_\alpha \in V$, $d(r_\alpha) \leq 3$, such that:*

- *For $S_0 = \{r_\alpha\}$, w.c.p., the run-time of adversarial RPULL is in $\Omega(\sqrt{n})$.*
- *For any non-empty $S_0 \subset V$, w.h.p., the run-time of randomized RPULL is in $O(\log^2 n)$.*

Proof (sketch). See Fig. 1 for a visualization of the graph defined here. We first introduce a special graph type that we call a k-leaf-connected tree (k-LCT). In simple words, a k-LCT is a binary tree with k leaves, but with its k leaves being fully interconnected, i.e., forming a clique. Propagation of the rumor from one node to all nodes of a k-LCT, happens in $O(\log k)$. This also holds true if we embed this k-LCT into a larger graph, as long as the degrees do not grow much in this manner. As long as the k-LCT is uninformed most requests of the leaves will target other leaf nodes, hence it is unlikely for a k-LCT to acquire the rumor through its leaves.

We construct G as follows: We let D_α and D_ζ be two n-LCTs, and we have m l-LCTs that we denote with $D_1, D_2, \ldots, D_m$, where $l := \sqrt{n}$ and $m := c\sqrt{n}$ for some $c > 1$. Their corresponding roots and leaf sets are denoted as $r_\alpha, r_\zeta, r_1, r_2, \ldots r_m$ and $L_\alpha, L_\zeta, L_1, L_2, \ldots L_m$ respectively, and with $l_{X,1}, l_{X,2}, \ldots$ we enumerate the leaves of leaf set L_X. Let $C_\alpha = \{c_1, \ldots c_m\}$ be an arbitrary m-sized subset of $D_\alpha \backslash L_\alpha$ and add the following edges:

- Between r and D_ζ: Add one edge from r to r_ζ
- Between r and D_α: For each $j \in [m \log n]$: add edge $\{r, l_{\alpha,j}\}$
- Between r and $D_1, \ldots, D_m$:   For each $i \in [m]$, $j \in [\log n]$: add edge $\{r, l_{i,j}\}$
- Between $D_1, \ldots, D_m$ and D_ζ: For each $i \in [m]$ add edge $\{r_i, l_{\zeta,i}\}$
- Between $D_1, \ldots, D_m$ and C_α: For each $i \in [m]$ add edge $\{l_{i,l}, c_i\}$.

The graph is built such that information propagation from D_ζ to the rest of the graph is quick, but not the other way round. In the random RPULL model, wherever the rumor starts, it reaches r quickly and from there r_ζ manages to pull in polylogarithmic time. Then the rumor quickly propagates through D_ζ, and from L_ζ to all LCTs $D_1, \ldots, D_m$ and afterwards to D_α.

In the adversarial RPULL model, as long as the rumor starts outside D_ζ, the rumor can quickly spread to r, D_α and a few of the D_is (i.e., the LCTs $D_1, \ldots, D_m$). But we let the adversary always prioritize a request at node r from a node in one of the D_is over a request from r_ζ to prevent that r_ζ gets the rumor. We can show that the number of informed D_is grows slowly and hence such requests exist w.h.p. as long as no node in D_ζ is informed. Also, with only few D_is informed, due to their high degrees, leaf nodes in L_ζ are unlikely to request from a D_i containing the rumor, and hence the progress of rumor propagation is stalled.

Theorem 1. *There is a graph $G = (V, E)$ of size $\Theta(n)$, such that for any $S_0 = \{s\} \subset V$:*

- *In expectation, the run-time of adversarial RPULL is in $\Omega(\sqrt{n})$.*
- *W.h.p., the run-time of randomized RPULL is in $O(\log^2 n)$.*

Proof. Let G' and G'' be duplicates of the graph G from Lemma 3, r'_α and r''_α being the respective duplicates of r_α. We set $G := G' \cup G''$ and add the edge $\{r'_\alpha, r''_\alpha\}$. Without loss of generality let $s \in V'$.

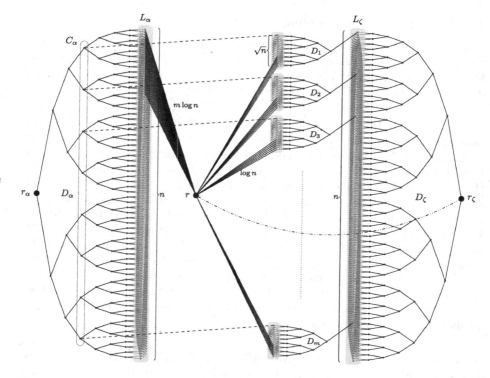

Fig. 1. The graph in Lemma 3. Grey areas indicate fully connected parts of the graph.

In the random version, the rumor propagates through all of G' in $O(\log^2 n)$ rounds. Due to its low degree, r''_α gets the rumor from r'_α within $O(\log n)$ time after G' is informed and again, in $O(\log^2 n)$ rounds G'' is informed completely.

In the adversarial version, G'' can only learn the rumor from G' through edge $\{r'_\alpha, r''_\alpha\}$. But once r''_α knows the rumor, we can apply Lemma 3 again to prove that now progress is stalled.

3   Comparison of PULL and Random RPULL

In this section we compare the two algorithms PULL and random RPULL on general graphs, i.e., we analyze how many rounds of random RPULL are enough to cover the progress of one round of PULL. More precisely, we show that w.h.p. the set of nodes informed after $O\left(\frac{\Delta}{\delta} \cdot \log n\right)$ rounds of random RPULL is a superset of the set of nodes informed in a single PULL round. We manage to do so by coupling both algorithms. At the end of the section we head out to prove that this bound is tight. Whenever we talk about RPULL in this section we mean random RPULL.

3.1   Dominance and Couplings

We begin with two examples of insufficient definitions of domination between two rumor spreading algorithms. Showing for two algorithms $\mathcal{A}$ and $\mathcal{A}'$ that $\mathbf{P}\left(u \in S^{\mathcal{A}}\right) \geq \mathbf{P}\left(u \in S^{\mathcal{A}'}\right)$ holds for all $u \in U$ is not enough to obtain a natural dominance definition of $\mathcal{A}$ over $\mathcal{A}'$, since due to dependencies for a set M with $|M| > 1$ it might still be true that $\mathbf{P}\left(M \subseteq S^{\mathcal{A}}\right) < \mathbf{P}\left(M \subseteq S^{\mathcal{A}'}\right)$.

Showing that $\mathbf{P}\left(M \subseteq S^{\mathcal{A}}\right) \geq \mathbf{P}\left(M \subseteq S^{\mathcal{A}'}\right)$ (*) holds for all $M \subseteq U$ is not enough either. Assume the following example: Let $U = \{a, b, c\}$ be the set of uninformed nodes. Assume that under $\mathcal{A}$ the probability that the set of newly informed nodes equals $\{a, b, c\}$, $\{a\}$, $\{b\}$ or $\{c\}$ is $1/8 + \epsilon$ each and the probability that it equals one of the sets $\{a, b\}$, $\{a, c\}$, $\{b, c\}$ or $\emptyset$ is $1/8 - \epsilon$ each. Under $\mathcal{A}'$ we inform any of those sets with probability $1/8$. A direct computation for all $M \subseteq \{a, b, c\}$, e.g., for $M = \{a\}$, $\mathbf{P}(\{a\} \subseteq S^{\mathcal{A}'}) = 1/2$ and $\mathbf{P}(\{a\} \subseteq S^{\mathcal{A}}) = \mathbf{P}(S^{\mathcal{A}} = \{a\}) + \mathbf{P}(S^{\mathcal{A}} = \{a, b\}) + \mathbf{P}(S^{\mathcal{A}} = \{a, c\}) + \mathbf{P}(S^{\mathcal{A}} = \{a, b, c\}) = 1/2$, shows that inequality (*) is fulfilled for any $M \subseteq U$, but the probability of the event *"at least 2 nodes are informed"* is by 2ϵ smaller for $\mathcal{A}$ than for $\mathcal{A}'$. In the following we introduce the classical method to relate stochastic processes, i.e., the notion of (first order) stochastic dominance. However, we show that (proper) stochastic dominance between RPULL and PULL does not exist and thus we weaken the notion afterwards.

Stochastic Dominance and Coupling. Let $(\mathcal{S}, \preceq_{\mathcal{S}})$ be a finite distributive lattice and let X_1 and X_2 be random variables with distributions $\mathbf{P}_1$ and $\mathbf{P}_2$ which take values in $\mathcal{S}$. A function $f : \mathcal{S} \to \mathbb{R}$ is called *increasing* if $A \preceq_{\mathcal{S}} B$ implies $f(A) \leq f(B)$.

Definition 1 (Stochastic Dominance). *We say that X_2 stochastically dominates X_1 if $E(f(X_2)) \geq E(f(X_1))$ holds for every increasing function $f : \mathcal{S} \to \mathbb{R}$, where $E(\cdot)$ denotes the expected value.*

Alternative to Definition 1, one can show that one process stochastically dominates a second process by defining a monotone coupling (cf. Theorem 2).

Definition 2 ((Monotone) Coupling). *A coupling of two random processes X_1 and X_2, taking values in $\mathcal{S}$ with distributions $\mathbf{P}_1$ and $\mathbf{P}_2$, is a joint distribution $\hat{\mathbf{P}}$ of a random process $(\hat{X}_1, \hat{X}_2)$ taking values in $\mathcal{S} \times \mathcal{S}$, such that its marginals equal the distributions of X_1 and X_2, respectively, i.e.,*

$$\sum_{B \in \mathcal{S}} \hat{\mathbf{P}}\left((\hat{X}_1, \hat{X}_2) = (A, B)\right) = \mathbf{P}_1(X_1 = A) \; \forall A \in \mathcal{S} \; and$$

$$\sum_{A \in \mathcal{S}} \hat{\mathbf{P}}\left((\hat{X}_1, \hat{X}_2) = (A, B)\right) = \mathbf{P}_2(X_2 = B) \; \forall B \in \mathcal{S}.$$

A coupling $\hat{\mathbf{P}}$ is called monotone (written $X_1 \leq X_2$) if also the following holds:

$$\forall A, B \in \mathcal{S} \; with \; \hat{\mathbf{P}}\left((\hat{X}_1, \hat{X}_2) = (A, B)\right) > 0 \; it \; follows \; that \; A \preceq_{\mathcal{S}} B. \tag{1}$$

Note that the choice of a coupling between two processes is generally not unique. The following theorem, Strassen's Theorem [17,18], shows an equivalence between stochastic dominance and the notion of monotone couplings.

Theorem 2 (Strassen [17,18]). *The following are equivalent:*

1. X_2 stochastically dominates X_1,
2. *There exists a monotone coupling between X_1 and X_2 such that $X_1 \leq X_2$,*
3. $\mathbf{P}(X_2 \in F) \geq \mathbf{P}(X_1 \in F)$ *holds for every monotone set $F \subseteq \mathcal{S}$.*[3]

Stochastic dominance/monotone couplings are the commonly known method to relate stochastic processes and we would like to show that $O(\frac{\Delta}{\delta} \log n)$ rounds of random RPULL stochastically dominate one round of PULL. This, however, is not possible as one can easily construct a graph in which some node u is informed with probability 1 in one round of PULL, but with probability less than 1 in $O(\frac{\Delta}{\delta} \log n)$ rounds of RPULL.[4] Therefore we introduce the notion of highly probable monotone couplings and – in analogy to the equivalencies from Strassen's Theorem – also the notion of highly probable stochastical dominance.

Definition 3. *A coupling $\hat{\mathbf{P}}$ of random processes X_1 and X_2 is called* ***monotone w.h.p.*** *(w.r.t. to n) (written $X_1 \leq_{w.h.p} X_2$) if for some $c > 1$ it satisfies*

$$\sum_{A \npreceq B} \hat{\mathbf{P}}\left((\hat{X}_1, \hat{X}_2) = (A, B)\right) \leq \frac{1}{n^c}. \tag{2}$$

We say X_2 stochastically dominates X_1 w.h.p. (w.r.t. n), if there exists a coupling between X_1 and X_2 that is monotone with high probability ($X_1 \leq_{w.h.p} X_2$).

In this paper we will set $\mathcal{S} = 2^U$ to be the power set of U, where $U \subseteq V$ is the set of uninformed nodes, $\preceq_\mathcal{S}$ equals the subset relation on U and X_2 and X_1 will be the respective random variables describing which nodes get informed in RPULL and PULL. In general, the parameter n in Definition 3 can be freely chosen; in our setting n will be the number of nodes of the communication network. With this choice of parameters a monotone coupling (w.h.p. w.r.t. n) is the desired relation of PULL and RPULL.

3.2   W.h.p. Monotone Coupling Between PULL and RPULL

Theorem 3. *W.h.p., for any set of informed nodes $S \subseteq V$, $T = O(\frac{\Delta}{\delta} \log n)$ rounds of random RPULL stochastically dominate a single round of PULL.*

Corollary 1. *If in a graph G with initially informed nodes $S \subseteq V$ the PULL algorithm informs all nodes in T rounds with probability p, then the random RPULL algorithm informs all nodes in time $O(T \cdot \frac{\Delta}{\delta} \cdot \log n)$ with probability $(1 - o(1))p$.*

---

[3] A set $F \subseteq \mathcal{S}$ is called monotone if $A \in F$ and $A \preceq_\mathcal{S} B$ implies $B \in F$.
[4] Figure 2 can be used to verify this.

By PUSH−RPULL we denote the combination of RPULL with a simultaneous execution of the classic PUSH protocol. The restriction of a single node to answer only a limited number of requests does not limit the progress of the PUSH algorithm when disseminating a rumor.

Corollary 2. *W.h.p., for any set of informed nodes $S \subseteq V$, $O\left(\frac{\Delta}{\delta} \log n\right)$ rounds of* PUSH − RPULL *stochastically dominate a single round of* PUSH − PULL.

To reduce dependencies between nodes which request from the same neighbor we introduce a new algorithm VPULL (virtual pull). VPULL is only introduced for the sake of analysis; difficulties that arise in an actual implementation of VPULL are not relevant. The proof of Theorem 3 is split into two parts such that it follows from the transitivity of the stochastical dominance relation:

Lemma 4: W.h.p., T rounds of RPULL stoch. dominate VPULL,
Lemma 6: VPULL stochastically dominates one round of PULL.

By RPULL$_T$ we denote the (randomized) process RPULL which runs for T rounds, by VPULL we denote one execution of VPULL and by PULL$_1$ we denote the process PULL which runs for one round only. The random variables S_T^{RPULL}, S^{VPULL} and S_1^{PULL} denote the respective sets of nodes that are informed after the corresponding number of rounds. The processes RPULL$_T$, VPULL and PULL$_1$ are not completely characterized by the random variables S_T^{RPULL}, S^{VPULL} and S_1^{PULL} – one has to include information about all requests and messages, that are sent by all nodes, to fully describe the random processes. Nevertheless, to show the desired result, it is sufficient to find a monotone coupling where condition (1) and (2), respectively, are fulfilled with regard to the subset relation of the set valued random variables S_T^{RPULL}, S^{VPULL} and S_1^{PULL}.

Definition of VPULL. Let us define *weakly connected* nodes $u \in U$ as nodes for which $d_S(u)/d(u) \leq 1/2$ and *strongly connected* otherwise. To introduce the process VPULL we need the following parameters, which we fix later.

$$K = \Theta\left(\frac{\Delta}{\delta} + \log n\right), \quad T' = O\left(\frac{\Delta}{\delta} \log n\right) \quad \text{and} \quad T = \Theta(T'), T \gg T'$$

An execution of VPULL consists of two phases. In the first phase nodes send tokens instead of the actual rumor and w.h.p. nodes who have received a token in the first phase are informed in the second phase. In an execution of VPULL we let $X_v(t)$ be the number of tokens which node v has sent up to round t. In a round t denote with $R_v(t)$ the set of nodes requesting from some informed node $v \in S$ and with $r_v(t) = |R_v(t)|$ its cardinality. R_v, r_v and X_v are random variables which describe certain properties of an execution of VPULL, where large values of X_v or r_v indicate the unlikely case in which the (strict) monotonicity of a tentative coupling between VPULL and RPULL$_T$ might break.

Definition 4 (Good, Bad Execution). *An execution of* VPULL *is called a* bad execution *if for some $v \in V$ or $1 \leq t \leq T$ it holds that $X_v(t) > K$ or $r_v(t) > K$, otherwise it is called a* good execution.

Here, we describe VPULL informally (for a formal defintion cf. Algorithm 1). An execution of VPULL is split into two phases – the first phase consists of T rounds and the second phase of one round. In the first phase an uninformed node requests the rumor uniformly at random from one of its neighbors and an informed node v decides with probability $\frac{r_v}{T'}$ whether to send out a token – in which case it selects, uniformly at random, one of its incoming requests as destination for the token. Nodes that get a token in those T rounds, stop requesting from neighbors, but are still unable to forward any information in consecutive rounds. In the second phase the limit to the number of requests that can be served by an informed node is stripped away. Then, in case of a bad execution (the first phase determines whether an execution is good or bad), all actions from the first phase are discarded and all uninformed nodes perform one round of PULL. In case of a good execution, all uninformed strongly connected nodes perform one round of PULL and afterwards all nodes holding a token are being informed. If we assume that tokens are as valuable as the information itself, in each of the T rounds of the first phase, VPULL differs from RPULL only in the fact that the selected incoming connection is established with probability $\frac{r_v}{T'}$ whereas it is established deterministically in RPULL. For an uninformed node $u \in U$, that chooses to request a neighbor $v \in S$, this normalizes the probability to get a token to $1/T'$, independent of the amount of other requesting nodes. Except for the second phase this algorithm is clearly dominated by RPULL.

W.h.p. Monotone Coupling Between RPULL and VPULL. First, we generate a monotone coupling between T rounds of RPULL and the first phase of VPULL as follows: For each round in the first phase both processes use the same randomness to decide on the outgoing calls of uninformed nodes (if in a round $t > 1$ a node is uninformed in VPULL but not in RPULL the process VPULL uses additional randomness; the contrary cannot happen). VPULL uses additional randomness to decide whether a node, which is contacted, sends out any message at all, confer line 13 from Algorithm 1. This simultaneous execution of both algorithms gives rise to a coupling of the first phase of VPULL and $RPULL_T$. Clearly, a node that is provided with a token in VPULL in any round is then also informed in RPULL, i.e., the coupling is monotone.

Lemma 4. $RPULL_T$ *stochastically dominates* VPULL *with high probability.*

Proof. Under the assumption that tokens are as valuable as the information itself we constructed a monotone coupling of the first phase of VPULL and S_T^{RPULL}. Then, it is sufficient to prove that in the second phase of VPULL, w.h.p., no node is informed, that has not been informed in the T rounds of RPULL: If neither ever any value r_v nor any X_v exceeded K, then only strongly connected nodes simulate one round of PULL in the second phase of the VPULL algorithm. We claim that, w.h.p., each strongly connected node has been informed in the first T rounds of RPULL: A strongly connected node $u \in U$ requests from an informed node $v \in S$ with probability at least $1/2$. In any given round due to Markov inequality with probability at least $1/2$ no more than $2\Delta/\delta$ nodes $u' \in U$

connect to v. The probability for u to get informed under RPULL is thus at least $\frac{\delta}{8\Delta}$. Choosing $T = O\left(\frac{\Delta}{\delta}\log n\right)$ big enough and a union bound gives us that, w.h.p., all strongly connected nodes are informed in process RPULL$_T$.

Algorithm 1. One execution of VPULL $((T+1)$-rounds)

Input: K – threshold for bad execution; T' – parameter to normalize probabilities
States: informed; uninformed
Oracle knowledge: $d_S(v)$ for every node v; $BE := \bigvee_{v \in V} BE_v$
Vars.: R_v set of nodes request. from v in the corresp. round $(r_v := |R_v|)$
$\quad\quad\quad$ BE_v boolean indicator for bad execution caused at node v
$\quad\quad\quad$ $tokenReceived_v$ indicates whether a node will be informed after T rounds
1: $BE_v \leftarrow$ **false**; $tokenReceived_v \leftarrow$ **false**
2: **for** T rounds **do**
3: $\quad$ **switch** $state_v$ **do**
4: $\quad\quad$ **case** uninformed
5: $\quad\quad\quad$ **if** $tokenReceived_v =$ **false then**
6: $\quad\quad\quad\quad$ send request for rumor uniformly at random
7: $\quad\quad\quad\quad$ **if** $msg = token$ **then**
8: $\quad\quad\quad\quad\quad$ $tokenReceived_v \leftarrow$ **true**
9: $\quad\quad$ **case** informed
10: $\quad\quad\quad$ **if** $r_v > K$ or $X_v > K$ **then**　　　// bad execution has been detected locally
11: $\quad\quad\quad\quad$ $BE_v \leftarrow$ **true**
12: $\quad\quad\quad$ **else**
13: $\quad\quad\quad\quad$ **with probability** r_v/T' **do**
14: $\quad\quad\quad\quad\quad$ send $token$ to uniformly at random chosen node in $R_v \neq \emptyset$
15: $\quad\quad\quad\quad\quad$ $X_v \leftarrow X_v + 1$
$\quad$ // Round $T+1$:
16: request $(BE, d_S(v))$ from global oracle
17: **if** $BE =$ **true then**　　　// bad execution has been detected globally
18: $\quad$ execute one round of PULL　　　// executed locally
19: **else**
20: $\quad$ **if** $d_S(v)/d(v) > 1/2$ **then**　　　// node is strongly connected
21: $\quad\quad$ execute one round of PULL　　　// executed locally
22: $\quad$ **else if** $tokenReceived_v =$ **true then**
23: $\quad\quad$ $state_v \leftarrow$ informed　　　// node learns rumor

---

W.h.p., $r_v \leq K$ in VPULL$_T$ for all v. For a fixed informed node v, in expectation, no more than $\frac{\Delta}{\delta}$ nodes can request from v. Using a Chernoff bound for a single round and a single node, $\mathbf{P}\left(r_v \geq \kappa\left(\frac{\Delta}{\delta} + \log n\right)\right) \leq n^{-\Theta(\kappa)}$ holds. With a union bound over all nodes and all rounds and κ large enough we obtain that, w.h.p., r_v never exceeds $\kappa K'$ and therefore neither K. A union bound over all nodes concludes the proof.

W.h.p., $X_v \le K$ in VPULL$_T$ for all v. For a fixed v, note that, w.h.p., in a single round no more than $\kappa K'$ nodes request from v, and therefore, X_v is increased at most with probability $\kappa K'/T'$ in any round. Over T rounds, in expectation, no more than $\kappa K' \frac{T}{T'}$ increments of X_v happen, and again a Chernoff bound gives us that X_v does not exceed $2\kappa K' \frac{T}{T'}$ with high probability. Choosing $c_{T,T'} = 2\kappa \frac{T}{T'}$ and a union bound over all nodes concludes the proof.

Stochastic Dominance Between VPULL and PULL. In a single round of PULL a node $u \in U$ is informed with probability $\frac{d_S(u)}{d(u)}$, independently from which other nodes are informed. For one execution of VPULL we can show that a node u is also informed at least with probability $\frac{d_S(u)}{d(u)}$, independently from which other nodes get informed (Lemma 5). Afterwards, we prove that Lemma 5 is sufficient to deduce the stochastic dominance of one execution of VPULL over PULL$_1$. For $u \in U$ and random process X, let $\mathcal{C}_u^X$ be the set of all conditions of the type $v \in X$ or $v \notin X$ where $v \ne u$.

Lemma 5. *In VPULL a node $u \in U$ is informed at least with probability $\frac{d_S(u)}{d(u)}$, independently from which other nodes are informed, i.e., for all sets of conditions $I \subseteq \mathcal{C}_u^{\mathrm{VPULL}}$ and $J \subseteq \mathcal{C}_u^{\mathrm{PULL}}$ with $\mathbf{P}(I), \mathbf{P}(J) > 0$ the following holds*

$$\mathbf{P}\left(u \in S_{T+1}^{\mathrm{VPULL}} \,\middle|\, I\right) \ge \frac{d_S(u)}{d(u)} = \mathbf{P}\left(u \in S_1^{\mathrm{PULL}}\right) = \mathbf{P}\left(u \in S_1^{\mathrm{PULL}} \,\middle|\, J\right). \quad (3)$$

Proof. If $u \in U$ is strongly connected, the result holds because VPULL executes one round of PULL for u in either way. In a bad execution, VPULL executes one round of PULL for any uninformed node and the claim holds trivially. Thus assume that u is weakly connected and we are in a good execution. Let $s = d_S(u)$ and $N_S(u) = \{v_1, \ldots, v_s\}$ be the neighbors of u in S. We call a node $v \in N_S(u)$ *busy* w.r.t. u in round t if it informs some node other than u. Let y_t be the number of busy nodes in round t w.r.t. u. In a good execution (which we denote by $\mathbb{G}$), any node in $N_S(u)$ can inform at most K nodes and hence there is the following constraint on the sum of all y_t's

$$\sum_{t=1}^{T} y_t \le s \cdot K. \quad (4)$$

We can ignore conditions in I corresponding to nodes which do not have a common neighbor with $N_S(u) \cup \{u\}$ as u can only get the rumor directly through S. The only negative effect on the probability that u gets informed by the conditions in I can be captured by the number of busy nodes w.r.t. u. Since the number of nodes which are informed per node in a good execution is small compared with T, there are sufficiently many rounds with sufficiently many non-busy nodes to inform u. More precisely, if u requests from a non-busy node it is informed at least with probability $\frac{1}{T'}$. Thus, the probability that u, conditioned on $I \wedge \mathbb{G}$

with $\mathbf{P}(I \wedge \mathbb{G}) > 0$, is not informed is smaller or equal to (with $c = T/T'$)

$$\prod_{t=1}^{T} \left(1 - \frac{s - y_t}{d(u) \cdot T'}\right) \le \left(1 - \frac{s\left(1 - \frac{K}{T}\right)}{d(u) \cdot T'}\right)^{T} \le e^{-c\left(1 - \frac{K}{T}\right)\frac{s}{d(u)}} \le 1 - \frac{d_S(u)}{d(u)}.$$

The first inequality holds because under constraint (4) the expression on the left hand side is maximized for $y_t = \frac{s \cdot K}{T}$. The last inequality holds due to $\frac{s}{d(u)} \le 1/2$, $c(1 - K/T) \ge 2$ and the fact that $e^{-2x} \le 1 - x$ for any $x \in [0, 1/2]$.

Lemma 5 is sufficient to show stochastic domination of VPULL over PULL$_1$. For a formal proof one can either directly construct a monotone coupling between the random variables or use the following theorem by Holley [19].

Theorem 4 (Holley Inequality, [19]). *Let $(\mathcal{S}, <)$ be a distributive lattice and let μ_1, μ_2 be measures on this lattice. The Holley criterion is satisfied if*

$$\mu_1(A \cap B)\mu_2(A \cup B) \ge \mu_1(A)\mu_2(B) \text{ holds for all } A, B \in \mathcal{S}. \quad (5)$$

If the Holley criterion is satisfied for μ_1 and μ_2 then

$$\sum_{A \in \mathcal{S}} \mu_1(A)f(A) \ge \sum_{A \in \mathcal{S}} \mu_2(A)f(A) \text{ holds for all increasing functions } f : \mathcal{S} \to \mathbb{R}.$$

Lemma 6. *One execution of VPULL stochastically dominates PULL$_1$.*

Proof. Let U be the uninformed nodes u with $0 < d_S(u) < d(u)$ and consider the distributive lattice $(\mathcal{S}, \preceq_\mathcal{S}) = (2^U, \subseteq)$. Every uninformed node u which is not contained in this set U has either no connection to S at all, i.e., it is not informed in either process, or $d_S(u) = d(u)$ holds, i.e., it is informed with probability one in either process because also VPULL executes one round of PULL for it. Hence it is sufficient to show stochastic domination of S^{VPULL} over S_1^{PULL} restricted to this set U. This choice of U provides $0 < \mathbf{P}(S_1^{\text{PULL}} = A), \mathbf{P}(S^{\text{VPULL}} = A) < 1$ for all $A \in \mathcal{S}$ and we define the *strictly positive* measures $\mu_1(F) := \mathbf{P}\left(S^{\text{VPULL}} \in F\right)$ and $\mu_2(F) := \mathbf{P}\left(S_1^{\text{PULL}} \in F\right)$ for $F \subseteq 2^U$. In [1, Claim xyz] we show that Lemma 5 implies the Holley criterion. Then Lemma 6 follows with Theorem 4 and the definition of the expected value of a function $f : \mathcal{S} \to \mathbb{R}$.

Proof (Theorem 3). The theorem follows with Lemmas 4 and 6.

3.3 The Round-by-Round Analysis is Tight

Lemma 7. *The time bound $T = O(\frac{\Delta}{\delta} \log n)$ from Theorem 3 is tight.*

Proof. We construct a graph G for which at least $T = \Omega(\frac{\Delta}{\delta} \log n)$ rounds of RPULL are necessary to guarantee that a node $a \in G$ is informed w.h.p. whereas a is informed with probability 1 in PULL (cf. Fig. 2).

We partition the set V into $V = A \cup B \cup T_{1,1} \cup \ldots \cup T_{k^2,k^2}$, where $k = n^{1/5}$, $A = \{a_1, \ldots, a_{k^2}\}$, $B = \{b_1, \ldots, b_{k^2}\}$. A and B form a complete bipartite graph

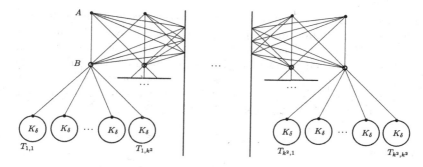

Fig. 2. Illustrating the graph in the proof of Lemma 7.

with edges running between A and B. For each $j \in [k^2]$, node b_i is connected to *one* node $t_{i,j} \in T_{i,j}$. Each $T_{i,j}$ forms a complete graph of size k. In this graph, $\delta = k - 1$ (acquired in $T_{i,j}$) and $\Delta = 2k^2$ (nodes in B), and therefore $\Delta/\delta \in \Theta(k)$. The total size of the graph is $|V| = n + o(n)$. Initially, we let $S_0 = B$.

Within one round of PULL, all nodes of A are informed with probability 1. Now, consider the same graph after $m \leq k^2/2$ rounds of RPULL and let us assume that some node $a \in A$ is still uninformed. It requests in this round from some node b_i. Let X_i be the number of requests at b_i. Within m rounds, each node b_i managed to inform at most m of its neighbors from $NB_i := \{t_{i,1}, \ldots, t_{i,m}\}$. Since $m \leq k^2/2$, at least half of all nodes in NB_i are still uninformed and thus, since they have degree k, $\mathbf{E}[X_i] \geq k/2$. Applying Chernoff, we get that w.h.p., $X_i \geq k/4$. In this scenario for a the probability to be chosen over one of its competitors is at most $4/k$, regardless of m, and therefore, $\mathbf{P}(a \in S_m^{\mathrm{RPULL}}) \leq \left(1 - \frac{4}{k}\right)^m$. For this to fall below $1/n$, m has to be in $\Theta\left(\frac{\Delta}{\delta}\log n\right)$.

4   Conclusions

Lemma 7 and Theorem 3 show that to simulate one round of PULL, $\Theta\left(\frac{\Delta}{\delta}\log n\right)$ rounds of RPULL are required. However, in case one wants stochastical dominance (w.h.p.) over $\mathcal{T} > 1$ rounds of PULL, the lower bound proof of Lemma 7 does not hold. We believe that for $\mathcal{T} = \Omega(\log n)$, on any graph G and any set of initially informed nodes $S \subseteq V$, $O\left(\mathcal{T}\left(\frac{\Delta}{\delta} + \log n\right)\right)$ or maybe even $O\left(\mathcal{T}\left(\frac{\Delta}{\delta}\right) + \log n\right)$ rounds of RPULL suffice to stochastically dominate $\mathcal{T}$ rounds of PULL. That proving this assumption might be a challenging task is underlined by a similar conjecture in [16], in which the authors do a coupling of synchronous and asynchronous PUSH-PULL. They obtain a similar multiplicative $O(\log n)$ factor and also conjecture that it can be improved to an additive $O(\log n)$ term.

A possible alternative restriction of the PUSH-PULL protocol could be given by the following algorithm. In each round, every node requests from an outgoing neighbor chosen uniformly at random. At each node, one of the incoming requests is chosen (e.g., uniformly at random) and a connection to the requesting node is established. Finally, over all established links between an informed and an uniformed node, the uninformed node learns the rumor. Note that unlike in the restricted PUSH-PULL variant described in our paper, here, also two informed nodes or two uninformed nodes could be paired. Such a PUSH-PULL variant can be analyzed in an analogous way to our analysis of the RPULL protocol and it can be shown that $O\left(\frac{\Delta}{\delta}\log n\right)$ rounds of this algorithm stochastically dominate a single round of the regular PUSH-PULL protocol.

References

1. Daum, S., Kuhn, F., Maus, Y.: Rumor spreading with bounded in-degree. CoRR, abs/1506.00828 (2015)
2. Demers, A., Greene, D., Hauser, C., Irish, W., Larson, J. Shenker, S., Sturgis, H., Swinehart, D., Terry, D.: Epidemic algorithms for replicated database management. In: Proceedings of the Symposium on Principles of Distributed Computing (PODC), pp. 1–12 (1987)
3. Feige, U., Peleg, D., Raghavan, P., Upfal, E.: Randomized broadcast in networks. Random Struct. Algorithms 1(4), 447–460 (1990)
4. Frieze, A.M., Grimmet, G.R.: The shortest-path problem for graphs with random arc-lengths. Discrete Appl. Math. 10(1), 57–77 (1985)
5. Chierichetti, F., Lattanzi, S., Panconesi, A.: Almost tight bounds for rumour spreading with conductance. In: Proceedings of the Symposium on Theory of Computing (STOC), pp. 399–408 (2010)
6. Giakkoupis, G.: Tight bounds for rumor spreading in graphs of a given conductance. In: Proceedings of the International Symposium on Theoretical Aspects of Computer Science (STACS), pp. 57–68 (2011)
7. Karp, R., Schindelhauer, C., Shenker, S., Vöcking, B.: Randomized rumor spreading. In: Proceedings of the Symposium on Foundations of Computer Science (FOCS), pp. 565–574 (2000)
8. Kempe, D., Dorba, A., Gehrke, J.: Gossip-based computation of aggregate information. In: Proceedings of the Symposium on Foundations of Computer Science (FOCS), pp. 482–491 (2003)
9. Mosk-Aoyama, D., Shah, D.: Fast distributed algorithms for computing separable functions. IEEE Trans. Inf. Theor. 54(7), 2997–3007 (2008)
10. Chen, J., Pandurangan, G.: Optimal gossip-based aggregate computation. In: Symposium on Parallelism in Algorithms and Architectures (SPAA), pp. 124–133 (2010)
11. Censor-Hillel, K., Haeupler, B., Kelner, J.A., Maymounkov, P.: Global computation in a poorly connected world: fast rumor spreading with no dependence on conductance. In: Proceedings of the Symposium on Theory of Computing (STOC), pp. 961–970 (2012)
12. Censor-Hillel, K., Shachnai, H.: Fast information spreading in graphs with large weak conductance. SIAM J. Comput. 41(6), 1451–1465 (2012)
13. Fountoulakis, N., Panagiotou, K., Sauerwald, T.: Ultra-fast rumor spreading in socal networks. In: Proceedings of the Symposium on Discrete Algorithms (SODA), pp. 1642–1660 (2012)
14. Giakkoupis, G., Sauerwald, T.: Rumor spreading and vertex expansion. In: Proceedings of the Symposium on Discrete Algorithms (SODA), pp. 1623–1641 (2012)

15. Giakkoupis, G.: Tight bounds for rumor spreading with vertex expansion. In: Proceedings of the Symposium on Discrete Algorithms (SODA), pp. 801–815 (2014)

16. Acan, H., Collevecchio, A., Mehrabian, A., Wormald, N.: On the push & pull protocol for rumour spreading. CoRR, abs/1411.0948 (2014)

17. Strassen, V.: The existence of probability measures with given marginals. Ann. Math. Stat. **36**(2), 423–439 (1965)

18. Domb, C., Lebowitz, J.L. (eds.): Phase Transitions and Critical Phenomena, vol. 18. Academic Press, London (2001)

19. Holley, R.: Remarks on the FKG inequalities. Commun. Math. Phys. **36**(3), 227–231 (1974)

Whom to Befriend to Influence People

Manuel Lafond[1], Lata Narayanan[2(✉)], and Kangkang Wu[2]

[1] Department of Computer Science and Operational Research,
Université de Montréal, Montréal, Canada
[2] Department of Computer Science and Software Engineering,
Concordia University, Montréal, Canada
lata@cs.concordia.ca

Abstract. Alice wants to join a new social network, and influence its members to adopt a new product or idea. Each person v in the network has a certain threshold $t(v)$ for *activation*, i.e. adoption of the product or idea. If v has at least $t(v)$ activated neighbors, then v will also become activated. If Alice wants to activate the entire social network, whom should she befriend? We study the problem of finding the minimum number of links that Alice should form to people in the network, in order to activate the entire social network. This *Minimum Links* Problem has applications in viral marketing and the study of epidemics. We show that the solution can be quite different from the related and widely studied Target Set Selection problem. We prove that the Minimum Links problem is NP-complete, in fact it is hard to approximate to within an $\epsilon \ln n$ factor for some constant ϵ, even for graphs with degree 3 and with threshold at most 2. In contrast, we give linear time algorithms to solve the problem for trees, cycles, and cliques, and give precise bounds on the number of links needed.

1 Introduction

The increasing popularity and proliferation of large online social networks, together with the availability of enormous amounts of data about customer bases, has contributed to the rise of *viral marketing* as an effective strategy in promoting new products or ideas. This strategy relies on the insight that once a certain fraction of a social network adopts a product, a larger cascade of further adoptions is predictable due to the *word-of-mouth* network effect [3,14,22]. Inspired by social networks and viral marketing, Domingos and Richardson [11,27] were the first to raise the following important algorithmic problem in the context of social network analysis: If a company can turn a subset of customers in a given network into early adopters, and the goal is to trigger a large cascade of further adoptions, which set of customers should they target?

We use the well-known threshold model to study the influence diffusion process in social networks from an algorithmic perspective. The social network is modelled by a node-weighted graph $G = (V, E, t)$ with $V(G)$ representing

Research supported by NSERC, Canada.

© Springer International Publishing AG 2016
J. Suomela (Ed.): SIROCCO 2016, LNCS 9988, pp. 340–357, 2016.
DOI: 10.1007/978-3-319-48314-6_22

individuals in the social network, $E(G)$ denoting the social connections, and t an integer-valued *threshold function*. Starting with a *target set*, that is, a subset $S \subseteq V$ of nodes in the graph, that are *activated* by some external incentive, influence propagates deterministically in discrete time steps, and *activates* nodes. For any unactivated node v, if the number of its activated neighbors at time step $t - 1$ is at least $t(v)$, then node v will be activated in step t. A node once activated stays activated. It is easy to see that if S is non-empty, then the process terminates after at most $|V| - 1$ steps. We call the set of nodes that are activated when the process terminates as the *activated set*. The problem proposed by Domingo and Richardson [11,27] can now be formulated as follows: Given a social network $G = (V, E, t)$, and an integer k, find a subset $S \subseteq V$ of size k so that the resulting activated set is as large as possible. In the context of viral marketing, the parameter k corresponds to the budget, and S is a target set that maximizes the size of the activated set. One question of interest is to find the cheapest way to activate the *entire network*, when possible. The optimization problem that results has been called the *Target Set Selection Problem*, and has been widely studied (see for eg. [1,4,25]): the goal is to find a minimum-sized set $S \subseteq V$ that activates the entire network (if such a set exists). In a certain sense, the elements of this minimum target set S are the most influential people in the network; if they are activated, the entire network will eventually be activated.

There are, however, two hidden flaws in the formulation of the target set problem. First, the nodes in the target set are assumed to be activated immediately by external incentives, *regardless of their own thresholds of activation*. This is not a realistic assumption; in the context of viral marketing, it is possible, perhaps even likely, that highly influential nodes have high thresholds, and cannot be activated by external incentives alone. Secondly, there is no possibility of giving *partial* external incentives; indeed the target set is activated *only* by external incentives, and the remaining nodes *only* by the internal network effect.

In this paper, we address the flaws mentioned above. We study a related but different problem. Suppose Alice wants to join a new social network, whom should she befriend if her goal is to influence the entire social network? In other words, to whom should Alice create links, so that she can activate the entire network? If Alice creates a link to a node v, the threshold of v is only effectively reduced by one, and so v in turn is activated only if its threshold is one. We call our problem the *Minimum Links* problem (Min-Links).

The Min-Links problem provides a new way to model a viral marketing strategy, which addresses the flaws described in the target set problem formulation. Indeed, Alice can represent the external initiator of a viral marketing strategy. The links added from the external node correspond to the external incentive given to the endpoints of these links. The nodes that are the endpoints of these new links may not be immediately completely activated, but their thresholds are effectively reduced; this corresponds to their receiving partial incentives. One way of seeing this is that every individual to whom we link is given a $10 coupon; for some people this may be enough for them to buy the product, for others, it reduces their resistance to buying it. Individuals with high thresholds cannot be activated only by external incentives. The Min-Links problem also has important applications in epidemiology or the spread of epidemics: in the spread of a new disease, where an infected person arrives from *outside* a community, the Min-Links problem corresponds to identifying the smallest set of people such that if the infected external person has contact with this set, the entire community could potentially be infected.

Observe that the solution to the Min-Links problem can be quite different from the solution to the Target Set Selection problem for a given network. For example, consider a star network, where the leaves all have threshold 1, while the central node has degree $n-1$ and has threshold n. The optimal target set is the central node, while the only solution to the Min-Links problem is to create links to all nodes in the network. Thus, a solution to the Min-Links problem can be arbitrarily larger than one to the Target Set Selection problem for the same social network. However, any solution to the Min-Links problem is clearly also a feasible solution to the Target Set Selection problem.

1.1   Our Results

We prove that the Min-Links problem is NP-hard, and is in fact, hard to approximate to within an $\epsilon \log n$ factor for some $\epsilon < 1$. In light of the hardness results, we study the complexity of the problem for social networks that can be represented as trees, cycles, and cliques. In each case, we give a necessary and sufficient condition for the feasibility of the Min-Links problem, based on the structural properties and an observation of the threshold function. We then give $O(|V|)$ algorithms to solve the Min-Links problem for all the studied graph topologies. Finally, we give exact bounds on the number of links needed to activate the entire network for all the above specific topologies, as a function of the threshold values.

1.2   Related Work

The problem of identifying the most influential nodes in a social network has received a tremendous amount of attention [2,5,12,15–18,23]. The algorithmic question of choosing the target set of size k that activates the most number of nodes in the context of viral marketing was first posed by Domingos and Richardson [11]. Kempe et al. [20] started the study of this problem as a discrete optimization problem, and studied it in both the probabilistic independent cascade model and the threshold model of the influence diffusion process. They showed the NP-hardness of the problem in both models, and showed that a natural greedy strategy has a $(1-1/e-\epsilon)$-approximation guarantee in both models; these results were generalized to a more general cascade model in [21].

In the Target Set Selection problem, the size of the target set is not specified in advance, but the goal is to activate the entire network. Chen [4] showed that it is hard to approximate the optimal Target Set to within a polylogarithmic factor, even when all nodes have majority thresholds, or have constant degrees and thresholds two. A polynomial-time algorithm for trees was given in the same paper. Ben-Zwi et al. [1] generalized the result on trees to show that target set selection can be solved in $n^{O(w)}$ time where w is the treewidth of the input graph. The effect of parameters such as diameter, vertex cover number etc. of the input graph on the complexity of the problem are studied in [25]. The Minimum Target Set has also been studied from the point of view of the spread of disease or epidemics. For eg., in [19], the case when all nodes have a threshold k is studied; the authors showed that the problem is NP-complete for fixed $k \geq 3$.

Influence diffusion under time window constraints were studied in [13]. Maximizing the number of nodes activated within a specified number of rounds has also been studied [9,24]. The problem of dynamos or dynamic monopolies in graphs (eg. [26]) is essentially the target set problem restricted to the case when every node's threshold is half its degree.

The paper closest to our work is [8], in which Demaine *et al.* introduce a model to *partially incentivize* nodes to maximize the spread of influence. Our work differs from theirs in several ways. First, they study the maximization of influence given a fixed budget, while we study in a sense the budget (number of links) needed to activate the entire network. Second, they consider thresholds chosen uniformly at random, while we study arbitrary thresholds. Finally, they allow arbitrary fractional influence to be applied externally on any node, while in our model, every node that receives a link has its threshold reduced by the same amount.

2   Notation and Preliminaries

Given a social network represented by an undirected graph $G = (V, E, t)$, we introduce a set of external nodes U that are assumed to be already activated. We assume that all edges have unit weight; this is generally called the *uniform weight assumption*, and has previously been considered in many papers [4,6,7,13]. A *link set* for (G, U) is a set S of links between nodes in U and nodes in V, i.e. $S \subseteq \{(u, v) \mid u \in U; v \in V\}$. For a link set S, we define $E(S) = \{v \in V \mid \exists (u, v) \in S\}$, that is, $E(S)$ is the set of V-endpoints of links in S. For a node v, define $r(v)$ to be the number of links in S for which v is an endpoint. Since the set of external nodes U is already activated, observe that adding the link set S to G is equivalent to reducing the threshold of the node v by $r(v)$. In the viral marketing scenario, the link set S represents giving v a partial incentive of $r(v)$.

Given a link set S for a graph G, we define $I(G, S)$ to be the set of nodes in G that are eventually activated as a result of adding the link set S, that is, by reducing the threshold of each node $v \in E(S)$ by $min\{r(v), t(v)\}$, and then running the influence diffusion process. See Fig. 1 for an illustration. Observe that in the target set formulation, this is the same as the set of nodes activated by using U as the target set in the graph G', the graph obtained from G by adding the set U to the vertex set and the set S to the set of edges.

A link set S such that $I(G, S) = V$, that is, S activates the entire network, is called a *pervading link set*. A pervading link set of minimum size is called an *optimal pervading link set*.

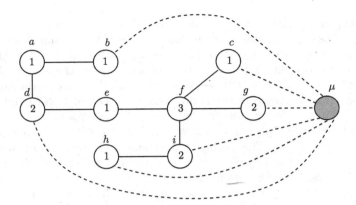

Fig. 1. Node μ is the external influencer and is assumed to be activated. Links in the link set are shown with dashed edges. The given link set activates the entire network and is an optimal pervading link set.

Definition 1. Minimum Links (Min-Links) **problem:** *Given a social network $G = (V, E, t)$, where t is the threshold function on V, and a set of external nodes U, find an optimal pervading link set for (G, U).*

In this paper, we consider the case of a *single influencer*, that is, $U = \{\mu\}$. In this case, a link given to a vertex v reduces its threshold by 1. Since μ must be an endpoint of each edge in the link set S, each such edge can be uniquely specified by a vertex in V. We therefore generally omit mention of μ in the rest of the paper. For each node $v \in E(S)$, we say we *give v a link*, or that v *receives* a link. If activating $X \subseteq V$ activates, directly or indirectly, the set of vertices Y, we write $X \sim Y$ (note that there may be vertices outside Y that X activates). We write $x \sim Y$ instead of $\{x\} \sim Y$. The minimum cardinality of a link set for a Min-Links instance G is denoted $ML(G)$.

Observe that for some graphs, a pervading link set may not exist; for example, consider a singleton node of threshold greater than 1. The existence of a feasible solution can be verified in $O(E)$ time by giving a link to every node in V, and simulating the influence diffusion process. The following simple observation stating two conditions under which no pervading link set exists, is used throughout the paper:

Observation 1. *A graph G does not have a pervading link set if it has a node v such that $t(v) > degree(v) + 1$, or if there is no node with threshold 1.*

3   NP-hardness

In this section, we prove that the Min-Links problem is NP-hard; in fact, it is almost as hard as Set-Cover to approximate, even if G has degree bounded by 3 and thresholds bounded by 2. Given a collection of n sets $\mathcal{S} = \{S_1, \ldots, S_n\}$

whose union is the universe $\mathcal{U}$ of cardinality m, with $n \leq m^k$ for some constant k, the Set-Cover problem is to find a minimum set cover, that is, a sub-collection of minimum cardinality $\mathcal{S}' \subseteq \mathcal{S}$ such that $\bigcup_{S \in \mathcal{S}'} S = \mathcal{U}$. The cardinality of $\mathcal{S}'$ is denoted $MSC(\mathcal{S})$. We shall make use of rooted binary trees. For such a tree T, denote the root by $r(T)$, and the set of leaves by $\mathcal{L}(T)$.

Constructing G from $\mathcal{S}$: Given a Set-Cover instance $\mathcal{S}$, we describe the construction of a corresponding Min-Links instance $G = (V, E, t)$ in polynomial time, which is used for our reduction. Figure 2 illustrates our construction. For each set in $\mathcal{S}$ and each element in $\mathcal{U}$, we introduce two binary trees in G, and then describe how to connect these trees. For each $S \in \mathcal{S}$, add to G a binary tree B_S with $|S|$ leaves $\mathcal{L}(B_S) = \{b_{S,u_1}, \ldots, b_{S,u_{|S|}}\}$, one for each element $u_i \in S$. Add another binary tree B'_S with $|S|$ leaves $\mathcal{L}(B'_S) = \{b'_{S,u_1}, \ldots, b'_{S,u_{|S|}}\}$, again one for each element $u_i \in S$. Then, add an edge between $r(B_S)$ and $r(B'_S)$.

The thresholds are $t(b) = 1$ for every $b \in V(B_S) \cup \mathcal{L}(B'_S)$, and $t(b') = 2$ for every internal node b' of $V(B'_S)$, that is for every $b' \in V(B'_S) \setminus \mathcal{L}(B'_S)$. Note that $\mathcal{L}(B'_S) \sim V(B'_S) \sim V(B_S)$.

Then for each element $u \in \mathcal{U}$, add a binary tree C_u with $|\mathcal{S}(u)|$ leaves, where $\mathcal{S}(u) = \{S \in \mathcal{S} : u \in S\}$ consists of the sets containing u. Denote $\mathcal{L}(C_u) = \{c_{u,S_1}, \ldots, c_{u,S_{|\mathcal{S}(u)|}}\}$, each leaf corresponding to a set S_i of $\mathcal{S}(u)$. Next, add yet another binary tree C'_u with $|\mathcal{S}(u)|$ leaves $\{c'_{u,S_1}, \ldots, c'_{u,S_{|\mathcal{S}(u)|}}\}$, again one for each $S_i \in \mathcal{S}(u)$. Add an edge between $r(C_u)$ and $r(C'_u)$. Every node $c \in V(C_u) \cup V(C'_u)$ has $t(c) = 1$.

We now define a gadget called a *heavy link*. Let x, y be two non-adjacent nodes with $t(x) = t(y) = 1$. Adding an $x - y$ heavy link consists of adding two nodes z_1, z_2 that are neighbors of x, then adding another node z_3 that is a neighbor of z_1, z_2 and y. We set the thresholds $t(z_1) = t(z_2) = 1$ and $t(z_3) = 2$. Note that the heavy link makes $x \sim y$ but not necessarily $y \sim x$ (thus adding an $x - y$ heavy link is different from adding a $y - x$ heavy link). Also notice that this operation increases the degree of x by 2 and of y by 1, and that z_1, z_2 and z_3 have degree bounded by 3.

To finish the construction, for every set $S \in \mathcal{S}$ and each element $u \in S$, add a $b_{S,u} - c_{u,S}$ heavy link, and a $c'_{u,S} - b'_{S,u}$ heavy link. Denote by H_S the set of nodes added to G by incorporating the heavy links to the B_S leaves, and by H'_u the set of heavy link nodes added to the C'_u leaves. It is not hard to see that G can be constructed in polynomial time. Note that for each $S \in \mathcal{S}$, the nodes of B_S are equivalent, in the sense that if one is activated, then they all get activated. The same holds for the nodes of C_u and C'_u, for every $u \in \mathcal{U}$. We will use their roots as *representatives*, meaning that we will implicitly use the fact that $r(B_S) \sim V(B_S)$ and $r(C_u) \sim V(C_u)$.

Lemma 1. *Let $\mathcal{S}$ be an instance of Set-Cover over universe $\mathcal{U}$, with $|\mathcal{S}| = n$ and $|\mathcal{U}| = m$, and let $G = (V, E, t)$ be the Min-Links instance constructed as above. Then all of the following conditions are met:*

1. $|V| \leq m^c$ *for some constant c;*
2. *each node of G has at most 3 neighbors;*
3. $t(v) \leq 2$ *for every node v of G.*

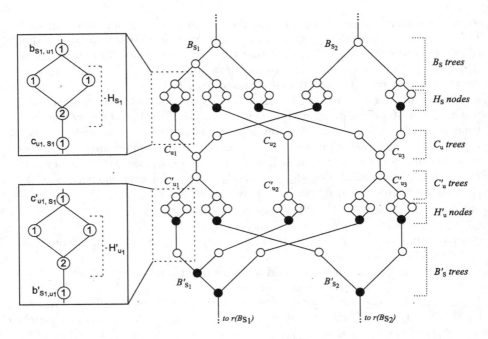

Fig. 2. The construction of G from $\mathcal{S}$ consisting of $S_1 = \{u_1, u_2, u_3\}$ and $S_2 = \{u_1, u_3\}$. White nodes have threshold 1, whereas black nodes have threshold 2.

Proof. For 1, there are $2n + 2m$ binary trees in G, which together contain at most $\ell = 2n \cdot m + 2m \cdot n = 4nm$ leaves. Thus the binary trees contain less than 2ℓ nodes in total. The heavy links account for at most 3ℓ nodes in total, and so $|V| \leq 5\ell \leq 20nm \leq m^c$ for some c (because $n \leq m^k$). To see that 1 holds, i.e. that the maximum degree is 3, observe that G consists of binary trees to which we add at most neighbor per root ($r(B_S)$ with $r(B'_S)$, and $r(C_u)$ with $r(C'_u)$), plus at most two neighbors per leaf (the heavy links). In the case that a node is both a root and a leaf (e.g. B_{S_i} is a single node because S_i has only one element), three neighbors are added to it, but it has zero neighbors initially. As for 1, it is easy to see that $t(v) \leq 2$ for every node $v \in V$ created. $\square$

We now show that both $\mathcal{S}$ and its corresponding instance G have the same optimality value.

Lemma 2. $MSC(\mathcal{S}) = ML(G)$.

Proof. First observe that for a given set $S \in \mathcal{S}$,

$$\bigcup_{u \in S} r(C_u) \sim \mathcal{L}(B'_S) \sim V(B'_S) \sim V(B_S)$$

which implies that

$$\bigcup_{u \in \mathcal{U}} r(C_u) \sim \bigcup_{S \in \mathcal{S}} V(B'_S) \sim \bigcup_{S \in \mathcal{S}} V(B_S)$$

and it follows that $\bigcup_{u \in \mathcal{U}} r(C_u) \sim V$.

To see that $MSC(\mathcal{S}) \geq ML(G)$, if $\mathcal{S}' \subseteq \mathcal{S}$ is a minimum set cover, then giving links to $V' = \bigcup_{S \in \mathcal{S}'} r(B_S)$ suffices to activate G since $V' \sim \bigcup_{u \in \mathcal{U}} r(C_u) \sim V$. Thus $MSC(\mathcal{S}) \geq ML(G)$.

It remains to show that $MSC(\mathcal{S}) \leq ML(G)$. Let $B = \{r(B_S) : S \in \mathcal{S}\}$. Let $V' \subseteq V$ be the set of endpoints of $E(\hat{S})$ for an optimal pervading link set $\hat{S}$ such that $|V' \cap B|$ is maximized among all possible choices. We divide this section of the proof into two claims.

Claim. $V' \subseteq B$.

Proof. First observe that we may assume that if $x \in V' \setminus B$, then there is no set S such that $r(B_S) \sim x$ (for otherwise, we can replace x by $r(B_S)$ in V', contradicting our choice of V'). But no such x can exist. If $x \in V(C_u)$ for some u, then $r(B_S) \sim x$ for any set S containing u. If x belongs to a $b_{S,u} - c_{u,S}$ heavy link, then $r(B_S) \sim x$. If x belongs to a $c'_{u,S} - b'_{S,u}$ heavy link, then again $r(B_S) \sim r(C_u) \sim x$. Finally if $x \in V(B'_S)$, then $r(B_S) \sim \bigcup_{u \in S} r(C_u) \sim \mathcal{L}(B'_S) \sim x$. We conclude that V' has only nodes from B. ☐

Claim. $\mathcal{S}' = \{S \in \mathcal{S} : r(B_S) \in V'\}$ is a set cover.

Proof. Suppose the claim is false, and let $w \in \mathcal{U}$ be an element not covered by $\mathcal{S}'$. Recall that $\mathcal{S}(w) = \{S_1, \ldots, S_{|\mathcal{S}(w)|}\}$ is the collection of sets containing w. Let $S_i \in \mathcal{S}(w)$. Then in B'_{S_i}, there is a leaf $b'_{S_i,w}$. Let P_{S_i} be the set of nodes lying on the unique $b'_{S_i,w} - r(B'_{S_i})$ shortest path in B'_{S_i} (inclusively). Define W as the node set that contains the C_w and C'_w nodes along with the heavy link nodes appended to $\mathcal{L}(C'_w)$, plus for each $S_i \in \mathcal{S}(w)$, the P_{S_i} nodes and the B_{S_i} nodes with the heavy link nodes appended to $\mathcal{L}(B_{S_i})$. Formally,

$$W = V(C_w) \cup V(C'_w) \cup H'_w \cup \left(\bigcup_{S_i \in \mathcal{S}(w)} (V(B_{S_i}) \cup H_{S_i} \cup P_{S_i}) \right)$$

We show that no node of W gets activated by V', contradicting the assertion that $\hat{S}$ is a pervading link set. Suppose instead that some W nodes do get activated. Let z be the first node of W activated by the propagation process (or if multiple nodes of W get simultaneously activated first, pick z arbitrarily among them). Then, since $V' \cap W = \emptyset$, z must have $t(z)$ neighbors outside of W that were activated and influenced it. Observe that the only nodes of W that have neighbors outside of W belong to either H_{S_i} or P_{S_i} for some $S_i \in \mathcal{S}(w)$. If $z \in H_{S_i}$, then the only heavy link node with neighbors outside of W is the threshold 2 node. But then, z has only one neighbor outside W (namely a c_{u,S_i} node for some u), which is not enough to activate z. Thus $z \notin H_{S_i}$. If $z \in P_{S_i}$, then $z \neq b'_{S_i,w}$ since $b'_{S_i,w}$ receives no influence from outside of W: it has two neighbors, one is in P_{S_i} and the other is in the $b'_{S_i,w} - c'_{w,S_i}$ heavy link, both of which are in W. If instead z is an interior node of the P_{S_i} path, then z has two neighbors in W (by the definition of a path). But $t(z) = 2$ and z has only three neighbors, i.e. only one outside of W, and so z cannot be activated only by influence from outside W. The last possible case is $z = r(B'_{S_i})$. But again, z has two neighbors in W: one is in P_{S_i} and the other is $r(B_{S_i})$, and the same argument applies. We conclude that z, and hence w, cannot exist, and that S' is a set cover.   □

Since V' yields a set cover $\mathcal{S}$ of size $ML(G)$, we deduce that $MSC(\mathcal{S}) \leq ML(G)$.   □

We can now state the main result of this section.

Theorem 1. *The decision version of* Min-Links *is NP-complete, even when restricted to instances with maximum degree 3 and maximum threshold 2. Moreover, there exists a constant $\epsilon > 0$ such that the optimization version of* Min-Links, *under the same restrictions, is NP-hard to approximate within a $\epsilon \ln n$ factor, where n is the number of nodes of the given graph.*

Proof. NP-completeness follows directly from Lemma 2, and observing that Min-Links is in NP, as it is easy to check that a given set V' is a pervading link set (because propagation must finish in a polynomial number of steps). As for the inapproximability result, let $\mathcal{S}$ be an instance of set cover over universe $\mathcal{U}$, $|\mathcal{S}| = n$ and $|\mathcal{U}| = m$, and let n' be the number of nodes of G constructed from $\mathcal{S}$ as described above, with $n' \leq m^c$ (c is the constant from Lemma 1). Dinur and Steurer showed that it is NP-hard to approximate set cover within a $d \ln m$ factor for any $0 < d < 1$ [10]. For our purposes, fix $0 < d < 1$, and suppose that some approximation algorithm $\mathcal{A}$ always finds a pervading link set of size at most $APP \leq \frac{d}{c} \ln(n') \cdot ML(G)$. Because $ML(G) = MSC(\mathcal{S})$, we have $APP \geq MSC(\mathcal{S})$, and in the other direction,

$$APP \leq \frac{d}{c} \ln(n') \cdot ML(G) \leq \frac{d}{c} \ln(m^c) \cdot ML(G) = d \ln(m) \cdot MSC(\mathcal{S})$$

and hence $\mathcal{A}$ can approximate Set-Cover to within a factor $d \ln(m)$ using the aforementioned reduction. Therefore, for $\epsilon = \frac{d}{c}$, it is hard to approximate the Min-Links problem within a $\epsilon \ln(n')$ factor.   □

4   Trees

In contrast to the NP-completeness of the Min-Links problem shown in the previous section, we now show that there is a linear time algorithm to solve the problem in trees. We start with a necessary and sufficient condition for a tree T to have a valid pervading link set.

Proposition 1. *Let T be a tree and let v be a leaf in T. Let $T' = T - \{v\}$ and T'' be the same as T' except that the threshold of w, the neighbor of v in T, is reduced by 1. Then T has a pervading link set if and only if (a) either $t(v) = 1$ and T'' has a pervading link set or (b) $t(v) = 2$ and T' has a pervading link set.*

We now prove a critical lemma that shows that for any node in the tree, there is an optimal solution that gives a link to that node.

Lemma 3. *Let T be a tree with n nodes that has a pervading link set, and let v be a node in T. Then there exists an optimal solution for Min-Links(T) in which v gets a link.*

Proof. We prove the lemma by induction on the number of nodes n in the tree. Clearly it is true if $n = 1$. Suppose $n > 1$, and let S be an optimal pervading link set for T. If v gets a link, we are done. If not, v must have a neighbor w that is activated before v, and that contributes to the activation of v. Let T_1 and T_2 be the two trees created by removing the edge between v and w, with T_1 containing w, and let S_1 (respectively S_2) be the links of S with an endpoint in T_1 (respectively T_2). Since T is a tree, and v is activated after w by S, none of the links in S_2 can contribute to the activation of nodes in T_1. It follows that S_1 is a pervading link set for T_1, and in fact is optimal, as a smaller solution for T_1 could be combined with S_2 to yield a better solution for T, contradicting the optimality of S. By the inductive hypothesis, there is an optimal solution S' for T_1 that gives a link to w. Note that $|S'| = |S_1|$, and $S' \cup S_2$ must also be an optimal solution for T. But clearly $S'' = S' \cup S_2 \cup \{(\mu, v)\} - \{(\mu, w)\}$ also activates the entire tree T , and since $|S''| = |S|$, we conclude that S'' is an optimal solution for T, that gives a link to v, as needed to complete the proof by induction. $\square$

The above lemma suggests a simple way to break up the Min-Links problem for a tree into subproblems that can be solved independently, which yields a linear-time greedy algorithm.

Theorem 2. *The Min-Links problem can be solved for trees in linear time.*

Proof. Given a tree T, let v be an arbitrary leaf in the tree. By Lemma 3, there is an optimal solution, say S; to the Min-Links problem for T that gives v a link. Suppose $t(v) = 2$, then the link to v is not enough to activate v, and therefore v's neighbor w must activate v. Also, v's activation cannot help in activating any other nodes in T. Thus $S - \{(\mu, v)\}$ must be an optimal solution to $T' = T - \{v\}$. Suppose instead that $t(v) = 1$. Then the link given to v activates it immediately. Consider the induced subgraph of T containing only nodes of threshold 1, and let C be the connected component (subtree) containing v in this subgraph. Then clearly $v \sim C$. Since S is optimal, S cannot contain any node in C except for v. Construct T' by removing C from T, and subtracting 1 from the threshold of any node x who is a neighbor of a node in C. Observe that any such node x can be a neighbor of exactly one node in C, since T is a tree. Then $S - \{(\mu, v)\}$ must be an optimal solution to T'; if instead there is a smaller-sized solution to T', we can add (μ, v) to that solution to obtain a smaller solution for T than S, contradicting the optimality of S.

The above argument justifies the correctness of the following simple greedy algorithm. Initialize $S = \emptyset$. Take a leaf v in the tree. If $t(v) > 2$ then there is no solution by Observation 1. If $t(v) = 2$, then put the link (μ, v) in S, remove v from the tree, and recursively solve the remaining tree. If $t(v) = 1$, then give a link to v, remove the subtree of T that is connected to v consisting only of nodes of degree 1, reduce the thresholds of all neighbors of the nodes in this subtree by 1, and recursively solve the resulting trees. It is easy to see that the algorithm can be implemented in linear time.   □

For the network in Fig. 1, assuming that leaves in the tree are always processed in alphabetical order, the greedy algorithm given in Theorem 2 first picks node b and adds a link to it. We then remove nodes b and a, and reduce the threshold of d by 1. Next we pick c, give it a link, remove it from the tree, and decrement $t(f)$ to 2. The next leaf that is picked and given a link is d; since d's threshold now is 1, we remove d and e from the tree, and reduce f's threshold to 1. Proceeding in this way, we arrive at the link set shown.

We now give an exact bound on $ML(T)$, the number of links required to activate the entire tree T:

Theorem 3. *Let T be a tree that has a pervading link set. Then $ML(T) = 1 + \sum_{v \in T}(t(v) - 1)$*

Proof. We give a proof by induction on the number of nodes n in the tree. Clearly if the tree consists of a single node x, there is a solution if and only if $t(x) = 1$, and the number of links needed is 1 which is equal to $1 + \sum_{v \in V}(t(v) - 1)$ as needed. Now consider a tree T with $n > 1$ nodes and let x be a leaf in the tree. Then by Lemma 3, there is an optimal solution S in which x gets a link. By Observation 1, there is a solution only if $t(x) = 1$ or $t(x) = 2$. Let $T' = T - \{x\}$ (all nodes keep the same thresholds as in T) and let T'' be the tree derived from T by removing x and reducing the threshold of w, the neighbor of x in T by 1.

First we consider the case when $t(x) = 2$. Then giving x a link is not sufficient to activate it. By the usual cut-and-paste argument, $S - \{(\mu, x)\}$ must be an optimal solution for tree T'.

$$ML(T) = 1 + ML(T')$$
$$= t(x) - 1 + (1 + \sum_{v \in T'}(t(v) - 1)) \text{ by the inductive hypothesis}$$
$$= 1 + \sum_{v \in T}(t(v) - 1)$$

Next we consider the case when $t(x) = 1$, and $t(w) > 1$. Then x is immediately activated by the link it receives in S, and the link given to x effectively reduces the threshold of w. Therefore, $S - \{(\mu, x)\}$ must be an optimal solution for the tree T'' in which the threshold of w is $t(w) - 1$. It follows that

$$
\begin{aligned}
ML(T) &= 1 + ML(T'') \\
&= 1 + (1 + \sum_{v \in T''} (t(v) - 1)) \text{ by the inductive hypothesis} \\
&= 2 + (t(w) - 2) + \sum_{v \in T'' - \{w\}} (t(v) - 1) \\
&= 1 + \sum_{v \in T} (t(v) - 1)
\end{aligned}
$$

Finally suppose $t(x) = t(w) = 1$. Then it is impossible that S contains w, as this would contradict the optimality of S. Therefore, we can move the link from node v to node w, to get a new optimal pervading link set S' for T. Furthermore, S' must also be an optimal pervading link set for T'. It follows that

$$
\begin{aligned}
ML(T) &= ML(T') \\
&= t(x) - 1 + (1 + \sum_{v \in T'} (t(v) - 1)) \text{ by the inductive hypothesis} \\
&= 1 + \sum_{v \in T} (t(v) - 1)
\end{aligned}
$$

$\square$

We remark that in contrast to the intuition for the optimal target set problem, where we would choose nodes of high degree or threshold to be in the target set, in the Min-Links problem, our algorithm gives links to leaves initially, though eventually nodes that were internal nodes in the tree may also receive links. That is, the best nodes to befriend might be the nodes with a single connection to other nodes in the tree!

5   Cycles

In this section, we give a solution for the Min-Links problem on cycles. Let $C_n = (V, E, t)$ be a cycle with n nodes, $V = \{0, 1, ..., n - 1\}$, $E = \{((i, i + 1) \bmod n) \mid 1 \le i \le n\}$, and $t : t(v) \to \mathcal{Z}^+$. We define $P_{i,j}$ $(i \ne j)$ to be the sub-path of C_n consisting of all nodes in $\{i, \ldots, j\}$ in the clockwise direction. We may use the $[i, j]$ notation to denote the vertices of $P_{i,j}$. By consecutive vertices of threshold 3, we mean two vertices i, j such that the only two vertices in $P_{i,j}$ with threshold 3 are i and j.

Proposition 2. A cycle has a pervading link set if and only there is at least one node of threshold 1, every node is of threshold at most 3, and between any two consecutive nodes of threshold 3, there is at least one node of threshold 1.

We note that a similar condition can be stated for paths, with the additional restriction that there must be a node of threshold 1 before (after) the first (last resp.) node of threshold 3.

We give a linear time algorithm for finding a minimum-sized link set for problem Min-Links(C_n). Essentially we reduce the problem to finding an optimal solution for an appropriate path.

Theorem 4. *The* Min-Links *problem for a cycle C_n can be solved in time $\Theta(n)$.*

Proof. By Observation 1, there is no solution if there is a node with threshold 4 or more. If there exists a node i such that $t(i) = 3$, then clearly i must get a link, and both of its neighbors must be activated before it. That is, i can play no role in activating any node in $P_{i+1,i-1}$. Therefore, $S = \{(\mu, i)\} \cup S'$ is an optimal solution to C_n where S' is an optimal solution to $P_{i+1,i-1}$. In this case, S' can be found in linear time using the tree algorithm of Theorem 2. If there is no node with threshold 3, a single node with threshold 2, and the remaining nodes all have threshold 1, then by giving a link to *any* of the nodes with threshold 1, we can activate the entire cycle.

It remains only to consider the case when there are no nodes of threshold 3, at least two nodes of threshold 2, and at least one node of threshold 1. Fix an arbitrary node i of threshold 1 in C_n. We define $c(i)$ and $cc(i)$ to be the first node with threshold 2 in i's clockwise direction and counter clockwise direction respectively, $c(i) \neq cc(i)$ (see Fig. 3). We also define $P_{c(i),cc(i)}$ be the path from $c(i)$ to $cc(i)$ where $t(c(i)) = t(cc(i)) = 2$; and $P'_{c(i),cc(i)}$ to be the same path as $P_{c(i),cc(i)}$ except that we set $t(c(i)) = t(cc(i)) = 1$.

We first claim that there exists an optimal solution that gives a link to i. To see this, let S be an optimal solution that does not give a link to node i. Since all nodes in C_n are activated by S, there must exist some node $j \in [cc(i), c(i)]$ that gets a link. If $t(j) = 1$, we can take the link given to j and give it instead to node i. Otherwise there exists $j \in \{c(i), cc(i)\}$ such that it gets a link and is activated before i, and eventually activates i. Again we can move the link from node j to node i, which clearly has the same effect of giving a link to node j. Therefore, we have a new solution of the same size as S that gives a link to node i.

Consider therefore an optimal solution S that gives a link to the node i. It is not hard to see that that $S - \{(\mu, i)\}$ must be an optimal solution to Min-Links($P'_{c(i),cc(i)}$), since activating i activates $[cc(i) + 1, c(i) - 1]$ and lowers the threshold of $cc(i)$ and $c(i)$. Again, since the Min-Links problem for a path can be solved in $\Theta(n)$ according to Theorem 2, we can construct an optimal solution for a cycle in $\Theta(n)$ time as well.   □

We give an exact bound on the number of links required to fully activate a cycle.

Theorem 5. *Given a cycle $C_n = (V, E, t)$ which has a pervading link set,*
$$ML(C_n) = \sum_{i=1}^{n}(t(i) - 1)$$

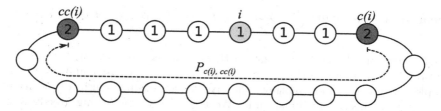

Fig. 3. A cycle with no threshold 3 vertices, illustrating the main components of the proof.

Proof. If there is a node i of threshold 3, then $ML(C_n) = 1 + ML(P_{i+1,i-1})$. Since by Theorem 3, $ML(P_{i+1,i-1}) = 1 + \sum_{j \neq i}(t(j) - 1)$, we have $ML(C_n) = 1 + (1 + \sum_{j \neq i}(t(j) - 1)) = (t(i) - 1) + \sum_{j \neq i}(t(j) - 1) = \sum_{j=1}^{n}(t(j) - 1)$ as needed. If there is no node of threshold 3 and a single node of threshold 2, then $ML(C_n) = 1 = \sum_{j=1}^{n}(t(j) - 1)$. Finally, if there is no node of threshold 3, and at least two nodes of threshold 2, and at least one of threshold 1, then $ML(C_n) = 1 + ML(P'_{cc(i),c(i)})$ where i is a node of threshold 1. Since the thresholds of $c(i)$ and $cc(i)$ have been reduced by 1 each in $P'_{cc(i),c(i)}$, by Theorem 3, we have $ML(P'_{cc(i),c(i)}) = -1 + \sum_{j \in [cc(i),c(i)]}(t(j) - 1)$. Therefore $ML(C_n) = 1 - 1 + \sum_{j \in [cc(i),c(i)]}(t(j) - 1) = \sum_{i=1}^{n}(t(i) - 1)$. $\qquad\square$

6 Cliques

In this section, we give an algorithm to solve the Min-Links problem on cliques. Let $K_n = (V, E, t)$ be a clique with n nodes, $V = \{1, 2, ..., n\}$ and $E = \{(i,j) : 1 \leq i < j \leq n\}$ and $t : t(v) \to \mathcal{Z}^+$. We first show a necessary and sufficient condition for the Min-Links problem to have a feasible solution:

Proposition 3. *Let K_n be a clique with $t(i) \leq t(i+1)$, for all $1 \leq i < n$. Then K_n has a pervading link set if and only if $t(i) \leq i$ for all $1 \leq i \leq n$.*

Proof. If $t(i) \leq i$ for all $1 \leq i \leq n$, it is easy to see that there exists a solution S by giving a link to every node i; we claim that node i is activated in or before round i. Since $t(1) \leq 1$, node 1 is activated in round 1. Inductively, node 1 to $i-1$ are already activated in round $i - 1$, the effective threshold of node i has been reduced to ≤ 1. Node i receives a link, therefore, node i must be activated in the ith round, if it is not already activated. Conversely, suppose there exist nodes j such that $t(j) > j$ and there exists a solution S to the Min-Links problem; let p be the smallest such node with $t(p) > p$. In order to activate any node $q \geq p$, at least p nodes have to be activated before q, since $t(q) \geq t(p) > p$. However, there are only $p - 1$ nodes that can be activated before any such node $q \geq p$. Thus no node q with $q \geq p$ can be activated, a contradiction. $\qquad\square$

We now give a greedy algorithm to solve the Min-Links problem on a clique.

Theorem 6. *The* Min-Links *problem for a clique K_n can be solved in time $\Theta(n)$.*

Proof. First sort the nodes in order of threshold. By Observation 1, there is no solution if any node has a threshold $> n$, therefore, we can use counting sort and complete the sorting in $\Theta(n)$ time. Clearly, the condition given in Proposition 3 can easily be checked in linear time. We now give the following greedy linear time algorithm for a clique which has a feasible solution: give a link to node 1, and let j be the maximum value such that $t(i) < i$ whenever $2 \leq i < j$. Remove all nodes in $\{1, \ldots, j-1\}$, decrement by $j-1$ the thresholds of all nodes $\geq j$, and solve the resulting graph recursively. It is easy to see that this algorithm can be implemented in linear time, in an iterative fashion as follows: we examine the nodes in order. When we process node i, if $t(i) < i$, we simply increment i and continue; if $t(i) = i$, we give a link to node i. We now show that the link set produced by this greedy algorithm is optimal.

First we show that there must be an optimal solution that contains the node 1. Consider an optimal solution S and let i be the smallest index of a node that receives a link in S. If $i = 1$, then we are done. If not, since there must always be a node with threshold 1 that receives a link, it must be that $t(i) = 1$. But then we can move the link from i to 1, to create a new solution S' which will activate node i in the next step. Since $|S'| = |S|$ and $I(K_n, S) = I(K_n, S')$, S' is an optimal solution to the Min-Links problem that contains the node 1. Thus, we can assume that the optimal solution S contains the node 1.

Next we claim that $S - \{1\}$ is an optimal solution to the clique C' which is the induced sub-graph on the nodes $\{j, j+1, \ldots, n\}$ where $j > 1$ is the smallest index with $t(j) = j$, and with thresholds of all nodes reduced by $j-1$. Suppose there is a smaller solution S' to C'. We claim that $S' \cup \{1\}$ activates all nodes in the clique K_n. Since for any node $1 < k < j$, we have $t(k) < k$, it can be seen inductively that the link given to node 1 suffices to activate node k. Thus, all nodes in $\{1, 2, \ldots j-1\}$ are activated. Furthermore, the thresholds of all nodes in $\{j, j+1, \ldots, n\}$ are effectively reduced by $j-1$. Thus using the links in S' suffices to activate them. Finally, since $|S'| < |S|-1$, $S' \cup \{1\}$ is a smaller solution than S to the clique K_n, contradicting the optimality of S for K_n. We conclude that the greedy algorithm described above produces a minimum sized solution to the Min-Links problem. $\square$

The following tight bound on the minimum number of links to activate an entire clique is immediate:

Theorem 7. *Given a clique K_n which has a feasible solution, $ML(K_n) = |\{j \mid t(j) = j\}|$*

The greedy algorithm from Theorem 6 can be extended to complete multi-partite graphs:

Theorem 8. *The* Min-Links *problem for a complete multi-partite graph G can be solved in time $O(|E(G)|)$.*

7   Discussion

In this paper, we introduced and studied the Min-Links problem: given a social network G where every node v has a threshold $t(v)$ to be activated, which minimum-sized set of nodes should an already activated external influencer μ befriend, so as to influence the entire network? We showed that the problem is NP-complete, in fact it is hard to approximate to within an $\epsilon \ln n$ factor (for some constant $0 < \epsilon < 1$) even for graphs with maximum degree 3, and with maximum threshold 2. In contrast, we show linear time algorithms for the problem for trees, cycles, cliques, and complete k-partite graphs, and give an exact bound (as a function of the thresholds) on the number of links needed for such graphs. This leaves open the question of a polynomial time algorithm for graphs of bounded treewidth, as well as the best possible approximation algorithm for general graphs. It would be interesting to generalize these algorithms to find the minimum number of links required to influence a specified fraction of the nodes. Other directions include studying the *multiple influencer case*, and the case with non-uniform weights on the edges. Clearly, the problem remains NP-complete in general, but the complexity for special classes of graphs remains open. Another interesting question is that of maximizing the number of activated nodes, given a fixed budget of k links.

References

1. Ben-Zwi, O., Hermelin, D., Lokshtanov, D., Newman, I.: Treewidth governs the complexity of target set selection. Discrete Optim. **8**, 702–715 (2011)
2. Borgs, C., Brautbar, M., Chayes, J., Lucier, B.: Maximizing social influence in nearly optimal time. In: Proceedings of the 25th Annual ACM-SIAM Symposium on Discrete Algorithms, SODA 2014, pp. 946–957 (2014)
3. Brown, J.J., Reingen, P.H.: Social ties and word-of-mouth referral behavior. J. Consum. Res. **14**(3), 350–362 (1987)
4. Chen, N.: On the approximability of influence in social networks. In: Proceedings of the 19th Annual ACM-SIAM Symposium on Discrete Algorithms, SODA 2008, pp. 1029–1037 (2008)
5. Chen, W., Wang, Y., Yang, S.: Efficient influence maximization in social networks. In: Proceedings of the 15th ACM SIGKDD International Conference on Knowledge Discovery and Data Mining, KDD 2009, pp. 199–208 (2009)
6. Cicalese, F., Cordasco, G., Gargano, L., Milanič, M., Peters, J.G., Vaccaro, U.: How to go viral: cheaply and quickly. In: Ferro, A., Luccio, F., Widmayer, P. (eds.) FUN 2014. LNCS, vol. 8496, pp. 100–112. Springer, Heidelberg (2014)
7. Cicalese, F., Cordasco, G., Gargano, L., Milanič, M., Vaccaro, U.: Latency-bounded target set selection in social networks. In: Bonizzoni, P., Brattka, V., Löwe, B. (eds.) CiE 2013. LNCS, vol. 7921, pp. 65–77. Springer, Heidelberg (2013)
8. Demaine, E.D., Hajiaghayi, M.T., Mahini, H., Malec, D.L., Raghavan, S., Sawant, A., Zadimoghadam, M.: How to influence people with partial incentives. In: Proceedings of the 23rd International Conference on World Wide Web, WWW 2014, pp. 937–948 (2014)

9. Dinh, T., Zhang, H., Nguyen, D., Thai, M.: Cost-effective viral marketing for time-critical campaigns in large-scale social networks. IEEE ACM Trans. Netw. **PP**(99), 1 (2014)

10. Dinur, I., Steurer, D.: Analytical approach to parallel repetition. In: Proceedings of the 46th Annual ACM Symposium on Theory of Computing, pp. 624–633. ACM (2014)

11. Domingos, P., Richardson, M.: Mining the network value of customers. In: Proceedings of the 7th ACM SIGKDD International Conference on Knowledge Discovery and Data Mining, KDD 2001, pp. 57–66 (2001)

12. Fazli, M.A., Ghodsi, M., Habibi, J., Jalaly Khalilabadi, P., Mirrokni, V., Sadeghabad, S.S.: On the non-progressive spread of influence through social networks. In: Fernández-Baca, D. (ed.) LATIN 2012. LNCS, vol. 7256, pp. 315–326. Springer, Heidelberg (2012)

13. Gargano, L., Hell, P., Peters, J., Vaccaro, U.: Influence diffusion in social networks under time window constraints. In: Moscibroda, T., Rescigno, A.A. (eds.) SIROCCO 2013. LNCS, vol. 8179, pp. 141–152. Springer, Heidelberg (2013)

14. Goldenberg, J., Libai, B., Muller, E.: Talk of the network: a complex systems look at the underlying process of word-of-mouth. Mark. Lett. **12**, 211–223 (2001)

15. Goyal, A., Bonchi, F., Lakshmanan, L., Venkatasubramanian, S.: On minimizing budget and time in influence propagation over social networks. Soc. Netw. Anal. Min. **3**, 179–192 (2013)

16. Goyal, A., Bonchi, F., Lakshmanan, L.: A data-based approach to social influence maximization. Proc. VLDB Endow. **5**, 73–84 (2011)

17. Goyal, A., Lu, W., Lakshmanan, L.: Celf++: optimizing the greedy algorithm for influence maximization in social networks. In: Proceedings of the 20th International Conference Companion on World Wide Web, WWW 2011, pp. 47–48 (2011)

18. He, J., Ji, S., Beyah, R., Cai, Z.: Minimum-sized influential node set selection for social networks under the independent cascade model. In: Proceedings of the 15th ACM International Symposium on Mobile Ad Hoc Networking and Computing, MobiHoc 2014, pp. 93–102 (2014)

19. Dreyer Jr., P., Roberts, F.: Irreversible -threshold processes: graph-theoretical threshold models of the spread of disease and of opinion. Discrete Appl. Math. **157**(7), 1615–1627 (2009)

20. Kempe, D., Kleinberg, J., Tardos, É.: Maximizing the spread of influence through a social network. In: Proceedings of the Ninth ACM SIGKDD International Conference on Knowledge Discovery and Data Mining, KDD 2003, pp. 137–146 (2003)

21. Kempe, D., Kleinberg, J.M., Tardos, É.: Influential nodes in a diffusion model for social networks. In: Caires, L., Italiano, G.F., Monteiro, L., Palamidessi, C., Yung, M. (eds.) ICALP 2005. LNCS, vol. 3580, pp. 1127–1138. Springer, Heidelberg (2005)

22. Leskovec, J., Adamic, L.A., Huberman, B.A.: The dynamics of viral marketing. In: Proceedings of the 7th ACM Conference on Electronic Commerce, pp. 228–237. ACM (2006)

23. Lu, W., Bonchi, F., Goyal, A., Lakshmanan, L.: The bang for the buck: fair competitive viral marketing from the host perspective. In: Proceedings of the 19th ACM SIGKDD International Conference on Knowledge Discovery and Data Mining, KDD 2013, pp. 928–936 (2013)

24. Lv, S., Pan, L.: Influence maximization in independent cascade model with limited propagation distance. In: Han, W., Huang, Z., Hu, C., Zhang, H., Guo, L. (eds.) APWeb 2014 Workshops. LNCS, vol. 8710, pp. 23–34. Springer, Heidelberg (2014)

25. Nichterlein, A., Niedermeier, R., Uhlmann, J., Weller, M.: On tractable cases of target set selection. Soc. Netw. Anal. Min. **3**(2), 233–256 (2012)
26. Peleg, D.: Local majority voting, small coalitions and controlling monopolies in graphs: a review. Theor. Comput. Sci. **282**, 231–257 (2002)
27. Richardson, M., Domingos, P.: Mining knowledge-sharing sites for viral marketing. In: Proceedings of the Eighth ACM SIGKDD International Conference on Knowledge Discovery and Data Mining, KDD 2002, pp. 61–70 (2002)

Approximating the Size of a Radio Network in Beeping Model

Philipp Brandes[1], Marcin Kardas[2]([✉]), Marek Klonowski[2],
Dominik Pająk[2], and Roger Wattenhofer[1]

[1] ETH Zürich, Zürich, Switzerland
[2] Department of Computer Science at the Faculty of Fundamental
Problems of Technology, Wrocław University of Science and Technology,
Wrocław, Poland
Marcin.Kardas@pwr.edu.pl

Abstract. In a single-hop radio network, nodes can communicate with each other by broadcasting to a shared wireless channel. In each time slot, all nodes receive feedback from the channel depending on the number of transmitters. In the Beeping Model, each node learns whether zero or at least one node have transmitted. In such a model, a procedure estimating the size of the network can be used for efficiently solving the problems of leader election or conflict resolution. We introduce a time-efficient uniform algorithm for size estimation of single-hop networks. With probability at least $1 - 1/f$ our solution returns $(1 + \varepsilon)$-approximation of the network size n within $\mathcal{O}\left(\log \log n + \log f / \varepsilon^2\right)$ time slots. We prove that the algorithm is asymptotically time-optimal for any constant $\varepsilon > 0$.

1 Introduction

The number of nodes in the network is a parameter that is necessary to effectively perform many fundamental protocols and is useful for network analysis, gathering statistics etc. However, in modern applications of communication networks we often cannot assume that the size of the network or even its constant-factor approximation in known. Hence, the problem of designing an algorithm to precisely and efficiently estimate the number of nodes in radio networks is an important challenge. This is particularly clear in the context of networks with strictly limited communication channel, wherein one needs a precise estimation of the number of nodes in order to avoid collisions of transmissions caused by several nodes broadcasting at the same time. As a consequence, the most efficient algorithms for classic problems in radio networks, like leader election, use the size approximation as a subroutine.

In our paper we consider the problem of size estimation in a communication model that is weaker than the classic Multiple Access Channel, namely in the Beeping Model.

This paper is supported by Polish National Science Center – decision number 2013/09/N/ST6/03440 (M. Kardas).

J. Suomela (Ed.): SIROCCO 2016, LNCS 9988, pp. 358–373, 2016.
DOI: 10.1007/978-3-319-48314-6_23

We consider a wireless network of n devices (nodes). The size n of the network is unknown to the nodes. The nodes have no identifiers or serial numbers that could be used to distinguish them. The aim is to estimate the network's size n by performing random transmissions and using the feedback of the communication channel. The main result of this paper is an asymptotically optimal (with respect to the time of execution) algorithm that returns a $(1 + \varepsilon)$-approximation of the number of nodes in the network with controllable error probability. As the second result we show the matching lower bound.

1.1 Model

We study a single-hop radio network of n nodes with the Beeping Model as a communication model [1,9]. The transmission of each node reaches all other nodes. That is, the network can be represented as a complete graph. We assume that the nodes are identical and indistinguishable and perform the same protocol. However, each node can independently sample any number of random bits. Randomization can be used freely, but the final result of the protocol needs to be deterministically computed based on the knowledge available to all the nodes. We ensure in this way that all the nodes upon completing the procedure obtain the same result, which could also be determined by a passive observer listening to the communication channel.

We assume that the time is discrete, i.e., it is divided into slots. We also assume that the nodes are synchronized as if they had access to a global clock. In every slot, each node independently decides whether to transmit to the channel or not. The nodes share a common communication channel and in every slot the channel can be in one of the two following states: NULL, when no node is transmitting and BEEP, if at least one node is transmitting (i.e., the channel is busy). All nodes receive the state of the channel immediately after each communication round.

The Beeping Model can be contrasted with the classical model of Radio Networks with Collision Detection where the channel can be in three states depending on whether zero, one, or more than one, node is transmitting. The third state is called "collision".

The result of any size estimation protocol is a random variable, an estimator $\hat{n}$ of true number of nodes n. We are interested in the probability of getting an approximation that differs from the true value by at most a constant multiplicative factor.

Definition 1. *For any $\varepsilon > 0$, we say that protocol $\mathcal{P}$ $(1 + \varepsilon)$-approximates the number of nodes with probability at least $1 - 1/f$, if for any n it returns $\hat{n}$ such that $\mathbb{P}\left(\hat{n}/(1 + \varepsilon) \leq n < (1 + \varepsilon)\hat{n}\right) \geq 1 - 1/f$.*

The time complexity of protocol is expressed as a function of three variables n, f and ε.

1.2   Related Work

There are many papers devoted to size approximation in radio networks. Most of them work in the model of Radio Networks with Collision Detection. In [2] Bordim et al. presented a size approximation protocol for the network of the (unknown) size n with execution time $\mathcal{O}\left((\log n)^2\right)$ that finds an approximation $\hat{n}$ of the real number of nodes such that $n/(16 \log n) < \hat{n} < 2n/\log n$ with probability at least $1 - \mathcal{O}\left(n^{-1.83}\right)$. The authors assume communication model with collision detection and aim at saving energy of the network. Greenberg et al. [13] proposed a size approximation algorithm working in time $\log n + \mathcal{O}(1)$ producing an estimate of n with mean approximately $0.914n$ and standard deviation of $0.630n$. Greenberg et al. [13] also showed that a size approximation algorithm can be used to efficiently schedule transmissions such that each node succeeds to transmit.

Some papers presented other, more complex protocols that use elaborated size-approximation algorithms as a sub-procedure (e.g. [20]). In [19] Nakano and Olariu presented an energy-efficient initialization algorithm which needs to know the number of nodes n or its fair approximation to work properly. In paper [24] Willard showed an algorithm for a selection problem that needs $O(\log \log n)$ steps on average with a respective lower bound. This result has been used extensively for many other papers about fast leader election and size approximation in the context of radio networks.

An energy-efficient size estimation algorithm is proposed in Jurdziński et al. [15] for a model without collision detection. The algorithm uses $\mathcal{O}\left(\log^{2+\alpha} n\right)$ time slots with nodes being awake for at most $\mathcal{O}\left((\log \log n)^\alpha\right)$ slots for any $\alpha > 0$. The algorithm is a c-approximation for some constant c (with respect to n). In [3] authors present approximation of the size of the network in a similar model. Their protocol designed for collision detection model works in $O(\log n \log \log n)$ steps and returns a 2-approximation. The second protocol for no-collision detection settings needs $O(\log^2 n)$ steps for a 3-approximation. Moreover, the authors of [3] take into account energy of nodes necessary for completing the protocol. All the results aforementioned in this paragraph hold with high probability.

The problem of size estimation has been extensively studied in the context of computer databases [5,10–12,23]. In that case, one is interested in estimating the cardinality (the number of distinct elements) of some multiset. Many protocols for size estimation have been proposed for radio networks [7,8,16]. In many cases (including [13]) the proposed solutions provide asymptotically unbiased estimator $\mathbb{E}(\hat{n}) = n(1 + o(1))$ that is **not** well concentrated, i.e. ($\mathbb{V}\mathrm{ar}(\hat{n}) = \Omega(n^2)$). In such case one can have $\mathbb{P}(|\hat{n} - n| \geq c \cdot n) = \Theta(1)$. Thus one cannot expect obtaining c-approximation with high probability. Moreover, in contrast to most of the previous work, we use a controllable parameter of algorithm's success f. This can be particularly important for small n.

Independently, the problem of estimation of cardinality of a set emerged in the research devoted to RFID (Radio Frequency IDentification) technologies. There are many significant papers including [14,17,18,21,22,25] presenting different methods for various settings offering also some extra features. The result closest to our contribution is included in [6] where authors present a protocol for the model wherein both RFID and a single distinguished device called *the reader* in each round can transmit $O(1)$ bits. Using recent communication complexity result [4] they prove that every Monte Carlo counting protocol with relative error $\epsilon \in [1/\sqrt{n}, 0.5]$ and probability of failure smaller than 0.2 needs $\Omega(\frac{1}{\epsilon^2 \log 1/\epsilon} + \log \log n)$ execution time. For the same range of ϵ they demonstrated how to construct a protocol with $O(\frac{1}{\epsilon^2} + \log \log n)$ running-time. The model of a single-hop radio network considered in our paper and models of RFID systems are seemingly completely different. It turns, however, that the results from [6] can be almost instantly applied to the settings investigated in our paper at least for some ranges of parameters. On the other hand their results holds with constant probability while we demand probability of failure limited by $1/f$. As authors of [6] suggested repeating the basic algorithm and choosing the median to obtain arbitrary small probability of failure. Nevertheless, such approach leads to $\Theta(\log f)$ multiplicative factor overhead.

1.3   Our Results

In Sect. 1.1 we recall our model and introduce some new definitions. In Sect. 2 we present a time-efficient uniform algorithm for computing a $(1+\varepsilon)$-approximation of the size of the network with probability $1 - 1/f$ (where f is a parameter of the protocol) and provide its analysis. Our protocol requires $\mathcal{O}\left(\log \log n + \log f/\varepsilon^2\right)$ time slots.

In Sect. 3 we give a lower bound for the number of slots that are necessary to get a linear size estimation. For n nodes and any $f \geq 2$ we show that $\Omega(\log \log n + \log f/\varepsilon)$ slots are required to get a $(1+\varepsilon)$-approximation with probability greater than $1 - 1/f$ in the beeping model.

2   Size Estimation Algorithm

In this section we present an algorithm for $(1 + \varepsilon)$-approximation of network size. The algorithm works in time $\mathcal{O}\left(\log \log n + \log f/\varepsilon^2\right)$ with probability at least $1 - 1/f$. With probability at most $1/f$ the algorithm may return a wrong estimate or work for a larger number of steps (or both). First in Subsect. 2.1 we present a procedure for 64-approximation and later in Subsect. 2.2 we show how to improve it to $(1 + \varepsilon)$-approximation, for any $\varepsilon > 0$. An important feature of our algorithm is its uniformity:

Definition 2. *A randomized distributed algorithm $\mathcal{A}$ is called **uniform** if, and only if, in round i every node that has not yet transmitted successfully, transmits independently with probability p_i (the same for all nodes).*

For k active nodes the probability that exactly j nodes transmit in the i-th round is $\binom{k}{j}(p_i)^j(1 - p_i)^{k-j}$. Note that p_i may depend on the state of the

Function 1 Broadcast($\hat{n}$)

transmit with probability $1/\hat{n}$
return the status of the channel

Function 2 Phase1()

$l \leftarrow 0$
$u \leftarrow 1$
while Broadcast(2^u) $\neq$ NULL do
    $u \leftarrow 2u$
while $l + 1 < u$ do
    $m \leftarrow \lceil (l + u)/2 \rceil$
    if Broadcast(2^m) = NULL then
        $u \leftarrow m$
    else
        $l \leftarrow m$
return u

Function 3 Phase2(u, L)

$\mathcal{M} \leftarrow [\,]$
for $k = 1$ to L do
    append u to $\mathcal{M}$
    $status \leftarrow$ Broadcast(2^u)
    if $status = $ NULL then
        $u \leftarrow \max(u - 3, 0)$
    else if $status = $ BEEP then
        $u \leftarrow u + 3$
return the most frequent value in $\mathcal{M}$

Algorithm 1 SizeApprox1(f)

$u \leftarrow$ Phase1()
$d \leftarrow \lceil (\log f + \log\log f + \log\log u + 5)/3 \rceil$
$L \leftarrow 100\log(2f) + \lceil 125d/4 \rceil + 13$
$u \leftarrow$ Phase2(u, L)
return 2^u

Fig. 1. The pseudocode of a 64-approximation algorithm.

communication channel in previous rounds. In general, p_i can be even chosen randomly from some distribution during the execution of the protocol (finally, all nodes have to use, however, the same value p_i). Due to their simplicity and robustness, uniform algorithms are commonly used [20, 24, 25].

2.1   64-Approximation

Phase 1 in the Algorithm is based on Leader Election Protocol by Nakano and Olariu [20]. Similarly, Phase 2 is a modification of a subprocedure used in [20]. Both phases make use of Broadcast function to determine (with a certain probability) if the current estimation of the network size is too high or too low. Intuitively, in Phase 1 nodes try to bound from above the network size by doubling the estimate until the status of the channel suggests that it is too high. In each round of Phase 2 nodes adjust the estimate by factor 8 according to the status of the channel. We should note here that the closer the estimate is to the real network size, the more probable it is that the decision based on an output of call to Broadcast is incorrect. Because of this, after Phase 2 we return the most common estimate. The following lemmas provide bounds on time complexity and accuracy of the returned estimator.

Lemma 1 (Nakano, Olariu [20]). *With probability exceeding $1 - \frac{1}{2f}$ Phase1 takes at most $\mathcal{O}\left(\log\log n + \log f\right)$ rounds after which the returned value, u, satisfies the double inequality*

$$\frac{n}{\ln(4(\lceil \log\log(4nf) \rceil + 1)f)} \leq 2^u \leq 4(\lceil \log\log(4nf) \rceil + 1)fn. \qquad (1)$$

Let us introduce the following notation (we assume that $n \geq 2$). Parameters $p_\alpha^{(N)}$, $p_\alpha^{(B)}$ will denote probabilities of NULL and BEEP conditioned that the broadcast probability in the current round is $\min\{\frac{1}{\alpha n}, 1\}$. If $\alpha \cdot n > 1$, then

$$p_\alpha^{(N)} = \mathbb{P}\left(\text{NULL} \mid 2^u = \alpha \cdot n\right) = \left(1 - \frac{1}{\alpha \cdot n}\right)^n,$$

$$p_\alpha^{(B)} = \mathbb{P}\left(\text{BEEP} \mid 2^u = \alpha \cdot n\right) = 1 - \left(1 - \frac{1}{\alpha \cdot n}\right)^n,$$

where $1/2^u$ is the probability of transmission for each node and n is the real number of nodes. Otherwise, with $\alpha n \leq 1$ we set $p_\alpha^{(N)} = 0$ and $p_\alpha^{(B)} = 1$. For any fixed α we can bound the values of $p_\alpha^{(N)}, p_\alpha^{(B)}$ using basic inequalities. The following Proposition can be easily verified.

Proposition 1. *For $n \geq 25$ we have:*

1. $p_{1/8}^{(N)} \leq 0.06$,
2. $p_8^{(B)} \leq 0.12$,
3. $p_{1/64}^{(B)} \geq 0.99$,
4. $p_{64}^{(N)} \geq 0.98$.

In the following Lemma we analyze Phase2 and show that Algorithm 1 is a 64-approximation.

Lemma 2. *If $n \geq 25$, then Algorithm 1 with probability at least $1 - 1/f$ returns value $\hat{n} = 2^u$ such that $n/64 \leq \hat{n} \leq 64 \cdot n$ in time $\mathcal{O}(\log\log n + \log f)$.*

Proof. Assume that u, after Phase1 satisfies the double inequality from Lemma 1. We want to show that, conditioned on such an event, the approximation returned by Algorithm 1 is a 64-approximation with probability at least $1 - \frac{1}{2f}$. Thus we need to analyze Phase2. The phase can be seen as a biased random walk of length L on a line, where points on the line correspond to the values of the estimator 2^u and transition probabilities equal $p_\alpha^{(N)}$ and $p_\alpha^{(B)}$ (see Fig. 2).

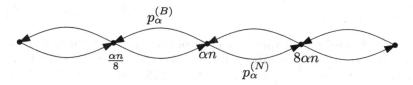

Fig. 2. An illustration of transition probabilities in Phase2.

Consider a sequence $\mathcal{U} = \{\ldots, u_{-2}, u_{-1}, u_0, u_1, u_2, \ldots\}$, such that $2^{u_0} \leq n < 2^{u_1}$ and $u_{i+1} = u_i + 3$ for all $i \in \mathbb{Z}$. Let $\mathcal{P} = \{u_{-1}, u_0, u_1, u_2\}$. Let us call a

- *good step* – a step that starts and ends inside $\mathcal{P}$,
- *improving step* – a step that start outside $\mathcal{P}$ moving towards $\mathcal{P}$ (a NULL or BEEP such that the estimator after the step is better),
- *bad step* – a step that is leaving $\mathcal{P}$ or the one that starts outside set $\mathcal{P}$ moving further from $\mathcal{P}$.

We want to show that the state with the maximum number of visits will be a state from set $\mathcal{P}$, and thus the returned estimator will be a 64-approximation. Observe that during a good step an estimate from set $\mathcal{P}$ is added to set $\mathcal{M}$.

Denote by G, B, I the number of good, bad and improving steps during L steps of Phase2. By Lemma 1 the probability of a bad step is at most 0.12. Clearly, steps are dependent, however all the bounds for each step hold independently from other steps. Thus we can limit B by the sum of stochastically independent 0–1 random variables and apply a Chernoff bound to get:

$$\mathbb{P}\left(B \geq 1.5 \cdot 0.12 \cdot L\right) \leq e^{1/12 \cdot 0.12 L} \leq \frac{1}{2f} \ .$$

Assume that $B < 0.12L$. Recall that d is the initial distance to set $\mathcal{P}$. Thus $G \leq B + d$. Since in a step (either good, bad or improving), the walk traverses an edge between two different states, the maximum number of visits to one state outside set $\mathcal{P}$ is at most

$$\left\lceil \frac{B}{2} \right\rceil + \left\lceil \frac{I}{2} \right\rceil \leq \frac{B+d}{2} + \frac{B}{2} + 2 = B + \frac{d}{2} + 2 \leq 0.18L + \frac{d}{2} + 2.$$

The total number of steps inside $\mathcal{P}$ is at least $L - I - B \leq L - (0.18L + d/2 + 2) = 0.82L - d/2 - 2$. Since $\mathcal{P}$ contains exactly four steps, there exists a step with at least $0.2L - d/6 - 2/3$ visits. Since the maximum number of visits to a state outside $\mathcal{P}$ is at most $0.18L + \frac{d}{2} + 2$, we need to show that

$$0.2L - d/8 - 1/4 \geq 0.18L + \frac{d}{2} + 2,$$

which is equivalent to

$$4L \geq 125d + 450.$$

We know from the definition of the algorithm that

$$L = 100 \log(2f) + 125d/4 + 13 = 100 \log f + 125d/4 + 113 > 125d/4 + 450/4.$$

Thus the state with the maximum number of visits is a state from set $\mathcal{P}$ which corresponds to a 64-approximation of the correct value of n. Now, by Lemma 1 with probability at least $1 - \frac{1}{2f}$, the total time of Phase1 is $\mathcal{O}(\log\log n)$ and the value of u after the phase satisfies the double inequality (1). Conditioned on this event, with probability at least $1 - \frac{1}{2f}$ Phase2 returns a 64-approximation. The time of Phase2 is always $\mathcal{O}(\log f + \log\log\log n)$. Thus overall our algorithm returns u such that 2^u is a 64-approximation of n in time $\mathcal{O}(\log\log n + \log f)$ with probability at least $1 - \frac{1}{f}$.

2.2 $(1 + \varepsilon)$-Approximation

We now describe how to enhance the algorithm from the previous section with an additional phase to obtain a $(1 + \varepsilon)$-factor approximation for any $\varepsilon > 0$. Intuitively, the procedure Vote checks whether the current estimate $\hat{n}$ is too big or too small. We let the nodes transmit with probability $1/\hat{n}$ for a fixed number of rounds. If our estimate is too small, a lot of nodes will transmit and there will not be enough silent rounds and thus we increase our estimate by a factor of $(1 + \varepsilon)$. Similarly, if our estimate is too large, too many rounds will be silent and thus we decrease our estimate by a factor of $(1 + \varepsilon)$.

Let $c = 1 + \varepsilon$, and denote $p_l = e^{-c}$ and $p_h = e^{-1/c}$.

Function 4 Vote($\hat{n}, c, f$)

$p_l \leftarrow e^{-c}, \quad p_h \leftarrow e^{-1/c}$
$\delta_c \leftarrow (p_h - p_l)/(p_h + p_l)$
$\ell \leftarrow \lceil 3 \cdot e^3 \cdot \log f/\delta_c^2 \rceil$
$nulls \leftarrow 0$
for $i = 1$ **to** ℓ **do**
  if Broadcast($\hat{n}$) = NULL **then**
    $nulls \leftarrow nulls + 1$
if $nulls < (1 + \delta_c) \cdot p_l \cdot \ell$ **then**
  return UNDERESTIMATED
else
  return OVERESTIMATED

Function 6 Phase3($\hat{n}, f$)

$f' \leftarrow 14f$
$initial \leftarrow$ Vote($\hat{n}, \sqrt{2}, f'$)
if $initial$=UNDERESTIMATED **then**
  $\phi \leftarrow \sqrt{2}$
else
  $\phi \leftarrow 1/\sqrt{2}$
for $i = 1$ **to** 13 **do**
  $\hat{n} \leftarrow \phi \cdot \hat{n}$
  if $initial \neq$ Vote($\hat{n}, \sqrt{2}, f'$) **then**
    return $\hat{n}$
return $\hat{n}$

Function 5 Refine($\hat{n}, c, f$)

$vote \leftarrow$ Vote($\hat{n}, c^{1/2}, f$)
if $vote =$ UNDERESTIMATED **then**
  return $c^{1/4} \cdot \hat{n}$
else
  return $c^{-1/4} \cdot \hat{n}$

Algorithm 2 SizeApprox2(f, c)

$\hat{n} \leftarrow$ SizeApprox1($4f$)
$\hat{n} \leftarrow$ Phase3($\hat{n}, 4f$)
$t \leftarrow \lceil \log_{4/3} \log_c 2 \rceil$
for $i = t$ **downto** 1 **do**
  $\hat{n} \leftarrow$ Refine($\hat{n}, c^{(4/3)^i}, 2^{i+1}f$)
return $\hat{n}$

Fig. 3. The pseudocode of c-approximation algorithm.

We have:

$$\mathbb{P}\left(\text{NULL}|\hat{n} \geq cn\right) \geq \left(1 - \frac{1}{cn}\right)^n \geq e^{-1/c}\left(1 - \frac{1}{cn}\right) \geq p_h/2, \qquad (2)$$

$$\mathbb{P}\left(\text{NULL}|\hat{n} \leq n/c\right) \leq \left(1 - \frac{c}{n}\right)^n \leq e^{-c} = p_l. \qquad (3)$$

Thus $p_h/2$ upper bounds the probability of NULL in a round under the condition that approximation $\hat{n}$ is c times too high. On the other hand p_l lowerbounds the probability of NULL in a round conditioned that $\hat{n}$ is c times too low.

Denote $\delta = \frac{p_h - p_l}{p_h + p_l}$, and observe that for such δ we have

$$p_h/2\,(1 - \delta) = p_l\,(1 + \delta). \tag{4}$$

Moreover since $p_h - p_l = e^{-1/c} - e^{-c} > 0$, then $\delta > 0$. Observe also that $\delta < 1/2$.

In the following lemmas we bound the probability that procedure Vote returns OVERESTIMATED and UNDERESTIMATED, assuming that estimator $\hat{n}$ deviates from n by factor c. We note that in all calls to Vote in the algorithm the inequality $c < 3$ holds.

Lemma 3. *If $\hat{n} < n/c$, then procedure Vote($\hat{n}, c, f$) returns UNDERESTI-MATED with probability at least $1 - \frac{1}{f}$.*

Proof. By (3), the probability that no node transmits is upperbounded by p_l. Let X_i denote the random variable that is 0 if at least one node transmits and 1 otherwise. Thus, if we let the nodes transmit ℓ times, we obtain as expected value for $X = \sum_{i=1}^{\ell} X_i$, $E[X] \leq \ell \cdot p_l$. Chernoff bound yields:

$$\mathbb{P}\,(X \geq (1 + \delta)\,p_l l) = \mathbb{P}\left(X \geq (1 + \delta)\left(1 + \frac{p_l l - E[X]}{E[X]}\right) E[X]\right)$$

$$\leq e^{-\frac{((1+\delta)p_l l - E[X])^2}{E[X]}}.$$

We know that $E[X] \leq p_l l$ hence $((1 + \delta)p_l l - E[X])^2 \geq (\delta p_l l)^2$. Since $\ell \geq \frac{3}{\delta^2} e^3 \log f$, then $\delta^2 p_l l \geq \log f$ hence $((1 + \delta)p_l l - E[X])^2 \geq E[X]\log f$ and $\mathbb{P}\,(X \geq (1 + \delta)\,p_l l) \leq \frac{1}{f}$. Thus, with probability at least $1 - 1/f$, variable *nulls* in procedure Vote satisfies *nulls* $< (1 + \delta) \cdot p_l \cdot \ell$. Thus Vote returns UNDER-ESTIMATED with probability at least $1 - 1/f$.

Lemma 4. *If $\hat{n} > cn$, then procedure Vote($\hat{n}, c, f$) returns OVERESTIMATED with probability at least $1 - \frac{1}{f}$.*

Proof. By (2), the probability that no node transmits is lowerbounded by $p_h/2$. Let X_i denote the random variable that is 0 if at least one node transmits and 1 otherwise. Thus, if we let the nodes transmit ℓ times, we obtain as expected value for $X = \sum_{i=1}^{\ell} X_i$, $E[X] \geq \ell \cdot p_h/2$. Chernoff bound yields:

$$\mathbb{P}\,(X \leq (1 + \delta)\,p_l l) = \mathbb{P}\,(X \leq (1 - \delta)\,p_h l/2) \leq \mathbb{P}\,(X \leq (1 - \delta)\,E[X])$$

$$\leq e^{-\frac{\delta^2}{2} E[X]} \leq \frac{1}{f}.$$

This holds for $\ell \geq \frac{3}{\delta^2} e^3 \log f$, since $p_h > e^{-1}$. Thus with probability at least $1 - 1/f$, variable *nulls* does not satisfy the condition after **if**, thus Vote returns OVERESTIMATED with probability at least $1 - 1/f$.

Lemma 5. *If $\hat{n}$ is a 64-approximation of the number of nodes n, then procedure Phase3($\hat{n}, f$) returns a 2-approximation of n with probability at least $1 - 1/f$ using $\mathcal{O}\,(\log f)$ slots.*

Proof. We call an execution of $\texttt{Vote}(\hat{n}, \sqrt{2}, 14f)$ successful if it:

- returns OVERESTIMATED when $\hat{n} \geq \sqrt{2}n$,
- returns UNDERESTIMATED when $\hat{n} \leq n/\sqrt{2}$.

Procedure $\texttt{Phase3}$ makes at most 14 calls to $\texttt{Vote}$ and by Lemmas 3 and 4 each call is successful with probability at least $1 - 1/(14f)$. Therefore the probability that all calls are successful is at least $1 - 1/f$.

We want to argue that if all calls to procedure Vote are successful, then we obtain 2-approximation. If $\hat{n} \geq \sqrt{2}n$, then the first call to $\texttt{Vote}$ returns OVERESTIMATED and we start decreasing the estimate. After at most $\log_{\sqrt{2}} 64 + 1 = 13$ iterations, the value $\hat{n}$ satisfies $\hat{n} \leq n/\sqrt{2}$ and $\texttt{Vote}$ returns UNDERESTIMATED. The returned estimator is a 2-approximation of n because we divide the estimator by $\sqrt{2}$ until it is at most $n/\sqrt{2}$ for the first time. We make similar argument if the initial estimate is too small, i.e., $\hat{n} \leq n/\sqrt{2}$. If the initial estimate is correct, then after making at most 2 increases we will obtain an estimate that is at least $\sqrt{2}$ times too big, thus the third call to $\texttt{Vote}$ returns OVERESTIMATED and we finish the procedure. Using the same argument as above we can show that the returned estimator is a 2-approximation. Similarly, if the initial value is correct, we make at most 2 decreases.

Each call to $\texttt{Vote}(\hat{n}, \sqrt{2}, 14f)$ requires $\mathcal{O}(\log f)$ slots.

Lemma 6. *If $\hat{n}$ is a c-approximation of the number of nodes then procedure Refine$(\hat{n}, c, f)$ returns $c^{3/4}$-approximation with probability at least $1 - 1/f$ using $\mathcal{O}\left(\log f/\varepsilon^2\right)$ slots.*

Proof. Observe that if $\hat{n}$ is already a $c^{1/2}$-approximation, then regardless of the output of Vote we obtain a $c^{3/4}$-approximation.

On the other hand if $cn \geq \hat{n} \geq c^{1/2}n$, then by Lemma 4, with probability at least $1 - 1/f$, procedure Vote returns OVERESTIMATED and we decrease the estimate by factor of $c^{1/4}$. Finally if $n/c \leq \hat{n} \leq c^{-1/2}n$, then with probability at least $1 - 1/f$, by Lemma 3 Vote returns UNDERESTIMATED and we increase the estimate by factor of $c^{1/4}$.

To bound the time complexity of procedure $\texttt{Refine}$ we need to bound the number of steps of procedure Vote. With $c = 1 + \varepsilon$ and $\varepsilon > 0$ we have

$$\delta_{1+\varepsilon} = \frac{e^{-\frac{1}{1+\varepsilon}} - e^{-(1+\varepsilon)}}{e^{-\frac{1}{1+\varepsilon}} + e^{-(1+\varepsilon)}} = \frac{e^{-1}}{e^{-1}} \frac{e^{\frac{\varepsilon}{1+\varepsilon}} - e^{-\varepsilon}}{e^{\frac{\varepsilon}{1+\varepsilon}} + e^{-\varepsilon}} \geq \frac{e^{\frac{\varepsilon}{1+\varepsilon}} - e^{-\frac{\varepsilon}{1+\varepsilon}}}{e^{\frac{\varepsilon}{1+\varepsilon}} + e^{-\frac{\varepsilon}{1+\varepsilon}}} = \tanh\left(\frac{\varepsilon}{1+\varepsilon}\right).$$

Therefore

$$\delta^{-2} \leq \coth^2\left(\frac{\varepsilon}{1+\varepsilon}\right) = 1 + \frac{1}{\sinh^2\left(\frac{\varepsilon}{1+\varepsilon}\right)} \leq \frac{1}{\varepsilon^2} + \frac{2}{\varepsilon} + 2,$$

where the last inequality is the result of $\sinh(x) \geq x$ for $x \geq 0$. Hence $\delta_\varepsilon^{-2} = O(\varepsilon^{-2})$ as $\varepsilon \to 0$. We call procedure Vote with $c^{1/2} = (1 + \varepsilon)^{1/2} \geq 1 + \varepsilon/4$ for $\varepsilon < 1$. Hence the complexity of a single execution of procedure Vote is $O(\varepsilon^{-2} \log f)$.

Theorem 1. *For any positive $\varepsilon > 0$ algorithm SizeApprox2$(f, 1 + \varepsilon)$ returns a $(1+\varepsilon)$-approximation of number of nodes with probability at least $1 - 1/f$ using $\mathcal{O}\left(\frac{\log f}{\varepsilon^2} + \log\log n\right)$ slots.*

Proof. With probability at least $1 - 1/(4f)$ call to SizeApprox1 returns 64-approximation, which we turn into 2-approximation with probability at least $1 - 1/(4f)$ by calling Phase3. Next, we refine the approximation using $t = \lceil\log_{4/3}\log_{1+\varepsilon} 2\rceil$ iterations. The probability of failure of the i-th iteration is at most $1/(2^{i+1}f)$, for $1 \leq i \leq t$. Therefore, by union bound, the probability of failure of the SizeApprox2 is at most

$$\frac{1}{4f} + \frac{1}{4f} + \frac{1}{2f} \cdot \sum_{i=1}^{t} 2^{-i} \leq \frac{1}{f}.$$

Assuming that none of the Vote calls failed we compute the quality of the resulting estimate. We can show by induction using Lemma 6 that after i-th iteration of the loop in algorithm SizeApprox2, the current estimate $\hat{n}$ is a $(1 + \varepsilon)^{(4/3)^{i-1}}$-approximation. Hence after t iterations we get a $(1+\varepsilon)$-approximation.

By Lemma 6 the number of slots used by t iterations is

$$\sum_{i=1}^{t} \mathcal{O}\left(\frac{\log(2^{i+1}f)}{\varepsilon^2(4/3)^{2i}}\right) \leq \sum_{i=1}^{\infty} \mathcal{O}\left(\frac{\log(2^{i+1}f)}{\varepsilon^2(4/3)^{2i}}\right) = \mathcal{O}\left(\frac{\log f}{\varepsilon^2}\right),$$

where the last inequality is justified by the fact that the $\mathcal{O}(\cdot)$ notation from Lemma 6 holds uniformly (i.e., the hidden constant is independent from f, i and ε).

Adding the slots used by SizeApprox1 and Phase3 we get the final time complexity.

3   Lower Bound

In this section we show that any (not necessarily uniform) size estimation algorithm returning a $(1+\varepsilon)$-approximation of the number of nodes with probability at least $1 - 1/f$ works in time $\Omega(\log\log n + \frac{\log f}{\varepsilon})$.

We start the analysis of beeping model by showing how the execution by different number of nodes relates to each other. Namely, we prove that (in probability) history of the channel state observed in case of n and m nodes performing any randomized protocol are similar for n close to m. We subscript symbol $\mathbb{P}$ with n to denote probability conditioned on the number of nodes running some algorithm, $\mathbb{P}_n(A) = \mathbb{P}(A \mid |\mathcal{N}| = n)$ for any event A. For a vector $h \in \{\text{NULL}, \text{BEEP}\}^t$ we write $\mathbb{P}(h)$ to denote the probability that during the first t slots of the execution of algorithm the global history of channel is h.

Lemma 7. *Let $\mathcal{A}$ be any randomized algorithm for a single-hop radio network with beeping communication model. For a global history of channel state, $h \in \{NULL, BEEP\}^*$ and $m \geq n \geq 1$, there is $\mathbb{P}_m(h) \geq (\mathbb{P}_n(h))^{m/n}$.*

Proof. We proceed with a coupling argument. Let $\mathcal{S} = \{s_1, \ldots, s_{nm}\}$ be a set consisting of nm nodes. Even though the nodes are indistinguishable, for the purpose of analysis we can identify them by the random sources they use. That is, we assume that node s_i has access to an infinite sequence of random bits $\boldsymbol{X}_i = X_i^{(1)}, X_i^{(2)}, \ldots$. Clearly, if $\boldsymbol{X}_i = \boldsymbol{X}_j$, then nodes s_i and s_j behave identically during an execution of any algorithm (of course $\mathbb{P}(\boldsymbol{X}_i = \boldsymbol{X}_j) = 0$ for $i \neq j$). We partition $\mathcal{S}$ in two different ways – into n independent networks $\mathcal{N}_1, \ldots, \mathcal{N}_n$ with m nodes each (called big networks) and m independent networks $\mathcal{N}'_1, \ldots, \mathcal{N}'_m$ with n nodes each (called small networks). We require that for each big network $\mathcal{N}_i$ there exists at least one small network $\mathcal{N}'_j$ such that $\mathcal{N}'_j \subseteq \mathcal{N}_i$. We assume that all networks are independent from each other, i.e., there are no interferences of communication channels. In these two settings, however, each node from $\mathcal{S}$ belongs to exactly one big and one small network and in both cases uses the same random source for making its decisions. Our goal is to compare the execution of algorithm $\mathcal{A}$ performed by the same nodes grouped into big and small networks.

Let $\boldsymbol{H}_1, \ldots, \boldsymbol{H}_n$ and $\boldsymbol{H}'_1, \ldots, \boldsymbol{H}'_m$ denote global histories of channel states during the executions of algorithm $\mathcal{A}$ by big and small networks, respectively. We are going to show by induction on h's length that if h is a prefix of channel histories of all small networks, $\boldsymbol{H}'_1, \ldots, \boldsymbol{H}'_m$, then it is also a prefix of channels histories of all big networks, $\boldsymbol{H}_1, \ldots, \boldsymbol{H}_n$. The base case of empty string $h = \varepsilon$ holds trivially. Therefore, let us assume that the statement is true for all global histories of length $t \geq 0$ and that $h = h_1, h_2, \ldots, h_t, h_{t+1}$ is a prefix of channel histories of small networks. By induction, $h_1, h_2, \ldots, h_t$ is a prefix of each $\boldsymbol{H}_1, \ldots, \boldsymbol{H}_n$. At the beginning of the $(t+1)$-st slot each node decides whether to transmit or not based on its random source, local history and the global history $h_1, h_2, \ldots, h_t$. However, in this case the local history is redundant as it can be reconstructed from $\boldsymbol{X}_i$ and the global history. Therefore, if in the $(t+1)$st slot the resulting channel states of each small network are $h_{t+1} = \text{NULL}$,

$$H_1'^{(t+1)} = \ldots = H_m'^{(t+1)} = \text{NULL},$$

then all nodes decided not to transmit and

$$H_1^{(t+1)} = \ldots = H_n^{(t+1)} = \text{NULL}.$$

Otherwise, if

$$H_1'^{(t+1)} = \ldots = H_m'^{(t+1)} = \text{BEEP},$$

then in every small network there is at least one node that decided to transmit during the $(t+1)$st slot. For each big network $\mathcal{N}_i$ there is some small network $\mathcal{N}'_j \subseteq \mathcal{N}_i$, hence $H_j'^{(t+1)} = \text{BEEP}$ implies $H_i^{(t+1)} = \text{BEEP}$. Therefore, h is a prefix of $\boldsymbol{H}_1, \ldots, \boldsymbol{H}_n$. Finally, all networks are independent, thus

$$(\mathbb{P}_n(h))^m = \mathbb{P}(\boldsymbol{H}'_1 \text{ starts with } h \wedge \ldots \wedge \boldsymbol{H}'_m \text{ starts with } h)$$
$$\leq \mathbb{P}(\boldsymbol{H}_1 \text{ starts with } h \wedge \ldots \wedge \boldsymbol{H}_n \text{ starts with } h)$$
$$= (\mathbb{P}_m(h))^n.$$

Lemma 8. *For any non-empty finite set of global histories of channel state* $H \subseteq \{NULL, BEEP\}^*$ *and* $m > n \geq 1$ *there is*

$$\mathbb{P}_m(H) \geq \frac{(\mathbb{P}_n(H))^{m/n}}{|H|^{m/n-1}}.$$

Proof. By Lemma 7 we get

$$\mathbb{P}_m(H) = \sum_{h \in H} \mathbb{P}_m(h) \geq \sum_{h \in H} (\mathbb{P}_n(h))^{m/n}.$$

Using Hölder inequality

$$\sum_{i=1}^{n} |x_i y_i| \leq \left(\sum_{i=1}^{n} |x_i|^p \right)^{1/p} \cdot \left(\sum_{i=1}^{n} |y_i|^q \right)^{1/q}$$

with $p = m/n$ and $q = m/(m-n)$ we obtain

$$\sum_{h \in H} (\mathbb{P}_n(h))^p = \frac{1}{|H|^{p/q}} \left(\sum_{h \in H} 1^q \right)^{p/q} \cdot \sum_{h \in H} (\mathbb{P}_n(h))^p \geq \frac{1}{|H|^{p/q}} \left(\sum_{h \in H} \mathbb{P}_n(h) \right)^p$$

$$= \frac{(\mathbb{P}_n(H))^{m/n}}{|H|^{m/n-1}}.$$

As we stated in Sect. 1.1, in any algorithm $\mathcal{A}$ the decision whether to stop the execution after the current slot and what estimation to return is based only on the global history of channel state. For any history $h \in \{NULL, BEEP\}^*$ that causes nodes to finish the execution of $\mathcal{A}$ we denote by $\mathcal{A}(h)$ the estimated network size returned by $\mathcal{A}$.

Theorem 2. *Let $\mathcal{A}$ be a size estimation algorithm for a single-hop radio network assuming the beeping communication model. If for any network size n algorithm $\mathcal{A}$ returns $(1 + \varepsilon)$-approximation with probability at least $1 - 1/f$ and within at most T_n time slots (T_n non-decreasing), then*

$$T_n \geq \max\left\{ \frac{\lg f + (1+\varepsilon)^2 \lg(1 - 1/f)}{(1+\varepsilon)^2 + 1/n - 1}, \lg\lg(1 + 2\varepsilon n + \varepsilon^2 n) - \lg\lg(1 + \varepsilon) - 1 \right\}.$$

Proof. For $k \in \mathbb{N}_+$ let be a set of all global histories of length at most T_k for which the value returned by algorithm $\mathcal{A}$ is a $(1+\varepsilon)$-approximation of k. Clearly, $\mathbb{P}_k(H_k) \geq 1 - 1/f$. Let $m = \lfloor (1+\varepsilon)^2 n + 1 \rfloor$, so that $m/(1+\varepsilon) > (1+\varepsilon)n$ and thus $H_n \cap H_m = \emptyset$. This way,

$$\mathbb{P}_m(H_n) \leq 1 - \mathbb{P}_m(H_m) \leq 1/f.$$

On the other hand by Lemma 8 there is

$$\mathbb{P}_m(H_n) \geq \frac{(\mathbb{P}_n(H_n))^{m/n}}{|H_n|^{m/n-1}} \geq \frac{(1 - 1/f)^{m/n}}{|H_n|^{m/n-1}}.$$

Therefore,

$$|H_n| \geq \left(f(1 - 1/f)^{m/n} \right)^{\frac{1}{m/n-1}} .$$

We know that set H_n contains words of length at most T_n and no word is a prefix of another, so $|H_n| \leq 2^{T_n}$. Finally, we get

$$T_n \geq \log_2 |H_n| \geq \frac{\log_2 f + \frac{m}{n} \log_2(1 - 1/f)}{m/n - 1} \geq \frac{\log_2 f + (1 + \varepsilon)^2 \log_2(1 - 1/f)}{(1 + \varepsilon)^2 + 1/n - 1} .$$

Now, let $a_1 = 1$ and

$$a_i = \lfloor (1 + \varepsilon)^2 a_{i-1} + 1 \rfloor \leq (1 + \varepsilon)^2 a_{i-1} + 1 \leq \frac{(1 + \varepsilon)^{2i} - 1}{(1 + \varepsilon)^2 - 1} .$$

All sets H_{a_i} must be non-empty and pairwise disjoint. Because T_n is non-decreasing, we have

$$\left| \bigcup_{i:\, a_i \leq n} H_{a_i} \right| \leq 2^{T_n} .$$

For

$$i \leq \frac{\log_2(((1 + \varepsilon)^2 - 1)n + 1)}{2 \log_2(1 + \varepsilon)}$$

there is $a_i \leq n$. Therefore,

$$T_n \geq \log_2 \log_2(1 + 2\varepsilon n + \varepsilon^2 n) - \log_2 \log_2(1 + \varepsilon) - 1.$$

Remark 1. For $\varepsilon \to 0$ and $f \geq 2$ we get

$$T_n = \Omega \left(\frac{\log f}{2\varepsilon + 1/n} + \log \log n \right).$$

For a constant ε (independent of n and f) there is

$$T_n = \Omega(\log f + \log \log n).$$

4   Final Remarks

We presented an algorithm for $(1 + \varepsilon)$-approximation of the size of a single-hop radio network with Beeping Model that needs $\mathcal{O}\left(\log \log n + \log f/\varepsilon^2 \right)$ time slots, wherein n is the real number of nodes and $1/f$ is the probability of failure. We also proved the matching lower bound for a constant ε. In some subprocedures we used quite big constants for the sake of technical simplicity of the analysis. As a future work we leave improving all those parameters. We believe that they can be significantly lowered to make the protocol practical for real-life scenarios already for moderate n.

References

1. Afek, Y., Alon, N., Bar-Joseph, Z., Cornejo, A., Haeupler, B., Kuhn, F.: Beeping a maximal independent set. Distrib. Comput. **26**(4), 195–208 (2013)
2. Bordim, J.L., Cui, J., Hayashi, T., Nakano, K., Olariu, S.: Energy-efficient initialization protocols for ad-hoc radio networks. In: Aggarwal, A., Pandu Rangan, C. (eds.) Algorithms and Computation. LNCS, vol. 1741, pp. 215–224. Springer, Heidelberg (1999). doi:10.1007/3-540-46632-0_23
3. Caragiannis, I., Galdi, C., Kaklamanis, C.: Basic computations in wireless networks. In: Deng, X., Du, D.-Z. (eds.) ISAAC 2005. LNCS, vol. 3827, pp. 533–542. Springer, Heidelberg (2005). http://dx.doi.org/10.1007/11602613_54
4. Chakrabarti, A., Regev, O.: An optimal lower bound on the communication complexity of gap-hamming-distance. In: Fortnow, L., Vadhan, S.P. (eds.) Proceedings of the 43rd ACM Symposium on Theory of Computing, STOC 2011, San Jose, CA, USA, 6–8 June 2011, pp. 51–60. ACM (2011). http://doi.acm.org/10.1145/1993636.1993644
5. Chassaing, P., Gerin, L.: Efficient estimation of the cardinality of large data sets. In: 4th Colloquium on Mathematics and Computer Science, DMTCS Proceedings, pp. 419–422 (2006)
6. Chen, B., Zhou, Z., Yu, H.: Understanding RFID counting protocols. In: Helal, S., Chandra, R., Kravets, R. (eds.) The 19th Annual International Conference on Mobile Computing and Networking, MobiCom 2013, Miami, FL, USA, 30 September–04 October 2013, pp. 291–302. ACM (2013). http://doi.acm.org/10.1145/2500423.2500431
7. Cichoń, J., Lemiesz, J., Szpankowski, W., Zawada, M.: Two-phase cardinality estimation protocols for sensor networks with provable precision. In: Proceedings of WCNC 2012, Paris, France. IEEE (2012)
8. Cichoń, J., Lemiesz, J., Zawada, M.: On size estimation protocols for sensor networks. In: Proceedings of the 51th IEEE Conference on Decision and Control, CDC 10–13, Maui, HI, USA. pp. 5234–5239, Proceedings of 51st Annual Conference on Decision and Control (CDC). IEEE, December 2012
9. Cornejo, A., Kuhn, F.: Deploying wireless networks with beeps. In: Lynch, N.A., Shvartsman, A.A. (eds.) DISC 2010. LNCS, vol. 6343, pp. 148–162. Springer, Heidelberg (2010)
10. Flajolet, P., Fusy, E., Gandouet, O., Meunier, F.: Hyperloglog: the analysis of a near-optimal cardinality estimation algorithm. In: Proceedings of the Conference on Analysis of Algorithms (AofA 2007), pp. 127–146 (2007)
11. Flajolet, P., Martin, G.N.: Probabilistic counting algorithms for data base applications. J. Comput. Syst. Sci. **31**(2), 182–209 (1985)
12. Giroire, F.: Order statistics and estimating cardinalities of massive data sets. Discrete Appl. Math. **157**(2), 406–427 (2009)
13. Greenberg, A.G., Flajolet, P., Ladner, R.E.: Estimating the multiplicities of conflicts to speed their resolution in multiple access channels. J. ACM **34**(2), 289–325 (1987). http://doi.acm.org/10.1145/23005.23006
14. Han, H., Sheng, B., Tan, C.C., Li, Q., Mao, W., Lu, S.: Counting RFID tags efficiently and anonymously. In: 29th IEEE International Conference on Computer Communications, Joint Conference of the IEEE Computer and Communications Societies, INFOCOM 2010, 15–19 March 2010, San Diego, CA, USA, pp. 1028–1036. IEEE (2010). http://dx.doi.org/10.1109/INFCOM.2010.5461944

15. Jurdziński, T., Kutyłowski, M., Zatopianski, J.: Energy-efficient size approximation of radio networks with no collision detection. In: Ibarra, O.H., Zhang, L. (eds.) COCOON 2002. LNCS, vol. 2387, pp. 279–289. Springer, Heidelberg (2002)
16. Kabarowski, J., Kutyłowski, M., Rutkowski, W.: Adversary immune size approximation of single-hop radio networks. In: Cai, J.-Y., Cooper, S.B., Li, A. (eds.) TAMC 2006. LNCS, vol. 3959, pp. 148–158. Springer, Heidelberg (2006)
17. Kodialam, M.S., Nandagopal, T.: Fast and reliable estimation schemes in RFID systems. In: Gerla, M., Petrioli, C., Ramjee, R. (eds.) Proceedings of the 12th Annual International Conference on Mobile Computing and Networking, MOBI-COM 2006, Los Angeles, CA, USA, 23–29 September 2006, pp. 322–333. ACM (2006). http://doi.acm.org/10.1145/1161089.1161126
18. Kodialam, M.S., Nandagopal, T., Lau, W.C.: Anonymous tracking using RFID tags. In: 26th IEEE International Conference on Computer Communications, Joint Conference of the IEEE Computer and Communications Societies, INFOCOM 2007, Anchorage, Alaska, USA, 6–12 May 2007, pp. 1217–1225. IEEE (2007). http://dx.doi.org/10.1109/INFCOM.2007.145
19. Nakano, K., Olariu, S.: Energy-efficient initialization protocols for single-hop radio networks with no collision detection. IEEE Trans. Parallel Distrib. Syst. **11**(8), 851–863 (2000)
20. Nakano, K., Olariu, S.: Uniform leader election protocols for radio networks. IEEE Trans. Parallel Distrib. Syst. **13**(5), 516–526 (2002). http://doi.ieeecomputersociety.org/10.1109/TPDS.2002.1003864
21. Qian, C., Ngan, H., Liu, Y., Ni, L.M.: Cardinality estimation for large-scale RFID systems. IEEE Trans. Parallel Distrib. Syst. **22**(9), 1441–1454 (2011). http://doi.ieeecomputersociety.org/10.1109/TPDS.2011.36
22. Shahzad, M., Liu, A.X.: Every bit counts: fast and scalable RFID· estimation. In: Akan, Ö.B., Ekici, E., Qiu, L., Snoeren, A.C. (eds.) The 18th Annual International Conference on Mobile Computing and Networking, Mobicom 2012, Istanbul, Turkey, 22–26 August 2012, pp. 365–376. ACM (2012). http://doi.acm.org/10.1145/2348543.2348588
23. Whang, K.Y., Zanden, B.T.V., Taylor, H.M.: A linear-time probabilistic counting algorithm for database applications. ACM Trans. Database Syst. **15**(2), 208–229 (1990)
24. Willard, D.E.: Log-logarithmic selection resolution protocols in a multiple access channel. SIAM J. Comput. **15**(2), 468–477 (1986). http://dx.doi.org/10.1137/0215032
25. Zheng, Y., Li, M.: ZOE: fast cardinality estimation for large-scale RFID systems. In: Proceedings of the IEEE INFOCOM 2013, Turin, Italy, 4–19 April 2013, pp. 908–916. IEEE (2013). http://dx.doi.org/10.1109/INFCOM.2013.6566879

An Approximation Algorithm for Path Computation and Function Placement in SDNs

Guy Even[1], Matthias Rost[2(✉)], and Stefan Schmid[3]

[1] School of Electrical Engineering, Tel Aviv University, Tel Aviv, Israel
[2] Department of Electrical Engineering and Computer Science,
TU Berlin, Berlin, Germany
`mrost@inet.tu-berlin.de`
[3] Department of Computer Science, Aalborg University, Aalborg, Denmark

Abstract. We consider the task of embedding multiple service requests in Software-Defined Networks (SDNs), i.e. computing (combined) mappings of network functions on physical nodes and finding routes to connect the mapped network functions. A single service request may either be fully embedded or be rejected. The objective is to maximize the sum of benefits of the served requests, while the solution must abide node and edge capacities.

We follow the framework suggested by Even *et al.* [5] for the specification of the network functions and routing of requests via processing-and-routing graphs (PR-graphs): a request is represented as a directed acyclic graph with the nodes representing network functions. Additionally, a unique source and a unique sink node are given for each request, such that any source-sink path represents a feasible chain of network functions to realize the service. This allows for example to choose between different realizations of the same network function. Requests are attributed with a global demand (e.g. specified in terms of bandwidth) and a benefit.

Our main result is a randomized approximation algorithm for path computation and function placement with the following guarantee. Let m denote the number of links in the substrate network, ε denote a parameter such that $0 < \varepsilon < 1$, and opt^* denote the maximum benefit that can be attained by a fractional solution (one in which requests may be partly served and flow may be split along multiple paths). Let $c_{\min}$ denote the minimum edge capacity, let $d_{\max}$ denote the maximum demand, and let $b_{\max}$ denote the maximum benefit of a request. Let $\Delta_{\max}$ denote an upper bound on the number of processing stages a request undergoes. If $c_{\min}/(\Delta_{\max} \cdot d_{\max}) = \Omega((\log m)/\varepsilon^2)$, then with probability at least $1 - \frac{1}{m} - exp(-\Omega(\varepsilon^2 \cdot \mathsf{opt}^*/(b_{\max} \cdot d_{\max})))$, the algorithm computes a $(1 - \varepsilon)$-approximate solution.

1 Introduction

Software Defined Networks (SDNs) and Network Function Virtualization (NFV) have been reinventing key issues in networking [8]. The key characteristics of these developments are: (i) separation between the data plane and the control

© Springer International Publishing AG 2016
J. Suomela (Ed.): SIROCCO 2016, LNCS 9988, pp. 374–390, 2016.
DOI: 10.1007/978-3-319-48314-6_24

plane, (ii) specification of the network control from a global view, (iii) introduction of network abstractions that provide a simple networking model, and (iv) programmability and virtualization of network components.

In this paper we focus on an algorithmic problem that an orchestrator needs to solve in an SDN + NFV setting, namely jointly optimizing the *path computation and function placement* (PCFP) [5]: In modern networks, networking is not limited to forwarding packets from sources to destinations. Requests can come in the form of flows (i.e., streams of packets from a source node to a destination node with a specified packet rate) that must undergo processing stages on their way to their destination. Examples of processing steps include: compression, encryption, firewall validation, deep packet inspection, etc. The crystal ball of SDN + NFV is the introduction of abstractions that allow one to specify, per request, requirements such as processing stages, valid locations for each processing stage, and allowable sets of links along which packets can be sent between processing stages. An important application for such goals is supporting security requirements that stipulate that unencrypted packets do not traverse untrusted links or reach untrusted nodes.

From an algorithmic point of view, the path computation and function placement problem combines two different optimization problems. Path computation alone (i.e., the case of pure packet forwarding without processing of packets) is an integral path packing problem. Function mapping alone (i.e., the case in which packets only need to be processed but not routed) is a load balancing problem.

To give a feeling of the problem, consider a special case of requests for streams, each of which needs to undergo the same sequence of k processing stages denoted by $w_1, w_2, \ldots, w_k$. This means that service of a request from s_i to t_i is realized by a concatenation of $k + 1$ paths: $s_i \overset{p_0}{\rightsquigarrow} v_1 \overset{p_1}{\rightsquigarrow} v_2 \overset{p_2}{\rightsquigarrow} \ldots \overset{p_{k-1}}{\rightsquigarrow} v_k \overset{p_k}{\rightsquigarrow} t_i$, where processing stage w_i takes place in node v_i. Note that the nodes $v_1, \ldots, v_k$ need not be distinct and the concatenated path $p_0 \circ p_1 \circ \cdots \circ p_k$ need not be simple. A collection of allocations that serve a set of requests not only incurs a forwarding load on the network elements, it also incurs a computational load on the nodes. The computational load is induced by the need to perform the respective processing stages for the requests.

Previous works. The opportunities introduced by the SDN/NFV paradigm, in terms of novel services which can be deployed quickly and on-demand, has inspired much research over the last years [3,4,12,13,18]. The main focus of these works is usually on the system aspects, while less attention has been given to the algorithmic challenges. Moreover, the existing papers which do deal with the algorithmic challenges, often resort to heuristics or non-polynomial algorithms. For example, in the seminal work on service chaining [18] as well as in [10,17,19], mixed-integer programming is employed (and heuristics are sketched), Hartert et al. [13] use constrained programming, and others propose fast heuristics without approximation guarantees [1,2], or ignore important aspects of the problem such as link capacity constraints [9]. The online version is studied in [5] in which also a new standby/accept service model is introduced, and in [9].

More generally, the problem of combined path computation and function place-ment is closely related to the virtual network embedding problem, for which many exponential-time and heuristic algorithms have been developed over the last years [7]. Indeed, only recently a first approximation scheme based on ran-domized rounding for the virtual network embedding problem was proposed in [16]. While the model is more general by allowing for cyclic request graphs, the proposed algorithms might violate node and edge capacities by a logarithmic factor and is only applicable on a limited class of request graphs.

Our starting point is the model of SDN requests presented in [5]. In this model, each request is represented by a special graph, called a processing-and-route graph (PR-graph, in short). The PR-graph represents both the routing requirement and the processing requirements that the packets of the stream must undergo. We also build on the technique of graph products for representing valid realizations of requests [5].

Raghavan [14] initiated the study of randomized rounding techniques for multi-commodity flows, where the LP has two types of constraints: capacity constraints and demand constraints. The joint capacity constraints are common to all the flows, while the demand constraints are per request. Raghavan proves that randomized rounding succeeds with high probability if the ratio of the minimum capacity to maximum demand is logarithmic.

Contribution and Techniques. To the best of our knowledge, we present the first polynomial-time algorithm which comes with provable approximation guarantees for the PCFP-problem, under reasonable assumptions (i.e., logarith-mic capacity-to-demand ratio, few processing stages per request, and sufficiently large optimal benefit). We begin by formulating a fractional relaxation of the problem. The fractional relaxation consists of a set of fractional single com-modity flows, each over a *different* product graph. Each flow is fractional in the sense that it may serve only part of a request and may split the flow among multiple paths. We emphasize that the fractional flows do not constitute a multi-commodity flow because they are over different graphs. The fractional problem is a general packing LP [14]. Namely, the LP can be formulated in the form $\max\{\mathbf{b}^T \cdot \mathbf{x} \mid A \cdot \mathbf{x} \leq \mathbf{c}, \mathbf{x} \geq \mathbf{0}\}$, where all the components of the vectors $\mathbf{b}, \mathbf{c}$ and the matrix A are nonnegative. However, this LP does not satisfy the logarith-mic ratio required in Raghavan's analysis of general packing problems (due to demand constraints).

Although randomized rounding is very well known and appears in many textbooks and papers, the version for the general packing problem appears only in half a page in the thesis by Raghavan [14, p. 41]. A special case with unit demands and unit benefits appears in [11]. One of the contributions of this paper is a full description of the analysis of randomized rounding for the case of multiple-commodity flows over different graphs with joint capacity constraints.

2   Modeling Requests in SDN

We model SDN/NFV requests as process-and-route graphs (PR-graphs) [5]. The model is quite general, and allows each request to have multiple sources and destinations, varying bandwidth demands based on processing stages, task specific capacities, prohibited locations of processing, and prohibited links for routing between processing stages, etc. We overview a simplified version of this model to concisely define the problem of path computation and function placement (PCFP).

2.1   The Substrate Network

The substrate network is a fixed network of servers and communication links. The network is represented by a graph $N = (V, E)$, where V is the set of *nodes* and E is the set of *edges*. Nodes and edges have *capacities*. The capacity of an edge e is denoted by $c(e)$, and the capacity of a node $v \in V$ is denoted by $c(v)$. Let $c_{\min}$ denote the minimum capacity. We note that the network is static and undirected (namely each edge represents a bidirectional communication link), but may contain parallel edges.

2.2   Requests and PR-Graphs

Each request is specified by a tuple $r_i = (G_i, d_i, b_i, U_i, s_i, t_i)$, where the components are as follows:

1. $G_i = (X_i, Y_i)$ is a directed (acyclic) graph called the process-and-route graph (PR-graph). There is a single source (respectively, sink) that corresponds to the source (resp. destination) of the request. We denote the source and sink nodes in G_i by s_i and t_i, respectively. The other vertices correspond to services or processing stages of a request. The edges of the PR-graph are directed and indicate precedence relations between PR-vertices. Any s_i-t_i path in G_i represents a valid realization of request i.
2. The demand of r_i is d_i and its benefit is b_i. By scaling, we may assume that $\min_i b_i = 1$.
3. $U_i : X_i \cup Y_i \to 2^V \cup 2^E$ where (1) $U_i(x) \subseteq V$ denotes a set of "allowed" nodes in the substrate N that can perform service x, and (2) $U_i(y) \subseteq E$ denotes the set of "allowed" edges of the substrate N along which the routing segment that corresponds to y may be routed.

Note that in the above definition the function $U_i(x)$ returns a set of substrate locations on which the function $x \in X_i$ can be executed. This allows to model network function *types*: If a substrate node $v \in V$ represents a specific hardware appliance (e.g. a firewall), then this node can only host this specific type of network function. Hence, if a virtualized network function $x \in X_i$ has the same type as $v \in V$, we include v in $U_i(x)$ and exclude v from $U_i(x)$ otherwise.

Given this understanding of the restriction function U_i, PR-graphs allow to model the selection of specific implementations of network functions. Assume e.g. that a request i is given that shall connect $v \in V$ to $u \in V$ such that the traffic passes through a firewall. The substrate network may offer two types of firewall implementations: a hardware-based (as hardware appliance) and a software-based (as virtual machine). Using the definition of PR-graphs, the selection of either of the choices can be modeled by setting $X_i = \{s_i, x_{\mathrm{hw}}, x_{\mathrm{sw}}, t_i\}$ and $Y_i = \{(s_i, x_{\mathrm{hw}}), (s_i, x_{\mathrm{sw}}), (x_{\mathrm{hw}}, t_i), (x_{\mathrm{sw}}, t_i)\}$ and restricting $U_i(x_{\mathrm{hw}})$ to all the hardware firewalls and $U_i(x_{\mathrm{sw}})$ to the set of all nodes that may host software firewalls. As any s_i-t_i path in G_i represents a valid realization of request i, a mapping of request i must select any of the options to realize the request (cf. [17] for a general discussion on decomposition opportunities).

We denote the maximum demand by any request as d_{max} and the maximum benefit of any request as b_{max}.

2.3   The Product Network

In [5] the concept of product graphs was introduced and we shortly revisit the definition. For each request r_i, the product network $\mathrm{pn}(\mathrm{N}, \mathrm{r}_i)$ is defined as follows. The node set of $\mathrm{pn}(\mathrm{N}, \mathrm{r}_i)$, denoted V_i, is defined as $V_i \triangleq \cup_{y \in Y_i} (V \times \{y\})$. We refer to the subset $V \times \{y\}$ as the y-layer in the product graph. Note that there is a layer for every edge y in the PR-graph. The edge set of $\mathrm{pn}(\mathrm{N}, \mathrm{r}_i)$, denoted E_i, consists of two types of edges $E_i = E_{i,1} \cup E_{i,2}$ defined as follows.

1. *Routing edges* connect vertices in the same layer.

$$E_{i,1} = \big\{ ((u, y), (v, y)) \mid y \in Y_i,\ (u, v) \in U_i(y) \big\} \ .$$

2. Directed *processing edges* connect two copies of the same network vertex in different layers.

$$E_{i,2} = \{((v, y), (v, y')) \mid y, y' \in Y_i \text{ with } y = (\cdot, x), y' = (x, \cdot) \text{ and } v \in U_i(x)\}.$$

To simplify the description of valid realizations, we add a super source $\hat{s}_i$ and a super sink $\hat{t}_i$ to the respective product networks. The super source $\hat{s}_i$ is connected to all vertices (v, y) such that $v \in U_i(s_i)$ and y emanates from s_i. Similarly, there is an edge to the super sink $\hat{t}_i$ from all vertices (v, y) such that $v \in U_i(t_i)$ and y enters t_i.

Remarks. The following remarks may help clarify the definition of the product network.

1. Consider an edge $y = (x_1, x_2)$ of a request. The y-layer in the product graph contains a copy of the substrate to compute a route from the vertex that performs the x_1 processing to the vertex that performs the x_2 processing.
2. Consider two edges $y_1 = (x_1, x_2)$ and $y_2 = (x_2, x_3)$ in the PR-graph. The only processing edges between the y_1-layer and the y_2-layer are edges of the form $(v, y_1) \to (v, y_2)$, where $v \in U_i(x_2)$.
3. If we coalesce each layer of the PR-graph to a single vertex, then the resulting graph is the line graph of the PR-graph.

2.4 Valid Realizations of SDN Requests

We use product graphs to define valid realizations of SDN requests. Consider a path $\tilde{p}_i$ in the product graph $\mathrm{pn}(\mathrm{N}, \mathrm{r}_i)$ that starts in the super source $\hat{s}_i$ and ends in the super sink $\hat{s}_i$. Such a path $\tilde{p}_i$ represents the routing of request r_i from its origin to its destination and the processing stages that it undergoes. The processing edges along $\tilde{p}_i$ represent nodes in which processing stages of r_i take place. The routing edges within each layer represent paths along which the request is routed between processing stages. (The edges incident to the super source and super sink are not important).

Definition 1. *A path $\tilde{p}$ in the product network* $\mathrm{pn}(\mathrm{N}, \mathrm{r}_i)$ *that starts in the super source and ends in the super sink is a* valid realization *of request r_i.*

2.5 The Path Computation and Function Placement Problem (PCFP)

Modeling SDN requests by product graphs helps in reducing SDN requests to path requests. The translation of paths in the product graph back to paths in the substrate network is called projection. This translation involves a loss due to multiple occurrences of the same substrate resource along a path in the product graph. We define projection and multiplicity before we present the formal definition of the PCFP-problem.

Projection of paths. Let $\tilde{p}_i$ denote a path in the product graph $\mathrm{pn}(\mathrm{N}, \mathrm{r}_i)$ from the super source to the super sink. The projection of $\tilde{p}_i$ to a path $p_i = \pi(\tilde{p}_i)$ in the substrate network N is simply the projection onto routing edges of $\tilde{p}_i$. Namely, each routing edge $((u, y), (v, y))$ in $\tilde{p}_i$ is projected to the edge (u, v) in the substrate. Hence, when projecting a path, we ignore the processing edges and the edges incident to the super source and super sink. Note that $p = \pi(\tilde{p}_i)$ may not be a simple path even if $\tilde{p}_i$ is simple.

Notation. The *multiplicity* of an edge or a vertex z in a path p is the number of times z appears in the path. We denote the multiplicity of z in p by $\mathsf{multiplicity}(z, p)$.

Capacity Constraints. Let $\tilde{P} = \{\tilde{p}_i\}_{i \in I'}$ denote a set of valid realizations for a subset of requests $\{r_i\}_{i \in I'}$ with $I' \subseteq I$. The set $\tilde{P}$ *satisfies the capacity constraints* if

$$\sum_{i \in I} d_i \cdot \mathsf{multiplicity}(e, \pi(\tilde{p}_i)) \le c(e), \quad \text{for every edge } e \in E$$

$$\sum_{i \in I} d_i \cdot \mathsf{multiplicity}(v, \pi(\tilde{p}_i)) \le c(v), \quad \text{for every vertex } v \in V$$

Definition of the PCFP-problem. The input in the PCFP-problem consists of (1) a substrate network $N = (V, E)$ with vertex and edge capacities, and (2) a set of requests $\{r_i\}_{i \in I}$. The goal is to compute valid realizations $\tilde{P} = \{\tilde{p}_i\}_{i \in I'}$

for a subset $I' \subseteq I$ such that: (1) $\tilde{P}$ satisfies the capacity constraints, and (2) the benefit $\sum_{i \in I'} b_i$ is maximum. We refer to the requests r_i such that $i \in I'$ as the *accepted* requests; requests r_i such that $i \in I \setminus I'$ are referred to as *rejected* requests.

3   The Approximation Algorithm for PCFP

The approximation algorithm for the PCFP-problem is described in this section. It is a variation of Raghavan's randomized rounding algorithm for general packing problems [14, Theorem 4.7, p. 41] (in which the approximation ratio is $\frac{1}{e} - \sqrt{\frac{2 \ln n}{\varepsilon \cdot e \cdot \text{opt}^*}}$ provided that $\frac{c_{\min}}{d_{\max}} \geq \frac{\ln n}{\varepsilon}$).

3.1   Fractional Relaxation of the PCFP-problem

We now define the fractional relaxation of the PCFP-problem. Instead of assigning a valid realization $\tilde{p}_i$ per accepted request r_i, we assign a fractional single commodity flow $\tilde{f}_i$ in the product graph $\text{pn}(N, r_i)$. The source of the flow $\tilde{f}_i$ is the super source $\hat{s}_i$. Similarly, the destination of $\tilde{f}_i$ is the super sink $\hat{t}_i$. The demand of $\tilde{f}_i$ is d_i. Hence the demand constraint is $|\tilde{f}_i| \leq d_i$.

The capacity constraints are accumulated across all requests' flows. Formally,

$$\sum_{i,y} \tilde{f}_i((u,y),(v,y)) \leq c(u,v)$$

$$\sum_{i,y,y'} \tilde{f}_i((v,y),(v,y')) \leq c(v).$$

Hence, the cumulative load on the copies of substrate edge $(u,v) \in E$ in the respective PR-graphs is upper bounded by the original edge capacity $c(u,v)$. Similarly, as the usage of processing edges $((v,y),(v,y'))$ in the PR-graphs denotes the processing on substrate node v, the cumulative load is bounded by $c(v)$.

The objective function of the LP relaxation is to maximize $\sum_i b_i \cdot |\tilde{f}_i|/d_i$.

We emphasize that this fractional relaxation is not a classic multi-commodity flow. The reason is that each flow $\tilde{f}_i$ is defined over a different product graph. However, the fractional relaxation is a general packing LP.

3.2   The Algorithm

The algorithm uses a parameter $1 > \epsilon > 0$. The algorithm proceeds as follows.

1. Divide all the capacities by $(1 + \varepsilon)$. Namely, $\tilde{c}(e) = c(e)/(1 + \varepsilon)$ and $\tilde{c}(v) = c(v)/(1 + \varepsilon)$.
2. Compute a maximum benefit fractional PCFP solution $\{\tilde{f}_i\}_i$.

3. Apply the randomized rounding procedure independently to each flow $\tilde{f}_i$ over the product network $\text{pn}(N, r_i)$ (See Appendix B for a description of the procedure.) Let $\tilde{p}_i$ denote the path in $\text{pn}(N, r_i)$ (if any) that is assigned to request r_i by the randomized rounding procedure. Let f_i denote a flow of amount d_i along the projection $\pi(\tilde{p}_i)$. Note that each f_i is an unsplittable all-or-nothing flow. The projection of p_i might not be a simple path in the substrate, hence the flow $f_i(e)$ along the edge e can be a multiple of the demand d_i.

3.3   Analysis of the Algorithm

Definition 2. *The* diameter *of G_i is the length of a longest path in G_i from the source s_i to the destination t_i. We denote the diameter of G_i by $\Delta(G_i)$.*

The diameter of G_i is well defined because G_i is acyclic for every request r_i. In all applications we are aware of, the diameter $\Delta(G_i)$ is bounded by a constant (i.e., e.g. less than 10).

Notation. Let $\Delta_{\max} \triangleq \max_{i \in I} \Delta(G_i)$ denote the maximum diameter of a request. Let $c_{\min}$ denote the minimum edge capacity, and let $d_{\max}$ denote the maximum demand. Let opt^* denote the maximum benefit achievable by a fractional PCFP solution (with respect to the original capacities $c(e)$ and $c(v)$). Let ALG denote the solution computed by the algorithm. Let $B(S)$ denote the benefit of a solution S. Define $\beta(\varepsilon) \triangleq (1 + \varepsilon) \ln(1 + \varepsilon) - \varepsilon$.

Our goal is to prove the following theorem.[1]

Theorem 1. *Assume that $\frac{c_{\min}}{\Delta_{\max} \cdot d_{\max}} \geq \frac{4.2 + \varepsilon}{\varepsilon^2} \cdot (1 + \varepsilon) \cdot \ln |E|$ and $\varepsilon \in (0, 1)$. Then,*

$$\mathbf{Pr}\,[\text{ALG} \textit{does not satisfy the capacity constraints}] \leq \frac{1}{|E|} \tag{1}$$

$$\mathbf{Pr}\left[B(\text{ALG}) < \frac{1 - \varepsilon}{1 + \varepsilon} \cdot B(\text{opt}^*)\right] \leq e^{-\beta(-\varepsilon) \cdot B(\text{opt}^*)/((1 + \varepsilon) \cdot b_{\max} \cdot d_{\max})}. \tag{2}$$

We remark in asymptotic terms, the theorem states that if $\frac{c_{\min}}{\Delta_{\max} \cdot d_{\max}} = \Omega(\frac{\log |E|}{\varepsilon^2})$, then ALG satisfies the capacity constraints with probability $1 - O(1/|E|)$ and attains a benefit of $(1 - O(\varepsilon)) \cdot B(\text{opt}^*)$ with probability $1 - \exp(-\Omega(\varepsilon^2) \cdot B(\text{opt}^*)/(b_{\max} \cdot d_{\max}))$.

Proof. The proof is based on the fact that randomized rounding is applied to each flow $\tilde{f}_i$ independently. Thus the congestion of an edge in ALG is the sum of independent random variables. The same holds for the $B(\text{ALG})$. The proof proceeds by applying Chernoff bounds.

---

[1] We believe there is a typo in the analogous theorem for integral MCFs with unit demands and unit benefits in [11, Theorem 11.2, p. 452] and that a factor of ε^{-2} is missing in their lower bound on the capacities.

Proof of Eq. 1. For the sake of simplicity we assume that there are no vertex capacities (i.e., $c(v) = \infty$). The proof is based on the Chernoff bound in Theorem 2. To apply the bound, fix a substrate edge $e \in E$, where $e = (u, v)$. Recall that the randomized rounding procedure decides which requests are supplied. A supplied request r_i is assigned a path $\tilde{p}_i$ in the product network $\text{pn}(\text{N}, \text{r}_i)$. The path $\tilde{p}_i$ is the support of a single commodity flow f_i' with flow amount $|f_i'| = d_i$. The projection of f_i' to the substrate network is denoted by f_i and its support is the projected path $\pi(\tilde{p}_i)$. The multiplicity of every edge in $\pi(\tilde{p}_i)$ is at most $\Delta_{\max}$. Hence, for every edge e, $f_i(e)$ is a multiple of d_i between 0 and $\Delta_{\max} \cdot d_i$.

Define the random variables X_i and the upper bounds μ_i on their expectation as follows (recall that $e = (u, v)$).

$$X_i \triangleq \frac{f_i(e)}{\Delta_{\max} \cdot d_{\max}}$$

$$\mu_i \triangleq \frac{\tilde{c}(e)}{\Delta_{\max} \cdot d_{\max}} \cdot \frac{\sum_y \tilde{f}_i((u, y), (v, y))}{\sum_{j, y} \tilde{f}_j((u, y), (v, y))}$$

The conditions of Theorem 2 are satisfied for the following reasons. The random variables X_i are independent and $0 \le X_i \le 1$ because $f_i(e) \in \{0, d_i, \ldots, \Delta_{\max} \cdot d_i\}$. Also, by Claim C (see Page 14) and linearity of expectation,

$$\mathbf{E}[X_i] = \frac{\sum_y \tilde{f}_i((u, y), (v, y))}{\Delta_{\max} \cdot d_{\max}}.$$

Since $\sum_{j, y} \tilde{f}_j((u, y), (v, y)) \le \tilde{c}(e)$, it follows that $\mathbf{E}[X_i] \le \mu_i$. Finally, $\mu \triangleq \sum_{i \in I} \mu_i = \tilde{c}(e)/(\Delta_{\max} \cdot d_{\max})$.

Let $\text{ALG}(e)$ denote the load incurred on the edge e by ALG. Namely $\text{ALG}(e) \triangleq \sum_{i \in I} f_i(e)$. Note that $\text{ALG}(e) \ge (1 + \varepsilon) \cdot \tilde{c}(e)$ iff

$$\sum_{i \in I} X_i \ge (1 + \varepsilon) \cdot \frac{\tilde{c}(e)}{\Delta_{\max} \cdot d_{\max}} = (1 + \varepsilon) \cdot \mu.$$

From Theorem 2 we conclude that:

$$\mathbf{Pr}[\text{ALG}(e) \ge (1 + \varepsilon) \cdot \tilde{c}(e)] = \mathbf{Pr}\left[\sum_{i \in I} X_i \ge (1 + \varepsilon) \cdot \mu\right]$$

$$\le e^{-\beta(\varepsilon) \cdot \mu}$$

$$= e^{-\beta(\varepsilon) \cdot \tilde{c}(e)/(\Delta_{\max} \cdot d_{\max})}$$

By scaling of capacities, we have $c(e) = (1 + \varepsilon) \cdot \tilde{c}(e)$. By Fact 4, $\beta(\varepsilon) \ge \frac{2\varepsilon^2}{4.2 + \varepsilon}$. By the assumption $\frac{\tilde{c}(e)}{\Delta_{\max} d_{\max}} \ge \frac{4.2 + \varepsilon}{\varepsilon^2} \cdot \ln |E|$. We conclude that

$$\mathbf{Pr}[\text{ALG}(e) \ge c(e)] \le \frac{1}{|E|^2}.$$

Equation 1 follows by applying a union bound over all the edges.

Proof of Eq. 2. The proof is based on the Chernoff bound stated in Theorem 3. To apply the bound, let

$$X_i \triangleq \frac{b_i \cdot |f_i|}{b_{\max} \cdot d_{\max}}$$

$$\mu_i \triangleq \frac{b_i \cdot |\tilde{f}_i|}{b_{\max} \cdot d_{\max}}.$$

The conditions of Theorem 3 are satisfied for the following reasons. Since $b_i \leq b_{\max}$ and $|f_i| \leq d_{\max}$, it follows that $0 \leq X_i \leq 1$. Note that $\sum_i X_i = B(\text{ALG})/(b_{\max} \cdot d_{\max})$. By Corollary 1, $\mathbf{E}[X_i] = \mu_i$. By linearity, $\sum_i b_i \cdot |\tilde{f}_i| = \text{opt}^*/(1+\varepsilon)$ and $\mu \triangleq \sum_i \mu_i = \frac{B(\text{opt}^*)}{(1+\varepsilon)b_{\max} \cdot d_{\max}}$. Hence,

$$\mathbf{Pr}\left[B(\text{ALG}) < \frac{1-\varepsilon}{1+\varepsilon} \cdot B(\text{opt}^*)\right] = \mathbf{Pr}\left[\sum_i X_i < (1-\varepsilon) \cdot \mu\right]$$

$$\leq e^{-\beta(-\varepsilon)\cdot\mu}$$

$$\leq e^{-\beta(-\varepsilon)\cdot B(\text{opt}^*)/((1+\varepsilon)b_{\max}\cdot d_{\max})},$$

and the theorem holds.   □

3.4   Unit Benefits

In the case of unit benefits (i.e., all the benefits equal one and hence $b_{\max} = 1$), Theorem 1 gives a fully polynomial randomized approximation scheme.

Corollary 1. *Suppose that $b_{\max} = 1$. Under the premises of Theorem 1, with probability $1 - O(1/Poly(|E|))$, the algorithm returns an all-or-nothing unsplittable multi-commodity flow whose benefit is at least $1 - O(\varepsilon)$ times the optimal benefit.*

Proof. If $B(\text{opt}^*) > c_{\min}$, then the large capacities assumption implies that $B(\text{opt}^*)/(d_{\max} \cdot b_{\max}) \geq c_{\min}/d_{\max} \geq \varepsilon^{-2} \cdot \ln|E|$. This implies that that $B(\text{ALG}) \geq (1 - O(\varepsilon)) \cdot B(\text{opt}^*)$ with probability at least $1 - 1/poly(|E|)$. By adding the probabilities of the two possible failures (i.e., violation of capacities and small benefit) and taking into account the prescaling of capacities, we obtain that with probability at least $1 - O(1/poly(|E|))$, randomized rounding returns an all-or-nothing unsplittable multi-commodity flow whose benefit is at least $1 - O(\varepsilon)$ times the optimal benefit.   □

4   Discussion

Theorem 1 provides an upper bound for the probability that ALG is not feasible and that $B(\text{ALG})$ is far from $B(\text{opt}^*)$. These bounds imply that our algorithm can be viewed as a version of an asymptotic PTAS in the following sense. Suppose that the parameters $b_{\max}$ and $d_{\max}$ are not a function of $|E|$. As the benefit of the optimal solution opt^* increases, the probability that $B(\mathrm{ALG}) \geq (1 - O(\varepsilon)) \cdot B(\mathsf{opt}^*)$ increases. On the other hand, we need the capacity-to-demand ratio to be logarithmic, namely, $c_{\min} \geq \Omega((\Delta_{\max} \cdot d_{\max} \cdot \ln|E|)/\varepsilon^2)$. We believe that the capacity-to-demand ratio is indeed large in realistic networks.

Acknowledgment. This research was supported by the EU project UNIFY FP7-IP-619609 as well as by the German BMBF Software Campus grant 01IS1205. Stefan Schmid is supported by the Aalborg University's inter-faculty Talent Program.

A   Multi-commodity Flows

Consider a directed graph $G = (V, E)$. Assume that edges have non-negative capacities $c(e)$. For a vertex $u \in V$, let $\mathsf{out}(u)$ denote the outward neighbors, namely the set $\{y \in V \mid (u, y) \in E\}$. Similarly, $\mathsf{in}(u) \triangleq \{x \in V \mid (x, u) \in E\}$. Consider two vertices s and t in V (called the *source* and *destination* vertices, respectively). A *flow* from s to t is a function $f : E \to \mathbb{R}^{\geq 0}$ that satisfies the following conditions:

(i) Capacity constraints: for every edge $(u, v) \in E$, $0 \leq f(u, v) \leq c(u, v)$.
(ii) Flow conservation: for every vertex $u \in V \setminus \{s, t\}$

$$\sum_{x \in \mathsf{in}(u)} f(x, u) = \sum_{y \in \mathsf{out}(u)} f(u, y).$$

The *amount* of flow delivered by the flow f is defined by

$$|f| \triangleq \sum_{y \in \mathsf{out}(s)} f(s, y) - \sum_{x \in \mathsf{in}(s)} f(x, s).$$

Consider a set ordered pairs of vertices $\{(s_i, t_i)\}_{i \in I}$. An element $i \in I$ is called a *commodity* as it denotes a request to deliver flow from s_i to t_i. Let $F \triangleq \{f_i\}_{i \in I}$ denote a set of flows, where each flow f_i is a flow from the source vertex s_i to the destination vertex t_i. We abuse notation, and let F denote the sum of the flows, namely $F(e) \triangleq \sum_{i \in I} f_i(e)$, for every edge e. Such a sequence is a *multi-commodity flow* if, in addition it satisfies *cumulative capacity constraints* defined by:

$$\text{for every edge } (u, v) \in E: \quad F(u, v) \leq c(u, v).$$

Demands are used to limit the amount of flow per commodity. Formally, let $\{d_i\}_{i \in I}$ denote a sequence of positive real numbers. We say that d_i is the *demand* of flow f_i if we impose the constraint that $|f_i| \leq d_i$. Namely, one can deliver at most d_i amount of flow for commodity i.

The *maximum benefit optimization problem* associated with multi-commodity flow is formulated as follows. The input consists of a (directed) graph $G = (V, E)$, edge capacities $c(e)$, a sequence source-destination pairs for commodities

$\{(s_i, t_i)\}_{i \in I}$. Each commodity has a nonnegative demand d_i and benefit b_i. The goal is to find a multi-commodity flow that maximizes the objective $\sum_{(u,v) \in E} b_i \cdot |f_i|$. We often refer to this objective as the *benefit* of the multi-commodity flow. When the demands are identical and the benefits are identical, the maximum benefit problem reduces to a maximum *throughput* problem.

A multi-commodity flow is *all-or-nothing* if $|f_i| \in \{0, d_i\}$, for every commodity $i \in I$. A multi-commodity flow is *unsplittable* if the support of each flow is a simple path. (The *support* of a flow f_i is the set of edges (u, v) such that $f_i(u, v) > 0$.) We often emphasize the fact that a multi-commodity flow is not all-or-nothing or not unsplittable by saying that it *fractional*.

B   Randomized Rounding Procedure

In this section we overview the randomized rounding procedure. The presentation is based on [11]. Given an instance $F = \{f_i\}_{i \in I}$ of a fractional multi-commodity flow with demands and benefits, we are interested in finding an all-or-nothing unsplittable multi-commodity flow $F' = \{f'_i\}_{i \in I}$ such that the benefit of F' is as close to the benefit of F as possible.

Observation 1. *As flows along cycles are easy to eliminate, we assume that the support of every flow $f_i \in F$ is acyclic.*

We employ a randomized procedure, called *randomized rounding*, to obtain F' from F. We emphasize that all the random variables used in the procedure are independent. The procedure is divided into two parts. First, we flip random independent coins to decide which commodities are supplied. Next, we perform a random walk along the support of the supplied commodities. Each such walk is a simple path along which the supplied commodity is delivered. We describe the two parts in detail below.

Deciding which commodities are supplied. For each commodity, we first decide if $|f'_i| = d_i$ or $|f'_i| = 0$. This decision is made by tossing a biased coin $bit_i \in \{0, 1\}$ such that

$$\mathbf{Pr}\left[bit_i = 1\right] \triangleq \frac{|f_i|}{d_i}.$$

If $bit_i = 1$, then we decide that $|f'_i| = d_i$ (i.e., commodity i is fully supplied). Otherwise, if $bit_i = 0$, then we decide that $|f'_i| = 0$ (i.e., commodity i is not supplied at all).

Assigning paths to the supplied commodities. For each commodity i that we decided to fully supply (i.e., $bit_i = 1$), we assign a simple path P_i from its source s_i to its destination t_i by following a random walk along the support of f_i. At each node, the random walk proceeds by rolling a dice. The probabilities of the sides of the dice are proportional to the flow amounts. A detailed description of the computation of the path P_i is given in Algorithm 1.

Algorithm 1. Algorithm for assigning a path P_i to flow f_i.

1: $P_i \leftarrow \{s_i\}$.
2: $u \leftarrow s_i$
3: **while** $u \neq t_i$ **do**   ▷ did not reach t_i yet
4:     $v \leftarrow$ *choose-next-vertex*(u).
5:     Append v to P_i
6:     $u \leftarrow v$
7: **end while**
8: **return** (P_i).
9: **procedure** *choose-next-vertex*(u, f_i) ▷ Assume that u is in the support of f_i
10:     Define a dice $C(u, f_i)$ with $|\text{out}(u)|$ sides. The side corresponding to an edge
          (u, v) has probability $f_i(u, v)/(\sum_{(u,v') \in \text{out}(u)} f_i(u, v'))$.
11:     Let v denote the outcome of a random roll of the dice $C(u, f_i)$.
12:     **return** (v)
13: **end procedure**

Definition of F'. Each flow $f'_i \in F'$ is defined as follows. If $bit_i = 0$, then f'_i is identically zero. If $bit_i = 1$, then f'_i is defined by

$$f'_i(u, v) \triangleq \begin{cases} d_i & \text{if } (u, v) \in P_i, \\ 0 & \text{otherwise.} \end{cases}$$

Hence, F' is an all-or-nothing unsplittable flow, as required.

C   Analysis of Randomized Rounding - Expected Flow per Edge

The presentation in this section is based on [11].

Claim. For every commodity i and every edge $(u, v) \in E$:

$$\mathbf{Pr}\left[(u, v) \in P_i\right] = \frac{f_i(u, v)}{d_i},$$
$$\mathbf{E}\left[f'_i(u, v)\right] = f_i(u, v).$$

Proof. Since

$$\mathbf{E}\left[f'_i(u, v)\right] = d_i \cdot \mathbf{Pr}\left[(u, v) \in P_i\right],$$

it suffices to prove the first part.

An edge (u, v) can belong to the path P_i only if $f_i(u, v) > 0$. We now focus on edges in the support of f_i. By Observation 1, the support is acyclic, hence we can sort the support in topological ordering. The claim is proved by induction on the position of an edge in this topological ordering.

The induction basis, for edges $(s_i, y) \in \mathsf{out}(s_i)$, is proved as follows. Since the support of f_i is acyclic, it follows that $f_i(x, s_i) = 0$ for every $(x, s_i) \in \mathsf{in}(s_i)$. Hence $|f_i| = \sum_{y \in \mathsf{out}(s_i, f_i)} f_i(s_i, y)$. Hence,

$$\mathbf{Pr}\left[(s_i, y) \in P_i\right] = \mathbf{Pr}\left[bit_i = 1\right] \cdot \mathbf{Pr}\left[\text{dice } C(s_i, f_i) \text{ selects } (s_i, y) \mid bit_i = 1\right]$$

$$= \frac{|f_i|}{d_i} \cdot \frac{f_i(s_i, y)}{\sum_{y \in \mathsf{out}(s_i, f_i)} f_i(s_i, y)}$$

$$= \frac{f_i(s_i, y)}{d_i},$$

and the induction basis follows.

The induction step, for an edge (u, v) in the support of f_i such that $u \neq s_i$, is proved as follows. Vertex u is in P_i if and only if P_i contains an edge whose head is u. We apply the induction hypothesis to these incoming edges, and use flow conservation to obtain

$$\mathbf{Pr}\left[u \in P_i\right] = \mathbf{Pr}\left[\bigcup_{x \in \mathsf{in}(u)} (x, u) \in P_i\right]$$

$$= \frac{1}{d_i} \cdot \sum_{x \in \mathsf{in}(u)} f_i(x, u)$$

$$= \frac{1}{d_i} \cdot \left(\sum_{y \in \mathsf{out}(u)} f_i(u, y)\right).$$

Now,

$$\mathbf{Pr}\left[(u, v) \in P_i\right] = \mathbf{Pr}\left[u \in P_i\right] \cdot \mathbf{Pr}\left[\text{dice } C(u, f_i) \text{ selects } (u, v) \mid u \in P_i\right]$$

$$= \frac{1}{d_i} \cdot \left(\sum_{y \in \mathsf{out}(u)} f_i(u, y)\right) \cdot \frac{f_i(u, v)}{\sum_{y \in \mathsf{out}(u)} f_i(u, y)}$$

$$= \frac{f_i(u, v)}{d_i},$$

and the claim follows. $\qquad\square$

By linearity of expectation, we obtain the following corollary.

Corollary 1. $\mathbf{E}\left[|f_i'|\right] = |f_i|$.

D   Chernoff Bounds

In this section we present material from Raghavan [15] and Young [20] about the Chernoff bounds used in the analysis of randomized rounding.

Fact 1. $e^x \geq 1 + x$ and $x \geq \ln(1 + x)$ for $x > -1$.

Fact 2. $(1+\alpha)^x \leq 1 + \alpha \cdot x$, for $0 \leq x \leq 1$ and $\alpha \geq -1$.

Fact 3 (Markov Inequality). *For a non-negative random variable X and $\alpha > 0$,*
$\mathbf{Pr}\,[X \geq \alpha] \leq \frac{\mathbf{E}[X]}{\alpha}$.

Definition 3. *The function $\beta : (-1, \infty) \to \mathbb{R}$ is defined by $\beta(\varepsilon) \triangleq (1+\varepsilon)\ln(1+\varepsilon) - \varepsilon$.*

Fact 4. *For ε such that $-1 < \varepsilon < 1$ we have $\beta(-\varepsilon) \geq \frac{\varepsilon^2}{2} \geq \beta(\varepsilon) \geq \frac{2\varepsilon^2}{4.2+\varepsilon}$.
Hence, $\beta(-\varepsilon) = \Omega(\varepsilon^2)$ and $\beta(\varepsilon) = \Theta(\varepsilon^2)$.*

Theorem 2 (Chernoff Bound). *Let $\{X_i\}_i$ denote a sequence of independent random variables attaining values in $[0,1]$. Assume that $\mathbf{E}\,[X_i] \leq \mu_i$. Let $X \triangleq \sum_i X_i$ and $\mu \triangleq \sum_i \mu_i$. Then, for $\varepsilon > 0$,*

$$\mathbf{Pr}\,[X \geq (1+\varepsilon) \cdot \mu] \leq e^{-\beta(\varepsilon) \cdot \mu}.$$

Proof. Let A denote the event that $X \geq (1+\varepsilon) \cdot \mu$. Let $f(x) \triangleq (1+\varepsilon)^x$. Let B denote the event that

$$\frac{f(X)}{f((1+\varepsilon) \cdot \mu)} \geq 1.$$

Because $f(x) > 0$ and $f(x)$ is monotonously increasing, it follows that $\mathbf{Pr}\,[A] = \mathbf{Pr}\,[B]$. By Markov's Inequality,

$$\mathbf{Pr}\,[B] \leq \frac{\mathbf{E}\,[f(X)]}{f((1+\varepsilon) \cdot \mu)}.$$

Since $X = \sum_i X_i$ is the sum of independent random variables,

$$
\begin{aligned}
\mathbf{E}\,[f(X)] &= \prod_i \mathbf{E}\,\left[(1+\varepsilon)^{X_i}\right] \\
&\leq \prod_i \mathbf{E}\,[1 + \varepsilon \cdot X_i] && (by\ Fact\ 2) \\
&\leq \prod_i (1 + \varepsilon \cdot \mu_i) \\
&\leq \prod_i e^{\varepsilon \cdot \mu_i} && (by\ Fact\ 1) \\
&= e^{\varepsilon \cdot \mu}
\end{aligned}
$$

We conclude that

$$
\begin{aligned}
\mathbf{Pr}\,[A] &\leq \frac{e^{\varepsilon \cdot \mu}}{f((1+\varepsilon) \cdot \mu)} \\
&= e^{-\beta(\varepsilon) \cdot \mu},
\end{aligned}
$$

and the theorem follows. $\qquad\square$

Using the same tools, the following theorem can be obtained for bounding the probability of the event that X is much smaller than μ (see [6] for the proof).

Theorem 3 (Chernoff Bound). *Under the same premises as in Theorem 2 except that* $\mathbf{E}\left[X_i\right] \geq \mu_i$, *it holds that, for* $1 > \varepsilon \geq 0$,

$$\Pr\left[X \leq (1 - \varepsilon) \cdot \mu\right] \leq e^{-\beta(-\varepsilon)\cdot\mu}.$$

References

1. Abujoda, A., Papadimitriou, P.: MIDAS: middlebox discovery and selection for on-path flow processing. In Proceedings of the COMSNETS Conference (2015)
2. Dietrich, D., Abujoda, A., Papadimitriou, P.: Network service embedding across multiple providers with nestor. In: IFIP Networking Conference (2015)
3. Gember-Jacobson, A., et al.: OpenNF: Enabling innovation in network function control. In: Proceedings of the ACM SIGCOMM (2014)
4. GSNFV ETSI: Network functions virtualisation (NFV); use cases. V1.1.1 (2013)
5. Even, G., Medina, M., Patt-Shamir, B.: Competitive path computation and function placement in SDNs. CoRR, abs/1602.06169 (2016)
6. Even, G., Rost, M., Schmid, S.: An approximation algorithm for path computation and function placement in SDNs. CoRR, abs/1603.09158 (2016)
7. Fischer, A., Botero, J., Beck, M.T., de Meer, H., Hesselbach, X.: Virtual network embedding: a survey. IEEE Commun. Surv. Tutorials **15**(4), 1888–1906 (2013)
8. Kreutz, D., Ramos, F.M.V., Verissimo, P.E., Rothenberg, C.E., Azodolmolky, S., Uhlig, S.: Software-defined networking: a comprehensive survey. Proc. IEEE **103**(1), 14–76 (2015)
9. Lukovszki, T., Schmid, S.: Online admission control and embedding of service chains. In: Scheideler, C. (ed.) Structural Information and Communication Complexity. LNCS, vol. 9439, pp. 104–118. Springer, Heidelberg (2015). doi:10.1007/978-3-319-25258-2_8
10. Mehraghdam, S., Keller, M., Karl, H.: Specifying and placing chains of virtual network functions. In: IEEE 3rd International Conference on Cloud Networking (CloudNet), pp. 7–13. IEEE (2014)
11. Motwani, R., Naor, J.S., Raghavan, P.: Randomized approximation algorithms in combinatorial optimization. In: Approximation Algorithms for NP-Hard Problems, pp. 447–481. PWS Publishing Co. (1996)
12. Skoldstrom, P., et al.: Towards unified programmability of cloud and carrier infrastructure. In: Proceedings of the European Workshop on Software Defined Networking (2014)
13. Hartert, R., et al.: Declarative and expressive approach to control forwarding paths in carrier-grade networks. In: Proceedings of the ACM SIGCOMM (2015)
14. Raghavan, P.: Randomized rounding, discrete ham-sandwich theorems: provably good algorithms for routing and packing problems. In: Report UCB/CSD 87/312. Computer Science Division, University of California Berkeley (1986)
15. Raghavan, P., Tompson, C.D.: Randomized rounding: a technique for provably good algorithms and algorithmic proofs. Combinatorica **7**(4), 365–374 (1987)
16. Rost, M., Schmid, S.: Service chain, virtual network embeddings: approximations using randomized rounding. CoRR, abs/1604.02180 (2016)

17. Sahhaf, S., Tavernier, W., Rost, M., Schmid, S., Colle, D., Pickavet, M., Demeester, P.: Network service chaining with optimized network function embedding supporting service decompositions. J. Comput. Net. (COMNET) **93**, 492–505 (2015)
18. Soulé, R., Basu, S., Marandi, P.J., Pedone, F., Kleinberg, R., Sirer, E.G., Foster, N.: Merlin: a language for provisioning network resources. In: Proceedings of the 10th ACM International on Conference on Emerging Networking Experiments and Technologies (CoNEXT), pp. 213–226 (2014)
19. Xia, M., Shirazipour, M., Zhang, Y., Green, H., Takacs, A.: Network function placement for NFV chaining in packet/optical datacenters. J. Lightwave Technol. **33**(8), 1565–1570 (2015)
20. Young, N.E.: Randomized rounding without solving the linear program. In: SODA, vol. 95, pp. 170–178 (1995)

Transiently Consistent SDN Updates: Being Greedy is Hard

Saeed Akhoondian Amiri[1], Arne Ludwig[1],
Jan Marcinkowski[2], and Stefan Schmid[3(✉)]

[1] Technical University Berlin, Berlin, Germany
saeed.amiri@tu-berlin.de, arne@inet.tu-berlin.de
[2] University of Wroclaw, Wroclaw, Poland
jasiekmarc@gmail.com
[3] Aalborg University, Aalborg, Denmark
schmiste@cs.aau.dk

Abstract. The software-defined networking paradigm introduces interesting opportunities to operate networks in a more flexible yet formally verifiable manner. Despite the logically centralized control, however, a Software-Defined Network (SDN) is still a distributed system, with inherent delays between the switches and the controller. Especially the problem of changing network configurations in a consistent manner, also known as the consistent network update problem, has received much attention over the last years. This paper revisits the problem of how to update an SDN in a transiently consistent, loop-free manner. First, we rigorously prove that computing a maximum ("greedy") loop-free network update is generally NP-hard; this result has implications for the classic maximum acyclic subgraph problem (the dual feedback arc set problem) as well. Second, we show that for special problem instances, fast and good approximation algorithms exist.

1 Introduction

By outsourcing and consolidating the control over multiple data-plane elements to a centralized software program, Software-Defined Networks (SDNs) introduce flexibilities and optimization opportunities. However, while a logically centralized control is appealing, an SDN still needs to be regarded as a distributed system, posing non-trivial challenges [3,9,16–19]. In particular, the communication channel between switches and controller exhibits non-negligible and varying delays [10,19], which may introduce inconsistencies during *network updates*.

Over the last years, the problem of how to consistently update routes in a (software-defined) network has received much attention, both in the systems

The research of Saeed Akhoondian Amiri has been supported by the European Research Council (ERC) under the European Union's Horizon 2020 research and innovation programme (ERC consolidator grant DISTRUCT, agreement No. 648527). The research of Arne Ludwig and Stefan Schmid was supported by the EU project UNIFY.

as well as in the theory community [9,15,17,19,21]. While in the seminal work by Reitblatt et al. [19], protocols providing strong, per-packet consistency guarantees were presented (using some kind of 2-phase commit technique), it was later observed that weaker ("relaxed"), but transiently consistent guarantees can be implemented more efficiently. In particular, Mahajan and Wattenhofer [17] proposed a first algorithm to update routes in a network in a transiently loop-free manner. Their approach is appealing as it does not require packet tagging (which comes with overheads in terms of header space and also introduces challenges in the presence of middleboxes [20] or multiple controllers [3]) or additional TCAM entries [3,19] (which is problematic given the fast growth of forwarding tables both in the Internet as well as in the virtualized datacenters [2]). Moreover, the relaxed notion of consistency also allows (parts of the) paths to become available sooner [17].

Concretely, to update a network in a transiently loop-free manner, an algorithm can proceed *in rounds* [15,17]: in each round, a "safe subset" of (so-called OpenFlow) switches is updated, such that, independently of the times and order in which the updates of this round take effect, the network is always consistent. The scheme can be implemented as follows: After the switches of round t have confirmed the successful update (e.g., using acknowledgements [12]), the next subset of switches for round $t + 1$ is scheduled.

It is easy to see that a simple update schedule always exists: we can update switches one-by-one, proceeding from the destination toward the source of a route. In practice, however, it is desirable that updates are fast and new routes become available quickly: Ideally, in order to be able to use as many new links as possible, one aims to maximize the number of concurrently updated switches [17]. We will refer to this approach as the *greedy approach*.

This paper revisits the problem of updating a maximum number of switches in a transiently loop-free manner. In particular, we consider the two different notions of loop-freedom introduced in [15]: *strong loop-freedom* and *relaxed loop-freedom*. The first variant guarantees loop-freedom in a very strict, topological sense: no single packet will ever loop. The second variant is less strict, and allows for a small bounded number of packets to loop during the update; however, at no point in time should newly arriving packets be pushed into a loop. It is known that by relaxing loop-freedom, in principle many more switches can be updated simultaneously.

Our Contributions. We rigorously prove that computing the maximum set of switches which can be updated simultaneously, without introducing a loop, is NP-hard, both regarding strong and relaxed loop-freedom. This result may be somewhat surprising, given the very simple graph induced by our network update problem. The result also has implications for the classic Maximum Acyclic Subgraph Problem (MASP), a.k.a. the dual Feedback Arc Set Problem (dFASP): The problem of computing a maximum set of switches which can be updated simultaneously, corresponds to the dFASP, on special graphs essentially describing two routes (the old and the new one). Our NP-hardness result shows that MASP/dFASP is hard even on such graphs. On the positive side, we identify network update problems which allow for optimal or almost optimal (with a provable approximation factor less than 2) polynomial-time algorithms, e.g., problem instances where the number of leaves is bounded or problem instances with bounded underlying undirected tree-width.

Organization. The remainder of this paper is organized as follows. Section 2 introduces preliminaries and presents our formal model. In Sect. 3, we prove that computing greedy updates is NP-hard, both for strong and for relaxed loop-freedom. Section 4 describes polynomial-time (approximation) algorithms. After reviewing related work in Sect. 5, we conclude and discuss future work in Sect. 6. Some technical details and longer discussions appear in the *arXiv Report* 1605.03158.

2 Model

We are given a network and two policies resp. *routes* π_1 (the *old policy*) and π_2 (the *new policy*). Both π_1 and π_2 are simple directed paths (*digraphs*). Initially, packets are forwarded (using the *old rules*, henceforth also called *old edges*) along π_1, and eventually they should be forwarded according to the new rules of π_2. Packets should never be delayed or dropped at a switch, henceforth also called *node:* whenever a packet arrives at a node, a matching forwarding rule should be present. Without loss of generality, we assume that π_1 and π_2 lead from a source s to a destination d.

We assume that the network is managed by a controller which sends out ·forwarding rule updates to the nodes. As the individual node updates occur in an asynchronous manner, we require the controller to send out simultaneous updates only to a "safe" subset of nodes. Only after these updates have been confirmed (*acked*), the next subset is updated.

We observe that nodes appearing only in one or none of the two paths are trivially updateable, therefore we focus on the network G induced by the nodes V which are part of *both* policies π_1 *and* π_2, i.e., $V = \{v : v \in \pi_1 \wedge v \in \pi_2\}$. We can represent the policies as $\pi_1 = (s = v_1, v_2, \ldots, v_\ell = d)$ and $\pi_2 = (s = v_1, \pi(v_2), \ldots, \pi(v_{\ell-1}), v_\ell = d)$, for some permutation $\pi : V \backslash \{s, d\} \to V \backslash \{s, d\}$ and some number ℓ. In fact, we can represent policies in an even more compact way: we are actually only concerned about the nodes $U \subseteq V$ which need to be updated. Let, for each node $v \in V$, $out_1(v)$ (resp. $in_1(v)$) denote the outgoing (resp. incoming) edge according to policy π_1, and $out_2(v)$ (resp. $in_2(v)$) denote the outgoing (resp. incoming) edge according to policy π_2. Moreover, let us extend these definitions for entire node sets S, i.e., $out_i(S) = \bigcup_{v \in S} out_i(v)$, for $i \in \{1, 2\}$, and analogously, for in_i. We define s to be the first node (say, on π_1) with $out_1(v) \neq out_2(v)$, and d to be the last node with $in_1(v) \neq in_2(v)$. We are interested in the set of to-be-updated nodes $U = \{v \in V : out_1(v) \neq out_2(v)\}$, and define $n = |U|$. Given this reduction, in the following, we will assume that V only consists of interesting nodes ($U = V$).

We require that paths be loop-free [17], and distinguish between *Strong Loop-Freedom* (SLF) and *Relaxed Loop-Freedom* (RLF) [15].

Strong Loop-Freedom. We want to find an *update schedule* $U_1, U_2, \ldots, U_k$, i.e., a sequence of subsets $U_t \subseteq U$ where the subsets form a partition of U (i.e., $U = U_1 \cup U_2 \cup \ldots \cup U_k$), with the property that for any round t, given that the updates $U_{t'}$ for $t' < t$ have been made, all updates U_t can be performed "asynchronously", that is, in an arbitrary order without violating loop-freedom. Thus, consistent paths will be maintained for any subset of updated nodes, independently of how long individual updates may take.

More formally, let $U_{<t} = \bigcup_{i=1,\ldots,t-1} U_i$ denote the set of nodes which have already been updated before round t, and let $U_{\leq t}$, $U_{>t}$ etc. be defined analogously. Since updates during round t occur asynchronously, an arbitrary subset of nodes $X \subseteq U_t$ may already have been updated while the nodes $\overline{X} = U_t \backslash X$ still use the old rules, resulting in a temporary forwarding graph $G_t(U, X, E_t)$ over nodes U, where $E_t = out_1(U_{>t} \cup \overline{X}) \cup out_2(U_{<t} \cup X)$. We require that the update schedule $U_1, U_2, \ldots, U_k$ fulfills the property that for all t and for any $X \subseteq U_t$, $G_t(U, X, E_t)$ is loop-free.

In the following we will call an edge (u, v) of the new policy π_2 *forward*, if v is closer (with respect to π_1) to the destination, resp. *backward*, if u is closer to the destination. It is also convenient to name nodes after their outgoing edges w.r.t. policy π_2 (e.g., *forward* or *backward*); similarly, it is sometimes convenient to say that we *update an edge* when we update the corresponding node.

While the initial network configuration consists of two paths, in later rounds, the already updated edges may no longer form a line from left to right, but rather an arbitrary directed tree, with tree edges directed towards the destination d. We will use the terms *forward* and *backward* also in the context of the tree: they are defined with respect to the direction of the tree root. However, there also emerges a third kind of edges: *horizontal edges* in-between two different branches of the tree.

Relaxed Loop-Freedom. *Relaxed Loop-Freedom* (RLF) is motivated by the practical observation that transient loops are not very harmful if they do not occur between the source s and the destination d. If relaxed loop-freedom is preserved, only a bounded number of packets can loop: we will never push new packets into a loop "at line rate". In other words, even if switches acknowledge new updates late (or never), new packets will not enter loops. Concretely, and similar to the definition of SLF, we require the update schedule to fulfill the property that for all rounds t and for any subset X, the temporary forwarding graph $G_t(U, X, E_t')$ is loop-free. The difference is that we only care about the subset E_t' of E_t consisting of edges *reachable from the source s*.

The Greedy Approach. Our objective is to update simultaneously as many nodes (or equivalently, edges) as possible: an objective initially studied in [17], which may also be seen as a greedy approach to minimize the number of rounds[1]. Note that in the first round, computing a maximum update set is trivial: All forward edges can be updated simultaneously, as they will never introduce a

---

[1] It is known however that in the worst case, a greedy approach can lead to an unnecessarily large number of rounds [15].

cycle, but no backward edge can be updated in the first round, as it can always induce a cycle; horizontal edges do not exist in the first round. Also observe that since all nodes lie on the path from source to destination, this holds for both strong and relaxed loop-freedom. However, as we will show in this paper, already in the second round, a computationally hard problem can arise.

3 Being Greedy is Hard

Interestingly, although the underlying graphs are very simple, and originate from just two (legal) paths, we now prove that the loop-free network update problem is NP-hard.

Theorem 1. *The greedy network update problem, the problem of selecting a maximum set of nodes which can be updated simultaneously, is NP-hard.*

Our reduction is from the NP-hard *Minimum Hitting Set* problem. This proof is similar for both consistency models: strong and relaxed loop-freedom, and we can present the two variants together. The inputs to the hitting set problem are:

1. A universe of m elements $\mathcal{E} = \{\varepsilon_1, \varepsilon_2, \ldots, \varepsilon_m\}$.
2. A set $S = \{S_1, S_2, S_3, \ldots, S_k\}$ of k subsets $S_i \subseteq \mathcal{E}$.

The objective is to find a subset $\mathcal{E}' \subseteq \mathcal{E}$ of minimal size, such that each set S_i includes at least one element from $\mathcal{E}'$: $\forall S_i \in S : S_i \cap \mathcal{E}' \neq \emptyset$. In the following, we will assume that elements are unique and can be ordered $\varepsilon_1 < \varepsilon_2 \ldots < \varepsilon_m$. The idea of the reduction is to create, in polynomial time, a legal network update instance where the problem of choosing a maximum set of nodes which can be updated concurrently is equivalent to choosing a minimum hitting set. While in the initial network configuration, essentially describing two paths from s to d, a maximum update set can be chosen in polynomial time (simply update all forwarding edges but no backward edges), we show in the following that already in the second round, the problem can be computationally hard.

More concretely, based on a hitting set instance, we aim to construct a network update instance of the following form, see Fig. 1. For each element $\varepsilon \in \mathcal{E}$, we create a pair of branches ε^{in} and ε^{out}, i.e., $2m$ branches in total. To model the relaxed loop-free case, in addition to the $\mathcal{E}$ branches, we add a source-destination branch, from s to d, depicted on the right in the figure. We will introduce the following to-be-updated new edges:

1. **Set Edges (SEs):** The first type of edges models sets. Let us refer to the (ordered) elements in a given set S_i by $\varepsilon_1^{(i)} < \varepsilon_2^{(i)} < \varepsilon_3^{(i)} \ldots$. For each set $S_i \in S$, we now create $m + 1$ edges from each $\varepsilon_j^{(i)}$ to $\varepsilon_{j+1}^{(i)}$, in a modulo fashion. That is, we also introduce $m + 1$ edges from the last element to the first element of the set. These edges start at the *out* branch of the smaller index and end at the *in* branch of the larger index. There are no requirements on how the edges of different sets are placed with respect to each other, as long as they are not mixed. Moreover, only one instance of multiple equivalent SEs arising in multiple sets must be kept.

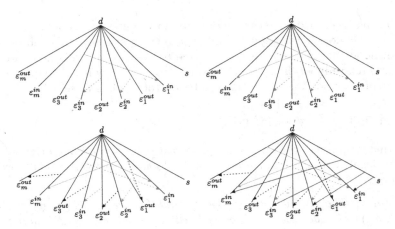

Fig. 1. Example: Construction of network update instance given a hitting set instance with $\mathcal{E} = \{1, 2, 3, \ldots, m\}$ and $S = \{\{1, 2, 3\}, \{1, m\}\}$. Each element $\varepsilon \in \mathcal{E}$ is represented by a pair of branches, one called outgoing (*out*) and one incoming (*in*). Moreover, we add a branch representing the $s - d$ path on the very right. The black branches represent already installed rules (either old or updated in the first round), and new rules (*dashed*) are situated between the branches. There are three types of to-be-updated, dashed edges: one type represents the sets (loosely and densely dashed grey), one type represents element selector edges (between in and out branch, loosely dashed black), and one type is required to connect the $s - d$ path to the elements (densely dashed grey). We prove that such a scenario can be reached after one update round where all (and only) forward edges are updated. **Top-left:** Each loosely dashed grey edge represents $m + 1$ edges, and is used to describe the set $\{1, 2, 3\}$:$(1, 2), (2, 3), (3, 1)$. **Top-right:** Each densely dashed grey edge represents $m + 1$ edges and is used for the set $\{1, m\}$: $(1, m), (m, 1)$. **Bottom-left:** The loosely dashed black edges are single edges and are the element selector edges, representing the decision if an element is part of $\mathcal{E}'$ or not. **Bottom-right:** Each densely dashed edge visualizes $m \cdot (m + 1)$ edges from the s-branch to the incoming branches of every $\varepsilon \in \mathcal{E}$.

2. **Anti-selector Edges (AEs):** These m edges constitute the decision problem of whether an element should be included in the minimum hitting set. AEs are created as follows: From the top of each *in* branch we create a *single* edge to the bottom of the corresponding *out* branch. That is, we ensure that an update of the edge from ε_i^{in} to ε_i^{out} is equivalent to $\varepsilon_i \notin \mathcal{E}'$, or, equivalently, every $\varepsilon_i \in \mathcal{E}'$ will not be included in the update set.

3. **Relaxed Edges (WEs):** These edges are only needed for the relaxed loop-free case. They connect the s-d branch to the other branches in such a way that no loops are missed. In other words, the edges aim to emulate a strong loop-free scenario by introducing artificial sources at the bottom of each branch. To achieve this, we create a certain number of edges from the s-branch to the bottom of every *in* branch. The precise amount will be explained at the detailed construction part of creating parallel edges. See Fig. 1 *bottom-left* for an example.

The rationale is as follows. If no *Anti-selector Edges* (AEs) are updated, all *Relaxed Edges* (WEs) as well as all *Set Edges* (SEs) can be updated simultaneously, without introducing a loop. However, since there are in total exactly m AEs but each set of SEs are $m + 1$ edges (hence they will all be updated), we can conclude that the problem boils down to selecting a maximum number of element AEs which do not introduce a loop. The set of non-updated AEs constitutes the selected sets, the hitting set: There must be at least one element for which there is an AE, preventing the loop. By maximizing the number of chosen AEs (maximum update set) we minimize the hitting set.

Let us consider an example: In Fig. 1 *bottom-right*, if for a set S_i every AE of $\varepsilon_i \in S_i$ is updated, a cycle is created: updating edges ε_1^{in} and ε_m^{in} results in a cycle with the $m + 1$ edges from ε_1^{out} and ε_m^{out}. Note that the resulting network update instance is of polynomial size (and can also be derived in polynomial time). In the remainder of the proof, we show that the described network update instance is indeed legal, e.g., we have a single path from source to destination, and this instance can actually be obtained after one update round.

3.1 Concepts and Gadgets

Before we describe the details of the construction, we first make some fundamental observations regarding greedy updates.

Introducing Forwarding Edges and Branches: First, a delayer concept is required to establish forwarding edges for the second round. Observe that every forwarding edge (a, b), with $a < b$, is always updated by a greedy algorithm in the first round. A delayer is used to construct a forward edge (a, b), with $a < b$, that is created in the second round. A *delayer* for edge (a, b) consists of two edges: an edge pointing backwards to a' from a with $a' < a$, plus an edge pointing from there to b. The forward edge (a', b) will be updated in the first round, which yields an edge (a, b) due to merging (see Fig. 2).

Fig. 2. Delayer concept: A forwarding edge $(a, a'b)$ can be created in round 2 using a helper node a'.

We next describe how to create the *in* and *out* branches as well as the s branch pointing to the destination d (recall Fig. 1). This can be achieved as follows: From a node close to the source s, we create a path of forward edges which ends at the destination. Each of these forward edges will be updated in the first round, and hence merged with its respective successor, which will be the destination for the very last forward edge. The nodes belonging to these forward edges will be called *branching nodes*. Every node in-between two *branching nodes* will be

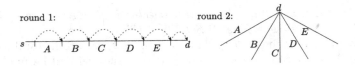

Fig. 3. Creating branches after a greedy update of forward edges.

part of a new branch pointing to the destination. See Fig. 3 for an example. The rightmost node before the *branching node* on the line will also be the topmost node on the branch after the first round update (as long as it has an outgoing backward edge, hence not being updated in the first round). We will use the terms right and high (rightmost-topmost) and left-low for the first and second round interchangeably.

Introducing Special Segments: In our construction, we split the line (old path) into disjoint segments which will become independent branches at the beginning of the second round. In addition to these segments, there will be two special segments, one at the beginning and one at the end. The first will not even become an independent branch at the beginning of the second round, but is merely used to realize the delayer edges. Behind the very last segment (ε_1^{in}) and just before d, there is a second special segment, which we call *relaxed*: it is needed to create the branch with the source s at the bottom and its connections to the other ε_i^{in} branches.

In our construction, SEs come in groups of $m + 1$ edges. These edges must eventually be part of a legal network update path, and must be connected in a loop-free manner. In other words, to create the desired problem instance, we need to find a way to connect two branches b_1 and b_2 with $m+1$ edges, such that there is a single complete path from s to d. Furthermore, these edges should not form a loop.

In the *arXiv Report* 1605.03158, we describe how parallel edges can be constructed.

3.2   Connecting the Pieces

Given these gadgets, we are able to complete the construction of our problem instance.

Realizing the Delayer: The first created segment, *temp*, serves for edges that are created using the *delayer* concept. This is due to our construction: every node that will be created in this interval in our construction will be a forward node and therefore updated in the first greedy round. The *temp* segment will be located right after the source s on the line.

Realizing the Branches: We create two segments for each $\varepsilon \in \mathcal{E}$, one *out* and one *in*, and sort them in descending global order (and depict them from left to right) w.r.t. $\varepsilon \in \mathcal{E}$, with the *out* segment closer to s than the *in* segment for each ε, i.e. $\varepsilon_m^{out}, \varepsilon_m^{in}, \ldots, \varepsilon_2^{out}, \varepsilon_2^{in}, \varepsilon_1^{out}, \varepsilon_1^{in}$ (Fig. 4).

s | $temp$ | ε_m^{out} | ε_m^{in} | $\ldots$ | ε_2^{out} | ε_2^{in} | ε_1^{out} | ε_1^{in} | $weak$ | d

Fig. 4. Illustration of how to split the old line into segments according to the amount of needed branches in the second round.

Connecting the Path: We will now create the new path from the source s to the destination d through all the different segments. This path requires additional edges. We will ensure that these edges can always be updated and hence do not violate the selector properties. Moreover, we ensure that they do not introduce a loop. In order to create a branch with s at the bottom (to ensure that the proof will also hold for relaxed loop-freedom), we start our path from the source s to a node $relaxed - bot$ on the very left part of the $relaxed$ segment. From here we need to create the $m \cdot (m + 1)$ connections to every other ε_i^{in} branch, more precisely to the very left of the top part of this branch ε_{i-t}^{in}: the relaxed Edges (WEs). Starting from $relaxed - bot$, we create the $m \cdot (m + 1)$ zigzag edges (described in detail in the technical report only) to the ε_1^{in} segment. Once this is done, we repeat this process for the remaining ε_i^{in} connecting them in the same order blockwise, as they are ordered on the line. See Fig. 5.

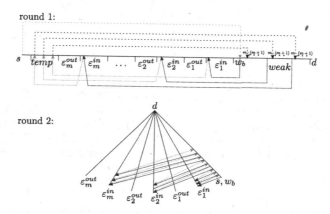

Fig. 5. Creating the branch with the source at the bottom and $m \cdot (m+1)$ connections to each ε_i^{in} segment of the line, as shown in Sect. 3.1. The $m \cdot (m + 1)$ connections are visualized as a single edge in the first round to enhance visibility.

At the beginning of the second round, we will now have a branch with the source s at the bottom and $m+1$ edges to each of the ε_i^{in} branches. The next step is to connect the out branches with the in branches (the Set Edges). For each set $S_j \in S$ and each pair $\varepsilon_i, \varepsilon_l \in S_j$ with no $\varepsilon' \in S_j, \varepsilon_i < \varepsilon' < \varepsilon_l$, we create $m+1$ edges from ε_i^{out} to ε_l^{in}, more precisely to the top part ε_{l-t}^{in} somewhere above the WEs. Each pair $\varepsilon_i, \varepsilon_l$ only needs to connect once with the $m + 1$ edges, even if it occurs in several different sets of S. The last element ε_i of a set S_j will additionally need to be connected to the first element of the set (the modulo edges).

After the $m + 1$ connections to ε_m^{in}, the path returns at the right most (or highest in the (s, d)-branch) node in the *relaxed* segment. From here we create a backward edge to the left part of ε_1^{out}. Here, we create $m + 1$ connections to every ε_i^{in}, which is the next larger element in any of the sets. An example is shown in Fig. 6.

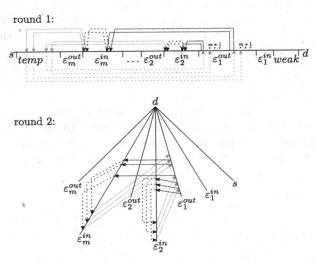

Fig. 6. Connecting the ε_1^{out} branch with the branches ε_2^{in}, ε_3^{in}, ε_m^{in}. This scenario would be created for the sets: $\{1, 2, \ldots\}, \{1, 3, \ldots\}, \{1, m\}$. The densely dashed black edges show the outgoing edges from ε_1^{out}. The loosely dashed black edges are the backward edges from the top part of a branch ε_i^{in} to its bottom part (ε_{i-t}^{in} to ε_{i-b}^{in}). The densely dashed grey edges are the way back from ε_i^{in} to ε_1^{out} and are needed to complete the path.

To complete the $m + 1$ connections for every pair, we proceed as follows: we connect the ε_1^{out} branch to all required in-branches, then add the edge from ε_1^{out} to the ε_2^{out} branch, then add the edges from the ε_2^{out} branch to all required in-branches, etc. Generally, we interleave adding the edges from the ε_i^{out} branch to all required in-branches and then add the i-out to $(i + 1)$-out edge. Until the path arrives at the end of the last out branch, ε_m^{out}:

- *Step A - Create the $m+1$ set specific edges:* Here we create $m+1$ connections to every successor in the respective sets (at most once per pair). If this element is the largest element in a set, it needs to be connected to the in part of the smallest element of this set again. Here the delayer concept needs to be used for the modulo edges.

– *Step B - Connecting the out branches:* In order to create the next $m + 1$ connections from the next out segment ε_{i+1}^{out}, we need to connect it from our current out segment ε_i^{out}. The edge therefore needs to point to the rightmost part of ε_{i+1}^{out}. Since this edge is always a backward edge in the first round (we start closer to the destination and move backward towards the source), it will turn out to be an edge which points to the very top of ε_{i+1}^{out} at the beginning of the second round. This assures that there are no loops created, since the only way is going directly towards the destination. From here we create an edge pointing to the very left side of ε_{i+1}^{out} (evolving to a backward rule from top to bottom of the branch in the second round, hence not being part of the update set in the first nor the second round).

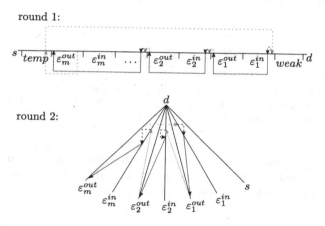

Fig. 7. Connecting the in and out branches of every ε_i, shown in densely dashed black. The edges shown in densely dashed grey are needed to keep the path complete and the backward edges in loosely dashed black are needed to ensure that only the destination can be reached from that point in the second round.

To finish the construction, we need to add the anti-selector edges (AEs), and connect the in and out branches of every single ε_i with each other. The goal is to create, for each given i, an edge from the top of each ε_i^{in} to the bottom of each ε_i^{out}. This way, if this edge is included in the update, a loop may be formed: as every incoming edge to ε_i^{in} arrives below the AEs start point and every outgoing edge on ε_i^{out} is above AE's destination. The decision to not include one of these edges is equivalent to $\varepsilon_i \in \mathcal{E}'$ in the minimum hitting set problem. In order to keep the path connected we will also need to include edges from ε_i^{out} to ε_{i+1}^{in}, compare Fig. 7. These edges will point to the top of ε_{i+1}^{in} and therefore do not create loops, since the only way is going directly to the destination. From here we create another backward edge to its left neighbor such that there is no possible other way than traversing towards d from this point. Without this backward edge loops may be created, since it introduces connections between branches which are not both in a set S_i of the hitting set problem. Therefore, an update of one of the additional connector edges will never lead to a loop, and the edges can all be included in the update set of the round 2.

The construction of these edges is straightforward. From the end of the current path which is located on the ε_m^{out} segment, we create a delayed edge (over $temp$) to the very right part of the ε_1^{in} segment. From here we construct the path as described with a short backward edge to its left neighbor and then to the very left part of the ε_i^{out} segment and again to the very right part of the ε_{i+1}^{in} segment afterwards, until we arrive at the very left part of the ε_m^{out} segment.

It remains to create the segments and branches for the second round. From ε_m^{out}, we create a backward edge to the $temp$ part. From here we use the branching concept and connect all horizontal nodes in-between the single parts that we created on the line (see Fig. 8).

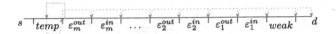

Fig. 8. Connecting the segments with forward edges. This creates a single branch from the destination for every segment due to the merging. The edge shown in loosely dashed grey is connecting this step with the step before.

In summary, we ensured that already after a single greedy first update round, we end up in a situation where choosing the maximum set of updateable nodes is equivalent to choosing the minimum hitting set.

4   Polynomial-Time Algorithms

While the computational hardness is disappointing, we can show that there exist several interesting specialized and approximate algorithms.

Optimal Algorithms. There are settings where an optimal solution can be computed quickly. For instance, it is easy to see that in the first round, in a configuration with two paths, updating all forward edges is optimal: Forward edges never introduce any loop, and at the same time we know that backward edges can never be updated in the first round, as any backward edge alone (i.e., taking effect in the first round), will immediately introduce a loop. In the following, we first present an optimal algorithm for SLF, for trees with only two leaves. We will then extend this algorithm to RLF. In the our arXiv report we will also show that optimal solutions can be computed efficiently if the underlying undirected graph is of bounded tree-width.

Lemma 1. *A maximum SLF update set can be computed in polynomial-time in trees with two leaves.*

Proof. Recall that there are three types of new edges in the graph (see also Fig. 9): forward edges (F), backward edges (B) and horizontal edges (H), hence $E = H \cup B \cup F$. Moreover, recall that forward edges can always be updated while backward edges can never be updated in SLF. Thus, the problem boils down to selecting a maximum subset of H, pointing from one branch to the other. If there is a simple loop $C \in G$ such that $H^C = E(C) \cap H \neq \emptyset$, then $|H^C| = 2$ and we say that the two edges $e_1, e_2 \in H^C$ cross each other, written $e_1 \times e_2$.

We observe that the different edge types can be computed efficiently. For illustration, suppose the policy graph $G = (V, E)$ (the union of old and new policy edges) is given as a straight line drawing Π in the 2-dimensional Euclidean plane, such that the old edges of the 2-branch tree form two disjoint segments which meet at the root of the tree (the destination), and such that each node is mapped to a unique location. Given the graph, such a drawing (including crossings) in the plane can be computed efficiently. Also note that there could be other edges which intersect w.r.t. the drawing Π, but those are not important for us.

Now create an auxiliary graph $G' = (V', E')$ where $V' = \{v_e \mid e \in H\}$, $E' = \{(v_{e_1}, v_{e_2}) \mid e_1, e_2 \in H : e_1 \times e_2\}$. The graph G' is bipartite, and therefore finding a minimum vertex cover $VC \in V(G)$ is equivalent to finding maximum matching, which can be done in polynomial time. Let $H' = \{e \mid e \in H : v_e \in VC\}$, then the set H' is a minimum size subset of H which is not updateable. Therefore the set $H \setminus H'$ is the maximum size subset of H which we can update in a SLF manner (the complement of H' is a maximum independent set in G' and therefore, by the definition of the collision graph G', a maximum updateable set).

We conclude the proof by observing that all these algorithmic steps can be computed in polynomial time.

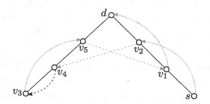

Fig. 9. Concept of horizontal edges shown in loosely dashed grey. Both horizontal edges (v_2, v_4) and (v_5, v_1) are crossing each other. The backward edge (v_4, v_3) is shown in loosely dashed black and the forward edges in densely dashed grey. Note that s does not necessarily have to be a leaf.

Lemma 2. *A maximum RLF update set can be computed in polynomial-time in trees with two leaves.*

For more details, see our technical report.

5   Related Work

In their seminal work, Reitblatt et al. [19] initiated the study of network updates providing strong, per-packet consistency guarantees, and the authors also presented a 2-phase commit protocol. This protocol also forms the basis of the distributed control plane implementation in [3]. Mahajan and Wattenhofer [17] started investigating a hierarchy of transient consistency properties— in particular also (strong) loop-freedom but for example also bandwidth-aware updates [1]—for destination-based routing policies. The measurement studies in [10] and [13] provide empirical evidence for the non-negligible time and high variance of switch updates, further motivating their and our work. In their paper, Mahajan and Wattenhofer proposed an algorithm to "greedily" select a maximum number of edges which can be used early during the policy installation process. This study was recently refined in [7,8], a parallel work to ours, where the authors also establish a hardness result for destination based routing (single- and multi-destination). Our work builds upon [17] and complements the results in [7,8]: We consider the scheduling complexity of updating *arbitrary routes* which are not necessarily destination-based. Interestingly, our results (using a different reduction) show that even with the requirement that the initial and the final routes are simple paths, the problem is NP-hard. Moreover, our results hold for both the strong SLF and the relaxed RLF loop-free problem variants introduced in [15]. The SLF can be seen as a special variant of the Dual Feedback Arc Set Problem (FASP) resp. Maximum Acyclic Subgraph Problem (MASP): important classic problems in approximation theory [11]. In particular, it is known that dual-FASP/MASP can be $1/2 + \varepsilon$ approximated on general graphs (for arbitrary small ε). The results presented in this paper also imply that better approximation algorithms and even optimal polynomial-time algorithms exist for special graph families, namely graph families describing network update problems; this may be of independent interest. The RLF variant is a new optimization problem, and to the best of our knowledge, existing bounds are not applicable to this problem. We should note that FASP is in FPT [4], and the hitting set problem is W[2]-hard [6]. In our hardness construction we actually find a reduction from hitting set to FASP for particular graph classes. But the reduction is not parameter preserving, so the W-hierarchy does not collapse. Finally, our model is orthogonal to the network update problems aiming at minimizing the number of interactions with the controller (the so-called *rounds*), which we have recently studied for single [15] and multiple [5] policies, also including additional properties, beyond loop-freedom, such as waypointing [14]. The two objectives conflict [15], a good approximation for the number of update edges yields a bad approximation for the number of rounds, and vice versa.

6   Open Problems

An interesting open question regards whether SLF and RLF can be approximated well or even solved optimally in polynomial time, in graphs of bounded tree width. See our accompanying arXiv report for a longer discussion.

References

1. Brandt, S., Förster, K.-T., Wattenhofer, R.: On consistent migration of flows in SDNs. In: Proceedings of IEEE INFOCOM (2016)
2. Bu, T., Gao, L., Towsley, D.: On characterizing BGP routing table growth. Comput. Netw. **45**(1), 45–54 (2004)
3. Canini, M., Kuznetsov, P., Levin, D., Schmid, S.: A distributed and robust SDN control plane for transactional network updates. In: Proceedings of IEEE INFOCOM (2015)
4. Chen, J., Liu, Y., Lu, S., O'sullivan, B., Razgon, I.: A fixed-parameter algorithm for the directed feedback vertex set problem. J. ACM **55**(5), 21:1–21:19 (2008)
5. Dudycz, S., Ludwig, A., Schmid, S.: Can't touch this: consistent network updates for multiple policies. In: Proceedings of 46th IEEE/IFIP International Conference on Dependable Systems and Networks (DSN) (2016)
6. Flum, J., Grohe, M.: Parameterized complexity theory. In: Texts in Theoretical Computer Science. An EATCS Series. Springer, Heidelberg (2006)
7. Förster, K.-T., Mahajan, R., Wattenhofer, R.: Consistent updates in software defined networks: on dependencies, loop freedom, and blackholes. In: Proceedings of 15th IFIP Networking (2016)
8. Förster, K.-T., Wattenhofer, R.: The power of two in consistent network updates: hard loop freedom, easy flow migration. In: Proceedings of 25th International Conference on Computer Communication and Networks (ICCCN) (2016)
9. Ghorbani, S., Godfrey, B.: Towards correct network virtualization. In: Proceedings of ACM HotSDN, pp. 109–114 (2014)
10. Jin, X., Liu, H., Gandhi, R., Kandula, S., Mahajan, R., Rexford, J., Wattenhofer, R., Zhang, M.: Dionysus: dynamic scheduling of network updates. In: Proceedings of ACM SIGCOMM (2014)
11. Kann, V.: On the approximability of NP-complete optimization problems. Ph.D. thesis (1992)
12. Kuzniar, M., Peresíni, P., Kostic, D.: Providing reliable FIB update acknowledgments in SDN. In: Proceedings of 10th ACM CoNEXT, pp. 415–422 (2014)
13. Kuźniar, M., Perešíni, P., Kostić, D.: What you need to know about SDN flow tables. In: Mirkovic, J., Liu, Y. (eds.) PAM 2015. LNCS, vol. 8995, pp. 347–359. Springer, Heidelberg (2015)
14. Ludwig, A., Dudycz, S., Rost, M., Schmid, S.: Transiently secure network updates. In: Proceedings of ACM SIGMETRICS (2016)
15. Ludwig, A., Marcinkowski, J., Schmid, S.: Scheduling loop-free network updates: it's good to relax! In: Proceedings of ACM Symposium on Principles of Distributed Computing (PODC) (2015)
16. Ludwig, A., Rost, M., Foucard, D., Schmid, S.: Good network updates for bad packets: waypoint enforcement beyond destination-based routing policies. In: Proceedings of ACM Workshop on Hot Topics in Networks (HotNets) (2014)
17. Mahajan, R., Wattenhofer, R.: On consistent updates in software defined networks. In: Proceedings of ACM HotNets (2013)
18. Padon, O., Immerman, N., Karbyshev, A., Lahav, O., Sagiv, M., Shoham, S.: Decentralizing SDN policies. In: Proceedings of 42nd Annual ACM SIGPLAN-SIGACT Symposium on Principles of Programming Languages (POPL), pp. 663–676 (2015)
19. Reitblatt, M., Foster, N., Rexford, J., Schlesinger, C., Walker, D.: Abstractions for network update. In: Proceedings of ACM SIGCOMM, pp. 323–334 (2012)

20. Qazi, Z., et al.: Simple-fying middlebox policy enforcement using SDN. In: Proceedings of ACM SIGCOMM (2013)
21. Zhou, W., (Kevin) Jin, D., Croft, J., Caesar, M., Godfrey, P.B.: Enforcing customizable consistency properties in software-defined networks. In: Proceedings of 12th USENIX Symposium on Networked Systems Design and Implementation (NSDI), pp. 73–85 (2015)

Author Index

Printed in the United States
By Bookmasters